$$\int_0^\infty \frac{\sin ax}{x}\, dx = \begin{cases} \frac{1}{2}\pi; & a > 0 \\ 0; & a = 0 \\ -\frac{1}{2}\pi; & a < 0 \end{cases}$$

$$\int_0^\pi \sin^2 ax\, dx = \int_0^\pi \cos^2 ax\, dx = \frac{\pi}{2}$$

$$\int_0^\pi \sin mx \sin nx\, dx = \int_0^\pi \cos mx \cos nx\, dx = 0; \quad m \ne n,\ m \text{ and } n \text{ integers}$$

$$\int_0^\pi \sin mx \cos nx\, dx = \begin{cases} 0; & m + n \text{ even} \\ \dfrac{2m}{m^2 - n^2}; & m + n \text{ odd} \end{cases}$$

A short table of trigonometric identities

$\sin(\alpha \pm \beta) = \sin\alpha \cos\beta \pm \cos\alpha \sin\beta$

$\cos(\alpha \pm \beta) = \cos\alpha \cos\beta \mp \sin\alpha \sin\beta$

$\cos(\alpha \pm 90°) = \mp \sin\alpha$

$\sin(\alpha \pm 90°) = \pm \cos\alpha$

$\cos\alpha \cos\beta = \frac{1}{2}\cos(\alpha + \beta) + \frac{1}{2}\cos(\alpha - \beta)$

$\sin\alpha \sin\beta = \frac{1}{2}\cos(\alpha - \beta) - \frac{1}{2}\cos(\alpha + \beta)$

$\sin\alpha \cos\beta = \frac{1}{2}\sin(\alpha + \beta) + \frac{1}{2}\sin(\alpha - \beta)$

$\sin 2\alpha = 2\sin\alpha \cos\alpha$

$\cos 2\alpha = 2\cos^2\alpha - 1 = 1 - 2\sin^2\alpha = \cos^2\alpha - \sin^2\alpha$

$\sin^2\alpha = \frac{1}{2}(1 - \cos 2\alpha)$

$\cos^2\alpha = \frac{1}{2}(1 + \cos 2\alpha)$

$\sin\alpha = \dfrac{e^{j\alpha} - e^{-j\alpha}}{j2}$

$\cos\alpha = \dfrac{e^{j\alpha} + e^{-j\alpha}}{2}$

$e^{\pm j\alpha} = \cos\alpha \pm j\sin\alpha$

$A\cos\alpha + B\sin\alpha = \sqrt{A^2 + B^2}\cos\left(\alpha + \tan^{-1}\dfrac{-B}{A}\right)$

Engineering Circuit Analysis

McGRAW-HILL SERIES IN ELECTRICAL ENGINEERING

Consulting Editor
Stephen W. Director, Carnegie-Mellon University

CIRCUITS AND SYSTEMS
COMMUNICATIONS AND SIGNAL PROCESSING
CONTROL THEORY
ELECTRONICS AND ELECTRONIC CIRCUITS
POWER AND ENERGY
ELECTROMAGNETICS
COMPUTER ENGINEERING
INTRODUCTORY
RADAR AND ANTENNAS
VLSI

Previous Consulting Editors

Ronald M. Bracewell, Colin Cherry, James F. Gibbons, Willis W. Harman,
Hubert Heffner, Edward W. Herold, John G. Linvill, Simon Ramo, Ronald A.
Rohrer, Anthony E. Siegman, Charles Susskind, Frederick E. Terman, John G.
Truxal, Ernst Weber, and John R. Whinnery

CIRCUITS AND SYSTEMS

Consulting Editor
Stephen W. Director, Carnegie-Mellon University

Engineering
Circuit Analysis Fourth Edition

William H. Hayt, Jr.
Professor of Electrical Engineering
Purdue University

Jack E. Kemmerly
Professor of Engineering, Emeritus
California State University, Fullerton

McGraw-Hill Book Company

New York St. Louis San Francisco Auckland Bogotá Hamburg
Johannesburg London Madrid Mexico Montreal New Delhi Panama
Paris São Paulo Singapore Sydney Tokyo Toronto

ENGINEERING CIRCUIT ANALYSIS

2 3 4 5 6 7 8 9 0 V N H V N H 8 9 8 7 6

ISBN 0-07-027397-9

This book was set in Times Roman by Bi-Comp, Incorporated. The editors were Sanjeev Rao, Steven Tenney, and Madelaine Eichberg; the designer was Hermann Strohbach; the production supervisor was Leroy A. Young. The drawings were done by J & R Services, Inc.
Von Hoffmann Press, Inc., was printer and binder.

Library of Congress Cataloging in Publication Data

Hayt, William Hart, 1920–
    Engineering circuit analysis.

    Includes index.
    1. Electric circuit analysis.   2. Electric
network analysis.   I. Kemmerly, Jack E., 1924–
II. Title.
TK454.H4   1986   621.319′2   85-9635
ISBN 0-07-027397-9
ISBN 0-07-027398-7 (solutions manual)

Contents

PART TWO/THE TRANSIENT CIRCUIT

PART THREE/SINUSOIDAL ANALYSIS

Preface

This book is intended for use with a first course in electrical engineering. In many colleges and universities such a course will be preceded or accompanied by an introductory physics course in which the basic concepts of electricity and magnetism are introduced, most often from the field aspect. Such a background is not a prerequisite, however. Instead, several of the requisite basic concepts of electricity and magnetism are discussed (or reviewed) in the first chapter. Only an introductory calculus course need be considered as a prerequisite, or possibly a corequisite, to the reading of the book. Circuit elements are introduced and defined here in terms of their circuit equations; only incidental comments are offered about the pertinent field relationships.

It is the authors' intention that this text be one from which students may teach themselves; it is written to the student and not to the instructor. If at all possible, each new term is clearly defined when it is first introduced. The basic material appears toward the beginning of each chapter and is explained carefully and in detail; numerical examples are usually used to introduce and suggest general results. Drill problems appear at the end of most sections; they are generally simple, and answers to the several parts are given in order. The more difficult problems appear at the ends of the chapters and follow the general order of presentation of the text material. These problems are occasionally used to introduce less important or more advanced topics through a guided step-by-step procedure, as well as to introduce topics which will appear in the following chapter. The introduction and resulting repetition are both important to the learning process. In all, there are 220 drill problems, each consisting of several parts, and 597 additional problems at the ends of the chapters. Most of these problems are new in this edition.

The general order of the material has been selected so that the student may learn as many of the techniques of circuit analysis as possible in the simplest context, namely, the resistive circuit which is the subject of the first part of this text. Fundamental laws, a few theorems, and some elementary network topology enable most of the basic analytical techniques to be developed. Numerous examples and problems are possible since the solutions are not mathematically complicated. The extension of these techniques to more advanced circuits in subsequent parts of the text affords the opportunity for both review and generalization. This first part of the text may be covered in four to six weeks, depending on the students' background and ability, and on the course intensity.

The second part of the text is devoted to the natural response and the complete response to dc excitation of the simpler *RL, RC,* and *RLC* circuits. Differential and integral calculus are necessary, but a background in differential equations is not required. The unit step is introduced as an important singularity function in this part, but the introduction of the unit impulse is withheld until transform techniques are introduced in Chapter 18.

The third part of the text introduces the frequency domain and initiates operations with complex numbers by concentrating on sinusoidal analysis in the steady state. Part 3 also includes a discussion of average power, rms values, and polyphase circuits, all of which are associated with the sinusoidal steady state.

In the fourth part of the book the complex-frequency concept is introduced, and its use in relating the forced response and the natural response is emphasized. The determination of the complete response of sinusoidally excited circuits begins to tie together the material of the first three parts.

The fifth part begins with a consideration of magnetic coupling, which is basically a two-port phenomenon, and logically leads into a consideration of two-port network analysis and the linear modeling of various electronic devices.

Part 6 of the text introduces more powerful techniques of network analysis. The first of these is the Fourier series description of periodic waveforms. This treatment is then extended to nonperiodic forcing and response functions by use of the Fourier transform. The final chapter covers the more important Laplace transform techniques and their use in obtaining the complete response of more complicated circuits.

It is felt that the material in this book is more than adequate for a two-semester course, but some selection may be made from the last four or five chapters. No material is included in the text which will not be of some value in the following term; thus, signal-flow graphs, the relationship of circuit theory to field theory, advanced topological concepts, state-variable techniques, and the details of computer-aided analysis are among those subjects which are relegated to subsequent courses.

A number of changes have been made in this edition. One of the most noticeable is the use of color, which adds emphasis to the textual material, and clarity and contrast to the illustrations. The operational amplifier is introduced as one of today's most important devices, and it is used to provide examples of circuits containing dependent sources, such as the voltage follower, the integrator and differentiator, the multiplier, the inverting amplifier, and circuits which can simulate lossless *LC* circuits and voltage-gain transfer functions.

Source transformations now precede the Thévenin and Norton theorems in Chapter 3 and furnish a logical introduction to those topics. Complex power is now represented by the standard symbol **S.** The special use of the wattmeter for three-phase power measurement has been removed from this edition.

In Chapter 13 the relationship between the poles of the transfer function and transient response is treated more carefully, and in the following chapter, the properties of parallel and series resonance have been summarized in tabular form. A new section has been added to Chapter 14 covering Bode diagrams. Chapter 16 has been improved by including input admittance in terms of generalized determinants, by using matrix notation for all two-port parameters, including an appendix on elementary properties of matrices, by investigating the series, parallel, and cascade interconnections of two ports, by adding a section on transmission parameters, and by including a table summarizing the interrelationships among the four sets of parameters studied.

The chapter on state variables has not been widely used, and it is omitted from this edition.

In Chapter 18, there is an additional example on convolution, emphasizing graphical techniques. Also, the one-sided Laplace transform has been specifically defined as an integral with a lower limit of $t = 0^-$. Several techniques for obtaining partial-fraction expansions are also outlined.

Added help is offered for both students and instructors. Appendix 5 now includes answers to every odd-numbered problem, and a Solution Manual is available for the instructor.

Throughout the book there is a logical trail leading from definition, through explanation, description, illustration, and numerical example, to problem-solving ability and to the obvious expectation of the authors that students continually ask themselves, ''Why does this happen? How is it related to last week's work? Where do we logically go next?'' There is a tremendous amount of enthusiastic momentum in a beginning engineering student, and this may be preserved by providing frequent drill problems whose successful solution confirms progress in the students' minds by integrating the various sections into a coherent whole, by pointing out future applications and more advanced techniques, and by maintaining interested, inquisitive attitudes in the students.

If the book occasionally appears to be informal, or even lighthearted, it is because the authors feel that it is not necessary to be dry or pompous to be educational. An amused smile on the face of a student is seldom an obstacle to his absorbing information. If the writing of the text had its entertaining moments, then why not the reading too?

Much of the material in the text is based on courses taught at California State University, Fullerton, at Fort Lewis College, Durango, Colorado, and at Purdue University.

We would also like to express our thanks for the many useful comments and suggestions provided by colleagues who reviewed this text during the course of its development, especially to Bennett Bashore, Oklahoma State University; William Beasley, Texas A&M University; Roger C. Conant, University of Illinois at Chicago Circle; James Delansky, Pennsylvania State

University; Steve Director, Carnegie-Mellon University; Robert Finch, South Dakota State University; Robert Hatch, Lawrence Institute of Technology; Russell Krackhardt, Worcester Polytechnic Institute; William Kuriger, University of Oklahoma; Evan Moustakas, San Jose State University; David G. Skitek. University of Missouri - Columbia, William J. Troop, New Jersey Institute of Technology; Gerald Tuma, University of Oklahoma; and Norman Weinberg, University of Miami.

We also thank Miss Tuyen Phung for her tireless and meticulous efforts in preparing the Solution Manual.

William H. Hayt, Jr.
Jack E. Kemmerly

PART ONE
THE RESISTIVE CIRCUIT

Chapter 1
Definitions and Units

The purpose of this text, as its title implies, is to provide material that should lead to some proficiency in the subject of engineering circuit analysis. This topic is extremely useful for virtually every type of engineer as well as for many physicists and applied mathematicians; it is also stimulating, challenging, and quite enjoyable. The uninitiated may well ask immediately, ''What *is* engineering circuit analysis?'' The question is a fair one, and we may answer it by taking a quick look at an available collegiate dictionary, which provides the following revelations:

engineering	a science by which the properties of matter and the sources of energy in nature are made useful to man.
circuit	an interconnection of simple electrical devices in which there is at least one closed path in which current may flow.
analysis	a (mathematical) study of a complex entity and the interrelationship of its parts.

Thus we might be inclined to decide that ''engineering circuit analysis'' is a mathematical study of some useful interconnection of simple electrical devices in which there is at least one closed current path. This definitior is essentially correct, although we cannot understand it fully until we clarify what is meant by ''current'' and ''electrical devices,'' a task which we shall undertake shortly.

Not long ago,[1] a textbook of this kind would have been viewed as a strictly electrical engineering text. Now, however, it has become increasingly common for students of civil engineering, mechanical engineering, and the other engineering disciplines, as well as an occasional student of applied mathematics, computer sciences, biology, or physics, to study introductory circuit analysis with the electrical engineer; moreover, courses based largely on this

---

[1] At least it seems that way to the two sweet old authors.

subject are taken by engineering students even before they have become identified with a particular branch of engineering.

If we have already entered or intend to enter an electrical engineering program, then circuit analysis may represent the introductory course in our chosen field. If we are associated with some other branch of engineering, then circuit analysis may represent a large fraction of our total study of electrical engineering, but it also enables us to continue our electrical work in electronics, instrumentation, and other areas. Most important, however, is the possibility given to us to broaden our educational base and become more informed members of a team which may be primarily concerned with the development of some electrical device or system. Effective communication within such a team can be achieved only if the language and definitions used are familiar to all.

Few of the engineering accomplishments of the recent past can be credited to a single individual. The era of the Edison-type inventor has passed, and a graduating engineer should expect to become part of a group consisting of many kinds of engineers, applied mathematicians, computer scientists, and physicists. The group effort will be coordinated by technically trained managers, and the technical products will be produced, sold, and often operated and maintained by men and women having scientific or engineering training. The engineering graduates of today are not all employed solely to work on the technical design aspects of engineering problems. Their efforts now extend beyond the creation of better steam irons and radar systems to vigorous efforts to solve socioeconomic problems such as air and water pollution, urban planning, mass transportation, the discovery of new energy sources, and the conservation of existing natural resources, particularly oil and natural gas.

In order to contribute to the solution of these engineering problems, an engineer must acquire many skills, one of which is a knowledge of electric-circuit analysis.

We shall begin this study by considering systems of units and several basic definitions and conventions. For those who have an elementary knowledge of basic electricity and magnetism, the material in this chapter should be rapid reading. After these introductory topics have been mastered, we can then turn our attention to a simple electric circuit.

1-2 Systems of units

We must first establish a common language. Engineers cannot communicate with one another in a meaningful way unless each term used is clear and definite. It is also true that little learning can be achieved from a textbook which does not define carefully each new quantity as it is introduced. If we speak in the vague generalities of a television commercial—''gets clothes up to 40 percent whiter''—and do not bother to define whiteness or to provide units by which it may be measured, then we shall certainly be deterred from success in engineering, although we might sell a lot of detergent.

In order to state the value of some measurable quantity, we must give both a *number* and a *unit,* such as "3 inches." Fortunately we all use the same number system and know it well. This is not as true for the units, and some time must be spent in becoming familiar with a suitable system of units. We must agree on a standard unit and be assured of its permanence and its general acceptability. The standard unit of length should not be defined in terms of the distance between two marks on a certain rubber band; this is not permanent, and furthermore everybody else is using another standard.

We shall also need to define each technical term at the time it is introduced, stating the definition in terms of previously defined units and quantities. Here the definition cannot always be as general as the more theoretically minded might wish. For instance, it will soon be necessary to define "voltage." We must either accept a very complete and general definition, which we can neither appreciate nor understand now, or else adopt a less general but simpler definition which will satisfy our purposes for the present. By the time a more general definition is needed, our familiarity with the simpler concepts will help our understanding at that time.

It will also become evident that many quantities are so closely related to each other that the first one defined needs a few subsequent definitions before it can be thoroughly understood. As an example, when the "circuit element" is defined, it is most convenient to define it in terms of *current* and *voltage,* and when current and voltage are defined, it is helpful to do so with reference to a circuit element. None of these three definitions can be well understood until all have been stated. Therefore, our first definition of the circuit element may be somewhat inadequate, but then we shall define current and voltage in terms of a circuit element and, finally, go back and define a circuit element more carefully.

We have very little choice open to us with regard to a system of units. The one we shall use was adopted by the National Bureau of Standards in 1964; it is used by all the major professional engineering societies and is the language in which today's textbooks are written. This is the *International System of Units* (abbreviated SI in all languages), adopted by the General Conference on Weights and Measures in 1960. The SI is built upon six basic units: the meter, kilogram, second, ampere, kelvin, and candela. This, of course, is a "metric system," some form of which is now in common use in most of the technologically advanced countries of the world, and its use is being strongly urged in the United States.

We shall look at the definitions of the meter, kilogram, second, and ampere in the following. Standard abbreviations for them and the other SI units that we shall be using are listed on the inside cover and will be used throughout our discussions.

In the late 1700s the meter was defined to be exactly one ten-millionth of the distance from the earth's pole to its equator. The distance was marked off by two fine lines on a platinum-iridium bar which had been cooled to zero degrees Celsius (°C) (formerly centigrade). Although more accurate surveys have shown since that the marks on the bar do not represent this fraction of the earth's meridian exactly, the distance between the marks was nonetheless accepted internationally as the definition of the standard meter until 1960. In that year the General Conference based a more accurate definition of the meter (m) on a multiple of the wavelength of radiation of the orange line of krypton 86. Then in 1983 the meter was defined even more accurately as the distance that light travels through space in 1/299 792 458 seconds (to be defined in a second or two).

The basic unit of mass, the kilogram (kg), was defined in 1901 as the mass of a platinum block kept with the standard meter bar at the International Bureau of Weights and Measures in Sèvres, France. This definition was reaffirmed in 1960. The mass of this block is approximately 0.001 times the mass of 1 m^3 of pure water at 4°C.

The third basic unit, the second (s), was defined prior to 1956 as 1/86 400 of a mean solar day. At that time it was defined as 1/31 556 925.9747 of the tropical year 1900. In 1964 it was defined more carefully as 9 192 631 770 periods of the transition frequency between the hyperfine levels $F = 4$, $m_F = 0$ and $F = 3$, $m_F = 0$ of the ground state $^2S_{1/2}$ of the atom of cesium 133, unperturbed by external fields. This latter definition is permanent and more reproducible than the former; it is also comprehensible only to atomic physicists. However, any of these definitions adequately describes the second with which we are all familiar.

The definition of the fourth basic unit, the ampere (A), will appear later in this chapter after we are more familiar with the basic properties of electricity. The remaining two basic units, the kelvin (K) and the candela (cd), are not of immediate concern to circuit analysts.[2]

The SI incorporates the decimal system to relate larger and smaller units to the basic unit and uses standard prefixes to signify the various powers of ten. These are:

atto-	(a-, 10^{-18})	deci-	(d-, 10^{-1})
femto-	(f-, 10^{-15})	deka-	(da-, 10^{1})
pico-	(p-, 10^{-12})	hecto-	(h-, 10^{2})
nano-	(n-, 10^{-9})	kilo-	(k-, 10^{3})
micro-	(μ-, 10^{-6})	mega-	(M-, 10^{6})
milli-	(m-, 10^{-3})	giga-	(G-, 10^{9})
centi-	(c-, 10^{-2})	tera-	(T-, 10^{12})

Those shown inside the block in the table above are the ones used most frequently by students of electric-circuit theory.

These prefixes are worth memorizing, for they will appear often, both in

---

[2] Complete definitions of all the basic units and a further discussion of the International System of Units may be found in C. H. Page et al., IEEE Recommended Practice for Units in Published Scientific and Technical Work, *IEEE Spectrum*, vol. 3, no. 3, pp. 169–173, March 1966. Accurate numerical values are given in chapter 2, National Bureau of Standards Applied Mathematics Series 55, "Handbook of Mathematical Functions," 10th printing, 1972.

this text and in other scientific work. Thus, a millisecond (ms) is 0.001 second, and a kilometer (km) is 1000 m. It is apparent now that the gram (g) was originally established as the basic unit of mass, and the kilogram then represented merely 1000 g. Now the kilogram is our basic unit, and we could describe the gram as a millikilogram if we wished to be confusing. Combinations of several prefixes, such as the millimicrosecond, are unacceptable; the term *nanosecond* should be used. Also officially frowned on is the use of micron for 10^{-6} m; the correct term is the micrometer (μm). The angstrom (Å), however, may be used for 10^{-10} m.

This power-of-10 relationship is not present in the so-called *British System of Units,* which unfortunately is in common use in this country. There are many occasions when the results of an engineering analysis must be transformed into the British System of Units for use in the shop or for clarity in discussions with others during the long interim period of transition to the popular use of the metric system.

Most of us have a better mental picture of 2 in than we do of 5 cm, although this great dependence on the older system is gradually changing.

The fundamental British units are defined in terms of the SI units as follows: 1 in is exactly 0.0254 m, 1 pound-mass (lbm) is exactly 0.453 592 37 kg, and the second is common to both systems.

As a last item in our discussion of units, we consider the three derived units used to measure force, work or energy, and power. The newton (N) is the fundamental unit of force,[3] and it is the force required to accelerate a 1-kg mass by one meter per second per second (1 m/s^2). A force of 1 N is equivalent to 0.224 81 pound of force (lbf), and the average nineteen-year-old male, having a mass of 68 kg, exerts a force of 667 N on the scales.

The fundamental unit of work or energy is the joule (J), defined as one newton-meter (N-m). The application of a constant 1-N force through a 1-m distance requires an energy expenditure of 1 J. The same amount of energy is required to lift this book, weighing about 10 N, a distance of approximately 10 cm. The joule is equivalent to 0.737 56 foot pound-force (ft-lbf). Other energy units include the calorie[4] (cal), equal to 4.1868 J, the British thermal unit (Btu), which is 1055.1 J, and the kilowatt-hour (kWh), equal to 3.6×10^6 J.

The last derived quantity with which we shall concern ourselves is power, the rate at which work is done or energy is expended. The fundamental unit of power is the watt (W), defined as 1 J/s. One watt is equivalent to 0.737 56 ft-lbf/s. It is also equivalent to 1/745.7 horsepower (hp), a unit which is now being phased out of engineering terminology.

NOTE: Throughout the text, drill problems appear following sections in which a new principle is introduced, in order to allow students to test their understanding of the basic fact itself. The problems are useful in gaining familiarization with new

---

[3] It is worth noting that all units named after famous scientists have *abbreviations* beginning with capital letters.
[4] The calorie used with food, drink, and exercise is really a kilocalorie, 4186.8 J.

terms and ideas and should all be worked. More general problems appear at the ends of the chapters. The answers to the drill problems are given in order.

Drill Problem

1-1. (*a*) What is the volume in dm³ of a rectangular parallelepiped with dimensions 8 cm × 10⁵ μm × 3 × 10⁸ nm? (*b*) If a bag of jelly doughnuts contains 5000 kcal and all that energy is converted to heat, how many Btu does it represent? (*c*) What force in newtons is required to give an acceleration of 200 in/s² to a mass of 3 lbm?   *Ans:* 2.40; 19 840; 6.91

**1-3
The unit of charge**

Before beginning a discussion of electricity and electric circuits, we may define in terms of an analogy the class of electrical phenomena which we are going to consider. When we hold a baseball out at arm's length and release it, we know that it falls toward the earth because of the gravitational force exerted on it. We can also describe precisely *how* it accelerates, *what* its velocity is at any given instant, *when* it reaches a given point, and *where* it will be at a given instant. Few of us understand, however, *why* it falls. Although we understand very well what gravitational forces do, we do not know what they are.

In an analogous way, electrical engineers are very familiar with forces, meter deflections, heating effects, and other measurable responses caused by electricity, but they are only rarely concerned with the theoretical (and philosophical) nature of electricity itself. Therefore, our goal is a competence in observing electrical phenomena, describing them mathematically, and putting them to a practical use. We shall be only incidentally concerned with their cause.

Suppose that we take a small piece of some light material such as pith and suspend it by a fine thread. If we now rub a hard rubber comb with a woolen cloth and then touch the pith ball with the comb, we find that the pith ball tends to swing away from it; a force of repulsion exists between the comb and the pith ball. After laying down the comb and then approaching the pith ball with the woolen cloth, we can see that there is a force of attraction present between the pith ball and the woolen cloth.

We explain both of these forces on the pith ball by saying that they are *electrical* forces caused by the presence of *electric charges* on the pith ball, the comb, and the woolen cloth. In an analogous way, we attribute the force on the baseball to a *gravitational* force caused by the presence of gravitational masses in the baseball and the earth.

Our experiment shows clearly that the electrical force may be one of either attraction or repulsion, and in this respect the gravitational analogy breaks down. As far as we know at the present time, a gravitational repulsive force does not exist.

We explain the existence of an electrical force of both attraction and repulsion by the hypothesis that there are two kinds of charge and that like charges repel and unlike charges attract. The two kinds of charge are called positive and negative, although we might have called them gold and black or vitreous and resinous (as they were termed many years ago). Arbitrarily, the type of charge originally present on the comb was called negative by Benjamin Franklin, and that on the woolen cloth, positive.

We may now describe our experiment in these new terms. By rubbing the comb with the cloth, a negative charge is produced on the comb and a positive charge on the cloth. Touching the pith ball with the comb transferred some of its negative charge to the pith ball, and the force of repulsion between the like kinds of charge on the pith ball and comb caused the ball to move away. As we brought the positively charged woolen cloth near the negatively charged pith ball, a force of attraction between the two different kinds of charge was evident.

We also know now that all matter is made up of fundamental building blocks called atoms and that the atoms, in turn, are composed of different kinds of fundamental particles. The three most important particles are the electron, the proton, and the neutron. The electron possesses a negative charge, the proton possesses an equal-magnitude positive charge, and the neutron is neutral, or has no charge at all. As we rubbed the rubber comb with the woolen cloth, the comb acquired its negative charge because some of the electrons on the wool were rubbed off onto the comb; the cloth then had an insufficient number of electrons to maintain its electrical neutrality and thus behaved as a positive charge.

The mass of each of the three particles named above has been determined experimentally and is $9.109\ 56 \times 10^{-31}$ kg for the electron and about 1840 times as large for the proton and the neutron.

Now we are ready to define the fundamental unit of charge, called the coulomb after Charles Coulomb, the first man to make careful quantitative measurements of the force between two charges. The coulomb can, of course, be defined in any way we wish as long as the definition is convenient, universally accepted, permanent, and does not contradict any previous definition. Again, this leaves us no freedom at all because the definition which is already universally accepted is as follows: two small, identically charged particles which are separated by one meter in a vacuum and repel each other with a force of $10^{-7}c^2$ N possess an identical charge of either plus or minus one coulomb (C). The symbol c represents the velocity of light, $2.997\ 925 \times 10^8$ m/s. In terms of this unit, the charge of an electron is a negative $1.602\ 19 \times 10^{-19}$ C, and 1 C (negative) therefore represents the combined charge of about 6.24×10^{18} electrons.

We shall symbolize charge by Q or q; the capital letter is reserved for a charge which does not change with time, or is a constant, and the lowercase letter represents the general case of a charge that may vary with time. We often call this the *instantaneous* value of the charge and may emphasize its time dependence by writing it as $q(t)$. Note that $q(t)$ might represent a constant as a special case. This same use of capital and lowercase letters will be carried over to all other electrical quantities as well.

In their handwriting, many students[5] do not distinguish between upper-

---

[5] Instructors are perfect.

case and lowercase letters. This can have serious consequences, very few of them beneficial. For example, in electronics these four collector currents all mean different things: i_c, i_C, I_c, and I_C. How about that?

Drill Problem

1-2. What charge (in fC) is represented by: (*a*) ten million protons? (*b*) 10^{-20} g of electrons? (*c*) 10^7 electrons plus 3×10^6 protons plus 10^8 neutrons?

Ans: 1602; -1759; -1122 fC

1-4
Current

The electrical phenomena discussed above belong to the field of electrostatics, which is concerned with the behavior of electric charges at rest. This is of interest to us only because it is a beginning and serves as a useful device to define charge.

One part of the experiment, however, departed from electrostatics, the process of transferring charge from the wool cloth to the comb or from the comb to the pith ball. This idea of "transfer of charge" or "charge in motion" is of vital importance to us in studying electric circuits, because, in moving a charge from place to place, we may also transfer energy from one point to another. The familiar cross-country power transmission line is a practical example.

Of equal importance is the possibility of varying the rate at which the charge is transferred in order to communicate or transfer intelligence. This process is the basis of communication systems such as radio, television, and telemetry.

Charge in motion represents a *current*, which we shall define more carefully below. The current present in a discrete path, such as a metallic wire, has both a magnitude and a direction associated with it; it is a measure of the rate at which charge is moving past a given reference point in a specified direction. We now consider a rather arbitrary example which, however, will lead us to the general definition of current as the time rate of change of charge, dq/dt.

Let us consider a discrete path along which charge can move and ask a number of questions about the manner in which charge is traveling along this lead or conductor. As a first-hand observer, we shall place a very small student at point *A* on the path and ask that the total amount of charge which has passed since some reference time $t = 0$ be recorded. We ask that data be taken every second[6] and then give these detailed instructions:

1 The positive direction is to your right.
2 If positive charge moves past you in the positive direction, add the magnitude of the charge.
3 If positive charge moves in the negative direction, subtract the charge magnitude.
4 If negative charge moves in the positive direction, also subtract the charge magnitude.

---

[6] The student is small and *quick*.

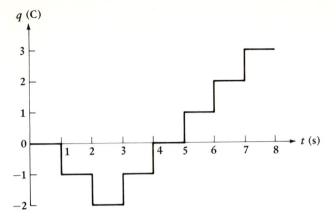

Fig. 1-1: A graph of the total charge q that has passed a given reference point since $t = 0$. The charge is measured over 1-s intervals.

Fig. 1-2: A graph of the instantaneous value of the total charge $q(t)$ which has passed a given reference point since $t = 0$.

5 If negative charge moves in the negative direction, add the charge magnitude.

The observer watches for 8 s, records data, and then hands us the graph, Fig. 1-1, explaining that q is the total charge which has moved past him since $t = 0$.

Now we see that there are many ways in which this record might be interpreted. For instance, in the first second either one unit positive charge moved past to the left or one unit negative charge moved past to the right. The same alternative is present for the second 1-s interval. As a matter of fact, in either of these intervals the observer might have had to count 100 unit positive charges moving to the right and 101 moving to the left. Perhaps positive and negative charges were in motion in both directions.

It is fortunate that we do not need to know which specific one of this infinite number of possibilities actually occurred; the electrical effects produced by each will be the same.

We now refine the data by making measurements much more often, and this requires that smaller and smaller elements of charge be counted. The limit is the amount of charge carried by a single electron. The graphical record now appears as a smooth curve, Fig. 1-2.

We are now ready to consider the *rate* at which charge is being transferred. In the time interval extending from t to $(t + \Delta t)$, the charge transferred past the reference point has increased from q to $(q + \Delta q)$. If the graph is decreasing in this interval, then Δq is a negative value. The rate at which charge is passing the reference point at time t is therefore very closely equal to $\Delta q/\Delta t$, and as the interval Δt decreases, the exact value of the rate is given by the derivative

$$\frac{dq}{dt} = \lim_{\Delta t \to 0} \frac{q(t + \Delta t) - q(t)}{\Delta t} = \lim_{\Delta t \to 0} \frac{\Delta q}{\Delta t}$$

We define the current at a specific point and flowing in a specified direction as the instantaneous rate at which net positive charge is moving past that point in the specified direction. Current is symbolized by I or i, and thus

$$i = \frac{dq}{dt} \tag{1}$$

The unit of current is the ampere (A), which corresponds to charge moving at the rate of 1 C/s. The ampere was named after A. M. Ampère, a French physicist of the early nineteenth century. It is often called an "amp," but this is informal and unofficial. The use of the lowercase letter i is again to be associated with an instantaneous value. Using the data of Fig. 1-2, the instantaneous current is given by the slope of the curve at every point. This current is plotted in Fig. 1-3.

The charge transferred between time t_0 and t may be expressed as a definite integral,

$$q\Big|_{t_0}^{t} = \int_{t_0}^{t} i \, dt$$

The total charge transferred over all time is obtained by adding $q(t_0)$, the charge transferred up to the time t_0, to the expression above,

$$q = \int_{t_0}^{t} i \, dt + q(t_0) \tag{2}$$

Several different types of current are illustrated in Fig. 1-4. A current which is constant is termed a direct current, or simply dc, and is shown by Fig. 1-4a. We shall find many practical examples of currents which vary sinusoidally with time, Fig. 1-4b; currents of this form are present in the normal household circuits. Such a current is often referred to as alternating current, or ac. Exponential currents and damped sinusoidal currents, sketched in Figs. 1-4c and d, will also be encountered later.

We establish a graphical symbol for current by placing an arrow next to the conductor. Thus, in Fig. 1-5a the direction of the arrow and the value

Fig. 1-3: The instantaneous current, $i = dq/dt$, where q is given in Fig. 1-2.

(a)   *(b)*

Fig. 1-5: Two methods of representation for the same current.

(a)   *(b)*

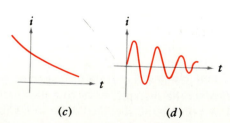

(c)   *(d)*

Fig. 1-4: Several types of current: (*a*) Direct current, or dc. (*b*) Sinusoidal current, or ac. (*c*) Exponential current. (*d*) Damped sinusoidal current.

Fig. 1-6: (a) and (b) Incomplete, improper, and incorrect definitions of a current. (c) The correct definition of $i_1(t)$.

$i_1(t)$ $i_1(t)$

(a)   (b)   (c)

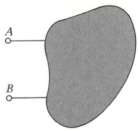

Fig. 1-7: A general circuit element is characterized by a pair of terminals to which other general circuit elements may be connected.

"3 A" indicate *either* that a net positive charge of 3 C/s is moving to the right or that a net negative charge of −3 C/s is moving to the left each second. In Fig. 1-5b there are again two possibilities: either −3 C/s is flowing to the left or +3 C/s is flowing to the right. All four of these statements and both figures represent currents which are equivalent in their electrical effects, and we say that they are equal.

It is convenient to think of current as the motion of positive charge even though it is known that current flow in metallic conductors results from electron motion. In ionized gases, in electrolytic solutions, and in some semiconductor materials, positively charged elements in motion constitute part or all of the current. Thus, any definition of current can agree only with the physical nature of conduction part of the time. The definition and symbolism we have adopted are standard.

It is essential that we realize that the current arrow does not indicate the "actual" direction of current flow, but is simply part of a convention that allows us to talk about "the current in the wire" in an unambiguous manner. The arrow is a fundamental part of the *definition* of a current! Thus, to talk about the value of a current $i_1(t)$ without specifying the arrow is to discuss an undefined entity. Therefore, Figs. 1-6a and b are meaningless representations of $i_1(t)$, whereas Fig. 1-6c is the proper definitive symbology. Remember:

The arrow is a fundamental part of the definition of a CURRENT i I

Drill Problem

1-3. The current $i_1(t)$ in Fig. 1-6c is given as $5e^{4t}$ A for $t < 0$, and $5e^{-4t}$ A for $t > 0$. Find: (a) $i_1(0.25)$; (b) the average value of $i_1(t)$ over the interval $-0.25 \le t \le 0.25$ s; (c) the total charge that has passed from left to right along the conductor in the interval $-0.25 \le t \le 0.25$ s.

Ans: 1.839 A; 3.16 A; 1.580 C

1-5
Voltage

We must now begin to refer to a circuit element, and we shall define it in very general terms. Such electrical devices as fuses, light bulbs, resistors, batteries, capacitors, generators, and spark coils can be represented by combinations of simple circuit elements. We shall begin by showing a very general circuit element as a shapeless object possessing two terminals at which connections to other elements may be made, Fig. 1-7. This simple picture may serve as the definition of a general circuit element. There are two paths by which current may enter or leave the element. Later we shall define particular circuit elements by describing the electrical characteristics which may be observed at their pairs of terminals.

Let us suppose that direct current is directed into terminal A of Fig. 1-7, through the general element, and out of B. Let us also assume that the passage of this charge through the element requires an expenditure of energy. We then say that an electrical *voltage* or a *potential difference* exists between the two terminals, or that there is a voltage or potential difference "across" the element. Thus, the voltage across a terminal pair is a measure of the work required to move charge through the element. Specifically, we shall define the voltage across the element as the work required to move a positive charge of 1 C from one terminal through the device to the other terminal. The sign of the voltage will be discussed below. The unit of voltage is the *volt* (V), which is 1 J/C, and voltage is represented by V or v. We are indeed fortunate that the full name of the eighteenth-century Italian physicist, Alessandro Giuseppe Antonio Anastasio Volta, is not used for our unit of potential difference.

A voltage or potential difference can exist between a pair of electrical terminals whether a current is flowing or not. An automobile battery, for example, has a voltage of 12 volts across its terminals even if nothing whatsoever is connected to the terminals.

The energy which is expended in forcing the charges through the element must appear somewhere else by the principle of conservation of energy. When we later meet specific circuit elements, we should note whether the energy is stored in some form which is readily available or whether it changes irreversibly into heat, acoustic energy, and so forth.

We must now establish a convention by which we can distinguish between energy supplied to the element by some external source and energy which may be supplied by the element itself to some external device. We do this by our choice of a sign for the voltage of terminal A with respect to terminal B. If a positive current is entering terminal A of the element and if an external source must expend energy to establish this current, then terminal A is positive with respect to terminal B. Alternatively, we may say also that terminal B is negative with respect to terminal A.

The sense of the voltage is indicated by a plus-minus pair of algebraic signs. In Fig. 1-8a, for example, the placement of the plus sign at terminal A indicates that terminal A is v volts positive with respect to terminal B. If we later find that v happens to have a numerical value of -5 V, then we may say either that A is -5 V positive with respect to B or that B is 5 V positive with

Fig. 1-8: In (a) and (b), terminal B is 5 V positive with respect to terminal A; in (c) and (d), terminal A is 5 V positive with respect to terminal B.

(a)   (b)   (c)   (d)

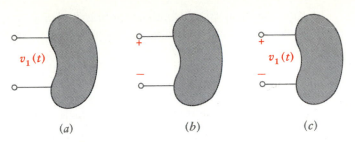

Fig. 1-9: (a) and (b) These are inadequate definitions of a voltage. (c) A correct definition includes both a symbol for the variable and a plus-minus symbol pair.

(a)   (b)   (c)

respect to A. Other cases are shown and described in Figs. 1-8b, c, and d.

Just as we noted in our definition of current, it is essential to realize that the plus-minus pair of algebraic signs does not indicate the "actual" polarity of the voltage, but is simply part of a convention that enables us to talk unambiguously about "the voltage across the terminal pair." The definition of any voltage must include a pair of plus-minus signs! Using a quantity $v_1(t)$ without specifying the location of the plus-minus-sign pair is using an undefined term. Figures 1-9a and b do not serve as definitions of $v_1(t)$; Fig. 1-9c does.

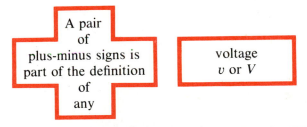

A pair of plus-minus signs is part of the definition of any   voltage v or V

Drill Problem

1-4. In Fig. 1-9c, let $v_1(t) = 100 \cos 250t$. Find: (a) $v_1(1 \text{ ms})$; (b) $v_1(8 \text{ ms})$; (c) the energy required to move 4 C from the lower to the upper terminal at $t = 4$ ms.   *Ans:* 96.9 V; -41.6 V; -216 J

**1-6
Power**

We now need to determine an expression for the power being absorbed by any circuit element in terms of the voltage across it and the current through it. Voltage has already been defined in terms of an energy expenditure, and power is the rate at which energy is expended. However, no statement can be made concerning energy transfer in any of the four cases shown in Fig. 1-8, for example, until the direction of the current is specified. Let us assume that a current arrow is placed by each upper lead, directed to the right, and labeled "+2 A"; then, since in both cases c and d terminal A is 5 V positive with respect to terminal B and since a positive current is entering terminal A, energy is being supplied to the element. In the remaining two cases, the element is delivering energy to some external device.

We have already defined power, and we shall represent it by P or p. If one joule of energy is expended in transferring one coulomb of charge through the device, then the rate of energy expenditure in transferring one coulomb of charge per second through the device is one watt. This absorbed power

Fig. 1-10: The power absorbed by the element is given by the product, $p = vi$.

must be proportional both to the number of coulombs transferred per second, or current, and to the energy needed to transfer one coulomb through the element, or voltage. Thus,

$$p = vi \qquad (3)$$

Dimensionally, the right side of this equation is the product of joules per coulomb and coulombs per second, which produces the expected dimension of joules per second, or watts.

With a current arrow placed by each upper lead of Fig. 1-8, directed to the right and labeled "2 A," 10 W is absorbed by the element in c and d and −10 W is absorbed (or 10 W is generated) in a and b.

The conventions for current, voltage, and power are summarized in Fig. 1-10. The sketch shows that if one terminal of the element is v volts positive with respect to the other terminal, and if a current i is entering the element through the first terminal, then a power $p = vi$ is being *absorbed by* or *delivered to* the element. When the current arrow is directed into the element at the plus-marked terminal, we satisfy the *passive sign convention*. This convention should be studied carefully, understood, and memorized. In other words, it says that if the current arrow and the voltage polarity signs are placed at the terminals of the element such that the current enters that end of the element marked with the positive sign, and if both the arrow and the sign pair are labeled with the appropriate algebraic quantities, then the power *absorbed by* the element can be expressed by the algebraic product of these two quantities. If the numerical value of the product is negative, then we say that the element is absorbing negative power, or that it is actually generating power and delivering it to some external element. For example, in Fig. 1-10 with $v = 5$ V and $i = -4$ A, then the element may be described as either absorbing −20 W or generating 20 W.

The three examples of Fig. 1-11 further illustrate these conventions.

Fig. 1-11: (a) A power, $p = (2)(3) = 6$ W, is absorbed by the element. (b) A power, $p = (-2)(-3) = 6$ W, is absorbed by the element. (c) A power, $p = (4)(-5) = -20$ W, is absorbed by the element, or 20 W is delivered by the element.

(a)   (b)   (c)

Fig. 1-12: See Drill Prob. 1-5.

(a)   (b)   (c)

Drill Problem

1-5. Find the power being: (*a*) delivered to the circuit element in Fig. 1-12*a* at *t* = 0.8 s; (*b*) generated by the circuit element in Fig. 1-12*b*; (*c*) absorbed by the circuit element in Fig. 1-12*c* at *t* = 0.

Ans: −34.68 W; 792 mW; −28.75 kW

1-7
Types of circuits and circuit elements

Using the concepts of current and voltage, it is now possible to be more specific in defining a circuit element.

It is important to differentiate between the physical device itself and the mathematical model of this device which we shall use to analyze its behavior in a circuit. Let us agree that we will use the expression "circuit element" to refer to the mathematical model. The choice of a particular model for any real device must be made on the basis of experimental data or experience; we shall usually assume that this choice has already been made. We must first learn the methods of analysis of idealized circuits.

Now let us distinguish a *general circuit element* from a *simple circuit element* by the statement that a general circuit element may be composed of more than one simple circuit element, but that a simple circuit element cannot be further subdivided into other simple circuit elements. For brevity, we shall agree that the term circuit element generally refers to a simple circuit element.

All the simple circuit elements that will be considered in the work that follows can be classified according to the relationship of the current through the element to the voltage across the element. For instance, if the voltage across the element is directly proportional to the current through it, or $v = ki$, we shall call the element a resistor. Other types of simple circuit elements have a terminal voltage which is proportional to the time derivative or the integral with respect to time of the current. There are also elements in which the voltage is completely independent of the current or the current is completely independent of the voltage; these are the *independent sources*. Furthermore, we shall need to define special kinds of sources in which the source voltage or current depends upon a current or voltage elsewhere in the circuit; such sources will be termed *dependent sources* or *controlled sources*.

By definition, a simple circuit element is the mathematical model of a two-terminal electrical device, and it can be completely characterized by its voltage-current relationship but cannot be subdivided into other two-terminal devices.

The first element which we shall need is an *independent voltage source*. It is characterized by a terminal voltage which is completely independent of the current through it. Thus, if we are given an independent voltage source and are notified that the terminal voltage is $50t^2$ V, we can be sure that at $t = 2$ s the voltage will be 200 V, regardless of the current that was flowing, is flowing, or is going to flow. The representation of an independent voltage source is shown in Fig. 1-13. The subscript *s* merely identifies the voltage as a "source" voltage.

Fig. 1-13: The circuit symbol of an independent voltage source. The circuit symbol of a dependent or controlled voltage source is shown in Fig. 1-16*a*.

A point worth repeating here is that the presence of the plus sign at the upper end of the symbol for the independent voltage source in Fig. 1-13 does not necessarily mean that the upper terminal is always positive with respect to the lower terminal. Instead, it means that the upper terminal is v_s volts positive with respect to the lower. If, at some instant, v_s happens to be negative, then the upper terminal is actually negative with respect to the lower at that instant.

If a current arrow labeled "i" is placed adjacent to the upper conductor of this source and directed to the left, then the current i is entering the terminal at which the positive sign is located and the source thus absorbs a power $p = v_s i$. More often than not, a source is expected to deliver power to a network and not to absorb it. Consequently, we might choose to direct the arrow to the right in order that $v_s i$ will represent the power delivered by the source. Either direction may be used.

The independent voltage source is an *ideal* source and does not represent exactly any real physical device, because the ideal source could theoretically deliver an infinite amount of energy from its terminals. Each coulomb passing through it receives an energy of v_s joules, and the number of coulombs per second is unlimited. This idealized voltage source does, however, furnish a reasonable approximation to several practical voltage sources. An automobile storage battery, for example, has a terminal voltage of 12 V that remains essentially constant as long as the current through it does not exceed a few amperes. The small current may flow in either direction through the battery. If it is positive and flowing out of the positively marked terminal, then the battery is furnishing power to the headlights, say, as it discharges; but if the current is positive and flowing into the positive terminal, then the battery is charging or absorbing energy from the generator, or possibly a battery charger. An ordinary household electrical outlet also approximates an independent voltage source providing the voltage $v_s = 115\sqrt{2} \cos 2\pi 60t$ V; the representation is valid for currents less than perhaps 20 A.

An independent voltage source which has a constant terminal voltage is often termed an independent dc voltage source[7] and is represented by either symbol shown in Fig. 1-14. Note in Fig. 1-14b that, when the physical plate

---

[7] Terms like "dc voltage source" and "dc current source" are commonly used. Literally, they mean "direct-current voltage source" and "direct-current current source," respectively. Although these terms may be redundant or misleading, the terminology is so widely used there's no point in fighting it.

Fig. 1-14: Alternative representations of a constant, or dc, independent voltage source. In (a) the source is delivering 12 W, and in (b) the battery is absorbing 12 W.

(a) (b)

Fig. 1-15: The circuit symbol of an independent current source.

structure of the battery is suggested, the longer plate is placed at the positive terminal; the plus and minus signs then represent redundant notation, but they are usually included anyway.

Another ideal source which we will need is the *independent current source*. Here, the current through the element is completely independent of the voltage across it. The symbol for an independent current source is shown in Fig. 1-15. If i_s is constant, we call the source an independent dc source.

Like the independent voltage source, the independent current source is at best a reasonable approximation for a physical element. In theory it can deliver infinite power from its terminals, because it produces the same finite current for any voltage across it, no matter how large that voltage may be. It is, however, a good approximation for many practical sources, particularly in electronic circuits. Also, the independent dc source represents very closely the proton beam of a cyclotron which is operating at a constant beam current of perhaps 1 μA and will continue to deliver 1 μA to almost any device placed across its "terminals" (the beam and the earth).

The two types of ideal sources that we have discussed up to now are called *independent* sources because the value of the source quantity is not affected in any way by activities in the remainder of the circuit. This is in contrast with yet another kind of ideal source, the *dependent* or *controlled source,* in which the source quantity is determined by a voltage or current existing at some other location in the electrical system under examination. To distinguish between independent and dependent sources, we introduce the diamond symbols shown in Fig. 1-16. Sources such as these will appear in the equivalent electrical models for many electronic devices, such as transistors, operational amplifiers, and integrated circuits. We shall see all of these in the following chapters.

Dependent and independent voltage and current sources are *active* elements; they are capable of delivering power to some external device. For the present we shall think of a *passive* element as one which is capable only of receiving power. However, we shall later see that several passive elements are able to store a finite amount of energy and then return it later to an external element, and since we shall still wish to call such an element passive, it will be necessary to improve upon our two definitions then.

The interconnection of two or more simple circuit elements is called an electrical *network*. If the network contains at least one closed path, it is also an electric *circuit*. Every circuit is a network, but not all networks are

Fig. 1-16: The diamond shape characterizes the circuit symbols for (*a*) the dependent voltage source and (*b*) the dependent current source.

v_s i_s

(*a*)   (*b*)

Fig. 1-17: (*a*) An electrical network which is not a circuit. (*b*) A network which is a circuit.

(*a*)   (*b*)

(*a*)

Fig. 1-18: (*a*) One possible physical configuration for an *npn* bipolar silicon transistor (not to scale). (*b*) The circuit symbol for the *npn* bipolar transistor. (*c*) A simple circuit model that is useful in the mid-frequency range. (*d*) A more accurate high-frequency model.

(*b*)   (*c*)

(*d*)

circuits. Figure 1-17*a* shows a network which is not a circuit, and Fig. 1-17*b* shows a network which is a circuit.

A network which contains at least one active element, such as an independent voltage or current source, is an *active network*. A network which does not contain any active elements is a *passive network*.

We have now defined what we mean by the term *circuit element,* and we have presented the definitions of several specific circuit elements, the independent and dependent voltage and current sources. Throughout the remainder of the book we will define only four additional circuit elements, all passive, called the *resistor,* the *inductor,* the *capacitor,* and a pair of *mutually coupled inductors.* These are all ideal elements. They are important because we may combine them into networks and circuits that represent real devices as accurately as we wish. Thus, the transistor whose physical construction is suggested by Fig. 1-18*a,* and whose electrical symbol is given in Fig. 1-18*b,* may be modeled by the single resistor and the single dependent current source of Fig. 1-18*c* if we need only approximate performance data at frequencies that are neither extremely high nor extremely low. Note that the dependent *current* source produces a current that depends on a *voltage* elsewhere in the circuit. Figure 1-18*d* illustrates a more accurate model for high-frequency applications that contains three resistors, two capacitors, and one dependent current source.

Transistors such as these may constitute only one small part of an integrated circuit that is possibly 2 mm square and 0.2 mm thick, and yet contains several thousand transistors plus thousands of resistors and capacitors. Thus we have a physical device that is about the size of one letter on this page, but that might require a model composed of ten thousand ideal simple circuit elements.

Suitable models for the various physical devices that have wide practical importance are studied in courses in electronics, energy conversion, antennas, and other subjects appearing later in engineering curricula.

Drill Problem

1-6. Find the power absorbed by each element in the circuit of Fig. 1-19.

Ans: −80; 100; 10; −12; −18 W

Fig. 1-19: See Drill Prob. 1-6.

Problems[8]

1 A famous mild-mannered reporter has a mass of 80 kg, can leap a tall (250-m) building at a single bound, and is as fast as a speeding (600 m/s) bullet. (*a*) What is his top speed in miles per hour? (*b*) What energy in joules must he impart to his leap to just clear the tall building? (*c*) For how many days would this energy power an electronic calculator requiring 100 mW? (*d*) What is the name of the reporter's girl friend?

2 A 70-kg man or woman burns up energy at an average rate of 120 W over a 24-h day. This metabolic rate drops to 75 W while sleeping, rises to 230 W while

[8] Answers are given at the back of the book for the odd-numbered problems.

walking and about 1000 W while running 10 mi/h. (*a*) What horsepower is expended running at 10 mi/h? (*b*) How many cans of light beer (100 kcal/can) should be consumed daily, just to keep the old body going?

3 When a certain electric clothes dryer is in full operation, it draws 5.2 kW. (*a*) How many "average horses" would have to be used to generate this power? (*b*) Assuming that fuel oil has a density of 50 lbm/ft³ and yields 18 500 Btu per pound-mass, how many liters (1 liter = 1000 cubic centimeters) of fuel oil would be used to operate the dryer for one hour?

4 The net charge that has moved to the right in a conductor at point x is given as a function of time by: $t \le -4$ s, $q = 0$; $-4 \le t \le -2$, $q = 4t + 16$; $-2 \le t \le 1$, $q = 2t^2$; $1 \le t \le 6$, $q = 5\sqrt{t + 3} - 8$; $t \ge 6$, $q = 7$ C. (*a*) Plot q vs. t. (*b*) Calculate i at $t = -4.5, -3.5, -2.5, \ldots , 5.5, 6.5$ s and plot i vs. t.

5 Given the plot of i vs. t of Fig. 1-20, calculate the total charge that has moved past the reference point in the time interval, $-2 < t < 5$ s.

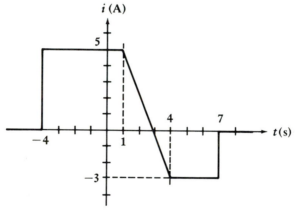

Fig. 1-20: See Prob. 5.

6 The waveform shown in Fig. 1-21 has a period of 5 ms. (*a*) What is the average value of the current over one period? (*b*) How much charge is transferred in the interval $1 \le t \le 4$ ms? (*c*) If $q(0) = 0$, sketch $q(t)$, $0 < t < 12$ ms.

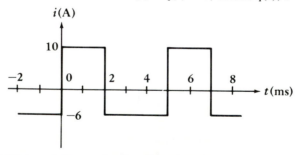

Fig. 1-21: See Probs. 6 and 10.

7 The total charge that has flowed to the right past point A in a certain conductor between $t = 0$ and t is identified as $q_A(t) = 100 \, e^{-200t} \cos 500t$ mC. (*a*) How much charge flows past A to the right between $t = 1$ ms and $t = 2$ ms? (*b*) What is the current to the right at A at $t = 1$ ms? (*c*) Now let the current directed to the right at A be $i_A(t) = 2(e^{-5000t} - e^{-8000t})$ A and find the amount of charge flowing to the right between $t = 10 \, \mu$s and $t = 80 \, \mu$s.

8 Determine the power being absorbed by each of the circuit elements shown in Fig. 1-22.

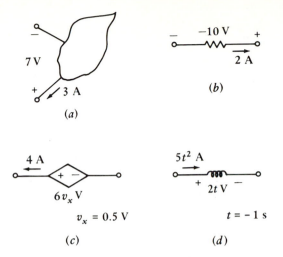

Fig. 1-22: See Prob. 8.

9 Let $i = 4e^{-50t}$ A and $v = 20 - 30e^{-50t}$ V for the circuit element of Fig. 1-10. (a) How much power is being absorbed by the device at $t = 10$ ms? (b) How much energy is delivered to the device in the interval, $0 < t < \infty$?

10 The current i shown in Fig. 1-21 enters the terminal with the (+) sign on it for a circuit element with a voltage v across it, where $v = 20 \sin 400\pi t$ V. (a) What is the maximum power absorbed by the element, and at what time does it occur? (b) What is the maximum power supplied by the element, and when does it occur? (c) How much energy is delivered to the circuit element in the time interval $0 < t < 5$ ms? (d) What average power is being absorbed by the element over this time interval?

11 In Fig. 1-10, let $v = 2t^2 - 8t + 6$ V. The power being absorbed by the circuit element is $p = 4(t^3 - 6t^2 + 11t - 6)$ W. (a) How much energy is delivered to the element from $t = 1$ to $t = 3$ s? (b) How much charge is delivered to the element in the time interval $1 < t < 3$ s?

12 Determine which of the five sources in Fig. 1-23 are being charged (absorbing positive power), and show that the algebraic sum of the five absorbed power values is zero.

Fig. 1-23: See Prob. 12.

Fig. 1-24: See Prob. 13.

13 Determine the power being delivered to each circuit element in Fig. 1-24.

14 In Fig. 1-10, let $v = 20e^{-0.2t}$ V. Find the power being absorbed by the element at $t = 4$ s if $i = $: (a) $0.1v$; (b) $0.1\, dv/dt$; (c) $0.1 \int_0^t v\, dt + 5$ A.

Chapter 2
Experimental Laws and
Simple Circuits

In the last chapter we became familiar with both independent and dependent voltage and current sources and were cautioned that they were idealized elements which could only be *approximated* in a real circuit. Another idealized element, the linear resistor, will be introduced in this chapter; we will also take a first look at the operational amplifier, a remarkable device which is an integral part of many of today's electric circuits.

We must next accept two fundamental laws as being axiomatic. With these laws, with the five simple circuit elements, and with the few definitions we already have, we may then begin to study simple electric circuits. This study will be restricted almost completely to *analysis*, which is the process by which the voltage and current associated with each element in a given circuit are determined. Fortunately, a complete analysis is not usually necessary, for often only a specific current, voltage, or perhaps power is needed.

After a proficiency in analysis has been achieved in this and other early courses, problems in *synthesis* may be considered. Here we are given a mathematical description of the desired behavior of a circuit and must determine the necessary elements and their interconnection in order to obtain the desired response. Synthesis problems often have more than one solution.

The final type of circuit problem, and the one for which engineering salaries are most often paid, is that of *design*. A real, physical, manufacturable, salable, economical, reliable device is the desired end product. Sometimes size, weight, temperature characteristics, environmental effects, or even eye appeal must be considered in the design. It is obvious that experience is a prerequisite for design proficiency; it is also evident that analysis and synthesis must come first.

This chapter and the following one are restricted to the analysis of simple circuits containing only current sources, voltage sources, and resistors; the sources may be independent or dependent. In analyzing these circuits we shall use several network transformations, network theorems, and mathematical methods which we shall later be able to apply, with only slight modifications, to circuits containing other types of passive elements excited by time-varying sources. We shall learn the methods useful in circuit analysis by applying them to the simplest possible case, the resistive circuit.

2-2 **Ohm's law**	The simplest passive element, the resistor, may be introduced by considering the work of an obscure German physicist, George Simon Ohm, who published a pamphlet in 1827 entitled "Die galvanische Kette mathematisch bearbeitet."[1] In it were contained the results of one of the first efforts to measure currents and voltages and to describe and relate them mathematically. One result was a statement of the fundamental relationship we now call Ohm's law, even though it has since been shown that this result was discovered 46 years earlier in England by Henry Cavendish, a brilliant semirecluse. However, no one, including Ohm, we shall hope, knew of the work done by Cavendish because it was not uncovered and published until long after both were dead.

Ohm's pamphlet received much undeserved criticism and ridicule for several years after its first publication, but it was later accepted and served to remove the obscurity associated with his name.

Ohm's law states that the voltage across many types of conducting materials is directly proportional to the current flowing through the material,

$$v = Ri \tag{1}$$

where the constant of proportionality R is called the *resistance*. The unit of resistance is the *ohm*, which is 1 V/A and customarily abbreviated by a capital omega, Ω.

When this equation is plotted on v versus i axes, it is a straight line passing through the origin. The equation is a linear equation, and we shall consider it as the definition of a *linear resistor*. Hence, if the ratio of the current and voltage associated with any simple current element is a constant, then the element is a linear resistor and has a resistance equal to the voltage-current ratio. Resistance is normally considered to be a positive quantity, although negative resistances may be simulated with special circuitry.

Again, it must be emphasized that the linear resistor is an idealized circuit element; it is a mathematical model of a physical device. "Resistors" may be easily purchased or manufactured, but it is soon found that the voltage-current ratio of this physical device is reasonably constant only within certain ranges of current, voltage, or power and depends also on temperature and other environmental factors. We shall usually refer to a linear resistor as simply a resistor, using the longer term only when the linear nature of the element needs emphasis. Any resistor which is nonlinear will always be described as such. Nonlinear resistors should not necessarily be considered as undesirable elements. Although it is true that their presence complicates an analysis, the performance of the device may depend on or be greatly

[1] "The Galvanic Circuit Investigated Mathematically."

Fig. 2-1: The circuit symbol for a resistor; $R = v/i$ and $p = vi = i^2R = v^2/R$.

improved by the nonlinearity. Zener diodes, tunnel diodes, and fuses are such elements.

Figure 2-1 shows the most common circuit symbol used for a resistor. In accordance with the voltage, current, and power conventions adopted in the last chapter, the product of v and i gives the power absorbed by the resistor. That is, v and i are selected to satisfy the passive sign convention. The absorbed power appears physically as heat and is always positive; a (positive) resistor is a passive element that cannot deliver power or store energy. Alternative expressions for the absorbed power are

$$p = vi = i^2R = \frac{v^2}{R} \tag{2}$$

One of the authors (who prefers not to be identified further)[2] had the unfortunate experience of inadvertently connecting a 100-Ω 2-W carbon resistor across a 110-V source. The ensuing flame, smoke, and fragmentation were rather disconcerting, demonstrating clearly that a practical resistor has definite limits to its ability to behave like the ideal linear model. In this case, the unfortunate resistor was called upon to absorb 121 W; since it was only designed to handle 2 W, its reaction was understandably violent.

The ratio of current to voltage is also a constant,

$$\frac{i}{v} = \frac{1}{R} = G \tag{3}$$

where G is called the *conductance*. The unit of conductance is the *mho*, 1 A/V, and is abbreviated by an inverted omega, ℧. The SI unit for conductance is the *siemen* (S), a term that is not as yet widely used in the United States. The same circuit symbol is used to represent both resistance and conductance. The absorbed power is again necessarily positive and may be expressed in terms of the conductance by

$$p = vi = v^2G = \frac{i^2}{G} \tag{4}$$

Thus a 2-Ω resistor has a conductance of $\frac{1}{2}$ ℧, and if a current of 5 A is flowing through it, a voltage of 10 V is present across the terminals and a power of 50 W is being absorbed.

All the expressions above have been written in terms of instantaneous current, voltage, and power, such as $v = Ri$ and $p = vi$. It is apparent that the current through and voltage across a resistor must both vary with time in the same manner. Thus, if $R = 10\,\Omega$ and $v = 2 \sin 100t$ V, then $i = 0.2 \sin 100t$ A; the power, however, is $0.4 \sin^2 100t$ W, and a simple sketch will illustrate the different nature of its variation with time. Although the current and voltage are each negative during certain time intervals, the absorbed power is never negative.

Resistance may be used as the basis for defining two commonly used

---

[2] Name gladly furnished upon written request to W.H.H.

terms, *short circuit,* and *open circuit.* We define a short circuit as a resistance of zero ohms; then, since $v = Ri$, the voltage across a short circuit must be zero, although the current may have any value. In an analogous manner, we define an open circuit as an infinite resistance. It follows then that the current must be zero, regardless of the voltage across the open circuit.

Remember, when any formula is written containing v's and i's, all of these quantities must be defined on a circuit diagram with their reference polarities and directions, respectively. Otherwise, $v = -Ri$ would be just as applicable to some resistor as $v = Ri$.

Drill Problem	**2-1.** If v and i are defined in Fig. 2-1, find: (*a*) R if $v = -8$ V and $i = -5$ mA; (*b*) the absorbed power if $i = -5$ A and $R = 2.2$ Ω; (*c*) i if $R = 8$ Ω and the resistor is absorbing 200 mW; (*d*) G if $v = 2.5$ V and $i = 100$ mA.

Ans: 1.60 kΩ; 55 W; ±158.1 mA; 40 m℧

2-3 Kirchhoff's laws

We are now ready to consider current and voltage relations in simple networks resulting from the interconnection of two or more simple circuit elements. The elements will be connected by electrical conductors, or leads, which have zero resistance, or are *perfectly conducting.* Since the network then appears as a number of simple elements and a set of connecting leads, it is called a *lumped-constant* network. A more difficult analysis problem arises when we are faced with a *distributed-constant* network, which essentially contains an infinite number of vanishingly small elements. Consideration of this latter type of network is mercifully relegated to later courses.

A point at which two or more elements have a common connection is called a *node.* Figure 2-2a shows a circuit containing three nodes. Sometimes networks are drawn so as to trap an unwary student into believing that there are more nodes present than is actually the case. This occurs when a node, such as node 1 in Fig. 2-2a, is shown as two separate junctions connected by a (zero-resistance) conductor, as in Fig. 2-2b. However, all that has been done is to spread the common point out into a common line. Thus, we must necessarily consider all of the perfectly conducting leads or portions of leads attached to the node as part of the node. Note also that every element has a node at each of its ends.

Suppose that we start at one node in a network and move through a simple

Fig. 2-2: (*a*) A circuit containing three nodes and five branches. (*b*) Node 1 is redrawn to look like two nodes; it is still one node.

(*a*)   (*b*)

element to the node at the other end, then continue from that node through a different element to the next node, and continue this movement until we have gone through as many elements as we wish. If no node was encountered more than once, then the set of nodes and elements that we have passed through is defined as a *path*. If the node at which we started is the same as the node on which we ended, then the path is, by definition, a *closed path* or a *loop*.

For example, in Fig. 2-2*a*, if we move from node 2 through the current source to node 1, and then through the upper right resistor to node 3, we have established a path; since we have not continued on to node 2 again, we do not have a closed path or a loop. If we proceeded from node 2 through the current source to node 1, down through the left resistor to node 2, and then up through the central resistor to node 1 again, we do not have a path, since a node (actually two nodes) was encountered more than once; we also do not have a loop, because a loop must be a path.

Another term whose use will prove convenient is a branch. We define a *branch* as a single path in a network, composed of one simple element and the node at each end of that element. Thus, a path is a particular collection of branches. The circuit shown in Figs. 2-2*a* and *b* contains five branches.

We are now ready to consider the first of the two laws named for Gustav Robert Kirchhoff (two h's and two f's), a German university professor who was born about the time Ohm was doing his experimental work.[3] This axiomatic law is called Kirchhoff's *current law* (or KCL for short), and it states that

> the algebraic sum of the currents entering any node is zero.

We shall not offer any proof of the law in this text. However, it simply represents a mathematical statement of the fact that charge cannot accumulate at a node. That is, if there were a net current into a node, then the rate at which coulombs were accumulating at the node would not be zero. However, a node is not a circuit element and it certainly cannot store, destroy, or generate charge. Hence, the currents must sum to zero.

Consider the node shown in Fig. 2-3. The algebraic sum of the four currents entering the node must be zero:

$$i_A + i_B - i_C - i_D = 0$$

Fig. 2-3: Kirchhoff's current law (KCL) enables us to write $i_A + i_B - i_C - i_D = 0$, $i_C + i_D - i_A - i_B = 0$, or $i_A + i_B = i_C + i_D$.

---

[3] Kirchhoff began his teaching career as a Privat-Dozent, or lecturer with official status but no salary, at the University of Berlin. This is not a popular option with today's young instructors.

It is evident that the law could be equally well applied to the algebraic sum of the currents *leaving* any node,

$$-i_A - i_B + i_C + i_D = 0$$

We might also wish to equate the sum of the currents having reference arrows directed into the node to the sum of those directed out of the node,

$$i_A + i_B = i_C + i_D$$

A compact expression for Kirchhoff's current law is

$$\sum_{n=1}^{N} i_n = 0 \tag{5}$$

and this is just a shorthand statement for

$$i_1 + i_2 + i_3 + \cdots + i_N = 0 \tag{6}$$

Whether (5) or (6) is used, it is understood that the N current arrows either are all directed toward the node in question or are all directed away from it.

It is sometimes helpful to interpret Kirchhoff's current law in terms of a hydraulic analogy. Water, like charge, cannot be stored at a point, and thus if we identify a junction of several pipes as a node, it is evident that the number of gallons of water entering the node every second must equal the number of gallons leaving the node each second.

We now turn to Kirchhoff's *voltage law* (KVL). This law states that

> the algebraic sum of the voltages around any closed path in a circuit is zero.

Again, we must accept this law as an axiom, even though it is developed in introductory electromagnetic theory.

Current is a variable that is related to the charge flowing *through* a circuit element, whereas voltage is a measure of potential energy difference *across* the element. There is a single unique value for this quantity in circuit theory. Thus, the energy required to move a unit charge from point A to point B in a circuit must have a value that is independent of the path taken from A to B.

In Fig. 2-4, if we carry 1 C from A to B through element 1, the reference polarity signs for v_1 show that we do v_1 joules of work. Now, if instead we

Fig. 2-4: The potential difference between points A and B is independent of the path selected, or $v_1 = v_2 - v_3$.

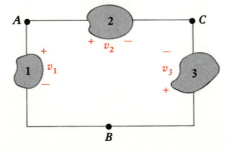

choose to proceed from A to B via C, then we expend $v_2 - v_3$ joules of energy. The work done, however is independent of the path in a circuit, and these values must be equal. Any route leads to the same value for the voltage. Thus,

$$v_1 = v_2 - v_3 \qquad (7)$$

It follows that if we trace out a closed path, the algebraic sum of the voltages across the individual elements around it must be zero. Thus, we may write

$$\sum_{n=1}^{N} v_n = 0 \qquad \text{or} \qquad v_1 + v_2 + v_3 + \cdots + v_N = 0 \qquad (8)$$

Kirchhoff's voltage law is a consequence of the conservation of energy and the conservative property of the electric circuit. This law can also be interpreted in terms of a gravitational analogy. If a mass is moved around any *closed* path in a conservative gravitational field, the total work done on the mass is zero.

We may apply KVL to a circuit in several different ways. One method that leads to fewer equation-writing errors than others consists of moving mentally around the closed path in a clockwise direction and writing down directly the voltage of each element whose + terminal is entered and writing down the negative of every voltage first met at the (−) sign. Applying this to the single loop of Fig. 2-4, we have

$$-v_1 + v_2 - v_3 = 0$$

which certainly agrees with our previous result, Eq. (7).

As a last example of the use of KVL, let us attack the circuit of Fig. 2-5. There are eight circuit elements, and voltages with plus-minus pairs are shown across each element. Suppose that we want to find v_{R2}, the voltage across R_2. We may do this by writing one KVL equation around the loop on the left:

$$4 - 36 + v_{R2} = 0$$

finding that

$$v_{R2} = 32 \quad V$$

Finally, let us assume that we want to determine the value of v_x. We might

Fig. 2-5: A circuit for which KVL shows that $v_x = 6$ V and $v_{R2} = 32$ V.

think of this as the (algebraic) sum of the voltages across the three elements on the right, or we might consider it as the voltage across an ideal voltmeter connected between points a and b, but it is just as easy to let it be the voltage across the gap from a to b. We apply KVL, beginning in the lower left corner, moving up and across the top to a, through v_x to b, and through the conducting lead to the starting point:

$$4 - 36 + 12 + 14 + v_x = 0$$

so that

$$v_x = 6 \quad V$$

Knowing v_{R2}, we might have taken the short cut through R_2,

$$-32 + 12 + 14 + v_x = 0$$

or $$v_x = 6 \quad V$$

once again.

Drill Problems

2-2. Determine the number of branches and nodes in each of the circuits of Fig. 2-6.   *Ans: 6,5; 6,4; 5,3*

2-3. Determine i_x in each of the circuits of Fig. 2-6.   *Ans: 3; −4; 13 A*

2-4. Determine v_x in each of the circuits of Fig. 2-6.   *Ans: −6; 50; 150 V*

(a)   (b)

(c)

Fig. 2-6: See Drill Probs. 2-2, 2-3, and 2-4.

2-4
Analysis of a single-loop circuit

Having established Ohm's and Kirchhoff's laws, we may flex our analytical muscles by applying these tools to the analysis of a simple resistive circuit. Figure 2-7a illustrates how the series connection of two batteries and two resistors might look as constructed by a youngster trying out a new soldering iron. Note that a connecting lead has been clamped onto the positive terminal of the left battery, and the other end of the wire has been soldered to one of the resistor leads. The terminal, the connecting leads, and the solder glob are all assumed to have zero resistance, and they constitute the node at the upper left-hand corner of the circuit diagram in Fig. 2-7b. Both batteries are replaced by ideal voltage sources; any resistances they may have are assumed to be small enough to neglect; if not, they may be incorporated into R_1 and R_2. The two resistors are assumed to be replaceable by ideal linear resistors.

We shall consider the resistance values and the source voltages of Fig. 2-7b to be known and attempt to determine the current through each element, the voltage across each element, and the power absorbed by each element.

Our first step in the analysis is the assumption of reference directions for the unknown currents. Arbitrarily, let us select a clockwise current i which flows out of the upper terminal of the left voltage source. This choice is indicated by an arrow labeled i at that point in the circuit, as shown in Fig. 2-7c. A trivial application of Kirchhoff's current law assures us that this same current must also flow through every other element in the circuit. We may emphasize this fact this one time by placing several other current symbols about the circuit.

By definition, all the elements that carry the *same* current are said to be connected in *series*. Note that elements may carry *equal* currents and not be in series; two 100-W lamp bulbs in neighboring houses may very well carry

Fig. 2-7: (a) A sketch of a physical single-loop circuit containing four elements. Solder connections and connecting wires are indicated. (b) The circuit model with source voltages and resistance values given. (c) Current and voltage reference signs have been added to the circuit.

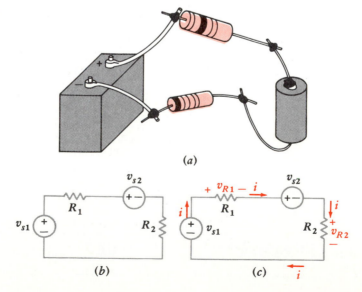

(a)

(b)

(c)

equal currents, but they do not carry the same current and are not in series.

Our second step in the analysis is the choice of a voltage reference for each of the two resistors. We have already found that the application of Ohm's law without a minus sign, $v = Ri$, demands that the sense of the current and voltage be selected so that the current enters the terminal at which the positive voltage reference is located. This is the passive sign convention. If the choice of the current direction is arbitrary, then the selection of the voltage sense is fixed if we intend to use Ohm's law in the form $v = Ri$. The voltages v_{R1} and v_{R2} are shown in Fig. 2-7c.

The third step is the application of Kirchhoff's voltage law to the single closed path present. Let us decide to move around the circuit in the clockwise direction, beginning at the lower left corner, and write down directly every voltage first met at its positive reference and write down the negative of every voltage encountered at the negative terminal. Thus,

$$-v_{s1} + v_{R1} + v_{s2} + v_{R2} = 0$$

Finally, we apply Ohm's law to the resistive elements,

$$v_{R1} = R_1 i \quad \text{and} \quad v_{R2} = R_2 i$$

and obtain

$$-v_{s1} + R_1 i + v_{s2} + R_2 i = 0$$

This equation is solved for i, and thus

$$i = \frac{v_{s1} - v_{s2}}{R_1 + R_2}$$

where all the quantities on the right side are known and enable us to determine i. The voltage or power associated with any element may now be obtained in one step by applying $v = Ri$, $p = vi$, or $p = i^2 R$.

Let us consider the numerical example illustrated in Fig. 2-8a. Two batter-

Fig. 2-8: (a) A given series circuit. (b) The circuit with current and voltage references assigned.

(a)   (b)

ies and two resistors are connected in a series circuit. A clockwise current and two resistor voltages are assigned to the circuit as indicated in Fig. 2-8b, and Kirchhoff's voltage law yields

$$-120 + v_{30} + 30 + v_{15} = 0$$

An application of Ohm's law to each resistor permits us to write

$$-120 + 30i + 30 + 15i = 0$$

from which

$$i = \frac{120 - 30}{30 + 15} = 2 \quad A$$

Thus, the voltages across the resistors are

$$v_{30} = 2(30) = 60 \quad V \qquad v_{15} = 2(15) = 30 \quad V$$

The power absorbed by each element is given by the product of the voltage across the element and the current flowing into the element terminal at which the positive voltage reference is located. For the 120-V battery, then, the power absorbed is

$$p_{120V} = 120(-2) = -240 \quad W$$

and thus 240 W is *delivered* to other elements in the circuit by this source. In a similar manner,

$$p_{30V} = 30(2) = 60 \quad W$$

and we find that this nominally active element is actually absorbing power (or being charged), delivered to it by the other battery.

The power absorbed by each (positive) resistor is necessarily positive and may be calculated by

$$p_{30} = v_{30}i = 60(2) = 120 \quad W$$

or by

$$p_{30} = i^2R = 2^2(30) = 120 \quad W$$

and

$$p_{15} = v_{15}i = i^2R = 60 \quad W$$

The results check because the total power absorbed must be zero, or in other words, the power delivered by the 120-V battery is exactly equal to the sum of the powers absorbed by the three other elements. A power balance is often a useful method of checking for careless mistakes.

Before leaving this example, it is important that we be convinced that our initial assumption of a current reference direction had nothing to do with the answers obtained. Let us assume a current i_x in a counterclockwise direction. Thus, $i_x = -i$. Both resistor voltages must now be assigned opposite polarities, and we obtain

$$-120 - 30i_x + 30 - 15i_x = 0$$

and $i_x = -2$ A, $v_{x30} = -60$ V, and $v_{x15} = -30$ V. Since each voltage and current reference is reversed and each quantity is the negative of the previously obtained value, it is evident that the results are the same. Each absorbed power will be the same.

Any random or convenient choice of current direction may be made, although clockwise currents are most often selected. Those who insist on positive answers may always go back and reverse the direction of the current arrow and rework the problem.

Now let us complicate the analysis slightly by letting one of the voltage

Fig. 2-9: A current i and voltage v_{30} are assigned in a single-loop circuit containing a dependent source.

sources be a dependent source, as exemplified by Fig. 2-9. We again assign a reference direction for the current i and the voltage v_{30}. There is no need to assign a voltage to the 15-Ω resistor, since the controlling voltage v_A for the dependent source is already available. It is worth noting, however, that the reference signs for v_A are reversed from those we would have assigned, and that Ohm's law for this element must thus be expressed as $v_A = -15i$. We apply Kirchhoff's voltage law around the loop,

$$-120 + v_{30} + 2v_A - v_A = 0$$

utilize Ohm's law twice,

$$v_{30} = 30i$$

$$v_A = -15i$$

and obtain

$$-120 + 30i - 30i + 15i = 0$$

$$i = 8 \quad \text{A}$$

The power relationships show that the 120-V battery supplies 960 W, the dependent source supplies 1920 W, and the two resistors together dissipate 2880 W.

More practical applications of the dependent source in equivalent circuits for the operational amplifier and the transistor will begin to appear in the following chapter.

Drill Problems

2-5. For the circuit shown in Fig. 2-10a, find: (a) i_a; (b) v_a; (c) the power supplied by the 15-V battery.   *Ans:* 7 mA; 8.4 V; -105 mW

2-6. In the circuit of Fig. 2-10b, find the power absorbed by each of the five elements in the circuit.
      Ans: CW from lower left: 576; -288; 144; -864; 432 W

Fig. 2-10: See Drill Probs. 2-5 and 2-6.

(a)   (b)

2-5
The single-node-pair circuit

The companion of the single-loop circuit discussed above is the single-node-pair circuit in which any number of simple elements are connected between the same pair of nodes. An example of such a circuit is shown in Fig. 2-11*a*. The two current sources and the conductance values are known, and we are to find the voltage, current, and power associated with each element once more.

Our first step is now to assume a voltage across any element, assigning an arbitrary reference polarity. Then Kirchhoff's voltage law forces us to recognize that the voltage across each branch is the same because a closed path proceeds through any branch from one node to the other and then is completed through any other branch. A total voltage of zero requires an identical voltage across every element. We shall say that elements having a common voltage across them are connected in *parallel*. Let us call this voltage v and arbitrarily select it as shown in Fig. 2-11*b*.

Two currents, flowing in the resistors, are then selected in conformance with the passive sign convention. These currents are also shown in Fig. 2-11*b*.

Our third step in the analysis of the single-node-pair circuit is the application of KCL to either of the two nodes in the circuit. It is usually clearer to apply it to the node at which the positive voltage reference is located, and thus we shall equate the algebraic sum of the currents leaving the upper node to zero,

$$-120 + i_{30} + 30 + i_{15} = 0$$

Finally, the current in each resistor is expressed in terms of v and the conductance of the resistor by Ohm's law,

$$i_{30} = 30v \qquad \text{and} \qquad i_{15} = 15v$$

and we obtain

$$-120 + 30v + 30 + 15v = 0$$

Thus,

$$v = 2 \quad \text{V}$$

$$i_{30} = 60 \quad \text{A} \qquad \text{and} \qquad i_{15} = 30 \quad \text{A}$$

The several values of absorbed power are now easily obtained. In the two resistors,

Fig. 2-11: (*a*) A single-node-pair circuit. (*b*) A voltage and two currents are assigned.

(*a*)   (*b*)

$$p_{30} = 30(2)^2 = 120 \quad W \qquad p_{15} = 15(2)^2 = 60 \quad W$$

and for the two sources,

$$p_{120A} = 120(-2) = -240 \quad W \qquad p_{30A} = 30(2) = 60 \quad W$$

Thus, the larger current source delivers 240 W to the other three elements in the circuit, and the conservation of energy is verified again.

The similarity of this example to the one previously completed, illustrating the solution of the series circuit with independent sources only, should not have gone unnoticed. The numbers are all the same, but currents and voltages, resistances and conductances, and "series" and "parallel" are interchanged. This is an example of *duality,* and the two circuits are said to be *exact duals* of each other. If the element values or source values were changed in either circuit, without changing the configuration of the network, the two circuits would be *duals,* although not exact duals. We shall study and use duality later, and at this time should only suspect that any result we obtain in terms of current, voltage, and resistance in a series circuit will have its counterpart in terms of voltage, current, and conductance for a parallel circuit.

Now let us try our skills on a single-node-pair circuit containing a dependent source, Fig. 2-12. The dependent current source is controlled by the current i_x in the 2-kΩ resistor. We assign a voltage v, arbitrarily with the positive reference at the top, and a current i_6 in the 6-kΩ resistance. The sum of the currents leaving the upper node is zero, so

$$i_6 - 2i_x - 0.024 - i_x = 0$$

We next apply Ohm's law to each resistor, noting that we are given the more common resistance values rather than conductances,

$$i_6 = \frac{v}{6000} \qquad \text{and} \qquad i_x = \frac{-v}{2000}$$

Therefore,

$$\frac{v}{6000} - 2\left(\frac{-v}{2000}\right) - 0.024 - \left(\frac{-v}{2000}\right) = 0$$

and

$$v = 600 \times 0.024 = 14.4 \quad V$$

Any other information we may want to find for this circuit is now easily obtained, usually in a single step. For example, the power supplied by the independent source is $p_{24} = 14.4(0.024) = 0.346$ W, and the current flowing to the right in the central conductor at the top is $i = -0.024 + (14.4/2000) = -0.0168$ A, or -16.8 mA.

Fig. 2-12: A voltage v and a current i_6 are assigned in a single-node-pair circuit containing a dependent source.

Drill Problems

2-7. For the single-node-pair circuit of Fig. 2-13a, find: (a) i_1; (b) i_2; (c) i_3.

Ans: -6; -9; 4 A

2-8. For the single-node-pair circuit shown in Fig. 2-13b, find the power absorbed by each of the five elements.

Ans: L to R: 15.4; 123.0; 0; -61.5; -76.9 W

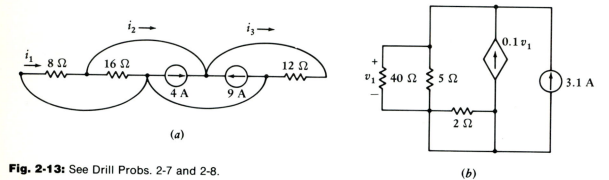

(a)

Fig. 2-13: See Drill Probs. 2-7 and 2-8.

(b)

2-6
Resistance and source combination

Some of the equation writing that we have been doing for the simple series and parallel circuits can be avoided. This is achieved by replacing relatively complicated resistor combinations by a single equivalent resistor whenever we are not specifically interested in the current, voltage, or power associated with any of the individual resistors in the combinations. All the current, voltage, and power relationships in the remainder of the circuit will be the same.

We first consider the series combination of N resistors, shown schematically in Fig. 2-14a. The shading surrounding the resistors is intended to suggest that they are enclosed in a "black box," or perhaps in another room, and we wish to replace the N resistors by a single resistor with resistance R_{eq} so that the remainder of the circuit, in this case only the voltage source, does not realize that any change has been made. The source current, power, and, of course, the voltage will be the same before and after.

We apply Kirchhoff's voltage law

$$v_s = v_1 + v_2 + \cdots + v_N$$

and Ohm's law

Fig. 2-14: (a) A circuit containing a series combination of N resistors. (b) A simpler equivalent circuit: $R_{eq} = R_1 + R_2 + \cdots + R_N$.

(a)

(b)

$$v_s = R_1 i + R_2 i + \cdots + R_N i = (R_1 + R_2 + \cdots + R_N)i$$

and then compare this result with the simple equation applying to the equivalent circuit shown in Fig. 2-14b,

$$v_s = R_{eq}i$$

Thus, the value of the equivalent resistance for N series resistors is

$$R_{eq} = R_1 + R_2 + \cdots + R_N \tag{9}$$

We are therefore able to replace a *two-terminal network* consisting of N resistors in series with a single *two-terminal element* R_{eq} that has the same *v-i* relationship. No measurements that we can make outside the "black box" can tell which network is which.

It should be emphasized again that we might be particularly interested in the current, voltage, or power of one of the original elements, as would be the case when the voltage of a dependent voltage source depends upon, say, the voltage across R_3. Once R_3 is combined with several series resistors to form an equivalent resistance, then it is gone and the voltage across it cannot be determined until R_3 is identified by removing it from the combination. It would have been better to have looked ahead and not made R_3 a part of the combination initially.

An inspection of the Kirchhoff voltage equation for a series circuit also shows two other possible simplifications. The order in which elements are placed in a series circuit makes no difference, and several voltage sources in series may be replaced by an equivalent voltage source having a voltage equal to the algebraic sum of the individual sources. There is usually little advantage in including a dependent voltage source in a series combination.

These simplifications may be illustrated by considering the circuit shown in Fig. 2-15a. We first interchange the element positions in the circuit, being careful to preserve the proper sense of the sources, and then combine the three voltage sources into an equivalent 90-V source and the four resistors into an equivalent 30-Ω resistance, as shown in Fig. 2-15b. Thus, instead of writing

$$-80 + 10i - 30 + 7i + 5i + 20 + 8i = 0$$

we have simply

$$-90 + 30i = 0$$

and

$$i = 3 \quad A$$

Fig. 2-15: (a) A given series circuit. (b) A simpler equivalent circuit.

(a)   (b)

Fig. 2-16: (a) A circuit containing N parallel resistors having conductances G_1, G_2, . . . , G_N. (b) A simpler equivalent circuit: $G_{eq} = G_1 + G_2 + \cdot \cdot \cdot + G_N$.

(a)   (b)

In order to calculate the power delivered to the circuit by the 80-V source appearing in the given circuit, it is necessary to return to that circuit with the knowledge that the current is 3 A. The desired power is 240 W.

It is interesting to note that no element of the original circuit remains in the equivalent circuit, unless we are willing to count the interconnecting wires as elements.

Similar simplifications can be applied to parallel circuits.[4] A circuit containing N conductances in parallel, as in Fig. 2-16a, leads to the KCL equation,

$$i_s = i_1 + i_2 + \cdot \cdot \cdot + i_N$$

or $$i_s = G_1v + G_2v + \cdot \cdot \cdot + G_Nv = (G_1 + G_2 + \cdot \cdot \cdot + G_N)v$$

whereas its equivalent in Fig. 2-16b gives

$$i_s = G_{eq}v$$

and thus

$$G_{eq} = G_1 + G_2 + \cdot \cdot \cdot + G_N$$

In terms of resistance instead of conductance,

$$\frac{1}{R_{eq}} = \frac{1}{R_1} + \frac{1}{R_2} + \cdot \cdot \cdot + \frac{1}{R_N}$$

or $$R_{eq} = \frac{1}{1/R_1 + 1/R_2 + \cdot \cdot \cdot + 1/R_N} \qquad (10)$$

This last equation is probably the most often used means of combining parallel resistive elements. The parallel combination is often indicated by writing $R_{eq} = R_1 \| R_2 \| R_3$, for example.

The special case of only two parallel resistors

$$R_{eq} = R_1 \| R_2 = \frac{1}{1/R_1 + 1/R_2} \quad \text{or} \quad R_{eq} = R_1 \| R_2 = \frac{R_1R_2}{R_1 + R_2} \qquad (11)$$

is needed very often. The last form is worth memorizing.

Parallel current sources may also be combined by algebraically adding the individual currents, and the order of the parallel elements may be rearranged as desired.

---

[4] Duality.

Fig. 2-17: (a) A given circuit. (b) A simplified equivalent circuit.

(a)   (b)

The various combinations described in this section are used to simplify the circuit of Fig. 2-17a. Let us suppose that we wish to know the power and voltage of the dependent source. We may just as well leave it alone, then, and combine the remaining two sources into one 2-A source. The resistances are combined by beginning with the parallel combination of the two 6-Ω resistors into a 3-Ω resistor, followed by the series combination of 3 Ω and 15 Ω. The 18-Ω and 9-Ω elements combine in parallel to produce 6 Ω, and this is as far as we can proceed profitably. Certainly 6 Ω in parallel with 3 Ω is 2 Ω, but the current i_3 on which the source depends then disappears.[5]

From the equivalent circuit in Fig. 2-17b, we have

$$-0.9i_3 - 2 + i_3 + \frac{v}{6} = 0$$

and
$$v = 3i_3$$

yielding

$$i_3 = \tfrac{10}{3} \quad \text{A}$$

$$v = 10 \quad \text{V}$$

Thus, the dependent source furnishes $v(0.9i_3) = 10(0.9 \times \tfrac{10}{3}) = 30$ W to the remainder of the circuit.

Now if we are belatedly asked for the power dissipated in the 15-Ω resistor, we must return to the original circuit. This resistor is in series with an equivalent 3-Ω resistor; a voltage of 10 V is across the 18-Ω total; thus, a current of $\tfrac{5}{9}$ A flows through the 15-Ω resistor and the power absorbed by this element is $(\tfrac{5}{9})^2(15)$, or 4.63 W.

To conclude the discussion of parallel and series element combinations, we should consider the parallel combination of two voltage sources and the series combination of two current sources. For instance, what is the equivalent of a 5-V source in parallel with a 10-V source? By the definition of a voltage source, the voltage across the source cannot change; by Kirchhoff's voltage law, then, 5 equals 10 and we have hypothesized a physical impossibility. Thus, voltage sources in parallel are permissible only when each has the same terminal voltage at every instant. Later, we shall see that *practical*

---

[5] Of course, we could have preserved it by using the given circuit to write $i_3 = v/3$, thus expressing i_3 in terms of variables appearing in the final circuit.

Fig. 2-18: (a) These two circuit elements are both in series and in parallel. (b) R_2 and R_3 are in parallel, and R_1 and R_8 are in series. (c) There are no circuit elements in series or in parallel.

voltage sources may be combined in parallel without any theoretical difficulty.

In a similar way, two current sources may not be placed in series unless each has the same current, including sign, for every instant of time.

A voltage source in parallel or series with a current source presents an interesting little intellectual diversion. The two possible cases are illustrated by Prob. 34 at the end of the chapter.

Three final comments on series and parallel combinations might be helpful. The first is illustrated by referring to Fig. 2-18a and asking, "Are v_s and R in series or in parallel?" The answer is clearly, "Both." The two elements carry the same current and are therefore in series; they also enjoy the same voltage and consequently are in parallel. This simple circuit is the only one for which this is true.

The second comment is a word of caution. Circuits can be drawn by inexperienced students or unscrupulous instructors in such a way as to make series or parallel combinations difficult to spot. In Fig. 2-18b, for example, the only two resistors in parallel are R_2 and R_3, while the only two in series are R_1 and R_8. Of course, v_s and R_7 are also in series.

The final comment is simply that a simple circuit element need not be in series or parallel with any other simple circuit element in a circuit. For example, R_4 and R_5 in Fig. 2-18b are not in series or parallel with any other simple circuit element, and there are no simple circuit elements in Fig. 2-18c that are in series or parallel with any other simple circuit element.

Drill Problems

2-9. An ohmmeter is an instrument that gives a value for the resistance seen between its two terminals. What is the correct reading if the instrument is

attached to the network of Fig. 2-19a at points: (a) ab; (b) ac; (c) bc?
Ans: 18.75; 24; 12.75 Ω

2-10. What conductance is measured at the terminals of the network in Fig. 2-19b if switch S is: (a) open; (b) closed; (c) replaced by a 10-m℧ conductance? *Ans:* 14; 20; 16.25 m℧

Fig. 2-19: See Drill Probs. 2-9 and 2-10.

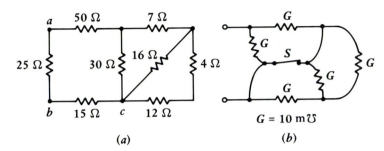

(a)   (b)

2-7
Voltage and current division

By combining resistances and sources, we have found one method of shortening the work of analyzing a circuit. Another useful shortcut is the application of the ideas of voltage and current division.

Voltage division is used to express the voltage across one of several series resistors in terms of the voltage across the combination. In Fig. 2-20, the voltage across R_2 is obviously

$$v_2 = R_2 i = R_2 \frac{v}{R_1 + R_2}$$

or

$$v_2 = \frac{R_2}{R_1 + R_2} v$$

and the voltage across R_1 is, similarly,

$$v_1 = \frac{R_1}{R_1 + R_2} v$$

If the network of Fig. 2-20 is generalized by replacing R_2 by R_2, R_3, . . . , and R_N in series, then we have the general result for voltage division across a string of N series resistors,

$$v_1 = \frac{R_1}{R_1 + R_2 + \cdots + R_N} v \tag{12}$$

Fig. 2-20: An illustration of voltage division, $v_2 = \frac{R_2}{R_1 + R_2} v.$

The voltage appearing across one of the series resistors is the total voltage times the ratio of its resistance to the total resistance. Voltage division and resistance combination may both be applied, as in the circuit shown in Fig.

Fig. 2-21: A numerical example illustrating resistance combination and voltage division. The wavy line within the source symbol indicates a sinusoidal variation with time.

2-21. We mentally combine the 3- and 6-Ω resistors, obtaining 2 Ω, and thus find that v_x is $\frac{2}{6}$ of 12 sin t, or 4 sin t V.

The dual of voltage division is current division. We now are given a total current supplied to several parallel conductances, as exemplified by the circuit of Fig. 2-22. The current flowing through G_2 is

$$i_2 = G_2 v = G_2 \frac{i}{G_1 + G_2}$$

or

$$i_2 = \frac{G_2}{G_1 + G_2} i$$

Fig. 2-22: An illustration of current division, $i_2 = \frac{G_2}{G_1 + G_2} i = \frac{R_1}{R_1 + R_2} i.$

and, similarly,

$$i_1 = \frac{G_1}{G_1 + G_2} i$$

Thus the current flowing through either of two parallel conductances is the total current times the ratio of that conductance to the total conductance.

Since we are given the value of the resistance more often than the conductance, a more important form of the last result is obtained by replacing G_1 by $1/R_1$ and G_2 by $1/R_2$,

$$i_2 = \frac{R_1}{R_1 + R_2} i \quad \text{and} \quad i_1 = \frac{R_2}{R_1 + R_2} i$$

Nature has not smiled on us here, for these last two equations have a factor which differs subtly from the factor used with voltage division, and some effort is going to be needed to avoid errors. Many students look on the expression for voltage division as "obvious" and that for current division as being "different." It also helps to realize that the larger of two parallel resistors always carries the smaller current.

We may also generalize these results by replacing G_2 in Fig. 2-22 by the parallel combination of $G_2, G_3, \ldots, G_N$. Thus, for N parallel conductances,

$$i_1 = \frac{G_1}{G_1 + G_2 + \cdots + G_N} i \qquad (13)$$

In terms of resistance values, the result becomes

$$i_1 = \frac{1/R_1}{1/R_1 + 1/R_2 + \cdots + 1/R_N} \, i \tag{14}$$

As an example of the use of both current division and resistance combination, let us return to the example of Fig. 2-21 and write an expression for the current through the 3-Ω resistor. The total current flowing into the 3- and 6-Ω combination is

$$i = \frac{12 \sin t}{4 + (6)(3)/(6 + 3)}$$

and thus the desired current is

$$i_3 = \frac{12 \sin t}{4 + (6)(3)/(6 + 3)} \frac{6}{6 + 3} = \tfrac{4}{3} \sin t$$

Unfortunately, current division is sometimes applied when it is not applicable. As one example, let us consider again the circuit shown in Fig. 2-18c, a circuit that we have already agreed contains no circuit elements that are in series or in parallel. Without parallel resistors, there is no way that current division can be applied. Even so, there are too many students who take a quick look at resistors R_A and R_B and try to apply current division, writing an incorrect equation such as

Remember, parallel resistors must be branches between the same pair of nodes.

Drill Problem

2-11. In the circuit of Fig. 2-23: (*a*) use resistance combination methods to find R_{eq}; (*b*) use current division to find i_1; (*c*) use voltage division to find v_2; (*d*) use current division to find i_3.   *Ans:* 30 Ω; 0.6 A; 9 V; 0.7 A

Fig. 2-23: See Drill Prob. 2-11.

2-8
A practical example: the operational amplifier

We have now been introduced to enough basic laws and simple analytical techniques that we should be able to apply them successfully to some interesting practical circuits. In this section we shall begin to consider an electrical device called the *operational amplifier,* or *op-amp* for short.

The first operational amplifiers were built in the 1940s using vacuum tubes

Fig. 2-24: (a) The circuit symbol for an operational amplifier. (b) Input voltages v_1, v_2, their difference v_i, and an output v_o are defined.

(a)   (b)

to perform the mathematical *operations* of addition, subtraction, multiplication, division, differentiation, and integration electrically, thus enabling the electrical solution of differential equations in the early analog computers.

Reduced to its basic essentials, the op-amp is just a voltage-controlled, dependent voltage source. The dependent voltage source appears at the output terminals of the op-amp, and the voltage on which it depends is applied to the input terminals.

The symbol commonly used for an op-amp is shown in Fig. 2-24a. Two input terminals are shown on the left, and a single output terminal appears at the right. There is also a common terminal or node called ''ground'' that is not usually shown explicitly as a terminal of the op-amp itself. In practical circuits there are often numerous elements connected to the metallic case or chassis on which a circuit is built, and this chassis is then connected through a good conductor to the earth. Thus, the metallic case becomes the ground node. The symbol for the ground node appears several times at the bottom of Fig. 2-24b.

Although we may consider a single voltage signal or voltage source to be applied directly between the pair of input terminals, we can provide for a greater number of applications by establishing a voltage between each input terminal and ground. The input terminal identified by a $(-)$ sign is the *inverting input,* and a voltage v_1 is defined between the inverting input and ground, as shown in Fig. 2-24b. The voltage v_2 is defined between the *noninverting input* and ground. We shall see shortly that the voltage $(v_1 - v_2)$ appears greatly amplified and reversed in polarity between the output terminal and ground. If $v_2 = 0$, then v_1 appears amplified and inverted at the output; if $v_1 = 0$, then v_2 appears amplified, without a sign change, between the output terminal and ground. The amplification factor ranges from 10^4 to 10^7 for different op-amps; a typical value is 10^5. The difference between v_1 and v_2 is the input voltage $v_i = v_1 - v_2$.

An op-amp may cost as little as twenty cents, for which we obtain an integrated-circuit (IC) chip containing about 25 transistors and a dozen resistors, all packaged in a small can or ceramic wafer with 8 or 10 terminal pins for connection to the external circuit. In some cases, the IC chip may contain several op-amps. In addition to the output pins and the two inputs, other pins enable voltages to be supplied for the transistors and external adjustments to be made to balance and compensate the op-amp. However, we are not concerned with the internal circuitry of the op-amp or the IC at this time, but only with the voltage and current relationships that exist at the external terminals. Thus, the connections shown in Fig. 2-24a are sufficient.

Fig. 2-25: A simple model for an op-amp contains one dependent source and several terminals.

Figure 2-25 shows a useful model for an op-amp. The resistance between the input terminals is so large (10^5 to 10^{15} Ω) that we can safely represent it by an open circuit, as shown. In later examples we shall connect such an *input resistance* R_i between the two input terminals, but at this time we shall work with a model that is closer to the ideal. A voltage-controlled dependent voltage source provides an output voltage equal to A times the difference of the two input voltages. More exact models of an op-amp include an *output resistance* R_o, ranging from 1 to 1000 Ω, in series with the dependent source and the output terminal.

Let us suppose that $v_1 = 1$ μV, $v_2 = 0.6$ μV, and $A = 10^5$. Then, $v_i = 10^{-6} - 0.6 \times 10^{-6} = 0.4 \times 10^{-6}$ V, and $v_o = -10^5 \times 0.4 \times 10^{-6} = -0.4$ V. Note the polarity signs on the dependent source. If we now connect the noninverting input to ground, then $v_2 = 0$, $v_i = 10^{-6}$ V, and $v_o = -10^5 \times 10^{-6} = -0.1$ V; with the inverting input grounded, $v_1 = 0$, $v_i = -0.6 \times 10^{-6}$ V, and $v_o = -10^5(-0.6 \times 10^{-6}) = 0.06$ V.

The operational amplifier is seldom used in the bare form suggested by the examples above. There are usually several circuit elements connected in parallel or series with the input or the output, or between the output and the input. We shall use many of these practical circuits as examples as we study circuit analysis in the chapters that follow.

The first example is the *voltage follower* shown in Fig. 2-26a. Here an input signal v_s is connected between the noninverting input and ground so that $v_2 = v_s$. A short circuit is connected directly from the output to the inverting input, and $v_1 = v_o$. Figure 2-26b shows the equivalent circuit, and it is seen to be a single-loop circuit even though there is an open circuit between the two input terminals. As a matter of fact, the loop current must be

Fig. 2-26: (a) An op-amp is connected as a voltage follower. (b) The equivalent circuit is treated as a single-loop circuit with an open circuit and zero loop current.

(a) (b)

zero, and there is therefore no current anywhere in the circuit. It follows that Kirchhoff's current law cannot give us any additional information, Ohm's law is worthless without any resistors, and we can only turn hopefully to Kirchhoff's voltage law for some information about the relationship between the output voltage v_o and the input voltage v_s. Around the loop, we have

$$-v_s - v_i - Av_i = 0$$

Also,

$$v_o = -Av_i$$

or

$$v_i = -\frac{v_o}{A}$$

Therefore

$$-v_s + \frac{v_o}{A}(1 + A) = 0$$

and

$$v_o = \frac{A}{1 + A} v_s \tag{15}$$

If $A = 10^5$, then $v_o = 0.999\,990\,v_s$. For all practical purposes, therefore, $v_o = v_s$, and the output voltage "follows" the input voltage. The advantage of such a unity-gain amplifier is that the input draws negligible current and power from the source, while the output can supply reasonable currents (10 to 20 mA) and power (100 to 500 mW) to a load connected across the output. Thus, the load has little effect on the source, and for this reason the voltage follower is also known as a *buffer amplifier*.

Some specific numerical values for the voltages can be helpful. If $v_s = 1$ V, then $v_o = 0.999\,990$ V when $A = 10^5$, and the input voltage $v_i = -9.9999\ \mu$V. The magnitude of the input voltage is quite small, and we often make an approximate analysis of an op-amp circuit by assuming that the input voltage $v_i = 0$, as well as the input current. If we do this for the voltage follower, then we can conclude immediately that $v_o = v_s$. This result is obtained more rigorously by letting the gain A approach infinity in Eq. (15).

Drill Problem

2-12. A voltage follower is constructed using an op-amp for which $A = 12\,000$. If $v_s = 1$ mV, find: (a) v_1; (b) v_2; (c) v_i; (d) v_o.

Ans: 0.999 917 mV; 1 mV; $-0.0833\ \mu$V; 0.999 917 mV

Problems

1 Find the resistance of a resistor if: (a) a constant current of 5 C/s delivers 10 kJ to the element in 4 min; (b) it dissipates 1 kW when connected across a 12-V battery. (c) What average power does the source $v_s = 80 \cos 500\pi t$ V deliver to a 5-Ω resistor over the time interval $0 \le t \le 12$ ms?

2 The current $i(t)$ in Fig. 2-1 is 4 A in the interval $0 < t < 1$ s, -4 A for $1 < t < 2$ s, and continues in this periodic manner. If $R = 100\ \Omega$, sketch $i(t)$, $v(t)$, and $p(t)$ for $0 < t < 4$ s.

3 The dc resistance of a conductor having a length l and a uniform cross-sectional area A is given by $R = \rho l/A = l/\sigma A$, where ρ (rho) is the electrical resistivity and

σ (sigma) is the electrical conductivity. If $\sigma = 5.8 \times 10^7$ Ʊ/m for copper: (*a*) what length (in feet) of #18 copper wire (diameter = 1.024 mm) has a resistance of 1 Ω? (*b*) If a circuit board has a copper-foil conducting ribbon 0.0013 in thick and 0.02 in wide that can carry 3 A safely at 60°C, what power does this current deliver to a conductor 6 in long?

4 In the circuit of Fig. 2-27, find R and G if the 5-A source is supplying 125 W.

Fig. 2-27: See Prob. 4.

5 Use Ohm's law and Kirchhoff's laws on the circuit of Fig. 2-28 to find: (*a*) v_{in}; (*b*) the power provided by the dependent source; (*c*) V_s.

Fig. 2-28: See Prob. 5.

6 (*a*) Use Ohm's and Kirchhoff's laws in a step-by-step procedure to determine all the voltages and currents in the circuit of Fig. 2-29. (*b*) Compute the power absorbed by every element and show that the sum is zero.

Fig. 2-29: See Prob. 6.

7 In the circuit shown in Fig. 2-30, find the power absorbed by the: (*a*) 4-A source; (*b*) 2-Ω resistor; (*c*) 12-Ω resistor; (*d*) 6-Ω resistor; (*e*) voltage source V_s.

Fig. 2-30: See Prob. 7.

8 A certain circuit contains six elements and four nodes, numbered 1, 2, 3, and 4. Each circuit element is connected between a different pair of nodes. The current directed from node 1 to node 2 in that branch is $i_{12} = 15$ A, while $i_{34} = -8$ A. Find i_{23}, i_{13}, and i_{41} if $i_{24} = :$ (a) 0; (b) 18 A; (c) -18 A.

9 Find the power being absorbed by element X in Fig. 2-31 if it is a: (a) 70-Ω resistor; (b) 2-V independent voltage source, + reference on left; (c) dependent voltage source, + reference on left, labeled $19i_x$; (d) 52-mA independent current source, arrow directed to right.

Fig. 2-31: See Prob. 9.

10 (a) Find i in the circuit shown in Fig. 2-32. (b) Let v_4 now extend across both the 4-Ω and 50-V elements, + reference on the left, and again find i.

Fig. 2-32: See Prob. 10.

11 For the battery charger modeled by Fig. 2-33: (a) find R so that a charging current of 2.5 A flows; (b) find R so that 25 W is delivered to the 11-V ideal voltage source; (c) find R so that 25 W is delivered to the combination of 0.04 Ω and 11 V.

Battery charger

Fig. 2-33: See Prob. 11.

Battery

Fig. 2-34: See Probs. 12 and 13.

12 In the circuit shown in Fig. 2-34, X is a simple circuit element. Assume that X is absorbing 100 W and: (a) find R if X is a resistor greater than 5 Ω; (b) find v_s, + reference at top, if X is an independent voltage source, $v_s < 5$ V.

13 Let the voltage across the 0.2-Ω resistor in Fig. 2-34 be 4 V, + reference at the left. Assume that X represents an unknown circuit element. (a) Find the absorbed power for each element in the circuit. (b) If X is a resistor, give its resistance R (which might be negative); if it is an independent voltage source, give v_s, including polarity; and if it is an independent current source, give i_s, including reference direction. (c) Repeat parts a and b if the + reference for the 4 V is at the right.

14 Find the power absorbed by each of the six elements in the circuit in Fig. 2-35.

Fig. 2-35: See Prob. 14.

15 Find the power absorbed by each element in the circuit illustrated in Fig. 2-36.

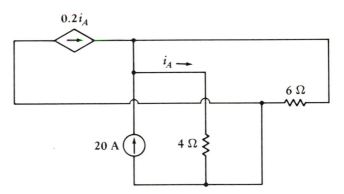

Fig. 2-36: See Prob. 15.

16 In Fig. 2-37, find the power supplied by each of the sources if the dependent source is labeled: (a) $3i_A$; (b) $3i_B$.

17 Find the power absorbed by each circuit element in the single-node-pair circuit of Fig. 2-38.

Fig. 2-38: See Prob. 17.

Fig. 2-37: See Prob. 16.

18 In the circuit of Fig. 2-39, X is a simple circuit element. Assume that X is absorbing 100 W and: (a) find R if X is a resistor larger than 50 Ω; (b) find i_s, reference arrow directed down, if X is an independent current source, $i_s > 2$ A.

Fig. 2-39: See Probs. 18 and 19.

19 Let the downward current through the 100-Ω resistor in Fig. 2-39 be 1.2 A. Assume that X represents an unknown circuit element. (a) Find the absorbed power for each element in the circuit. (b) If X is a resistor, give its resistance R (which might be negative); if it is an independent current source, give i_s, including reference direction; and if it is an independent voltage source, give v_s, including polarity. (c) Repeat parts a and b if the 1.2-A current is directed upward.

20 (a) Apply the techniques used in single-node-pair analysis to the upper right node

Fig. 2-40: See Prob. 20.

(a)

(b)

(c)

Fig. 2-41: See Prob. 21.

in Fig. 2-40 and find i_x. (b) Now work with the upper left node and find v_5. (c) How much power is the 16-A source generating?

21 Find R_{eq} for each of the three resistive networks shown in Fig. 2-41.

22 For the network of Fig. 2-42: (a) find R_{eq} if $R = 14\ \Omega$; (b) find R if $R_{eq} = 14\ \Omega$.

Fig. 2-42: See Prob. 22.

23 Find R_{eq} for each network shown in Fig. 2-43.

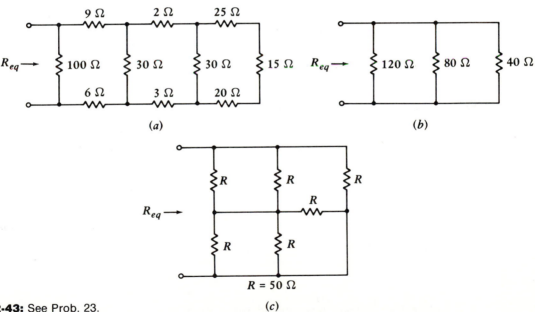

(a)

(b)

$R = 50\ \Omega$

(c)

Fig. 2-43: See Prob. 23.

24 By combining sources and resistances in the circuit of Fig. 2-44, find v_x and i_x.

Fig. 2-44: See Prob. 24.

25 Use source and resistance combinations to simplify the circuit of Fig. 2-45 and then find i_1 and v_2.

Fig. 2-45: See Prob. 25.

26 Given 10 resistors, each having a resistance of 20 Ω: (*a*) what is the smallest (nonzero) equivalent resistance that they could be assembled to produce? (*b*) What is the largest (finite) equivalent resistance? Show how some of these resistors might be combined to produce an equivalent resistance of: (*c*) 29 Ω; (*d*) 6 Ω.

27 Find the equivalent conductance for each of the networks of Fig. 2-46.

(*a*)

(*b*)

Fig. 2-46: See Prob. 27.

28 By combining resistors and using current and voltage division on the circuit of Fig. 2-47, write an expression for: (*a*) i_4; (*b*) v_2; (*c*) i_o.

Fig. 2-47: See Prob. 28.

29 Use resistance combination and voltage or current division to find i_x in each circuit shown in Fig. 2-48.

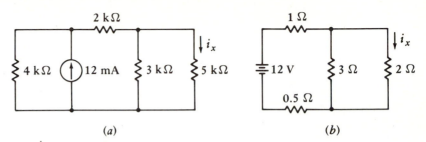

Fig. 2-48: See Prob. 29.

(a)   (b)

30 (a) Write a one-line expression, using resistance combination and voltage division, to find v_{57} in the circuit of Fig. 2-49a. (b) Write a one-line expression, using resistance combination and current division, to find i_{57} for the circuit of Fig. 2-49b.

(a)   (b)

Fig. 2-49: See Prob. 30.

31 The circuit shown in Fig. 2-50 contains a three-section resistive ladder network. (a) Assume $i_o = 1$ A and work one step at a time from the right side of the network toward the source and find V_s. (b) What is V_s if $i_o = 0.4$ A? (c) If $V_s = 100$ V, find i_o.

Fig. 2-50: See Prob. 31.

32 Find the value of R in the circuit of Fig. 2-51.

Fig. 2-51: See Prob. 32.

33 (*a*) Use voltage division, current division, and resistance combination to find i_y in the circuit of Fig. 2-52. (*b*) To what value should the 3-Ω resistance be changed to make $i_y = 1$ A?

Fig. 2-52: See Prob. 33.

Fig. 2-53: See Prob. 34.

34 The circuit shown in Fig. 2-53 shows several examples of independent current and voltage sources in series and parallel. (*a*) Find the power absorbed by each source. (*b*) To what value should the 4-V source be changed to reduce the power supplied by the -5-A source to zero?

35 A certain op-amp has a gain A of 20 000 and a resistance, $R_i = 50$ kΩ, inside the op-amp between the inverting and noninverting inputs. The op-amp is connected as a voltage follower with $v_s = 1$ V. Find: (*a*) v_o; (*b*) v_i; (*c*) the power supplied by v_s.

36 (*a*) Find v_o for the voltage follower shown in Fig. 2-54a if A is large. (*b*) The network shown in Fig. 2-54b is connected to the follower output. Find v_x.

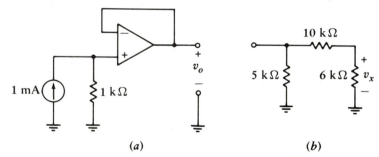

Fig. 2-54: See Prob. 36.

(*a*)

(*b*)

37 (*a*) Find v_x for the circuit shown in Fig. 2-55 if A is large. (*b*) If the op-amp is removed from the circuit and points *a* and *b* are connected together, what is v_x?

Fig. 2-55: See Prob. 37.

Chapter 3
Some Useful Techniques of
Circuit Analysis

We should now be familiar with Ohm's law and Kirchhoff's laws and their application in the analysis of simple series and parallel resistive circuits. When it will produce results more easily we should be able to combine resistors or sources in series or parallel and be able to use the principles of voltage and current division. Most of the circuits on which we have been practicing are simple and of questionable practical importance; they are useful in helping us learn to apply the fundamental laws. Now we must begin analyzing more complicated circuits.

Physical systems which we shall want to analyze and design in the coming years will include electric and electronic control circuits, communication systems, energy converters such as motors and generators, power distribution systems, interconnecting circuits for commercially available integrated circuits, and entertainment or other devices which are now unknown. Many of us will be confronted with allied problems involving heat flow, fluid flow, and the behavior of various mechanical systems. In the analysis of any of these cases it is often helpful to replace the system with an equivalent electric circuit. As an example, we might consider a transistor amplifier, an electronic device which is part of many communication systems and control circuits. Transistors, along with resistors and other passive circuit elements, are used to amplify or magnify electrical signals (voltages or currents) and to direct the amplified signals to desired loads. They are also widely used as components of the high-speed electrical switches and logic circuits that make up the digital computers that pervade our technological society. It is possible to replace the transistor, the resistors, the other passive circuit elements, the signal source, and the load by combinations of simple circuit elements, such as current sources, voltage sources, and resistors. The problem solutions are then achieved by circuit analysis methods which we either know already or will meet in this chapter.

When we are able to describe mathematically the behavior of fluid-flow and heat-flow systems, the dynamic response of aircraft control surfaces, and other nonelectrical phenomena, we shall see that the resultant equations are often precisely analogous to those describing current and voltage relationships in electric circuits. We may decide, then, that it is much easier and cheaper to construct the analogous electric circuit than it is to build a prototype of the actual physical system. The electric circuit may then be used to predict the performance of the other system as various elements are changed and may help achieve a better final design. This is the basis on which the electronic analog computer operates.

It is evident that one of the primary goals of this chapter must be learning methods of simplifying the analysis of more complicated circuits. Among these methods will be superposition, loop, mesh, and nodal analysis. We shall also try to develop the ability to select the most convenient analysis method. Most often we are interested only in the detailed performance of an isolated portion of a complex circuit; a method of replacing the remainder of the circuit by a greatly simplified equivalent is then very desirable. The equivalent is often a single resistor in series or parallel with an ideal source. Thévenin's and Norton's theorems will enable us to effect this replacement.

We shall begin studying methods of simplifying circuit analysis by considering a powerful general method, that of nodal analysis.

3-2
Nodal analysis

In the previous chapter we considered the analysis of a simple circuit containing only two nodes. We found then that the major step of the analysis was taken as we obtained a single equation in terms of a single unknown quantity, the voltage between the pair of nodes. We shall now let the number of nodes increase, and correspondingly provide one additional unknown quantity and one additional equation for each added node. Thus, a three-node circuit should have two unknown voltages and two equations; a ten-node circuit will have nine unknown voltages and nine equations; and an N-node circuit will need $(N - 1)$ voltages and $(N - 1)$ equations.

We consider the mechanics of nodal analysis in this section, but the justification for our methods will not be developed until later in this chapter.

As an example, let us consider the three-node circuit shown in Fig. 3-1a. We may emphasize the locations of the three nodes by redrawing the circuit, as shown in Fig. 3-1b, where each node is identified by a number. We would now like to associate a voltage with each node, but we must remember that a voltage must be defined as existing between *two nodes* in a network. We thus select one node as a reference node, and then define a voltage between each remaining node and the reference node. Hence, we note again that there will be only $(N - 1)$ voltages defined in an N-node circuit.

We choose node 3 as the reference node. Either of the other nodes could have been selected, but a little simplification in the resultant equations is obtained if the node to which the greatest number of branches is connected is identified as the reference node. If there is a ground node, it is usually most convenient to select it as the reference node. More often than not, the ground node appears as a common lead across the bottom of a circuit diagram.

Fig. 3-1: (*a*) A given three-node circuit. (*b*) The circuit is redrawn to emphasize the three nodes, and each node is numbered. (*c*) A voltage, including polarity reference, is assigned between each node and the reference node. (*d*) The voltage assignment is simplified by eliminating the polarity references; it is understood that each voltage is sensed positive relative to the reference node.

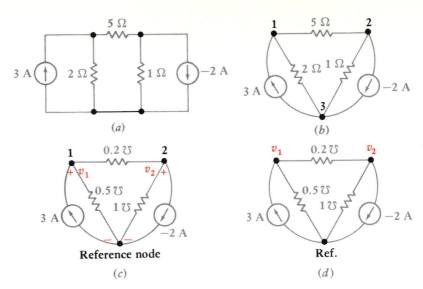

The voltage of node 1 relative to the reference node 3 is defined as v_1, and v_2 is defined as the voltage of node 2 with respect to the reference node. These two voltages are sufficient, and the voltage between any other pair of nodes may be found in terms of them. For example, the voltage of node 1 with respect to node 2 is $(v_1 - v_2)$. The voltages v_1 and v_2 and their reference signs are shown in Fig. 3-1*c*. In this figure the resistance values have also been replaced with conductance values.

The circuit diagram is finally simplified in Fig. 3-1*d* by eliminating all voltage reference symbols. A reference node is plainly marked, and the voltage placed at each remaining node is *understood* to be the voltage of that node with respect to the reference node. This is the only situation for which we should use voltage symbols without the associated plus-minus-sign pair, except for the battery symbol that was defined in Fig. 1-14*b*.

We must now apply Kirchhoff's current law to nodes 1 and 2. We do this by equating the total current leaving the node through the several conductances to the total source current entering the node. Thus,

$$0.5v_1 + 0.2(v_1 - v_2) = 3$$

or
$$0.7v_1 - 0.2v_2 = 3 \tag{1}$$

At node 2 we obtain

$$1v_2 + 0.2(v_2 - v_1) = 2$$

or
$$-0.2v_1 + 1.2v_2 = 2 \tag{2}$$

Equations (1) and (2) are the desired two equations in two unknowns, and they may be solved easily. The results are:

$$v_1 = 5 \quad V$$

$$v_2 = 2.5 \quad V$$

Fig. 3-2: (*a*) A circuit containing four nodes and eight branches. (*b*) The same circuit redrawn with node voltages assigned.

Also, the voltage of node 1 relative to node 2 is ($v_1 - v_2$), or 2.5 V, and any current or power in the circuit may now be found in one step. For example, the current directed downward through the 0.5-℧ conductance is $0.5v_1$, or 2.5 A.

Now let us increase the number of nodes by one. A new circuit is shown in Fig. 3-2*a*, and it is redrawn in Fig. 3-2*b*, with the nodes identified, a convenient reference node chosen, and the node voltages specified. We next sum the currents leaving node 1:

$$3(v_1 - v_2) + 4(v_1 - v_3) - (-8) - (-3) = 0$$

or
$$7v_1 - 3v_2 - 4v_3 = -11 \tag{3}$$

At node 2:

$$3(v_2 - v_1) + 1v_2 + 2(v_2 - v_3) - 3 = 0$$

or
$$-3v_1 + 6v_2 - 2v_3 = 3 \tag{4}$$

and at node 3:

$$4(v_3 - v_1) + 2(v_3 - v_2) + 5v_3 - 25 = 0$$

or
$$-4v_1 - 2v_2 + 11v_3 = 25 \tag{5}$$

Equations (3) through (5) may be solved by a simple process of elimination of variables, or by Cramer's rule and determinants.[1] Using the latter method, we have

$$v_1 = \frac{\begin{vmatrix} -11 & -3 & -4 \\ 3 & 6 & -2 \\ 25 & -2 & 11 \end{vmatrix}}{\begin{vmatrix} 7 & -3 & -4 \\ -3 & 6 & -2 \\ -4 & -2 & 11 \end{vmatrix}}$$

[1] Appendix 1 provides a short review of determinants and the solution of a system of simultaneous linear equations by Cramer's rule.

Expanding the numerator and denominator determinants by minors along their first columns leads to

$$v_1 = \frac{-11 \begin{vmatrix} 6 & -2 \\ -2 & 11 \end{vmatrix} - 3 \begin{vmatrix} -3 & -4 \\ -2 & 11 \end{vmatrix} + 25 \begin{vmatrix} -3 & -4 \\ 6 & -2 \end{vmatrix}}{7 \begin{vmatrix} 6 & -2 \\ -2 & 11 \end{vmatrix} - (-3) \begin{vmatrix} -3 & -4 \\ -2 & 11 \end{vmatrix} + (-4) \begin{vmatrix} -3 & -4 \\ 6 & -2 \end{vmatrix}}$$

$$= \frac{-11(62) - 3(-41) + 25(30)}{7(62) + 3(-41) - 4(30)} = \frac{-682 + 123 + 750}{434 - 123 - 120}$$

$$= \frac{191}{191} = 1 \quad V$$

Similarly,

$$v_2 = \frac{\begin{vmatrix} 7 & -11 & -4 \\ -3 & 3 & -2 \\ -4 & 25 & 11 \end{vmatrix}}{191} = 2 \quad V$$

and

$$v_3 = \frac{\begin{vmatrix} 7 & -3 & -11 \\ -3 & 6 & 3 \\ -4 & -2 & 25 \end{vmatrix}}{191} = 3 \quad V$$

The denominator determinant is common to each of the three evaluations above. For circuits that do not contain either voltage sources or dependent sources (i.e., circuits containing only independent current sources), this denominator is the determinant of a matrix[2] that is defined as the *conductance matrix* of the circuit:

$$\mathbf{G} = \begin{bmatrix} 7 & -3 & -4 \\ -3 & 6 & -2 \\ -4 & -2 & 11 \end{bmatrix}$$

It should be noted that the nine elements of the matrix are the ordered array of the coefficients of Eqs. (3), (4), and (5), each of which is a conductance value. The first row is composed of the coefficients of the KCL equation at the first node, the coefficients being given in the order of v_1, v_2, and v_3. The second row applies to the second node, and so on.

The conductance matrix is symmetrical about the major diagonal (upper left to lower right), and all elements not on this diagonal are negative, whereas all elements on it are positive. This is a general consequence of the systematic way in which we assigned variables, applied KCL, and ordered the equations, as well as of the reciprocity theorem, which we shall discuss

---

[2] We shall not begin to manipulate matrices mathematically until Chap. 16; at that time a rudimentary knowledge of linear algebra is presumed.

in Chap. 16. For the present, we merely acknowledge the symmetry in these circuits that have only independent current sources and accept the check that it provides us in discovering errors we may have committed in writing circuit equations.

We still must see how voltage sources and dependent sources affect the strategy of nodal analysis. We now investigate the consequences of including a voltage source.

As a typical example, consider the circuit shown in Fig. 3-3. Our previous four-node circuit has been changed by replacing the 2-℧ conductance between nodes 2 and 3 by a 22-V voltage source. We still assign the same node-to-reference voltages, v_1, v_2, and v_3. Previously, the next step was the application of KCL at each of the three nonreference nodes. If we try to do that once again, we see that we shall run into some difficulty at both nodes 2 and 3, for we do not know what the *current* is in the branch with the *voltage* source. There is no way by which we can express the current as a function of the voltage, for the definition of a voltage source is exactly that the voltage is independent of the current.

There are two ways out of these difficulties. The more difficult is to assign an unknown current to the branch with the voltage source, proceed to apply KCL three times, and then apply KVL once between nodes 2 and 3; the result is four equations in four unknowns for this example.

The easier method is to agree that we are primarily interested in the node voltages, so that we may avoid the current in the voltage-source branch that is causing our problems. We do this by treating node 2, node 3, and the voltage source together as a sort of supernode and applying KCL to both nodes at the same time. This is certainly possible, because, if the total current leaving node 2 is zero and the total current leaving node 3 is zero, then the total current leaving the totality of the two nodes is zero.

The supernode is indicated by the shaded region enclosed by the broken line in Fig. 3-3, and we shall set the sum of the six currents leaving the supernode equal to zero. Beginning with the 3-℧ conductance branch and working clockwise, we have

$$3(v_2 - v_1) - 3 + 4(v_3 - v_1) - 25 + 5v_3 + 1v_2 = 0$$

or

$$-7v_1 + 4v_2 + 9v_3 = 28$$

Fig. 3-3: The 2-℧ conductance in the circuit of Fig. 3-2 is replaced by an independent voltage source. Kirchhoff's current law is used on the supernode enclosed by the broken line, and the source voltage is set equal to $v_3 - v_2$.

The KCL equation at node 1 is unchanged from (3):

$$7v_1 - 3v_2 - 4v_3 = -11$$

We need one additional equation since we have three unknowns, and it must utilize the fact that there is a 22-V voltage source between nodes 2 and 3,

$$v_3 - v_2 = 22$$

Rewriting these last three equations,

$$7v_1 - 3v_2 - 4v_3 = -11$$

$$-7v_1 + 4v_2 + 9v_3 = 28$$

$$-v_2 + v_3 = 22$$

the determinant solution for v_1 is

$$v_1 = \frac{\begin{vmatrix} -11 & -3 & -4 \\ 28 & 4 & 9 \\ 22 & -1 & 1 \end{vmatrix}}{\begin{vmatrix} 7 & -3 & -4 \\ -7 & 4 & 9 \\ 0 & -1 & 1 \end{vmatrix}} = \frac{-189}{42} = -4.5 \quad \text{V}$$

Note the lack of symmetry about the major diagonal in the denominator determinant as well as the fact that not all of the off-diagonal elements are negative. This is the result of the presence of the voltage source. Note also that it would not make much sense to call the denominator the determinant of the *conductance* matrix, because the bottom row comes from the equation $-v_2 + v_3 = 22$, and this equation does not depend on conductances in any way.

The presence of a voltage source thus reduces by one the number of nonreference nodes at which we must apply KCL, regardless of whether the voltage source extends between two nonreference nodes or is connected between a node and the reference.

Now let us consider a circuit containing a dependent source. As our example, we select the op-amp connected as a voltage follower, Fig. 3-4a.

Fig. 3-4: (a) A voltage follower feeds a finite load R_L. (b) The op-amp is replaced with an equivalent circuit that includes a noninfinite R_i and a nonzero R_o. Three node-to-reference voltages are assigned, and one supernode is indicated.

(a) (b)

This is the same circuit we investigated in the last section of Chapter 2, except that a finite load resistance $R_L = 1 \text{ k}\Omega$ now appears between the output terminal and ground. We represent the op-amp by a model that includes a noninfinite input resistance, $R_i = 50 \text{ k}\Omega$, and a nonzero output resistance, $R_o = 2 \text{ k}\Omega$, as shown in the circuit of Fig. 3-4b; we assume a typical value of $A = 10^4$.

Ground is selected as the reference node, and the three nonreference nodes are assigned the voltages, v_2, v_o, and v_3. We note that the independent voltage source v_2 causes the v_2 node and the reference node to form a supernode; moreover, the dependent voltage source forces us to consider the v_3 node and the reference node as a supernode also. Thus, the v_2, v_o, and reference nodes form one large supernode, shown by the shaded, broken-line enclosure in Fig. 3-4b. Since the supernode includes the reference node, we will not write a KCL equation for it. The only KCL equation to be written is that at the v_o node. It is

$$\frac{v_o - v_2}{50\,000} + \frac{v_o - v_3}{2000} + \frac{v_o}{1000} = 0 \tag{6}$$

Let us set the independent source $v_2 = 1$ V. There are now two unknown node voltages, v_o and v_3 in (6), and no other (independent) KCL equations can be written. However, we must still express the voltage of each voltage source extending from node to node (and thus inside the broken-line enclosure) in terms of the node voltages, and we must also express the control of the dependent source (here, the voltage v_i) in terms of the node voltages.

First, we look at all the voltage sources inside the supernode. The source v_2 has been set equal to 1 V, and furthermore the node voltage itself was designated v_2. If we had been foolish enough to call it v_A, for example, then we should have to write the trivial equation, $v_A = v_2$. Next is the source Av_i. Since it is connected between node 3 and ground, we have

$$v_3 = -10^4 v_i$$

Finally, we must relate the currents or voltages on which the controlled sources depend to the node voltages. Here, v_i is defined across R_i, and

$$v_i = v_o - v_2 = v_o - 1$$

To solve (6) for v_o, we let $v_3 = -10^4(v_o - 1)$, obtaining one equation in one unknown,

$$\frac{v_o - 1}{50\,000} + \frac{v_o + 10^4(v_o - 1)}{2000} + \frac{v_o}{1000} = 0$$

We find that $v_o = 0.999\,700$ V, and thus the output voltage closely equals the input even for an op-amp that has relatively low gain, low input resistance, and high output resistance.

In passing, we note that the negative of the first term in (6) is the current supplied by the 1-V source, here $(1 - v_o)/50\,000 = 6.00$ nA, an exceedingly small value that is not apt to disturb the most delicate of sources. In contrast, the output current is the third term in (6), $v_o/1000 = 1.000$ mA, over 10^5 times as large. Thus the voltage follower can deliver much more current and power to the load than it draws from the source. In doing so it is not violating the conservation of energy, but merely drawing power from the dc supplies, which are usually not shown.

Let us summarize the method by which we may obtain a set of nodal equations for any resistive circuit:

1. Make a neat, simple, circuit diagram. Indicate all element and source values. Each source should have its reference symbol.

2. Assuming that the circuit has N nodes, choose one of these nodes as a reference node. Then write the node voltages $v_1, v_2, \ldots, v_{N-1}$ at their respective nodes, remembering that each node voltage is understood to be measured with respect to the chosen reference.

3. If the circuit contains only current sources, apply Kirchhoff's current law at each nonreference node. To obtain the conductance matrix if a circuit has only independent current sources, equate the total current leaving each node through all conductances to the total source current entering that node, and order the terms from v_1 to v_{N-1}. For each dependent current source present, relate the source current and the controlling quantity to the variables $v_1, v_2, \ldots, v_{N-1}$, if they are not already in that form.

4. If the circuit contains voltage sources, form a supernode about each one by enclosing the source and its two terminals within a broken-line enclosure, thus reducing the number of nodes by one for each voltage source that is present. The assigned node voltages should not be changed. Using these assigned node-to-reference voltages, apply KCL at each of the nodes or supernodes (that do not contain the reference node) in this modified circuit. Relate each source voltage to the variables $v_1, v_2, \ldots, v_{N-1}$, if it is not already in that form.

With these suggestions in mind, let us consider the circuit displayed in Fig. 3-5, one which contains all four types of sources and has five nodes. We

Fig. 3-5: A five-node circuit containing all of the four different types of sources.

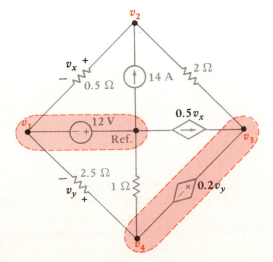

select the central node as the reference, and assign v_1 to v_4 in a clockwise direction starting from the left node.

After establishing a supernode about each voltage source, we see that we need to write KCL equations only at node 2 and at the supernode containing both nodes 3 and 4 and the dependent voltage source. No extra equation need be written for the supernode which contains node 1 and the independent voltage source; it is obvious that $v_1 = -12$ V.

At node 2,

$$\frac{v_2 - v_1}{0.5} + \frac{v_2 - v_3}{2} = 14$$

while at the 3-4 supernode,

$$\frac{v_3 - v_2}{2} - 0.5v_x + \frac{v_4}{1} + \frac{v_4 - v_1}{2.5} = 0$$

We next relate the source voltages to the node voltages:

$$v_1 = -12$$

$$v_3 - v_4 = 0.2v_y = 0.2(v_4 - v_1)$$

And finally we express the dependent current source in terms of the assigned variables:

$$0.5v_x = 0.5(v_2 - v_1)$$

Thus, we obtain four equations in the four node voltages:

$$-2v_1 + 2.5v_2 - 0.5v_3 = 14$$

$$v_1 = -12$$

$$0.1v_1 - v_2 + 0.5v_3 + 1.4v_4 = 0$$

$$0.2v_1 + v_3 - 1.2v_4 = 0$$

to which the solutions are

$$v_1 = -12 \quad \text{V}$$

$$v_2 = -4 \quad \text{V}$$

$$v_3 = 0$$

$$v_4 = -2 \quad \text{V}$$

Drill Problems

3-1. Use nodal analysis to find v_y in the circuit shown in Fig. 3-6 if element A is: (a) a 2-A current source, arrow pointing right; (b) an 8-Ω resistor; (c) a 10-V voltage source, the positive reference on the right.

Ans: -20; -16; -6.91 V

3-2. Use nodal analysis to find v_y in the circuit of Fig. 3-6 if element A is: (a) a dependent current source, $0.2v_x$, arrow pointing left; (b) a dependent voltage source, $3i_Q$, the positive reference on the left; (c) an open circuit; (d) a short circuit.

Ans: 49.3; -14.14; -22.7; -8.73 V

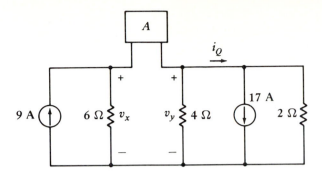

Fig. 3-6: See Drill Probs. 3-1 and 3-2.

3-3
Mesh Analysis

The technique of nodal analysis described in the preceding section is completely general and can always be applied to any electrical network. This is not the only method for which a similar claim can be made, however. In particular, we shall meet a generalized nodal analysis method and a technique known as *loop analysis* in the concluding sections of this chapter.

First, however, let us consider a method known as *mesh analysis*. Even though this technique is not applicable to every network, it can be applied to most of the networks we shall need to analyze, and it is probably used more often than it should be; other methods are often simpler. Mesh analysis is applicable only to those networks which are planar, a term we hasten to define.

If it is possible to draw the diagram of a circuit on a plane surface in such a way that no branch passes over or under any other branch, then that circuit is said to be a *planar* circuit. Thus, Fig. 3-7a shows a planar network, Fig. 3-7b shows a nonplanar network, and Fig. 3-7c shows a planar network, although it is drawn in such a way as to make it *appear* nonplanar at first glance.

Fig. 3-7: (a) A planar network can be drawn on a plane surface without crossovers. (b) A nonplanar network cannot be drawn on a plane surface without at least one crossover. (c) A planar network can be drawn so that it may look nonplanar.

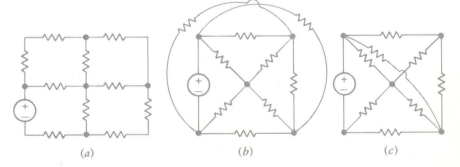

(a)   (b)   (c)

In the second chapter, the terms *path*, *closed path*, and *loop* were defined. Before we define a mesh, let us consider the sets of branches drawn with heavy lines in Fig. 3-8. The first set of branches is not a path since four branches are connected to the center node, and it is of course also not a loop. The second set of branches does not constitute a path since it is traversed

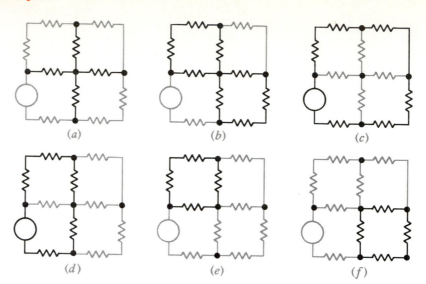

Fig. 3-8: (a) The set of branches identified by the heavy lines is neither a path nor a loop. (b) The set of branches here is not a path, since it can be traversed only by passing through the central node twice. (c) This path is a loop but not a mesh, since it encloses other loops. (d) This path is also a loop but not a mesh. (e) and (f) Each of these paths is both a loop and a mesh.

only by passing through the central node twice. The remaining four paths are all loops. The circuit contains 11 branches.

The mesh is a property of a planar circuit and is not defined for a non-planar circuit. We define a *mesh* as a loop which does not contain any other loops within it. Thus, the loops indicated in Fig. 3-8c and d are not meshes, whereas those of e and f are meshes. Once a circuit has been drawn neatly in planar form, it often has the appearance of a multipaned window; each pane in the window may be considered to be a mesh.

If a network is planar, mesh analysis can be used to accomplish its analysis. This technique involves the concept of a *mesh current,* which we shall introduce by considering the analysis of the two-mesh circuit of Fig. 3-9.

As we did in the single-loop circuit, we shall begin by assuming a current through one of the branches. Let us call the current flowing to the right through the 6-Ω resistor i_1. We intend to apply Kirchhoff's voltage law around each of the two meshes, and the resulting two equations are sufficient to determine two unknown currents. Therefore we select a second current i_2 flowing to the right in the 4-Ω resistor. We might also choose to call the current flowing downward through the central branch i_3, but it is evident from Kirchhoff's current law that i_3 may be expressed in terms of the two previously assumed currents as $(i_1 - i_2)$. The assumed currents are shown in Fig. 3-9.

Fig. 3-9: Two currents, i_1 and i_2, are assumed in a two-mesh circuit.

Following the method of solution for the single-loop circuit, we now apply Kirchhoff's voltage law to the left-hand mesh,

$$-42 + 6i_1 + 3(i_1 - i_2) = 0$$

or

$$9i_1 - 3i_2 = 42 \qquad (7)$$

and then to the right-hand mesh,

$$-3(i_i - i_2) + 4i_2 - 10 = 0$$

or

$$-3i_1 + 7i_2 = 10 \qquad (8)$$

Equations (7) and (8) are independent equations; one cannot be derived from the other.[3] There are two equations and two unknowns, and the solution is easily obtained: i_1 is 6 A, i_2 is 4 A, and $(i_1 - i_2)$ is therefore 2 A. The voltage and power relationships may be quickly obtained if desired.

If our circuit had contained M meshes, then we should have had to assume M branch currents and write M independent equations.[4] The solution in general may be systematically obtained through the use of determinants.

Now let us consider this same problem in a slightly different manner by using mesh currents. We define a *mesh current* as a current which flows only around the perimeter of a mesh. If we label the left-hand mesh of our problem as mesh 1, then we may establish a mesh current i_1 flowing in a clockwise direction about this mesh. A mesh current is indicated by a curved arrow that almost closes on itself and is drawn inside the appropriate mesh, as shown in Fig. 3-10. The mesh current i_2 is established in the remaining mesh, again in a clockwise direction. Although the directions are arbitrary, we shall always choose clockwise mesh currents because a certain error-minimizing symmetry then results in the equations.

Fig. 3-10: A clockwise mesh current is assigned to each mesh of a planar circuit.

We no longer have a current or current arrow shown directly on each branch in the circuit. The current through any branch must be determined by considering the mesh currents flowing in every mesh in which that branch appears. This is not difficult because it is obvious that no branch can appear in more than two meshes. For example, the 3-Ω resistor appears in both meshes, and the current flowing downward through it is $(i_1 - i_2)$. The 6-Ω resistor appears only in mesh 1, and the current flowing to the right in that branch is equal to the mesh current i_1.

A mesh current may often be identified as a branch current, as i_1 and i_2 are

---

[3] It will be shown in Sec. 3-8 that mesh equations are always independent.
[4] The proof of this statement will be found in Sec. 3-8.

identified above. This is not always true, however, for consideration of a square nine-mesh network soon shows that the central mesh current cannot be identified as the current in any branch.

One of the greatest advantages in the use of mesh currents is the fact that Kirchhoff's current law is automatically satisfied. If a mesh current flows into a given node, it obviously flows out of it also.

We therefore may turn our attention to the application of KVL to each mesh. For the left-hand mesh,

$$-42 + 6i_1 + 3(i_1 - i_2) = 0$$

while for the right-hand mesh,

$$3(i_2 - i_1) + 4i_2 - 10 = 0$$

and these two equations are equivalent to (7) and (8).

Let us next consider the five-node, seven-branch, three-mesh circuit shown in Fig. 3-11. The three required mesh currents are assigned as indicated, and we methodically apply KVL about each mesh:

$$-7 + 1(i_1 - i_2) + 6 + 2(i_1 - i_3) = 0$$

$$1(i_2 - i_1) + 2i_2 + 3(i_2 - i_3) = 0$$

$$2(i_3 - i_1) - 6 + 3(i_3 - i_2) + 1i_3 = 0$$

Simplifying,

$$3i_1 - i_2 - 2i_3 = 1$$

$$-i_1 + 6i_2 - 3i_3 = 0$$

$$-2i_1 - 3i_2 + 6i_3 = 6$$

and Cramer's rule leads to the formulation for i_3:

$$i_3 = \frac{\begin{vmatrix} 3 & -1 & 1 \\ -1 & 6 & 0 \\ -2 & -3 & 6 \end{vmatrix}}{\begin{vmatrix} 3 & -1 & -2 \\ -1 & 6 & -3 \\ -2 & -3 & 6 \end{vmatrix}} = \frac{117}{39} = 3 \text{ A}$$

The other mesh currents are $i_1 = 3$ A, and $i_2 = 2$ A.

Fig. 3-11: Mesh currents i_1, i_2, and i_3 are assumed in a five-node, seven-branch, three-mesh circuit.

Again we notice that we have a denominator determinant that is symmetrical about the major diagonal and has positive terms on the major diagonal and zero or negative terms off it. This occurs for circuits that contain only independent voltage sources when clockwise mesh currents are assigned, where the elements appearing in the first row of the determinant are the ordered coefficients of $i_1, i_2, \ldots, i_M$ in the KVL equation about the first mesh, where the second row corresponds to the second mesh, and so on. This symmetrical array appearing in the denominator is the determinant of the *resistance matrix* of the network,

$$\mathbf{R} = \begin{bmatrix} 3 & -1 & -2 \\ -1 & 6 & -3 \\ -2 & -3 & 6 \end{bmatrix}$$

How must we modify this straightforward procedure when a current source is present in the network? Taking our lead from nodal analysis (and duality), we should feel that there are two possible methods. First, we could assign an unknown voltage across the current source, apply KVL around each mesh as before, and then relate the source current to the assigned mesh currents. This is generally the more difficult approach.

A better technique is one that is quite similar to the supernode approach in nodal analysis. There we formed a supernode, completely enclosing the voltage source inside the supernode and reducing the number of nonreference nodes by one for each voltage source. Now we create a kind of "supermesh" from two meshes that have a current source as a common element; the current source is in the interior of the supermesh. We thus reduce the number of meshes by one for each current source present. If the current source lies on the perimeter of the circuit, then the single mesh in which it is found is ignored.[5] Kirchhoff's voltage law is applied only to those meshes or supermeshes in the modified network.

As an example of this procedure, consider the network shown in Fig. 3-12, in which a 7-A independent current source is in the common boundary of two meshes. Mesh currents $i_1, i_2,$ and i_3 are assigned, and the current source

---

[5] Such a current source is a common element with the mesh that "encloses" the outside of the entire circuit. Just as we do not write a nodal equation at the reference node, we do not write a KVL equation for this external mesh.

Fig. 3-12: Mesh analysis is applied to this circuit containing a current source by writing the KVL equation about the loop: 7 V, 1 Ω, 3 Ω, 1 Ω.

causes us to create a supermesh whose interior is that of meshes 1 and 3. Applying KVL about this loop,

$$-7 + 1(i_1 - i_2) + 3(i_3 - i_2) + 1i_3 = 0$$

or
$$i_1 - 4i_2 + 4i_3 = 7 \qquad (9)$$

and around mesh 2,

$$1(i_2 - i_1) + 2i_2 + 3(i_2 - i_3) = 0$$

or
$$-i_1 + 6i_2 - 3i_3 = 0 \qquad (10)$$

Finally, the source current is related to the assumed mesh currents,

$$i_1 - i_3 = 7 \qquad (11)$$

Solving (9) through (11), we have

$$i_3 = \frac{\begin{vmatrix} -1 & 6 & 0 \\ 1 & -4 & 7 \\ 1 & 0 & 7 \end{vmatrix}}{\begin{vmatrix} -1 & 6 & -3 \\ 1 & -4 & 4 \\ 1 & 0 & -1 \end{vmatrix}} = \frac{28}{14} = 2 \quad \text{A}$$

We may also find that $i_1 = 9$ A and $i_2 = 2.5$ A.

The presence of one or more dependent sources merely requires each of these source quantities and the variable on which it depends to be expressed in terms of the assigned mesh currents. In Fig. 3-13, for example, we note that both a dependent and an independent current source are included in the network. Three mesh currents are assigned and KVL is applied to mesh 2:

$$1(i_2 - i_1) + 2i_2 + 3(i_2 - i_3) = 0$$

Fig. 3-13: The presence of two current sources in this three-mesh circuit makes it necessary to apply KVL only once, around mesh 2.

The current sources appear in meshes 1 and 3. Since the 15-A source is located on the perimeter of the circuit, we may eliminate mesh 1 from consideration. Then, the dependent current source is on the perimeter of the modified network, and thus we avoid any equation writing for mesh 3. Only mesh 2 remains, and we already have an equation for it. We therefore turn our attention to the source quantities, obtaining

$$i_1 = 15$$

and $$\tfrac{1}{9}v_x = i_3 - i_1 = \tfrac{1}{9}[3(i_3 - i_2)]$$

Thus, $$-i_1 + 6i_2 - 3i_3 = 0$$

$$i_1 = 15$$

$$-i_1 + \tfrac{1}{3}i_2 + \tfrac{2}{3}i_3 = 0$$

from which we have $i_1 = 15$, $i_2 = 11$, and $i_3 = 17$ A. We might note that we wasted a little time in assigning a mesh current i_1 to the left mesh; we should simply have indicated a mesh current and labeled it 15 A.

Let us summarize the method by which we may obtain a set of mesh equations for a resistive circuit:

1 Make certain that the network is a planar network. If it is nonplanar, mesh analysis is not applicable.
2 Make a neat, simple, circuit diagram. Indicate all element and source values. Resistance values are preferable to conductance values. Each source should have its reference symbol.
3 Assuming that the circuit has M meshes, assign a clockwise mesh current in each mesh, $i_1, i_2, \ldots, i_M$.
4 If the circuit contains only voltage sources, apply Kirchhoff's voltage law around each mesh. To obtain the resistance matrix if a circuit has only independent voltage sources, equate the clockwise sum of all the resistor voltages to the counterclockwise sum of all the source voltages, and order the terms, from i_1 to i_M. For each dependent voltage source present, relate the source voltage and the controlling quantity to the variables $i_1, i_2, \ldots, i_M$, if they are not already in that form.
5 If the circuit contains current sources, create a supermesh for each current source that is common to two meshes by applying KVL around the larger loop formed by the branches that are not common to the two meshes; KVL need not be applied to a mesh containing a current source that lies on the perimeter of the entire circuit. The assigned mesh currents should not be changed. Relate each source current to the variables $i_1, i_2, \ldots, i_M$, if it is not already in that form.

Drill Problems

3-3. Use mesh analysis to find v_A in the circuit of Fig. 3-14 if element A is: (a) a 4-V voltage source, the positive reference at the top; (b) a 9-Ω resistor; (c) a 600-mA current source, arrow directed downward.

Ans: 23.7; 23.6; 23.9 V

Fig. 3-14: See Drill Probs. 3-3 and 3-4.

3-4. Use mesh analysis to find i_C in Fig. 3-14 if element A is: (*a*) a dependent voltage source, $0.2v_A$, the positive reference at the top; (*b*) a dependent current source, $0.5i_B$, arrow directed downward.   *Ans:* 0.462; −0.281 A

**3-4
Linearity and
superposition**

All the circuits which we have analyzed up to now (and which we shall analyze later) are linear circuits. At this time we must be more specific in defining a linear circuit. Having done this, we can then consider the most important consequence of linearity, the principle of superposition. This principle is very basic and will appear repeatedly in our study of linear circuit analysis. As a matter of fact, the nonapplicability of superposition to nonlinear circuits is the reason they are so difficult to analyze.

The principle of superposition states that the response (a desired current or voltage) at any point in a linear circuit having more than one independent source can be obtained as the sum of the responses caused by each independent source acting alone. In the following discussion, we shall investigate the meaning of "linear" and "acting alone." We shall also take note of a slightly broader form of the theorem.

Let us first define a *linear element* as a passive element that has a linear voltage-current relationship. By a "linear voltage-current relationship" we shall mean simply that multiplication of the time-varying current through the element by a constant K results in the multiplication of the time-varying voltage across the element by the same constant K. At this time, only one passive element has been defined, the resistor, and its voltage-current relationship

$$v(t) = Ri(t)$$

is obviously linear. As a matter of fact, if $v(t)$ is plotted as a function of $i(t)$, the result is a straight *line*. We shall see in Chap. 4 that the defining voltage-current equations for inductance and capacitance are also linear relationships, as is the defining equation for mutual inductance presented in Chap. 15.

We must also define a *linear dependent source* as a dependent current or voltage source whose output current or voltage is proportional only to the first power of some current or voltage variable in the circuit or to the sum of such quantities. That is, a dependent voltage source, $v_s = 0.6i_1 - 14v_2$, is linear, but $v_s = 0.6i_1^2$ and $v_s = 0.6i_1v_2$ are not.

We may now define a *linear circuit* as a circuit composed entirely of independent sources, linear dependent sources, and linear elements. From this definition, it is possible to show[6] that "the response is proportional to

---

[6] The proof involves first showing that the use of nodal analysis on the linear circuit can produce only linear equations of the form:

$$a_1v_1 + a_2v_2 + \cdots + a_Nv_N = b$$

where the a_i are constants (combinations of resistance or conductance values, constants appearing in dependent source expressions, 0, or ± 1), the v_i are the unknown node voltages (responses), and b is an independent source value or a sum of independent source values. Given a set of such equations, if we multiply all the b's by K, then it is evident that the solution of this new set of equations will be the node voltages $Kv_1, Kv_2, \ldots, Kv_N$.

Fig. 3-15: A three-node circuit, containing two forcing functions, used to illustrate the superposition principle.

the source," or that multiplication of *all independent* source voltages and currents by a constant K increases all the current and voltage responses by the same factor K (including the dependent source voltage or current outputs).

The most important consequence of linearity is superposition. Let us develop the superposition principle by considering first the circuit of Fig. 3-15, which contains two independent sources, the current generators which force the currents i_a and i_b into the circuit. Sources are often called *forcing functions* for this reason, and the voltages which they produce between node 1 or 2 and the reference node may be termed *response functions*, or simply *responses*. Both the forcing functions and the responses may be functions of time.

The two nodal equations for this circuit are

$$0.7v_1 - 0.2v_2 = i_a \tag{12}$$

$$-0.2v_1 + 1.2v_2 = i_b \tag{13}$$

Now let us perform experiment x. We change the two forcing functions to i_{ax} and i_{bx}; the two unknown voltages will now be different, and we shall let them be v_{1x} and v_{2x}. Thus,

$$0.7v_{1x} - 0.2v_{2x} = i_{ax} \tag{14}$$

$$-0.2v_{1x} + 1.2v_{2x} = i_{bx} \tag{15}$$

We next perform experiment y by changing the source currents to i_{ay} and i_{by} and by letting the responses be v_{1y} and v_{2y},

$$0.7v_{1y} - 0.2v_{2y} = i_{ay} \tag{16}$$

$$-0.2v_{1y} + 1.2v_{2y} = i_{by} \tag{17}$$

These three sets of equations describe the same circuit with different source currents. Let us *add* or *superpose* the last two sets of equations. Adding (14) and (16),

$$(0.7v_{1x} + 0.7v_{1y}) - (0.2v_{2x} + 0.2v_{2y}) = i_{ax} + i_{ay} \tag{18}$$

$$0.7v_1 \qquad - \qquad 0.2v_2 \qquad = \qquad i_a \tag{12}$$

and adding (15) and (17),

$$-(0.2v_{1x} + 0.2v_{1y}) + (1.2v_{2x} + 1.2v_{2y}) = i_{bx} + i_{by} \tag{19}$$

$$-0.2v_1 \qquad + \qquad 1.2v_2 \qquad = \qquad i_b \tag{13}$$

where (12) has been written immediately below (18), and (13) below (19) for easy comparison.

The linearity of all these equations allows us to compare (18) with (12) and (19) with (13) and draw an interesting conclusion. If we select i_{ax} and i_{ay} such that their sum is i_a, select i_{bx} and i_{by} such that their sum is i_b, then the desired responses v_1 and v_2 may be found by *adding* v_{1x} to v_{1y} and v_{2x} to v_{2y}, respectively. In other words, we may perform experiment x and note the responses, perform experiment y and note the responses, and finally add the corresponding responses. These are the responses of the original circuit to independent sources which are the sums of the independent sources used in experiments x and y. This is the fundamental concept involved in the superposition principle.

It is evident that we may extend these results by breaking up either source current into as many pieces as we wish; there is no reason why we cannot perform experiments z and q also. It is only necessary that the algebraic sum of the pieces be equal to the original current.

The *superposition theorem* usually appears in a form similar to the following:

> In any linear resistive network containing several sources, the voltage across or the current through any resistor or source may be calculated by adding algebraically all the individual voltages or currents caused by each independent source acting alone, with all other independent voltage sources replaced by short circuits and all other independent current sources replaced by open circuits.

Thus if there are N independent sources, we perform N experiments. Each independent source is active in only one experiment, and only one independent source is active in each experiment. An inactive independent voltage source is identical with a short circuit, and an inactive independent current source is an open circuit. Note that *dependent* sources are in general active in *every* experiment.

The circuit used as an example above, however, should indicate that a much stronger theorem might be written; a group of independent sources may be made active and inactive collectively, if we wish. For example, suppose there are three independent sources. The theorem above states that we may find a given response by considering each of the three sources acting alone and adding the three results. Alternatively, we may find the response due to the first and second sources operating with the third inactive, and then add to this the response caused by the third source acting alone. This amounts to treating several sources collectively as a sort of supersource.

There is also no reason that an independent source must assume only its given value or a zero value in the several experiments; it is only necessary for the sum of the several values to be equal to the original value. An inactive source almost always leads to the simplest circuit, however.

Let us illustrate the application of the superposition principle by considering an example in which both types of independent source are present. For the circuit of Fig. 3-16, let us use superposition to write an expression for the

Fig. 3-16: A circuit, containing both an independent current and voltage source, which is easily analyzed by the superposition principle.

unknown branch current i_x. We may first set the current source equal to zero and obtain the portion of i_x due to the voltage source as 0.2 A. Then if we let the voltage source be zero and apply current division, the remaining portion of i_x is seen to be 0.8 A. We might write the answer in detail as

$$i_x = i_x|_{i_s=0} + i_x|_{v_s=0} = \frac{3}{6+9} + 2\frac{6}{6+9} = 0.2 + 0.8 = 1.0 \quad A$$

As an example of the application of the superposition principle to a circuit containing a dependent source, consider Fig. 3-17. We seek i_x, and we first open-circuit the 3-A source. The single mesh equation is

$$-10 + 2i'_x + 1i'_x + 2i'_x = 0$$

so that

$$i'_x = 2$$

Next, we short-circuit the 10-V source and write the single-node equation,

$$\frac{v''}{2} + \frac{v'' - 2i''_x}{1} = 3$$

and relate the dependent-source-controlling quantity to v'',

$$v'' = -2i''_x$$

We find

$$i''_x = -0.6$$

and, thus,

$$i_x = i'_x + i''_x = 2 - 0.6 = 1.4$$

It usually turns out that little if any time is saved in analyzing a circuit containing one or more dependent sources by use of the superposition principle, for there must always be at least two sources in operation: one independent source and all the dependent sources.

We must constantly be aware of the limitations of superposition. It is

Fig. 3-17: Superposition may be used to analyze this circuit by first replacing the 3-A source by an open circuit and then replacing the 10-V source by a short circuit. The dependent voltage source is always active (unless $i_x = 0$).

applicable only to linear responses, and thus the most common nonlinear response—power—is not subject to superposition. For example, consider two 1-V batteries in series with a 1-Ω resistor. The power delivered to the resistor is obviously 4 W, but if we mistakenly try to apply superposition we might say that each battery alone furnished 1 W and thus the total power is 2 W. This is incorrect.

Drill Problem

3-5. Use superposition to find i_x in each of the circuits shown in Fig. 3-18.

Ans: -175; 200; 50 mA

Fig. 3-18: See Drill Prob. 3-5.

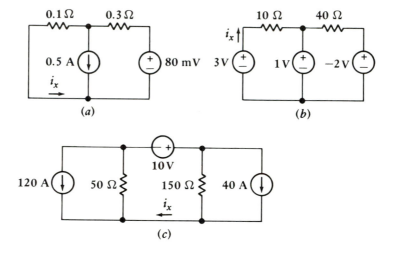

(a)   (b)

(c)

3-5
Source transformations

In all our previous work we have been making continual use of *ideal* voltage and current sources; it is now time to take a step closer to reality by considering *practical* sources. These sources will enable us to make more realistic representations of physical devices. Having defined the practical sources, we shall then study methods whereby practical current and voltage sources may be interchanged without affecting the remainder of the circuit. Such sources will be called *equivalent* sources. Our methods will be applicable for both independent and dependent sources, although we shall see later that they are not as useful for dependent sources.

The ideal voltage source was defined as a device whose terminal voltage is independent of the current through it. A 1-V dc source produces a current of 1 A through a 1-Ω resistor and a current of 1 000 000 A through a 1-μΩ resistor; it may provide an unlimited amount of power. No such device exists practically, of course, and we agreed that a real physical voltage source might be represented by an ideal voltage source only as long as relatively small currents, or powers, were drawn from it. For example, an automobile storage battery may be approximated by an ideal dc voltage source if its current is limited to a few amperes. However, anyone who has ever tried to start an automobile with the headlights on must have observed that the lights dimmed perceptibly when the battery was asked to deliver the heavy starter current, 100 A or more, in addition to the headlight current. Under these conditions, an ideal voltage source may be a very poor representation of the storage battery.

The ideal voltage source must be modified to account for the lowering of its terminal voltage when large currents are drawn from it. Let us suppose that we observe experimentally that a storage battery has a terminal voltage of 12 V when no current is flowing through it and a reduced voltage of 11 V when 100 A is flowing. Thus, a more accurate model might be an ideal voltage source of 12 V in series with a resistor across which 1 V appears when 100 A flows through it. The resistor must be 0.01 Ω, and the ideal voltage source and this series resistor comprise a *practical voltage source*. Thus, we are using the series combination of two ideal circuit elements, an independent voltage source and a resistor, to model a real device.

We should not expect to find such an arrangement of ideal elements inside our storage battery, of course. Any real device is characterized by a certain current-voltage relationship at its terminals, and our problem is to develop some combination of ideal elements that can furnish a similar current-voltage characteristic, at least over some useful range of current, voltage, or power.

This particular practical voltage source is shown connected to a general load resistor R_L in Fig. 3-19a. The terminal voltage of the practical source is the same as the voltage across R_L and is marked v_L.

Figure 3-19b shows a plot of load current i_L vs. load voltage v_L for this practical source. The KVL equation for the circuit of Fig. 3-19a may be written in terms of i_L and v_L:

$$12 = 0.01 i_L + v_L$$

and thus

$$i_L = 1200 - 100 v_L$$

This is a linear equation in i_L and v_L, and the plot in Fig. 3-19b is a straight line. Each point on the line corresponds to a different value of R_L. For example, the midpoint of the straight line is obtained when the load resistance is equal to the internal resistance of the practical source, $R_L = 0.01\ \Omega$. Here, the load voltage is only one-half the ideal source voltage.

Fig. 3-19: (a) A practical source, which approximates the behavior of a certain 12-V storage battery, is shown connected to a load resistor R_L. (b) The relationship between i_L and v_L is linear.

When $R_L = \infty$ and no current whatsoever is being drawn by the load, the practical source is open-circuited and the terminal voltage, or open-circuit voltage, is $v_{Loc} = 12$ V. If, on the other hand, $R_L = 0$, thereby short-circuiting the load terminals, then a load current or short-circuit current, $i_{Lsc} = 1200$ A, would flow. In practice, such an experiment would probably result in the destruction of the short circuit, the battery, and any measuring instruments incorporated in the circuit.

Since the plot of i_L vs. v_L is a straight line for this practical voltage source, we should note that the values of v_{Loc} and i_{Lsc} uniquely determine the entire i_L-v_L curve.

The vertical broken line shows the i_L-v_L plot for an ideal voltage source; the terminal voltage remains constant for any value of load current. For the practical voltage source, the terminal voltage has a value near that of the ideal source only when the load current is relatively small.

Let us now consider a general practical voltage source, as shown in Fig. 3-20a. The voltage of the ideal source is v_s, and a resistance R_{sv}, called an *internal resistance* or *output resistance,* is placed in series with it. Again, we must note that the resistor is not one that is really present, or that we would wire or solder into the circuit, but merely serves to account for a terminal voltage which decreases as the load current increases. Its presence enables us to model the behavior of a physical voltage source more closely.

The linear relationship between v_L and i_L is

$$v_L = v_s - R_{sv} i_L \tag{20}$$

and this is plotted in Fig. 3-20b. The open-circuit voltage and short-circuit current are

$$v_{Loc} = v_s \tag{21}$$

$$i_{Lsc} = \frac{v_s}{R_{sv}} \tag{22}$$

Once again, these values are the intercepts for the straight line on Fig. 3-20b, and they serve to define it completely.

An ideal current source is also nonexistent in the real world; there is no physical device which will deliver a constant current, regardless of the load

Fig. 3-20: (a) A general practical voltage source connected to a load resistor R_L. (b) The terminal voltage decreases as i_L increases and $R_L = v_L/i_L$ decreases.

Fig. 3-21: (a) A general practical current source connected to a load resistor R_L. (b) The load current provided by the practical current source is shown as a function of the load voltage.

resistance to which it is connected or the voltage across its terminals. Certain transistor circuits will deliver a constant current to a wide range of load resistances, but the load resistance can always be made sufficiently large that the current through it becomes very small. Infinite power is simply never available.

A practical current source is defined as an ideal current source in parallel with an internal resistance R_{si}. Such a source is shown in Fig. 3-21a, and the current i_L and voltage v_L associated with a load resistance R_L are indicated. It is apparent that

$$i_L = i_s - \frac{v_L}{R_{si}} \tag{23}$$

which is again a linear relationship. The open-circuit voltage and the short-circuit current are:

$$v_{Loc} = R_{si}i_s \tag{24}$$

$$i_{Lsc} = i_s \tag{25}$$

The variation of load current with changing load voltage may be investigated by changing the value of R_L as shown in Fig. 3-21b. The straight line is traversed from the short-circuit or northwest end to the open-circuit termination at the southeast by increasing R_L from zero to infinite ohms. The midpoint occurs for $R_L = R_{si}$. It is evident that the load current i_L and the ideal source current i_s are approximately equal only for small values of load voltage, obtained with values of R_L that are small compared with R_{si}.

Having defined both practical sources, we are now ready to discuss their equivalence. We shall define two sources as being *equivalent* if they produce identical values of v_L and i_L when they are connected to identical values of R_L, no matter what the value of R_L may be. Since $R_L = 0$ and $R_L = \infty$ are two such values, then equivalent sources provide the same open-circuit voltage and short-circuit current. In other words, if we are given two equivalent sources, one a practical voltage source and the other a practical current source, each enclosed in a black box with only a single pair of terminals on it, then there is no way in which we can differentiate between the boxes by measuring current or voltage in a resistive load.

The conditions for equivalence are now quickly established. Since the open-circuit voltages must be equal, from Eqs. (21) and (24) we have

$$v_{Loc} = v_s = R_{si}i_s \tag{26}$$

The short-circuit currents are also equal, and Eqs. (22) and (25) lead to

$$i_{Lsc} = \frac{v_s}{R_{sv}} = i_s$$

It follows that

$$R_{sv} = R_{si} = R_s \tag{27}$$

and

$$v_s = R_s i_s \tag{28}$$

where we now let R_s represent the internal resistance of either practical source.

As an example of the use of these ideas, consider the practical current source shown in Fig. 3-22a. Since its internal resistance is 2 Ω, the internal resistance of the equivalent practical voltage source is also 2 Ω; the voltage of the ideal voltage source contained within the practical voltage source is (2)(3) = 6 V. The equivalent practical voltage source is shown in Fig. 3-22b.

To check the equivalence, let us visualize a 4-Ω resistor connected to each source. In both cases a current of 1 A, a voltage of 4 V, and a power of 4 W are associated with the 4-Ω load. However, we should note very carefully that the ideal current source is delivering a total power of 12 W, while the ideal voltage source is delivering only 6 W. Furthermore, the internal resistance of the practical current source is absorbing 8 W, whereas the internal resistance of the practical voltage source is absorbing only 2 W. Thus we see that the two practical sources are only equivalent with respect to what transpires at the load terminals; they are *not* equivalent internally!

A very useful power theorem may be developed with reference to a practical voltage or current source. For the practical voltage source (Fig. 3-20a with $R_{sv} = R_s$), the power delivered to the load R_L is

$$p_L = i_L{}^2 R_L = \frac{v_s{}^2 R_L}{(R_s + R_L)^2}$$

To find the value of R_L that absorbs a maximum power from the given practical source, we differentiate with respect to R_L:

$$\frac{dp_L}{dR_L} = \frac{(R_s + R_L)^2 v_s{}^2 - v_s{}^2 R_L(2)(R_s + R_L)}{(R_s + R_L)^4}$$

and equate the derivative to zero, obtaining,

$$2R_L(R_s + R_L) = (R_s + R_L)^2$$

or

$$R_s = R_L$$

Since the values, $R_L = 0$ and $R_L = \infty$, both give a minimum ($p_L = 0$), and since we have already developed the equivalence between practical voltage

Fig. 3-22: (a) A given practical current source. (b) The equivalent practical voltage source: $R_{sv} = R_{si} = R_s$; $v_s = R_s i_s$.

and current sources, we have therefore proved the following *maximum power transfer* theorem:

> An independent voltage source in series with a resistance R_s or an independent current source in parallel with a resistance R_s delivers a maximum power to that load resistance R_L for which $R_L = R_s$.

Thus, the maximum power transfer theorem tells us that a 2-Ω resistor draws the greatest power (4.5 W) from either practical source of Fig. 3-22, whereas a resistance of 0.01 Ω receives the maximum power (3.6 kW) in Fig. 3-19.

Drill Problems

3-6. (a) Let $R_L = 80$ Ω in Fig. 3-23a and find i_L. (b) Transform the practical current source into a practical voltage source and again find i_L if $R_L = 80$ Ω. (c) Find the power supplied by the ideal source in each case. (d) What value of R_L will absorb a maximum power, and what is the value of that power?
Ans: 20 mA; 20 mA; 160 mW, 40 mW; 20 Ω, 50 mW

Fig. 3-23: See Drill Probs. 3-6 and 3-7.

3-7. Transform both 24-V practical sources in Fig. 3-23b into practical current sources, combine resistors and ideal current sources, then transform the resultant practical current source back into a practical voltage source, and combine ideal voltage sources. (a) If $R_x = 2$ Ω, find the power delivered to R_x. (b) What is the maximum power that can be delivered to any R_x? (c) What two values of R_x will have exactly 5 W delivered to them?
Ans: 4.5 W; 6 W; 0.280 and 1.587 Ω

3-6
Thévenin's and
Norton's theorems

Now that we have the superposition principle, it is possible to develop two more theorems which will greatly simplify the analysis of many linear circuits. The first of these theorems is named after M. L. Thévenin, a French engineer working in telegraphy, who first published a statement of the theorem in 1883; the second may be considered a corollary of the first and is credited to E. L. Norton, a scientist formerly with the Bell Telephone Laboratories.

Let us suppose that we need to make only a partial analysis of a circuit; perhaps we wish to determine the current, voltage, and power delivered to a single load resistor by the remainder of the circuit, which may consist of any number of sources and resistors; or perhaps we wish to find the response for different values of the load resistance. Thévenin's theorem then tells us that it is possible to replace everything except the load resistor by an equivalent circuit containing only an independent voltage source in series with a resistor; the response measured at the load resistor will be unchanged. Using Norton's theorem, we obtain an equivalent composed of an independent current source in parallel with a resistor.

It should thus be apparent that one of the main uses of Thévenin's and Norton's theorems is the replacement of a large part of a network, often a complicated and uninteresting part, by a very simple equivalent. The new, simpler circuit enables us to make rapid calculations of the voltage, current, and power which the original circuit is able to deliver to a load. It also helps us to choose the best value of this load resistance. In a transistor power amplifier, for example, the Thévenin or Norton equivalent enables us to determine the maximum power that can be taken from the amplifier and the type of load that is required to accomplish a maximum transfer of power or to obtain maximum practical voltage or current amplification.

As an introductory example, consider the circuit shown in Fig. 3-24. The shaded regions separate the circuit into networks A and B; we shall assume that our main interest is in network B, which consists only of the load resistor R_L. Network A may be simplified by making repeated source transformations. We first treat the 12-V source and the 3-Ω resistor as a practical voltage source and replace it with a practical current source consisting of a 4-A source in parallel with 3 Ω. The parallel resistances are then combined into 2 Ω, and the practical current source which results is transformed back into a practical voltage source. The steps are indicated in Fig. 3-25, the final result appearing in Fig. 3-25d. From the viewpoint of the load resistor R_L, this circuit (the Thévenin equivalent) is equivalent to the original circuit; from our viewpoint, the circuit is much simpler and we can now easily compute the power delivered to the load. It is

$$p_L = \left(\frac{8}{9 + R_L}\right)^2 R_L$$

Furthermore we can see from the equivalent circuit that the maximum voltage which can be obtained across R_L is 8 V when $R_L = \infty$; a quick transformation of network A to a practical current source (the Norton equivalent) indicates that the maximum current which may be delivered to the load is $\frac{8}{9}$ A

Fig. 3-24: A simple resistive circuit is divided into network A, in which we have no detailed interest, and network B, a load resistor with which we are fascinated.

Network A   Network B

Fig. 3-25: The source transformations and resistance combinations involved in simplifying network A are shown in order. The result, given in (d), is the Thévenin equivalent.

(a) Network A   (b) Network A

(c) Network A   (d) Network A   Network B

for $R_L = 0$; and the maximum power transfer theorem shows that a maximum power is delivered to R_L when R_L is 9 Ω. None of these facts is readily apparent from the original circuit.

If network A had been more complicated, the number of source transformations and resistance combinations necessary to obtain the Thévenin or Norton equivalent could easily become prohibitive; also, with dependent sources present, the method of source transformation is usually inapplicable. Thévenin's and Norton's theorems allow us to find the equivalent circuit much more quickly and easily, even in more complicated circuits.

Let us now state *Thévenin's theorem* formally:

Given any linear circuit, rearrange it in the form of two networks A and B that are connected together by two resistanceless conductors. If either network contains a dependent source, its control variable must be in that same network. Define a voltage v_{oc} as the open-circuit voltage which would appear across the terminals of A if B were disconnected so that no current is drawn from A. Then all the currents and voltages in B will remain unchanged if A is killed (all independent voltage sources and independent current sources in A replaced by short circuits and open circuits, respectively) and an independent voltage source v_{oc} is connected, with proper polarity, in series with the dead (inactive) A network.

The terms *killed* and *dead* are a little bloodthirsty, but they are descriptive and concise, and we shall use them in a friendly way. Moreover, it is possible that network *A* may only be sleeping, for it may still contain *dependent* sources which come to life whenever their controlling currents or voltages are nonzero.

Let us see if we can apply Thévenin's theorem successfully to the circuit we considered in Fig. 3-24. Disconnecting R_L, voltage division enables us to determine that v_{oc} is 8 V. Killing the *A* network, that is, replacing the 12-V source by a short circuit, we see looking back into the dead *A* network a 7-Ω resistor connected in series with the parallel combination of 6 Ω and 3 Ω. Thus the dead *A* network can be represented here by simply a 9-Ω resistor. This agrees with the previous result.

The equivalent circuit we have obtained is completely independent of the *B* network, because we have been instructed first to remove the *B* network and measure the open-circuit voltage produced by the *A* network, an operation which certainly does not depend on the *B* network in any way, and then to place the inactive *A* network in series with a voltage source v_{oc}. The *B* network is mentioned in the theorem and proof only to indicate that an equivalent for *A* may be obtained *no matter what arrangement of elements is connected to the A network;* the *B* network represents this general network.

A proof of Thévenin's theorem in the form in which we have stated it is rather lengthy, and therefore it has been placed in Appendix 3 where the curious or rigorous may peruse it.

There are several points about the theorem which deserve emphasis. First, the only restriction that we must impose on *A* or *B*, other than requiring that the original circuit composed of *A* and *B* be a linear circuit, is that all the dependent sources in *A* have their control variables in *A,* and similarly for *B*. No restrictions were imposed on the complexity of *A* or *B*; either one may contain any combination of independent voltage or current sources, linear dependent voltage or current sources, resistors, or any other circuit elements which are linear. The general nature of the theorem (and its proof) will enable it to be applied to networks containing inductors and capacitors, which are linear passive circuit elements to be defined in the following chapter. At this time, however, resistors are the only passive circuit elements which have been defined, and the application of Thévenin's theorem to resistive networks is a particularly simple special case. The dead *A* network can be represented by a single equivalent resistance, R_{th}. If *A* is an active *resistive* network, then it is obvious that the inactive *A* network may be replaced by a single equivalent resistance, which we shall also call the Thévenin resistance, since it is once again the resistance viewed at the terminals of the inactive *A* network.

Norton's theorem bears a close resemblance to Thévenin's theorem, another consequence of duality. As a matter of fact, the two statements will be used as an example of dual language when the duality principle is discussed in the following chapter.

Norton's theorem may be stated as follows:

> Given any linear circuit, rearrange it in the form of two networks A and B that are connected together by two resistanceless conductors. If either network contains a dependent source, its control variable must be in that same network. Define a current i_{sc} as the short-circuit current which would appear at the terminals of A if B were short-circuited so that no voltage is provided by A. Then all the voltages and currents in B will remain unchanged if A is killed (all independent current sources and independent voltage sources in A replaced by open circuits and short circuits, respectively) and an independent current source i_{sc} is connected, with proper polarity, in parallel with the dead (inactive) A network.

The Norton equivalent of an active resistive network is the Norton current source i_{sc} in parallel with the Thévenin resistance R_{th}.

There is an important relationship between the Thévenin and Norton equivalents of an active *resistive* network. The relationship may be obtained by applying a source transformation to either equivalent network. For example, if we transform the Norton equivalent, we obtain a voltage source $R_{th}i_{sc}$ in series with the resistance R_{th}; this network is in the form of the Thévenin equivalent, and thus

$$v_{oc} = R_{th}i_{sc} \qquad (29)$$

In resistive circuits containing *dependent* sources as well as independent sources, we shall often find it more convenient to determine either the Thévenin or Norton equivalent by finding both the open-circuit voltage and the short-circuit current and then determining the value of R_{th} as their quotient. It is therefore advisable to become adept at finding both open-circuit voltages and short-circuit currents, even in the simple problems which follow. If the Thévenin and Norton equivalents are determined independently, (29) may serve as a useful check.

Let us consider four examples of the determination of a Thévenin or Norton equivalent circuit. The first is shown in Fig. 3-26*a*; the Thévenin and Norton equivalents are desired for the network faced by the 1-kΩ resistor. That is, network B is this resistor, and network A is the remainder of the given circuit.

This circuit contains no dependent sources, and the easiest way to find the Thévenin equivalent is to determine R_{th} for the dead network directly, followed by a calculation of either v_{oc} or i_{sc}. We first kill both independent sources to determine the form of the dead A network. With the 4-V source short-circuited and the 2-mA source open-circuited, the result is the series combination of a 2-kΩ and 3-kΩ resistor, or the equivalent, a 5-kΩ resistor. The open-circuit voltage is easily determined by superposition. With only the 4-V source operating, the open-circuit voltage is 4 V; when only the 2-mA source is on, the open-circuit voltage is also 4 V; with both independent

Fig. 3-26: (a) A given circuit in which the 1-kΩ resistor is identified as network B. (b) The Thévenin equivalent is shown for network A. (c) The Norton equivalent is shown for network A.

sources operating, we see that $v_{oc} = 4 + 4 = 8$ V. This determines the Thévenin equivalent, shown in Fig. 3-26b, and from it the Norton equivalent of Fig. 3-26c can be drawn quickly. As a check, let us determine i_{sc} for the given circuit. We use superposition and a little current division:

$$i_{sc} = i_{sc}|_{4V} + i_{sc}|_{2mA} = \frac{4}{2 + 3} + 2\frac{2}{2 + 3} = 0.8 + 0.8 = 1.6 \quad \text{mA}$$

which completes the check.

As the second example, we consider the network A shown in Fig. 3-27, which contains both independent and dependent sources. The presence of the dependent source prevents us from determining R_{th} directly for the inactive network through resistance combination; instead, we find both v_{oc} and i_{sc}. To find v_{oc} we note that $v_x = v_{oc}$, and that the dependent source current

Fig. 3-27: (a) A given network whose Thévenin equivalent is desired. (b) A possible, but rather useless, form of the Thévenin equivalent. (c) The best form of the Thévenin equivalent for this linear resistive network.

must pass through the 2-kΩ resistor since there is an open circuit to the right. Summing voltages around the outer loop:

$$-4 + 2 \times 10^3 \left(\frac{-v_x}{4000}\right) + 3 \times 10^3(0) + v_x = 0$$

and
$$v_x = 8 = v_{oc}$$

By Thévenin's theorem, then, the equivalent could be formed with the dead A network in series with an 8-V source, as shown in Fig. 3-27b. This is correct, but is not very simple and not very helpful; in the case of linear resistive networks, we should certainly show a much simpler equivalent for the inactive A network, namely R_{th}. We therefore seek i_{sc}. Upon short-circuiting the output terminals in Fig. 3-27a, it is apparent that $v_x = 0$ and the dependent current source is zero. Hence, $i_{sc} = 4/(5 \times 10^3) = 0.8$ mA. Thus, $R_{th} = v_{oc}/i_{sc} = 8/(0.8 \times 10^{-3}) = 10$ kΩ, and the accepted Thévenin equivalent of Fig. 3-27c is obtained.

Another example for which we should find both v_{oc} and i_{sc} appears in Fig. 3-28. The circuit happens to be an op-amp connected as a voltage follower

Fig. 3-28: The Thévenin equivalent of this voltage follower is desired for terminals *a-b*.

with $v_s = 5$ V, $R_i = \infty$, $A = 10^4$, and $R_o = 2$ kΩ. To find v_{Loc} we set $R_L = \infty$, or we simply consider it to be removed from the circuit. There now can be no current through the 2-kΩ resistor, and therefore

$$v_{Loc} = -10^4 v_i$$

where
$$v_i = v_{Loc} - 5$$

Therefore
$$v_{Loc} = \frac{10^4(5)}{10\ 001} = 5.00 \quad V$$

Next we need a value for i_{Lsc}; so we replace R_L by a short circuit. Around the right mesh, KVL gives

$$10^4 v_i + 2000 i_{Lsc} = 0$$

while the application of KVL around the perimeter of the circuit gives

$$-5 - v_i = 0$$

Therefore

$$10^4(-5) + 2000i_{Lsc} = 0$$

and

$$i_{Lsc} = 25.0 \quad \text{A}$$

We find R_{th} by the quotient,

$$R_{th} = \frac{v_{Loc}}{i_{Lsc}} = \frac{5.00}{25.0} = 0.200 \quad \Omega$$

Thus, the Thévenin equivalent presented to R_L at a-b by the voltage follower is 5.00 V in series with a very low value of resistance, 0.200 Ω.

As our final example, let us consider a network having a dependent source but no independent source, such as that shown in Fig. 3-29a. The network therefore qualifies already as the dead A network, and $v_{oc} = 0$. We thus seek the value of R_{th} represented by this two-terminal network. However, we cannot find v_{oc} and i_{sc} and take their quotient, for there is no independent source in the network and both v_{oc} and i_{sc} are zero. Let us, therefore, be a little tricky. We apply a 1-A source *externally*, measure the resultant voltage, and then set $R_{th} = v/1$. Referring to Fig. 3-29b, we see that $i = -1$ and

$$\frac{v - 1.5(-1)}{3} + \frac{v}{2} = 1$$

so that

$$v = 0.6 \quad \text{V}$$

and

$$R_{th} = 0.6 \quad \Omega$$

The Thévenin equivalent is shown in Fig. 3-29c.

Fig. 3-29: (a) A network, containing no independent sources, whose Thévenin equivalent is desired. (b) R_{th} is numerically equal to v. (c) The Thévenin equivalent of (a).

We have now looked at four examples in which we determined a Thévenin or Norton equivalent circuit. The first example (Fig. 3-26) contained only independent sources and resistors, and we were able to use several different methods on it. One involved calculating R_{th} for the dead network and then v_{oc} for the live network. We could also have found R_{th} and i_{sc}, or v_{oc} and i_{sc}.

In the second and third examples (Figs. 3-27 and 3-28), both independent and dependent sources were present, and the only method we used required us to find v_{oc} and i_{sc}. We could not easily find R_{th} for the dead network because the dependent source could not be made inactive.

The last example did not contain any independent sources, and we found R_{th} by applying 1 A and finding $v = 1 \times R_{th}$. We could also apply 1 V and determine $i = 1/R_{th}$. These Thévenin and Norton equivalents do not contain an independent source.

These important techniques and the types of circuits to which they may be applied most readily are indicated in Table 3-1.

Table 3-1: Recommended methods of finding Thévenin or Norton equivalents

	Circuit contains		
Methods			
R_{th} and v_{oc} or i_{sc}	✔	—	—
v_{oc} and i_{sc}	Possible	✔	—
$i = 1$ A or $v = 1$ V	—	—	✔

All possible methods do not appear in the table. We have used repeated source transformations on several networks when there were no dependent sources, and this is certainly an easy technique for networks that do not contain too many elements. Another method has a certain appeal because it can be used for any of the three types of networks tabulated in the foregoing. Simply replace the B network by a voltage source v_s, define the current leaving its positive terminal as i, then analyze the A network to obtain i, and put the equation in the form $v_s = ai + b$. Evidently, $a = R_{th}$ and $b = v_{oc}$. Of course, we could also apply a current source i_s, let its voltage be v, and then determine $v = ai_s + b$. This procedure is universally applicable, but some other method is usually easier and more rapid.

Although we are devoting our attention almost entirely to the analysis of *linear* circuits, its is enlightening to know that Thévenin's and Norton's theorems are both valid if network B is nonlinear; only network A must be linear.

Drill Problems

3-8. Find the Thévenin equivalent for the network of: (*a*) Fig. 3-30*a*; (*b*) Fig. 3-30*b*.　　　　　*Ans:* 356 V, 40 Ω; $v_s/(1 - g_m R_1)$, $(R_1 + R_2)/(1 - g_m R_1)$

3-9. Find the Norton equivalent for the network of: (*a*) Fig. 3-30*c*; (*b*) Fig. 3-30*d*.　　　　　*Ans:* 0 A, 14.6 Ω; 0.75 A, 26.7 Ω

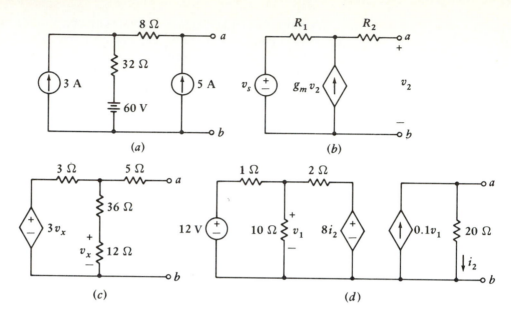

Fig. 3-30: See Drill
Probs. 3-8 and 3-9.

3-7

Trees and general nodal analysis

In this section we shall generalize the method of nodal analysis that we have come to know and love. Since nodal analysis is applicable to any network, we cannot promise that we shall be able to solve a wider class of circuit problems. We can, however, look forward to being able to select a general nodal analysis method for any particular problem that may result in fewer equations and less work.

We must first extend our list of definitions relating to network topology. We begin by defining *topology* itself as a branch of geometry which is concerned with those properties of a geometrical figure which are unchanged when the figure is twisted, bent, folded, stretched, squeezed, or tied in knots, with the provision that no parts of the figure are to be cut apart or to be joined together. A sphere and a tetrahedron are topologically identical, as are a square and a circle. In terms of electric circuits, then, we are not now concerned with the particular types of elements appearing in the circuit, but only with the way in which branches and nodes are arranged. As a matter of fact, we usually suppress the nature of the elements and simplify the drawing of the circuit by showing the elements as lines. The resultant drawing is called a *linear graph,* or simply a *graph.* A circuit and its graph are shown in Fig. 3-31. Note that all nodes are identified by heavy dots in the graph.

Since the topological properties of the circuit or its graph are unchanged when it is distorted, the three graphs shown in Fig. 3-32 are all topologically identical with the circuit and graph of Fig. 3-31.

---

[7] This and the following section may be postponed if desired. They introduce analysis methods that are a little more general than those using node-to-reference voltages and mesh currents. Their use can lead to fewer equations or more useful current and voltage variables.

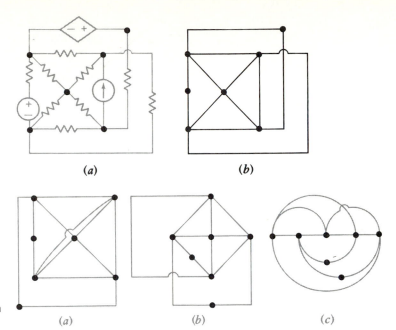

Fig. 3-31: (a) A given circuit. (b) The linear graph of this circuit.

(a)   (b)

Fig. 3-32: The three graphs shown are topologically identical to each other and to the graph of Fig. 3-31b, and each is a graph of the circuit shown in Fig. 3-31a.

(a)   (b)   (c)

Topological terms which we already know and have been using correctly are:

node: a point at which two or more elements have a common connection.
path: a set of elements that may be traversed in order without passing through the same node twice.
branch: a single path, containing one simple element, which connects one node to any other node.
loop: a closed path.
mesh: a loop which does not contain any other loops within it.
planar circuit: a circuit which may be drawn on a plane surface in such a way that no branch passes over or under any other branch.
nonplanar circuit: any circuit which is not planar.

The graphs of Fig. 3-32 each contain 12 branches and 7 nodes.

Three new properties of a linear graph must now be defined, a tree, a cotree, and a link. We define a *tree* as any set of branches which does not contain any loops and yet connects every node to every other node, not necessarily directly. There are usually a number of different trees which may be drawn for a network, and the number increases rapidly as the complexity of the network increases. The simple graph shown in Fig. 3-33a has eight possible trees, four of which are shown by heavy lines in Figs. 3-33b, c, d, and e.

Fig. 3-33: (a) The linear graph of a three-node network. (b), (c), (d), and (e) Four of the eight different trees which may be drawn for this graph are shown by the heavy lines.

(a)   (b)   (c)   (d)   (e)

Fig. 3-34: (*a*) A linear graph. (*b*) A possible tree for this graph. (*c*) and (*d*) These sets of branches do not satisfy the definition of a tree.

(*a*)   (*b*)   (*c*)   (*d*)

In Fig. 3-34*a* a more complex graph is shown. Figure 3-34*b* shows one possible tree, and Figs. 3-34*c* and *d* show sets of branches which are *not* trees because neither set satisfies the definition above.

After a tree has been specified, those branches that are not part of the tree form the *cotree*, or complement of the tree. The lightly drawn branches in Figs. 3-33*b* to *e* show the cotrees that correspond to the heavier trees.

Once we understand the construction of a tree and its cotree, the concept of the link is very simple, for a *link* is any branch belonging to the cotree. It is evident that any particular branch may or may not be a link, depending on the particular tree which is selected.

The number of links in a graph may be related to the number of branches and nodes very simply. If the graph has N nodes, then exactly $(N - 1)$ branches are required to construct a tree because the first branch chosen connects two nodes and each additional branch includes one more node. Thus, given B branches, the number of links L must be

$$L = B - (N - 1)$$

or

$$L = B - N + 1 \tag{30}$$

There are L branches in the cotree and $(N - 1)$ branches in the tree.

In any of the graphs shown in Fig. 3-33, we note that $3 = 5 - 3 + 1$, and in the graph of Fig. 3-34*b*, $6 = 10 - 5 + 1$. A network may be in several disconnected parts, and (30) may be made more general by replacing $+1$ with $+S$, where S is the number of separate parts. However, it is also possible to connect two separate parts by a *single* conductor, thus causing two nodes to form one node; no current can flow through this single conductor. This process may be used to join any number of separate parts, and thus we shall not suffer any loss of generality if we restrict our attention to circuits for which $S = 1$.

We are now ready to discuss a method by which we may write a set of nodal equations that are independent and sufficient. The method will enable us to obtain many different sets of equations for the same network, and all the sets will be valid. However, the method does not provide us with *every* possible set of equations. Let us first describe the procedure, illustrate it by three examples, and then point out the reason that the equations are independent and sufficient.

Given a network, we should:

1 Draw a graph and then identify a tree.
2 Place all voltage sources in the tree.
3 Place all current sources in the cotree.

4 Place all control-voltage branches for voltage-controlled dependent sources in the tree, if possible.

5 Place all control-current branches for current-controlled dependent sources in the cotree, if possible.

These last four steps effectively associate voltages with the tree and currents with the cotree.

We now assign a voltage variable (with its plus-minus pair) across each of the $(N - 1)$ branches in the tree. A branch containing a (dependent or independent) voltage source should be assigned that source voltage, and a branch containing a controlling voltage should be assigned that controlling voltage. The number of new variables that we have introduced is therefore equal to the number of branches in the tree $(N - 1)$, reduced by the number of voltage sources in the tree, and reduced also by the number of control voltages we were able to locate in the tree. In the third example below, we shall find that the number of new variables required may be zero.

Having a set of variables, we now need to write a set of equations that are sufficient to determine these variables. The equations are obtained through the application of KCL. Voltage sources are handled in the same way as in our earlier attack on nodal analysis; each voltage source and the two nodes at its terminals comprise a supernode or a part of a supernode. Kirchhoff's current law is then applied at all but one of the remaining nodes and supernodes. We set the sum of the currents leaving the node in all of the branches connected to it equal to zero. Each current is expressed in terms of the voltage variables we just assigned. One node may be ignored, just as was the case earlier for the reference node. Finally, in case there are current-controlled dependent sources, we must write an equation for each control current that relates it to the voltage variables; this also is no different from the procedure used before with nodal analysis.

Let us try out this process on the circuit shown in Fig. 3-35a. It contains four nodes and five branches, and its graph is shown in Fig. 3-35b. In accor-

Fig. 3-35: (a) A circuit used as an example for general nodal analysis. (b) The graph of the given circuit. (c) The voltage source and the control voltage are placed in the tree, while the current source goes in the cotree. (d) The tree is completed and a voltage is assigned across each tree branch.

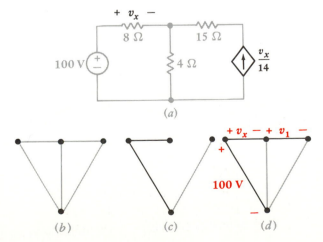

dance with steps 2 and 3 of the tree-drawing procedure, we place the voltage source in the tree and the current source in the cotree. Following step 4, we see that the v_x branch may also be placed in the tree since it does not form any loop which would violate the definition of a tree. We have now arrived at the two tree branches and the single link shown in Fig. 3-35c, and we see that we do not yet have a tree since the right node is not connected to the others by a path through tree branches. The only possible way to complete the tree is shown in Fig. 3-35d. The 100-V source voltage, the control voltage v_x, and a new voltage variable v_1 are next assigned to the three tree branches as shown.

We therefore have two unknowns, v_x and v_1, and it is obvious that we need to obtain two equations in terms of them. There are four nodes, but the presence of the voltage source causes two of them to form a single supernode. Kirchhoff's current law may be applied at any two of the three remaining nodes or supernodes. Let's attack the right node first. The current leaving to the left is $-v_1/15$, while that leaving downward is $-v_x/14$. Thus, our first equation is

$$-\frac{v_1}{15} - \frac{v_x}{14} = 0$$

The central node at the top looks easier than the supernode, so we set the sum of the current to the left, $-v_x/8$, the current to the right, $v_1/15$, and the downward current through the 4-Ω resistor equal to zero. This latter current is given by the voltage across the resistor divided by 4 Ω, but there is no voltage labeled on that link. However, when a tree is constructed according to the definition, there is a path through it from any node to any other node. Then, since every branch in the tree is assigned a voltage, we may express the voltage across any link in terms of the tree-branch voltages. This downward current is therefore $(-v_x + 100)/4$, and we have the second equation,

$$-\frac{v_x}{8} + \frac{v_1}{15} + \frac{-v_x + 100}{4} = 0$$

The simultaneous solution of these two nodal equations gives $v_1 = -60$ V, $v_x = 56$ V.

Fig. 3-36: (a) The circuit of Fig. 3-5 is repeated. (b) A tree is chosen such that both voltage sources and both control voltages are tree branches.

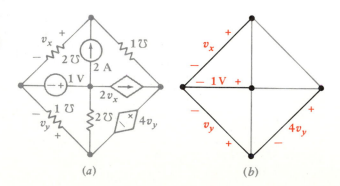

(a)   (b)

As a second example, let us reconsider a more complex circuit that we first analyzed by defining all node voltages with respect to a reference node. The circuit is that of Fig. 3-5, repeated as Fig. 3-36a. We draw a tree so that both voltage sources and both control voltages appear as tree-branch voltages and, hence, as assigned variables. As it happens, these four branches constitute a tree, Fig. 3-36b, and tree-branch voltages v_x, 1, v_y, and $4v_y$ are chosen, as shown.

Both voltage sources define supernodes and we apply KCL twice, once to the top node,

$$2v_x + 1(v_x - v_y - 4v_y) = 2$$

and once to the supernode consisting of the right node, the bottom node, and the dependent voltage source,

$$1v_y + 2(v_y - 1) + 1(4v_y + v_y - v_x) = 2v_x$$

Instead of the four equations we had previously, we have only two, and we find easily that $v_x = \frac{26}{9}$ V and $v_y = \frac{4}{3}$ V, both values agreeing with the earlier solution.

For a final example we consider the circuit of Fig. 3-37a. The two voltage sources and the control voltage establish the three-branch tree shown in Fig. 3-37b. Since the two upper and the lower right node all join to form one supernode, we need write only one KCL equation. Selecting the lower left node, we have

$$-1 - \frac{v_x}{4} + 3 + \frac{-v_x + 30 + 6v_x}{5} = 0$$

and it follows that $v_x = -\frac{32}{3}$ V. In spite of the apparent complexity of this circuit, the use of general nodal analysis has led to an easy solution. Employing mesh currents or node-to-reference voltages would require more equations and more effort.

We shall discuss the problem of finding the best analysis scheme in the following section.

If we should need to know some other voltage, current, or power in the previous example, one additional step would give the answer. For example, the power provided by the 3-A source is $3(-30 - \frac{32}{3}) = -122$ W.

Fig. 3-37: (a) A circuit for which only one general nodal equation need be written. (b) The tree and the tree-branch voltages used.

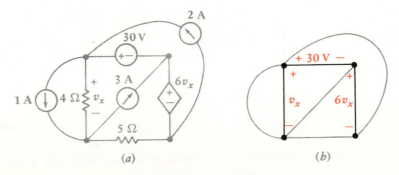

Let us conclude by discussing the sufficiency of the assumed set of tree-branch voltages and the independence of the nodal equations. If these tree-branch voltages are *sufficient,* then the voltage of every branch in either the tree or the cotree must be obtainable from a knowledge of the values of all the tree-branch voltages. This is certainly true for those branches in the tree. For the links we know that each link extends between two nodes, and, by definition, the tree must also connect those two nodes. Hence, every link voltage may also be established in terms of the tree-branch voltages.

Once the voltage across every branch in the circuit is known, then all the currents may be found by using either the given value of the current if the branch consists of a current source, Ohm's law if it is a resistive branch, or by using KCL and these current values if the branch happens to be a voltage source. Thus, all the voltages and currents are determined and sufficiency is demonstrated.

To demonstrate independency, let us satisfy ourselves by assuming the situation where the only sources in the network are independent current sources. As we have noticed earlier, independent voltage sources in the circuit result in fewer equations, while dependent sources usually necessitate a greater number of equations. For independent current sources only, there will then be precisely $(N - 1)$ nodal equations written in terms of $(N - 1)$ tree-branch voltages. To show that these $(N - 1)$ equations are *independent,* visualize the application of KCL to the $(N - 1)$ different nodes. Each time we write the KCL equation, there is a new tree branch involved—the one which connects that node to the remainder of the tree. Since that circuit element has not appeared in any previous equation, we must obtain an independent equation. This is true for each of the $(N - 1)$ nodes in turn, and hence we have $(N - 1)$ independent equations.

Drill Problem

3-10. (*a*) How many trees may be constructed for the circuit of Fig. 3-38 that follow all five of the tree-drawing suggestions listed earlier? (*b*) Draw a suitable tree, write two equations in two unknowns, and find v_x. (*c*) What power is supplied by the dependent source? *Ans:* 1; 250 V; 20.5 kW

Fig. 3-38: See Drill Prob. 3-10.

3-8
Links and loop analysis

Now we shall consider the use of a tree to obtain a suitable set of loop equations. In some respects this is the dual of the method of writing nodal equations. Again it should be pointed out that, although we are able to guarantee that any set of equations we write will be both sufficient and independent, we should not expect that the method will lead directly to every possible set of equations.

We again begin by constructing a tree, and we use the same set of rules as we did for general nodal analysis. The objective either for nodal or loop analysis is to place voltages in the tree and currents in the cotree; this is a mandatory rule for sources and a desirable rule for controlling quantities.

Now, however, instead of assigning a voltage to each branch in the tree, we assign a current (including reference arrow, of course) to each element in the cotree or to each link. If there were ten links, we would assign exactly 10 link currents. Any link that contains a current source is assigned that source current as the link current. Note that each link current may also be thought of as a loop current, for the link must extend between two specific nodes, and there must also be a path between those same two nodes through the tree. Thus, with each link there is associated a single loop that includes that one link and a unique path through the tree. It is evident that the assigned current may be thought of either as a loop current or as a link current. The link connotation is most helpful at the time the currents are being defined, for one must be established for each link; the loop interpretation is more convenient at equation-writing time, because we shall apply KVL around each loop.

Let us try out this process of defining link currents by reconsidering an earlier example we worked using mesh currents. The circuit is shown in Fig. 3-12 and redrawn in Fig. 3-39a. The tree selected is one of several that might be constructed for which the voltage source is in a tree branch and the current source is in a link. Let us first consider the link containing the current source. The loop associated with this link is the left-hand mesh, so we show our link current flowing about the perimeter of this mesh, Fig. 3-39b. An obvious choice for the symbol for this link current is ''7 A.'' Remember that no other current can flow through this particular link, and thus its value must be exactly the strength of the current source.

We next turn our attention to the 3-Ω resistor link. The loop associated with it is the upper right-hand mesh, and this loop (or mesh) current is defined as i_A and also shown in Fig. 3-39b. The last link is the lower 1-Ω resistor, and the only path between its terminals through the tree is around the perimeter of the circuit. That link current is called i_B, and the arrow indicating its path and reference direction appears in Fig. 3-39b. It is not a mesh current.

Fig. 3-39: (a) The circuit of Fig. 3-12 is shown again. (b) A tree is chosen such that the current source is in a link and the voltage source is in a tree branch.

Note that each link has only one current present in it, but a tree branch may have any number from one to the total number of link currents assigned. The use of long, almost closed, arrows to indicate the loops helps to indicate which loop currents flow through which tree branch and what their reference directions are.

A KVL equation must now be written around each of these loops. The variables used are the assigned link currents. Since the voltage across a current source cannot be expressed in terms of the source current, and since we have already used the value of the source current as the link current, we discard any loop containing a current source.

For the example of Fig. 3-39, we first traverse the i_A loop, proceeding clockwise from its lower left corner. The current going our way in the 1-Ω resistor is $(i_A - 7)$, in the 2-Ω element it is $(i_A + i_B)$, and in the link it is simply i_A. Thus

$$1(i_A - 7) + 2(i_A + i_B) + 3i_A = 0$$

For the i_B link, clockwise travel from the lower left corner leads to

$$-7 + 2(i_A + i_B) + 1i_B = 0$$

Traversal of the loop defined by the 7-A link is not required. Solving, we have $i_A = 0.5$ A, $i_B = 2$ A, once again. The solution has been achieved with one less equation than before.

An example containing a dependent source appears in Fig. 3-40a. This circuit contains six nodes, and its tree therefore must have five branches. Since there are eight elements in the network, there are three links in the

Fig. 3-40: (a) A circuit for which i_1 may be found with one equation using general loop analysis. (b) The only tree that satisfies the rules outlined in Sec. 3-7. (c) The three link currents are shown with their loops.

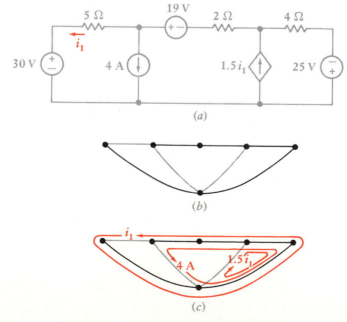

(a)

(b)

(c)

cotree. If we place the three voltage sources in the tree and the two current sources and the current control in the cotree, we are led to the tree shown in Fig. 3-40b. The source current of 4 A defines a loop as shown in Fig. 3-40c. The dependent source establishes the loop current $1.5i_1$ around the right mesh, and the control current i_1 gives us the remaining loop current about the perimeter of the circuit. Note that all three currents flow through the 4-Ω resistor.

We have only one unknown quantity, i_1, and, after discarding the loops defined by the two current sources, we apply KVL around the outside of the circuit:

$$-30 + 5(-i_1) + 19 + 2(-i_1 - 4) + 4(-i_1 - 4 + 1.5i_1) - 25 = 0$$

Besides the three voltage sources, there are three resistors in this loop. The 5-Ω resistor has one loop current in it since it is also a link, the 2-Ω resistor contains two loop currents, and the 4-Ω resistor has three. A carefully drawn set of loop currents is a necessity if errors in skipping currents, utilizing extra ones, or erring in the correct direction are to be avoided. The foregoing equation is guaranteed, however, and it leads to $i_1 = -12$ A.

How may we demonstrate sufficiency? Let us visualize a tree. It contains no loops and therefore contains at least two nodes to each of which only one tree branch is connected. The current in each of these two branches is easily found from the known link currents by applying KCL. If there are other nodes at which only one tree branch is connected, these tree-branch currents may also be immediately obtained. In the tree shown in Fig. 3-41, we thus have found the currents in branches a, b, c, and d. Now we move along the branches of the tree, finding the currents in the tree branches e and f; the process may be continued until all the branch currents are determined. The link currents are therefore sufficient to determine all branch currents. It is helpful to look at the situation where an incorrect ''tree'' has been drawn which contains a loop. Even if all the link currents were zero, a current might still circulate about this ''tree loop.'' Hence, the link currents could not determine this current, and they would not represent a sufficient set. Such a ''tree'' is by definition impossible.

To demonstrate independence, let us satisfy ourselves by assuming the situation where the only sources in the network are independent voltage sources. As we have noticed earlier, independent current sources in the circuit result in fewer equations, while dependent sources usually necessi-

Fig. 3-41: A tree which is used as an example to illustrate the sufficiency of the link currents.

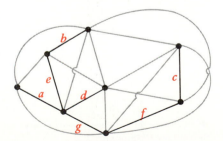

tate a greater number of equations. For independent voltage sources only, there will then be precisely $(B - N + 1)$ loop equations written in terms of the $(B - N + 1)$ link currents. To show that these $(B - N + 1)$ loop equations are independent, it is only necessary to point out that each represents the application of KVL around a loop which contains one link not appearing in any other equation. We might visualize a different resistance $R_1, R_2, \ldots, R_{B-N+1}$ in each of these links, and it is then apparent that one equation can never be obtained from the others since it contains one coefficient not appearing in any other equation.

Hence, the link currents are sufficient to enable a complete solution to be obtained, and the set of loop equations which we use to find the link currents is a set of independent equations.

Having looked at both general nodal analysis and loop analysis, we should now consider the advantages and disadvantages of each method so that an intelligent choice of a plan of attack can be made.

The nodal method in general requires $(N - 1)$ equations, but this number is reduced by one for each independent or dependent voltage source in a tree branch, and increased by one for each dependent source that is voltage controlled by a link voltage or current controlled.

The loop method basically involves $(B - N + 1)$ equations. However, each independent or dependent current source in a link reduces this number by one, while each dependent source that is current controlled by a tree-branch current or is voltage controlled increases the number by one.

As a grand finale for this discussion, let us inspect the T-equivalent-circuit model for a transistor, shown in Fig. 3-42, to which are connected a sinusoidal source, 4 sin 1000t mV, and a 10-kΩ load. We select typical values for the emitter resistance, $r_e = 50\ \Omega$; for the base resistance, $r_b = 500\ \Omega$; for the collector resistance, $r_c = 20\ \text{k}\Omega$; and for the common-base forward-current-transfer ratio, $\alpha = 0.98$. Suppose that we wish to find the input (emitter) current i_e and the load voltage v_L.

Although the details are requested in Drill Probs. 3-12 and 3-13 below, we should see readily that the analysis of this circuit might be accomplished by drawing trees requiring three general nodal equations $(N - 1 - 1 + 1)$ or two loop equations $(B - N + 1 - 1)$. We might also note that three equations are required in terms of node-to-reference voltages, as are three mesh equations.

Fig. 3-42: A sinusoidal voltage source and a 10-kΩ load are connected to the T-equivalent circuit of a transistor. The common connection between the input and output is at the base terminal of the transistor and the arrangement is called the common-base configuration.

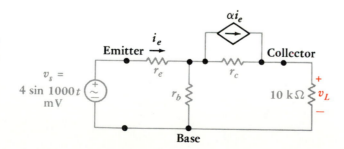

No matter which method we choose, these results are obtained for this specific circuit:

$$i_e = 18.14 \sin 1000t \quad \mu A$$

$$v_L = 119.6 \sin 1000t \quad mV$$

and we therefore find that this transistor circuit provides a voltage gain (v_L/v_s) of 29.9, a current gain ($v_L/10\,000 i_e$) of 0.659, and a power gain equal to the product, 29.9 (0.659) = 19.70. Higher gains could be secured by operating this transistor in a common-emitter configuration, as illustrated by the equivalent circuit of Prob. 42.

Drill Problems

3-11. Draw a suitable tree and use general loop analysis to find i_A in the circuit of: (a) Fig. 3-43a by writing just one equation with i_A as the variable; (b) Fig. 3-43b by writing just two equations with i_A and i_B as the variables.
Ans: 1.6; 9.39 A

3-12. For the transistor amplifier equivalent circuit shown in Fig. 3-42, let $r_e = 50\,\Omega$, $r_b = 500\,\Omega$, $r_c = 20\,k\Omega$, $\alpha = 0.98$, and find both i_e and v_L by drawing a suitable tree and using: (a) two loop equations; (b) three nodal equations with a common reference node for the voltages; (c) three nodal equations without a common reference node.
Ans: 18.14 sin 1000t mA; 119.6 sin 1000t mV

3-13. Determine the Thévenin and Norton equivalent circuits presented to the 10-kΩ load in Fig. 3-42 by finding: (a) the open-circuit value of v_L; (b) the (downward) short-circuit current; (c) the Thévenin equivalent resistance. All circuit values are given in Drill Prob. 3-12.
Ans: 146.2 sin 1000t mV; 65.6 sin 1000t μA; 2.23 kΩ

Fig. 3-43: See Drill Prob. 3-11.

(a)   (b)

Problems

1 (a) Find v_2 if $v_1 + 2v_2 + 3v_3 = 20$, $v_1 - 7v_2 - 5v_3 = -5$, and $v_3 + 3v_2 + 4v_1 - 10 = 0$. (b) Evaluate the determinant:

$$\begin{vmatrix} 1 & 2 & 3 & 4 \\ 2 & 3 & 4 & 1 \\ 3 & 4 & 1 & -2 \\ 4 & -1 & 2 & 3 \end{vmatrix}$$

2 Use nodal analysis to find v_P in the circuit shown in Fig. 3-44.

Fig. 3-44: See Prob. 2.

3 Use nodal analysis on the circuit given in Fig. 3-45 to find: (*a*) v_3; (*b*) the power being supplied by the 5-A source.

Fig. 3-45: See Prob. 3.

4 Make use of nodal analysis to find v_x and the power delivered to the 50-Ω resistor in the circuit of Fig. 3-46.

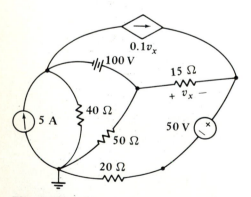

Fig. 3-46: See Prob. 4.

Fig. 3-47: See Prob. 5.

5 Set up nodal equations for the circuit illustrated in Fig. 3-47, and then find the power supplied by the 5-V source.

6 Use nodal analysis to find v_x in the circuit of Fig. 3-48.

Fig. 3-48: See Probs. 6, 12, and 28.

7 Analyze the circuit of Fig. 3-49 using node voltages and find the power being supplied by the 6-A source.

Fig. 3-49: See Prob. 7.

8 In Fig. 3-50, find v_2 through the use of nodal analysis.

Fig. 3-50: See Prob. 8.

9 In the circuit of Fig. 3-51, use mesh analysis to: (a) find the power delivered to the 4-Ω resistor. (b) To what voltage should the 100-V battery be changed so that no power is delivered to the 4-Ω resistor?

Fig. 3-51: See Prob. 9.

10 Use mesh analysis on the circuit shown in Fig. 3-52 to find the power supplied by the 4-V battery.

11 In Fig. 3-53, every resistance is 6 Ω and every battery voltage is 12 V. Find i_A.

Fig. 3-52: See Prob. 10.

Fig. 3-53: See Prob. 11.

12 In the circuit of Fig. 3-48, change the right-hand element to an 8-A independent current source, arrow directed upward, and use mesh analysis to find the power absorbed by the 3-Ω resistor.

13 Use mesh analysis on the circuit of Fig. 3-54 to find the values of all the mesh currents.

14 In the circuit of Fig. 3-4b, use mesh analysis to find i_o, the current flowing downward in R_L if $v_2 = 1.234\ 321$ V.

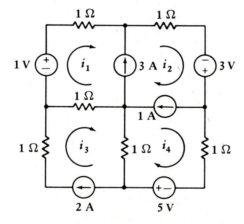

Fig. 3-54: See Prob. 13.

Fig. 3-55: See Prob. 15.

15 In the circuit shown in Fig. 3-55: (*a*) if $v_A = 20$ V and $v_B = 0$, $i_3 = 1.5$ A; find i_3 if $v_A = 50$ V and $v_B = 0$; (*b*) if $v_A = 20$ V and $v_B = 50$ V, $i_4 = 2$ A, while $i_4 = -1$ A if $v_A = 50$ V and $v_B = 20$ V; find i_4 if $v_A = 30$ V and $v_B = 100$ V.

16 Use superposition with the circuit of Fig. 3-56 to find i_x.

Fig. 3-56: See Prob. 16.

17 With reference to the circuit outlined in Fig. 3-57, when $v_s = 120$ V, it is found that $i_1 = 3$ A, $v_2 = 50$ V, and the power delivered to R_3 is 60 W. If Con Ed reduces v_s to 105 V, find new values for i_1, v_2, and the power delivered to R_3.

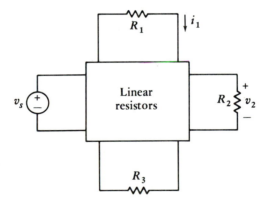

Fig. 3-57: See Prob. 17.

18 Use the superposition theorem on the circuit shown in Fig. 3-58 to find i.

Fig. 3-58: See Prob. 18.

19 The circuit illustrated in Fig. 3-59 contains a dependent source. Use the superposition theorem to find the current I.

Fig. 3-59: See Probs. 19 and 29.

20 In the circuit of Fig. 3-60, use superposition to help find the power absorbed by the: (a) 6-V source; (b) 3-A source; (c) 12-V source; (d) 2-A source.

Fig. 3-60: See Prob. 20.

21 A Radio Shack type 222 flashlight bulb is intended for use with two 1.55-V penlight (AA) batteries in series. The bulb, however, is marked "2.25 V, 0.25 A." Assuming all markings are correct, what practical voltage source might be used to model one of the batteries?

22 By making repeated source transformations and resistance combinations for each network in Fig. 3-61, replace the network to the left of terminals a-b with the series combination of a single independent voltage source and a single resistor.

Fig. 3-61: See Prob. 22. (a) (b)

23 In the circuit of Fig. 3-62, what value of R_L: (a) will absorb a maximum power from this network and what is $p_{L,max}$? (b) will have the maximum voltage across it and what is $v_{L,max}$? (c) will have a maximum current through it and what is $i_{L,max}$?

Fig. 3-62: See Prob. 23.

24 Consider the practical voltage source to the left of terminals *a-b* in Fig. 3-63. (*a*) What must be the value of R_L to cause the maximum possible power to be drawn from the *practical source?* (*b*) What is the value of this maximum power?

Fig. 3-63: See Prob. 24.

25 (*a*) Use three separate analyses to find v_{oc}, i_{sc}, and R_{th} with respect to terminals *a-b* for the circuit shown in Fig. 3-64. (*b*) Draw the Thévenin and Norton equivalents as seen at *a-b*.

Fig. 3-64: See Prob. 25.

26 Find the Thévenin equivalent circuit with respect to terminals *a-b* for the circuit shown in Fig. 3-65.

Fig. 3-65: See Probs. 26 and 30.

27 (*a*) Determine the Thévenin and Norton equivalent circuits as seen at terminals *a-b* for the network of Fig. 3-66. (*b*) Replace the 5-A source with a dependent voltage source labeled $5i_x$ (+ reference at the right) and again find the Thévenin and Norton equivalents.

Fig. 3-66: See Prob. 27.

28 In the circuit of Fig. 3-48, find the Thévenin equivalent of the network: (*a*) to the left of the 8-V source; (*b*) to the right of the 6-V source.

29 (*a*) Refer to the circuit shown in Fig. 3-59 and find the Thévenin equivalent circuit faced by the 3-Ω resistor. (*b*) Find I. (*c*) Change the 3-Ω resistor to 13 Ω and again find I.

30 If the voltage v_L across or the current $i_L = v_L/R_L$ through a general load resistance R_L is known, then the Thévenin or Norton equivalent may be determined easily since $v_{oc} = \lim_{R_L \to \infty} v_L$ and $i_{sc} = \lim_{R_L \to 0} i_L$. (*a*) Find v_L (+ sign at terminal "*a*") for the circuit of Fig. 3-65 if a resistance R_L is connected between *a* and *b*. (*b*) Use The expressions above to determine v_{oc} and i_{sc}.

31 Find the Thévenin equivalent circuit for the network shown in Fig. 3.67.

Fig. 3-67: See Prob. 31.

32 Determine the Thévenin equivalent circuit of the network shown in Fig. 3-68.

Fig. 3-68: See Prob. 32.

33 The voltage follower shown in Fig. 3-28 is modified by inserting a finite $R_i = 10$ kΩ between the terminals across which v_i is defined. Find the new Thévenin equivalent.

34 (*a*) Construct all possible trees for the linear graph shown in Fig. 3-69. (*b*) If the top two branches are voltage sources and the left branch is a current source, show all possible trees.

Fig. 3-69: See Prob. 34.

Fig. 3-70: See Prob. 35.

35 For the circuit shown in Fig. 3-70, construct a tree in which v_1 and v_2 are tree-branch voltages, write nodal equations, and solve for v_1.

36 Construct a suitable tree for the circuit of Fig. 3-71, assign tree-branch voltages, write KCL and control equations, and find i_2.

Fig. 3-71: See Prob. 36.

37 Use nodal analysis with tree-branch voltages on the circuit of Fig. 3-72 to determine what value of V_2 will cause $v_z = 0$.

Fig. 3-72: See Prob. 37.

38 For the circuit shown in Fig. 3-73, construct a tree in which i_1 and i_2 are link currents, write loop equations, and evaluate i_1 and i_2.

Fig. 3-73: See Prob. 38.

39 (a) Construct a suitable tree for the circuit of Fig. 3-74 and write the single equation necessary to find i_1. (b) Find i_1.

Fig. 3-74: See Prob. 39.

40 Devise a tree for the circuit shown in Fig. 3-75 for which all loop currents flow through the 1-Ω resistor. Find i.

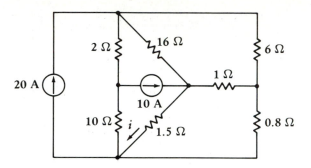

Fig. 3-75: See Prob. 40.

41 Use general loop analysis on the nonplanar circuit of Fig. 3-76 to find i_x.

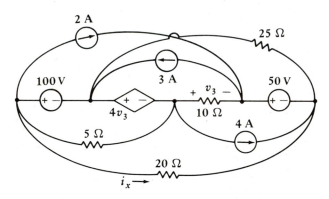

Fig. 3-76: See Prob. 41.

42 Figure 3-77 shows one form of the equivalent circuit for a transistor amplifier. Determine the open-circuit value of v_2 and the output resistance (R_{th}) of the amplifier.

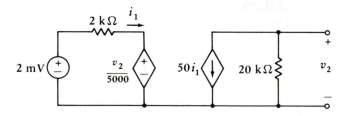

Fig. 3-77: See Prob. 42.

PART TWO
THE TRANSIENT CIRCUIT

Chapter 4
Inductance and Capacitance

4-1
Introduction

We are now ready to begin the second major portion of our study of circuits. In this chapter we shall introduce two new simple circuit elements whose voltage-current relationships involve the rate of change of a voltage or current. Before we begin this new study, it will be worthwhile to pause for a moment and look back upon our study of the analysis of resistive circuits. A little bit of philosophical review will aid in our understanding of the coming work.

After setting up a satisfactory system of units, we began our discussion of electric circuits by defining current, voltage, and five simple circuit elements. The independent and dependent voltage and current sources were called active elements and the linear resistor was termed a passive element, although our definitions of "active" and "passive" are still slightly fuzzy and need to be brought into sharper focus. We define an *active element* as one which is capable of furnishing an average power greater than zero to some external device, where the average is taken over an infinite time interval; the ideal sources are active elements. The operational amplifier is also an active device. A *passive element,* however, is defined as one that cannot supply an average power that is greater than zero over an infinite time interval; the resistor falls into this category. The energy it receives is usually transformed into heat.

Each of these elements was defined in terms of the restrictions placed on its voltage-current relationship. In the case of the independent voltage source, for example, the terminal voltage must be completely independent of the current drawn from its terminals. We then considered circuits composed of the different building blocks. In general, we used only constant voltages and currents, but now that we have gained a familiarity with the basic analytical techniques by treating only the resistive circuit, we may begin to consider the much more interesting and practical circuits in which inductance and capacitance may be present and in which both the forcing functions and the responses usually vary with time.

4-2
The inductor

Both the inductor, which is the subject of this and the following section, and the capacitor, which is discussed later in the chapter, are passive elements which are capable of storing and delivering finite amounts of energy. Unlike an ideal source, they cannot provide an unlimited amount of energy or a finite average power over an infinite time interval.

Although we shall define an inductor and inductance strictly from a circuit point of view, that is, by a voltage-current equation, a few comments about the historic development of the magnetic field may provide a better understanding of the definition. In the early 1800s the Danish scientist, Oersted, showed that a current-carrying conductor produced a magnetic field, or that compass needles were affected in the presence of a current-carrying conductor. In France, shortly thereafter, Ampère made some careful measurements which demonstrated that this magnetic field was linearly related to the current which produced it. The next step occurred some twenty years later when the English experimentalist, Michael Faraday, and the American inventor, Joseph Henry, discovered almost simultaneously[1] that a changing magnetic field could induce a voltage in a neighboring circuit. They showed that this voltage was proportional to the time rate of change of the current which produced the magnetic field. The constant of proportionality we now call the *inductance*, symbolized by L, and therefore

$$v = L \frac{di}{dt} \qquad (1)$$

where we must realize that v and i are both functions of time. When we wish to emphasize this, we may do so by using the symbols $v(t)$ and $i(t)$.

The circuit symbol for the inductor is shown in Fig. 4-1, and it should be noted that the passive sign convention is used, just as it was with the resistor. The unit in which inductance is measured is the *henry*[2] (H), and the defining equation shows that the henry is just a shorter expression for a volt-second per ampere.

The inductor whose inductance is defined by (1) is a mathematical model; it is an ideal element which we may use to approximate the behavior of a real device. A physical inductor may be constructed by winding a length of wire into a coil. This serves effectively to increase the current which is causing the magnetic field and also to increase the "number" of neighboring circuits into which Faraday's voltage may be induced. The result of this twofold effect is that the inductance of a coil is approximately proportional to the square of the number of complete turns made by the conductor out of which it is formed. An inductor, or "coil," which has the form of a long helix of very small pitch is found to have an inductance of $\mu N^2 A/s$, where A is the cross-sectional area, s is the axial length of the helix, N is the number of complete turns of wire, and μ (mu) is a constant of the material inside the

Fig. 4-1: The reference signs for voltage and current are shown on the circuit symbol for an inductor: $v = L \, di/dt$.

[1] Faraday won.
[2] An empty victory.

helix, called the *permeability*. For free space (and very closely for air), $\mu = \mu_0 = 4\pi \times 10^{-7}$ H/m.

Physical inductors should be on view in an accompanying laboratory course, and the topics concerned with the magnetic flux, permeability, and the methods of using the characteristics of the physical coil to calculate a suitable inductance for the mathematical model are treated in both physics courses and courses in electromagnetic field theory.

It is also possible to assemble electronic networks that do not contain any inductors and yet can provide the *v-i* relationship of (1) at their input terminals. We shall look at one example of this in Sec. 7-8.

Let us now scrutinize (1) to determine some of the electrical characteristics of this mathematical model. This equation shows that the voltage across an inductor is proportional to the time rate of change of the current through it. In particular, it shows that there is no voltage across an inductor carrying a constant current, regardless of the magnitude of this current. Accordingly, we may view an inductor as a ''short circuit to dc.'' Another fact which is evidenced by this equation is related to an infinite rate of change of the inductor current, such as that caused by an abrupt change in current from one finite value to some other finite value. This sudden or discontinuous change in the current must be associated with an infinite voltage across the inductor. In other words, if we wish to produce an abrupt change in an inductor current, we must apply an infinite voltage. Although an infinite-voltage forcing function might be acceptable theoretically, it can never be a part of the phenomena displayed by a real physical device. As we shall see shortly, an abrupt change in the inductor current also requires an abrupt change in the energy stored in the inductor, and this sudden change in energy requires infinite power at that instant; infinite power is again not a part of the real physical world. In order to avoid infinite voltage and infinite power, an inductor current must not be allowed to jump instantaneously from one value to another. If an attempt is made to open-circuit a physical inductor through which a finite current is flowing, an arc may appear across the switch. The stored energy is dissipated in ionizing the air in the path of the arc. This is useful in the ignition system of an automobile, where the current through the spark coil is interrupted by the distributor, and the arc appears across the spark plug.

We shall not consider any circuits at the present time in which an inductor is suddenly open-circuited. It should be pointed out, however, that we shall remove this restriction later when we hypothesize the existence of a voltage forcing function or response which does become infinite instantaneously.

Equation (1) may also be interpreted (and solved, if necessary) by graphical methods. Let us assume a current (Fig. 4-2a) which is zero prior to $t = -1$ s, increases linearly to 1 A in the next second, remains at 1 A for 2 s, and then decreases to zero in the next second, remaining zero thereafter. If this current is present in a 3-H inductor, and if the voltage and current senses are assigned to satisfy the passive sign convention, then we may use (1) to obtain the voltage waveform. Since the current is zero and constant for $t < -1$, the

Fig. 4-2: (a) The current waveform in a 3-H inductor. (b) The corresponding voltage waveform, $v = 3\,di/dt$.

Fig. 4-3: (a) The time required for the current of Fig. 4-2a to change from 0 to 1 and from 1 to 0 is decreased by a factor of 10. (b) The resultant voltage waveform. Note that the pulse widths are exaggerated slightly for clarity.

voltage is zero in this interval. The current then begins to increase at the linear rate of 1 A/s, and thus a constant voltage of 3 V is produced. During the following 2-s interval, the current is constant and the voltage is therefore zero. The final decrease of the current causes a negative 3 V and no response thereafter. The voltage waveform is sketched in Fig. 4-2b on the same time scale.

Let us now investigate the effect of a more rapid rise and decay of the current between the zero and 1-A values. If the intervals required for the rise and fall are decreased to 0.1 s, then the derivative must be ten times as great in magnitude. This condition is shown in the current and voltage sketches of Figs. 4-3a and b. In the voltage waveforms of Figs. 4-2b and 4-3b, it is interesting to note that the area under each voltage pulse is 3 V-s.

A further decrease in the length of these two intervals will produce a proportionally larger voltage magnitude, but only within the interval in which the current is increasing or decreasing. An abrupt change in the current will cause the infinite voltage "spikes" (each having an area of 3 V-s) that are suggested by the waveforms of Figs. 4-4a and b; or, from the equally

Fig. 4-4: (a) The time required for the current of Fig. 4-2a to change from 0 to 1 and from 1 to 0 is decreased to zero; the rise and fall are abrupt. (b) The associated voltage across the 3-H inductor consists of a positive and negative infinite spike.

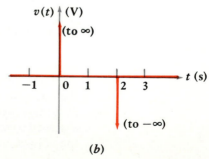

valid but opposite point of view, these infinite voltage spikes are required to produce the abrupt changes in the current. It will be convenient later to provide such infinite voltages (and currents), and we shall then call them "impulses"; for the present, however, we shall stay closer to physical reality by not permitting infinite voltage, current, or power. An abrupt change in the inductor current is therefore temporarily outlawed.

Drill Problems

4-1. For the circuit of Fig. 4-5a, find: (a) i_1; (b) i_2; (c) i_3.

Ans: 0.6; 0; 0.25 A

4-2. The current through a 40-mH inductor is shown as a function of time in Fig. 4-5b. Find $v_L(t)$ at $t =$: (a) 1; (b) 3; (c) 5 ms.

Ans: 40; -160; 160 mV

(a)

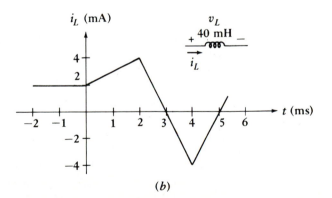

Fig. 4-5: See Drill Probs. 4-1 and 4-2.

(b)

4-3
Integral relation-ships for the inductor

We have defined inductance by a simple differential equation

$$v = L \frac{di}{dt} \qquad (2)$$

and we have been able to draw several conclusions about the characteristics of an inductor from this relationship. For example, we have found that we may consider an inductor as a short circuit to direct current, and we have agreed that we cannot permit an inductor current to change abruptly from one value to another because this would require that an infinite voltage and power be associated with the inductor. The defining equation for inductance still contains more information, however. Rewritten in a slightly different form,

$$di = \frac{1}{L} v \, dt$$

it invites integration. Let us first consider the limits to be placed on the two integrals. We desire the current i at time t, and this pair of quantities therefore provides the upper limits on the integrals appearing on the left and right side of the equation, respectively; the lower limits may also be kept general by merely assuming that the current is $i(t_0)$ at time t_0. Thus

$$\int_{i(t_0)}^{i(t)} di = \frac{1}{L} \int_{t_0}^{t} v \, dt$$

or

$$i(t) - i(t_0) = \frac{1}{L} \int_{t_0}^{t} v \, dt$$

and

$$i(t) = \frac{1}{L} \int_{t_0}^{t} v \, dt + i(t_0) \tag{3}$$

Equation (2) expresses the inductor voltage in terms of the current, whereas (3) gives the current in terms of the voltage. Other forms are also possible for this latter equation. We may write the integral as an indefinite integral and include a constant of integration k,

$$i(t) = \frac{1}{L} \int v \, dt + k \tag{4}$$

We may assume that we are solving a realistic problem in which the selection of t_0 as $-\infty$ ensures no current or energy in the inductor. Thus, if $i(t_0) = i(-\infty) = 0$, then

$$i(t) = \frac{1}{L} \int_{-\infty}^{t} v \, dt \tag{5}$$

Let us consider the use of these several integrals by working a simple example. Suppose that the voltage across a 2-H inductor is known to be $6 \cos 5t$ V; what information is then available about the inductor current? From Eq. (3),

$$i(t) = \frac{1}{2} \int_{t_0}^{t} 6 \cos 5t \, dt + i(t_0)$$

or $i(t) = \frac{1}{2} \cdot \frac{6}{5} \sin 5t - \frac{1}{2} \cdot \frac{6}{5} \sin 5t_0 + i(t_0) = 0.6 \sin 5t - 0.6 \sin 5t_0 + i(t_0)$

The first term indicates that the inductor current varies sinusoidally; the second and third terms together merely represent a constant which becomes known when the current is numerically specified at some instant of time. Let us assume that the statement of our example problem also shows us that the current is 1 A at $t = -\pi/2$ s. We thus identify t_0 as $-\pi/2$, $i(t_0)$ as 1, and find that

$$i(t) = 0.6 \sin 5t - 0.6 \sin (-2.5\pi) + 1$$

or

$$i(t) = 0.6 \sin 5t + 1.6$$

We may obtain the same result from (4). We have

$$i(t) = 0.6 \sin 5t + k$$

and we establish the numerical value of k by forcing the current to be 1 A at $t = -\pi/2$,

$$1 = 0.6 \sin (-2.5\pi) + k$$

or

$$k = 1 + 0.6 = 1.6$$

and

$$i(t) = 0.6 \sin 5t + 1.6$$

once more.

Equation (5) is going to cause trouble with this particular voltage. We based the equation on the assumption that the current was zero when $t = -\infty$. To be sure, this must be true in the real physical world, but we are working in the land of the mathematical model; our elements and forcing functions are all idealized. The difficulty arises after we integrate,

$$i(t) = 0.6 \sin 5t \Big|_{-\infty}^{t}$$

and attempt to evaluate the integral at the lower limit,

$$i(t) = 0.6 \sin 5t - 0.6 \sin (-\infty)$$

The sine of $\pm\infty$ is indeterminate; we might just as well represent it by an unknown constant,

$$i(t) = 0.6 \sin 5t + k$$

and we see that this result is identical with that which we obtained when we assumed an arbitrary constant of integration in (4).

We should not make any snap judgments, based on this example, as to which single form we are going to use forever; each has its advantages, depending on the problem and the application. Equation (3) represents a long, general method, but it shows clearly that the constant of integration is a current. Equation (4) is a somewhat more concise expression of (3), but the nature of the integration constant is suppressed. Finally, (5) is an excellent expression since no constant is necessary; however, it applies only when the current is zero at $t = -\infty$ and when the analytical expression for the current is not indeterminate there.

Let us now turn our attention to power and energy. The absorbed power is given by the current-voltage product

$$p = vi = Li \frac{di}{dt} \qquad \text{W}$$

The energy w_L accepted by the inductor is stored in the magnetic field around the coil and is expressed by the integral of the power over the desired time interval,

$$\int_{t_0}^{t} p \, dt = L \int_{t_0}^{t} i \frac{di}{dt} \, dt = L \int_{i(t_0)}^{i(t)} i \, di = \tfrac{1}{2}L\{[i(t)]^2 - [i(t_0)]^2\}$$

and thus

$$w_L(t) - w_L(t_0) = \tfrac{1}{2}L\{[i(t)]^2 - [i(t_0)]^2\} \qquad \text{J} \tag{6}$$

where we have again assumed that the current is $i(t_0)$ at time t_0. In using the energy expression, it is customary to assume that a value of t_0 is selected at which the current is zero; it is also customary to assume that the energy is zero at this time. We then have simply

$$w_L(t) = \tfrac{1}{2}Li^2 \tag{7}$$

where we now understand that our reference for zero energy is any time at which the inductor current is zero. At any subsequent time at which the current is zero, we also find no energy stored in the coil. Whenever the current is not zero, and regardless of its direction or sign, energy is stored in the inductor. It follows, therefore, that power must be delivered to the inductor for a part of the time and recovered from the inductor later. All the stored energy may be recovered from an ideal inductor; there are no storage charges or agent's commissions in the mathematical model. A physical coil, however, must be constructed out of real wire and thus will always have a resistance associated with it. Energy can no longer be stored and recovered without loss.

These ideas may be illustrated by a simple example. In Fig. 4-6 a 3-H inductor is shown in series with a 0.1-Ω resistor and a sinusoidal current source. The resistor may be interpreted, if we wish, as the resistance of the wire which must be associated with the physical coil.

The voltage across the resistor is given by

$$v_R = Ri = 1.2 \sin \frac{\pi}{6} t$$

and the voltage across the inductor is found by applying the defining equation for inductance,

$$v_L = L \frac{di}{dt} = 3 \frac{d}{dt} \left(12 \sin \frac{\pi}{6} t \right) = 6\pi \cos \frac{\pi}{6} t$$

Fig. 4-6: A sinusoidal current is applied as a forcing function to a series *RL* circuit.

The energy stored in the inductor is

$$w_L = \frac{1}{2} Li^2 = 216 \sin^2 \frac{\pi}{6} t$$

and it is apparent that this energy increases from zero at $t = 0$ to 216 J at $t = 3$ s. During the next 3 s, the energy leaves the inductor completely. Let us see what price we have paid in this coil for the privilege of storing and removing 216 J in these few seconds. The power dissipated in the resistor is easily found as

$$p_R = i^2 R = 14.4 \sin^2 \frac{\pi}{6} t$$

and the energy converted into heat in the resistor within this 6-s interval is therefore

$$w_R = \int_0^6 p_R \, dt = \int_0^6 14.4 \sin^2 \frac{\pi}{6} t \, dt$$

or

$$w_R = \int_0^6 14.4 \left(\frac{1}{2}\right) \left(1 - \cos \frac{\pi}{3} t\right) dt = 43.2 \quad \text{J}$$

This represents 20 percent of the maximum stored energy and is a reasonable value for many coils having this large an inductance. For coils having an inductance of about 100 μH, we should expect a figure closer to 2 or 3 percent. In Chap. 14 we shall formalize this concept by defining a quality factor Q that is proportional to the ratio of the maximum energy stored to the energy lost per period.

Let us now recapitulate by listing several characteristics of an inductor which result from its defining equation:

1 There is no voltage across an inductor if the current through it is not changing with time. An inductor is therefore a short circuit to dc.
2 A finite amount of energy can be stored in an inductor even if the voltage across the inductor is zero, such as when the current through it is constant.
3 It is impossible to change the current through an inductor by a finite amount in zero time, for this requires an infinite voltage across the inductor. It will be advantageous later to *hypothesize* that such a voltage may be generated and applied to an inductor, but for the present we shall avoid such a forcing function or response. An inductor resists an abrupt change in the current through it in a manner analogous to the way a mass resists an abrupt change in its velocity.
4 The inductor never dissipates energy, but only stores it. Although this is true for the mathematical model, it is not true for a physical inductor.

Drill Problem

4-3. Let $L = 50$ mH for the inductor of Fig. 4-1. Find v at $t = 10$ ms if: (a) $i = 20(1 - e^{-100t})$ A for $t > 0$; (b) the energy stored by the inductor is $0.4t$ J and $i(0.01) < 0$. (c) Find i at $t = 10$ ms if $v = 20t^2$ V for $t > -0.1$ ms and $i(-0.02) = 5$ mA. *Ans:* 36.8 V; -1 V; 6.2 mA

4-4

The capacitor

Our next passive circuit element is the capacitor. We shall define *capacitance C* by the voltage-current relationship

$$i = C \frac{dv}{dt} \tag{8}$$

where v and i satisfy the conventions for a passive element, as shown in Fig. 4-7. From (8), we may determine the unit of capacitance as an ampere-second per volt, or coulomb per volt, but we shall now define the *farad* (F) as one coulomb per volt.

The capacitor whose capacitance is defined by (8) is again a mathematical model of a real device. The construction of the physical device is suggested by the circuit symbol shown in Fig. 4-7, in much the same way that the helical symbol used for the inductor represents the coiled wire in that physical element. A capacitor, physically, consists of two conducting surfaces on which charge may be stored, separated by a thin insulating layer which has a very large resistance. If we assume that this resistance is sufficiently large that it may be considered infinite, then equal and opposite charges placed on the capacitor "plates" can never recombine, at least by any path *within* the element. Let us visualize some external device, such as a current source, connected to this capacitor and causing a positive current to flow into one plate of the capacitor and out of the other plate. Equal currents are entering and leaving the two terminals of the element, and this is no more than we expect for any circuit element. Now let us examine the interior of the capacitor. The positive current entering one plate represents positive charge moving toward that plate through its terminal lead; this charge cannot pass through the interior of the capacitor, and it therefore accumulates on the plate. As a matter of fact, the current and the increasing charge are related by the familiar equation

$$i = \frac{dq}{dt}$$

Now let us pose ourselves a troublesome problem by considering this plate as an overgrown node and applying Kirchhoff's current law. It apparently does not hold; current is approaching the plate from the external circuit, but it is not flowing out of the plate into the "internal circuit." This dilemma bothered a famous Scottish scientist about a century ago, and the unified electromagnetic theory which James Clerk Maxwell then developed hypothesizes a "displacement current" which is present wherever an electric field or voltage is varying with time. The displacement current flowing internally between the capacitor plates is exactly equal to the conduction current flow-

Fig. 4-7: The current and voltage reference marks are shown on the circuit symbol for a capacitor so that $i = C\, dv/dt$.

ing in the capacitor leads; Kirchhoff's current law is therefore satisfied if we include both conduction and displacement currents. However, circuit analysis is not concerned with this internal displacement current, and since it is fortunately equal to the conduction current, we may consider Maxwell's hypothesis as relating the conduction current to the changing voltage across the capacitor. The relationship is linear, and the constant of proportionality is obviously the capacitance C,

$$i_{\text{disp}} = i = C \frac{dv}{dt}$$

A capacitor constructed of two parallel conducting plates of area A, separated by a distance d, has a capacitance $C = \epsilon A/d$, where ϵ is the permittivity, a constant of the insulating material between the plates, and where the linear dimensions of the conducting plates are all very much greater than d. For air or vacuum, $\epsilon = \epsilon_0 = 8.854$ pF/m.

The concepts of the electric field, displacement current, and the generalized form of Kirchhoff's current law are more appropriate subjects for courses in physics and electromagnetic field theory, as is the determination of a suitable mathematical model to represent a specific physical capacitor.

Several important characteristics of our new mathematical model can be discovered from the defining equation (8). A constant voltage across a capacitor requires zero current passing through it; a capacitor is thus an "open circuit to dc." This fact is certainly represented by the capacitor symbol. It is also apparent that a sudden jump in the voltage requires an infinite current. Just as we outlawed abrupt changes in inductor currents and the associated infinite voltages on physical grounds, we shall not permit abrupt changes in capacitor voltage; the infinite current (and infinite power) which results is nonphysical. We shall remove this restriction at the time we assume the existence of the current impulse.

The capacitor voltage may be expressed in terms of the current by integrating (8). We first obtain

$$dv = \frac{1}{C} i \, dt$$

and then integrate between the times t_0 and t and between the corresponding voltages $v(t_0)$ and $v(t)$,

$$v(t) = \frac{1}{C} \int_{t_0}^{t} i \, dt + v(t_0) \tag{9}$$

Equation (9) may also be written as an indefinite integral plus a constant of integration,

$$v(t) = \frac{1}{C} \int i \, dt + k \tag{10}$$

Finally, in many real problems, t_0 may be selected as $-\infty$ and $v(-\infty)$ as zero,

$$v(t) = \frac{1}{C} \int_{-\infty}^{t} i \, dt \tag{11}$$

Since the integral of the current over any time interval is the charge accumulated in that period on the capacitor plate into which the current is flowing, it is apparent that capacitance might have been defined as

$$q = Cv$$

The similarity between the several integral equations introduced in this section and those appearing in our discussion of inductance is striking and suggests that the duality we observed between mesh and nodal equations in resistive networks may be extended to include inductance and capacitance as well. The principle of duality will be presented and discussed later in this chapter.

As an illustration of the use of the several integral equations displayed above, let us find the capacitor voltage which is associated with the current shown graphically in Fig. 4-8a. We shall assume that the single 20-mA rectangular pulse of 2-ms duration is applied to a 5-μF capacitor. Interpreting (9) graphically, we know that the difference between the values of the voltage at t and t_0 is proportional to the area under the current curve between these same two values of time. The proportionality constant is $1/C$. The area can be obtained from Fig. 4-8a by inspection for desired values of t_0 and t. Thus, if $t_0 = -0.5$ and $t = 0.5$ (in ms),

$$v(0.5) = 2 + v(-0.5)$$

or, if $t_0 = 0$ and $t = 3$,

$$v(3) = 8 + v(0)$$

We may express our results in more general terms by dividing the interesting range of time into several intervals. Let us select our starting point t_0 prior to zero time. Then the first interval of t is selected between t_0 and zero,

$$v(t) = 0 + v(t_0) \qquad t_0 \le t \le 0$$

Fig. 4-8: (a) The current waveform applied to a 5-μF capacitor. (b) The resultant voltage waveform, easily obtained by integrating graphically.

(a)

(b)

and since our waveform implies that no current has ever been applied to this capacitor since the Creation,

$$v(t_0) = 0$$

and, thus,

$$v(t) = 0 \qquad t \le 0$$

If we now consider the time interval represented by the rectangular pulse, we obtain

$$v(t) = 4000t \qquad 0 \le t \le 2 \text{ ms}$$

For the semi-infinite interval following the pulse, we have

$$v(t) = 8 \qquad t \ge 2 \text{ ms}$$

The results for these three intervals therefore provide us with analytical expressions for the capacitor voltage at any time after $t = t_0$; the time t_0, however, may be selected as early as we wish. The results are expressed much more simply in a sketch than by these analytical expressions, as shown in Fig. 4-8b.

The power delivered to a capacitor is

$$p = vi = Cv \frac{dv}{dt}$$

and the energy stored in its electric field is therefore

$$\int_{t_0}^{t} p \, dt = C \int_{t_0}^{t} v \frac{dv}{dt} \, dt = C \int_{v(t_0)}^{v(t)} v \, dv = \frac{1}{2} C\{[v(t)]^2 - [v(t_0)]^2\}$$

and thus

$$w_C(t) - w_C(t_0) = \tfrac{1}{2}C\{[v(t)]^2 - [v(t_0)]^2\} \tag{12}$$

where the stored energy is $w_C(t_0)$ and the voltage is $v(t_0)$ at t_0. If we select a zero-energy reference at t_0, implying that the capacitor voltage is also zero at that instant, then

$$w_C(t) = \tfrac{1}{2}Cv^2 \tag{13}$$

Let us consider a simple numerical example. As sketched in Fig. 4-9, we shall assume a sinusoidal voltage source in parallel with a 1-MΩ resistor and a 20-μF capacitor. The parallel resistor may be assumed to represent the

Fig. 4-9: A sinusoidal voltage source is applied to a parallel RC network.

resistance of the insulator or dielectric between the plates of the physical capacitor. The current through the resistor is

$$i_R = \frac{v}{R} = 10^{-4} \sin 2\pi t$$

and the current through the capacitor is

$$i_C = C\frac{dv}{dt} = 20 \times 10^{-6}\frac{d}{dt}(100 \sin 2\pi t) = 4\pi \times 10^{-3} \cos 2\pi t$$

We next obtain the energy stored in the capacitor,

$$w_C = \tfrac{1}{2}Cv^2 = 0.1 \sin^2 2\pi t$$

and see that the energy increases from zero at $t = 0$ to a maximum of 0.1 J at $t = \frac{1}{4}$ s and then decreases to zero in another $\frac{1}{4}$ s. During this $\frac{1}{2}$-s interval, the energy dissipated in the resistor is

$$w_R = \int_0^{0.5} p_R\,dt = \int_0^{0.5} 10^{-2} \sin^2 2\pi t\,dt = 2.5 \quad \text{mJ}$$

Thus, an energy equal to 2.5 percent of the maximum stored energy is lost in the process of storing and removing the energy in the ideal capacitor. Much smaller values are possible in "low-loss" capacitors, but these smaller percentages are customarily associated with much smaller capacitors.

Some of the important characteristics of a capacitor are now apparent:

1 The current through a capacitor is zero if the voltage across it is not changing with time. A capacitor is therefore an open circuit to dc.
2 A finite amount of energy can be stored in a capacitor even if the current through the capacitor is zero, such as when the voltage across it is constant.
3 It is impossible to change the voltage across a capacitor by a finite amount in zero time, for this requires an infinite current through the capacitor. It will be advantageous later to *hypothesize* that such a current may be generated and applied to a capacitor, but for the present we shall avoid such a forcing function or response. A capacitor resists an abrupt change in the voltage across it in a manner analogous to the way a spring resists an abrupt change in its displacement.
4 The capacitor never dissipates energy, but only stores it. Although this is true for the mathematical model, it is not true for a physical capacitor.

It is interesting to anticipate our discussion of duality by rereading the previous four statements with certain words replaced by their "duals." If capacitor and inductor, capacitance and inductance, voltage and current, across and through, open circuit and short circuit, spring and mass, and displacement and velocity are interchanged (in either direction), the four statements previously given for inductors are obtained.

We conclude our introduction to the ideal capacitor by seeing how it might be used in conjunction with an ideal op-amp. Up to now we have only discussed the op-amp as it is used as a voltage follower, a device having a very low output resistance and an output voltage that is virtually *equal* to the input voltage. We will now show that an ideal capacitor and an ideal resistor can be combined with an op-amp to form a device having an output voltage that is proportional to the *time integral* of the input.

To create this integrator, we ground the noninverting input of an op-amp that has infinite input resistance and zero output resistance, install an ideal capacitor as a feedback element from the output back to the inverting input, and connect a signal source v_s to the inverting input through an ideal resistor, as shown in Fig. 4-10a. With $R_i = \infty$ and $R_o = 0$, we obtain the equivalent circuit shown in Fig. 4-10b.

The output voltage is related to the voltage v_i by

$$v_o = -Av_i$$

so that

$$v_i = -\frac{v_o}{A} \tag{14}$$

Let us relate the output v_o to the signal voltage v_s by first assuming that A is infinite. It follows from (14) that $v_i = 0$. Therefore, we see from Fig. 4-10b that

$$i = \frac{v_s}{R}$$

Also, with $v_i = 0$, the capacitor voltage v_C is equal to $(-v_o)$, and

$$v_C = -v_o = \frac{1}{C}\int_0^t i\, dt + v_C(0) = \frac{1}{C}\int_0^t \frac{v_s}{R}\, dt + v_C(0)$$

or

$$v_o = -\frac{1}{RC}\int_0^t v_s\, dt - v_C(0) \tag{15}$$

We therefore have combined a resistor, a capacitor, and an op-amp to

Fig. 4-10: (*a*) An op-amp is connected as an integrator. (*b*) The equivalent circuit, assuming $R_i = \infty$ and $R_o = 0$.

Fig. 4-11: The inclusion of a normally-closed switch and an ideal voltage source enables the value of $v_o(0^+)$ to be set at $-v_C(0)$.

form an *integrator*. Note that the first term of the output is $1/RC$ times the *negative* of the integral of the input from $t = 0$ to t, and the second term is the negative of the initial value of v_C. The value of $(RC)^{-1}$ can be made equal to unity if we wish by choosing $R = 1$ MΩ and $C = 1$ μF, for example; or other selections may be made that will increase or decrease the output voltage. The sign change is often convenient when the integrator is used for the simulation of engineering systems. However, the sign of the output may also be changed by using the inverting amplifier whose circuit appears in Chapter 7.

The initial voltage $v_C(0)$ appearing in (15) may be included in the integrator circuit by adding a battery and a normally closed switch, as indicated in Fig. 4-11. In practical circuits, both the switch and the initial voltage are usually electronic devices, such as transistors or other op-amps.

If we do not make the assumption that A is infinite, then we may write one KVL equation around the perimeter of the circuit in Fig. 4-10b,

$$-v_s + Ri + \frac{1}{C}\int_0^t i\,dt + v_C(0) + v_o = 0$$

use $i = (v_s - v_i)/R$ to eliminate i, and then (14) to eliminate v_i. The result is

$$\left(1 + \frac{1}{A}\right)v_o = -\frac{1}{RC}\int_0^t \left(v_s + \frac{v_o}{A}\right)dt - v_C(0) \tag{16}$$

As $A \rightarrow \infty$, this results in an equation identical to (15).

Before we leave the integrator circuit, we might anticipate a question from an inquisitive reader, "Could we use an inductor in place of the capacitor and obtain a differentiator?" Indeed we could, but circuit designers usually avoid the use of inductors whenever possible because of their size, weight, cost, and associated resistance and capacitance. Instead, it is possible to interchange the positions of the resistor and capacitor in Fig. 4-10a and obtain a differentiator. The analysis of this circuit is the subject of Prob. 10.

Drill Problems

4-4. Let the value of C in Fig. 4-7 be 0.05 μF. Find $i(0.02)$ if: (*a*) $v = 25e^{-40t} \sin(20\pi t + 0.1\pi)$ V; (*b*) the energy stored in C is $10e^{-50t}$ J and $v(0.02) > 0$. (*c*) Find $v(0.02)$ if $i = 2t/(t^2 + 10^{-4})$ μA and $v(0) = 12$ V.

Ans: -22.5 μA; -15.16 mA; 44.2 V

4-5. In the circuit of Fig. 4-10a, let v_s be a periodic sequence of 1-V pulses, rising instantaneously from 0 to 1 V at $t = 0, 1, 2, \ldots$ s, and falling instantaneously back to zero at $t = 0.1, 1.1, 2.1, \ldots$ s. Let $R = 2$ MΩ and $C = 1$ μF, and assume that $v_C(0^-) = 0$. At what instant of time does v_o become equal to: (*a*) -0.01 V; (*b*) -0.12 V; (*c*) -5 V?

Ans: 0.02; 2.04; 99.1 s

4-5
Inductance and capacitance combinations

Now that we have added the inductor and capacitor to our list of passive circuit elements, we need to decide whether or not the methods we have developed for resistive circuit analysis are still valid. It will also be convenient to learn how to replace series and parallel combinations of either of

Fig. 4-12: (a) A circuit containing N inductors in series. (b) The desired equivalent circuit, in which $L_{eq} = L_1 + L_2 + \cdots + L_N$.

these elements with simpler equivalents, just as we did with resistors in Chap. 2.

We look first at Kirchhoff's two laws, both of which are axiomatic. However, when we hypothesized these two laws, we did so with no restrictions as to the types of elements constituting the network. Both, therefore, remain valid.

Now we may extend the procedures we have derived for reducing various combinations of resistors into one equivalent resistor to the analogous cases of inductors and capacitors. We shall first consider an ideal voltage source applied to the series combination of N inductors, as shown in Fig. 4-12a. We desire a single equivalent inductor, with inductance L_{eq}, which may replace the series combination such that the source current $i(t)$ is unchanged. The equivalent circuit is sketched in Fig. 4-12b. For the original circuit

$$v_s = v_1 + v_2 + \cdots + v_N$$

$$= L_1 \frac{di}{dt} + L_2 \frac{di}{dt} + \cdots + L_N \frac{di}{dt}$$

$$= (L_1 + L_2 + \cdots + L_N) \frac{di}{dt}$$

or, written more concisely,

$$v_s = \sum_{n=1}^{N} v_n = \sum_{n=1}^{N} L_n \frac{di}{dt} = \frac{di}{dt} \sum_{n=1}^{N} L_n$$

But for the equivalent circuit we have

$$v_s = L_{eq} \frac{di}{dt}$$

and thus the equivalent inductance is

$$L_{eq} = L_1 + L_2 + \cdots + L_n \qquad \text{or} \qquad L_{eq} = \sum_{n=1}^{N} L_n$$

The inductor which is equivalent to several inductors connected in series is one whose inductance is the sum of the inductances in the original circuit. This is exactly the same result we obtained for resistors in series.

The combination of a number of parallel inductors is accomplished by writing the single nodal equation for the original circuit, shown in Fig. 4-13a,

Fig. 4-13: (a) The parallel combination of N inductors. (b) The equivalent circuit, where $L_{eq} = 1/[(1/L_1) + (1/L_2) + \cdots + (1/L_N)]$.

(a)   (b)

$$i_s = \sum_{n=1}^{N} i_n = \sum_{n=1}^{N} \left[\frac{1}{L_n} \int_{t_0}^{t} v \, dt + i_n(t_0) \right]$$

$$= \left[\sum_{n=1}^{N} \frac{1}{L_n} \right] \int_{t_0}^{t} v \, dt + \sum_{n=1}^{N} i_n(t_0)$$

and comparing it with the result for the equivalent circuit of Fig. 4-13b,

$$i_s = \frac{1}{L_{eq}} \int_{t_0}^{t} v \, dt + i_s(t_0)$$

Since Kirchhoff's current law demands that $i_s(t_0)$ be equal to the sum of the branch currents at t_0, the two integral terms must also be equal; hence,

$$L_{eq} = \cfrac{1}{\cfrac{1}{L_1} + \cfrac{1}{L_2} + \cdots + \cfrac{1}{L_N}}$$

For the special case of two inductors in parallel,

$$L_{eq} = \frac{L_1 L_2}{L_1 + L_2}$$

and we note that inductors in parallel combine exactly as do resistors in parallel.

In order to find a capacitor which is equivalent to N capacitors in series, we use the circuit of Fig. 4-14a and its equivalent Fig. 4-14b to write

$$v_s = \sum_{n=1}^{N} v_n = \sum_{n=1}^{N} \left[\frac{1}{C_n} \int_{t_0}^{t} i \, dt + v_n(t_0) \right]$$

$$= \left[\sum_{n=1}^{N} \frac{1}{C_n} \right] \int_{t_0}^{t} i \, dt + \sum_{n=1}^{N} v_n(t_0)$$

Fig. 4-14: (a) A circuit containing N capacitors in series. (b) The desired equivalent, $C_{eq} = 1/[(1/C_1) + (1/C_2) + \cdots + (1/C_N)]$.

(a)   (b)

Fig. 4-15: (a) The parallel combination of N capacitors. (b) The equivalent circuit, where $C_{eq} = C_1 + C_2 + \cdots + C_N$.

and
$$v_s = \frac{1}{C_{eq}} \int_{t_0}^{t} i \, dt + v_s(t_0)$$

However, Kirchhoff's voltage law establishes the equality of $v_s(t_0)$ and the sum of the capacitor voltages at t_0; thus

$$C_{eq} = \frac{1}{\dfrac{1}{C_1} + \dfrac{1}{C_2} + \cdots + \dfrac{1}{C_N}}$$

and capacitors in series combine as do conductances in series, or resistors in *parallel*.

Finally, the circuits of Fig. 4-15 enable us to establish the value of the capacitance of the capacitor which is equivalent to N parallel capacitors as

$$C_{eq} = C_1 + C_2 + \cdots + C_N$$

and it is no great source of amazement to note that capacitors in parallel combine in the same manner in which we combine resistors in series, that is, by simply adding all the individual capacitances.

As an example in which some simplification may be achieved by combining like elements, consider the network of Fig. 4-16a. The 6- and 3-μF capacitors are first combined into a 2-μF equivalent, and this capacitor is then combined with the 1-μF element with which it is in parallel to yield an equivalent capacitance of 3 μF. In addition, the 3- and 2-H inductors are replaced by an equivalent 1.2-H inductor which is then added to the 0.8-H element to give a total equivalent inductance of 2 H. The much simpler (and probably less expensive) equivalent network is shown in Fig. 4-16b.

The network shown in Fig. 4-17 contains three inductors and three capaci-

Fig. 4-16: (a) A given LC network. (b) A simpler equivalent circuit.

Fig. 4-17: An LC network in which no series or parallel combinations of either the inductors or capacitors can be made.

tors, but no series or parallel combinations of either the inductors or the capacitors can be achieved. Simplification of this network cannot be accomplished at this time.

Next let us turn to mesh, loop, and nodal analysis. Since we already know that we may safely apply Kirchhoff's laws, we should have little difficulty in writing a set of equations that are both sufficient and independent. They will be constant-coefficient linear integrodifferential equations, however, which are hard enough to pronounce, let alone solve. Consequently, we shall write them now to gain familiarity with the use of Kirchhoff's laws in *RLC* circuits and discuss the solution of the simpler cases in the following chapters.

Let us attempt to write nodal equations for the circuit of Fig. 4-18. Node voltages are chosen as indicated, and we sum currents leaving the central node,

$$\frac{1}{L}\int_{t_0}^{t}(v_1 - v_s)\,dt + i_L(t_0) + \frac{v_1 - v_2}{R} + C_2\frac{dv_1}{dt} = 0$$

where $i_L(t_0)$ is the value of the inductor current at the time the integration begins, or the initial value. At the right node,

$$C_1\frac{d(v_2 - v_s)}{dt} + \frac{v_2 - v_1}{R} - i_s = 0$$

Rewriting these two equations, we have

$$\frac{v_1}{R} + C_2\frac{dv_1}{dt} + \frac{1}{L}\int_{t_0}^{t}v_1\,dt - \frac{v_2}{R} = \frac{1}{L}\int_{t_0}^{t}v_s\,dt - i_L(t_0)$$

$$-\frac{v_1}{R} + \frac{v_2}{R} + C_1\frac{dv_2}{dt} = C_1\frac{dv_s}{dt} + i_s$$

These are the promised integrodifferential equations, and we may note several interesting points about them. First, the source voltage v_s happens to enter the equations as an integral and as a derivative, but not simply as v_s. Since both sources are specified for all time, we should be able to evaluate the derivative or integral. Secondly, the initial value of the inductor current, $i_L(t_0)$, acts as a (constant) source current at the center node.

Fig. 4-18: A four-node *RLC* circuit with node voltages assigned.

We shall not attempt the solution of these equations at this time. It is worthwhile pointing out, however, that when the two voltage forcing functions are sinusoidal functions of time it will be possible to define a voltage-current ratio (called impedance) or a current-voltage ratio (called admittance) for each of the three passive elements. The factors operating on the two node voltages in the equations above will then become simple multiplying factors, and the equations will be linear *algebraic* equations once again. These we may solve by determinants or a simple elimination of variables as before.

Drill Problem

4-6. (a) Find L_{eq} in Fig. 4-19a. (b) Find C_{eq} in Fig. 4-19b. (c) Find i_s in Fig. 4-19c if $i_L = 0.03 \sin 5000t$ A.   *Ans:* 2.62 H; 6.91 μF; 0.015 sin 5000t A

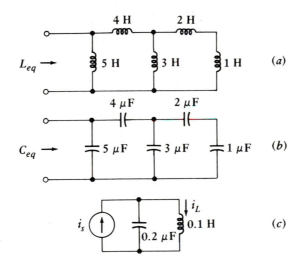

Fig. 4-19: See Drill Prob. 4-6.

4-6
Duality

Duality has been mentioned earlier in connection with resistive circuits and more recently in the discussion of inductance and capacitance; the comments made were introductory and, like the man who tried to pet the alligator, a little offhand. Now we may make an exact definition and then use the definition to recognize or construct dual circuits and thus avoid the labor of analyzing both a circuit and its dual.

We shall define duality in terms of the circuit equations. Two circuits are *duals* if the mesh equations that characterize one of them have the same mathematical form as the nodal equations that characterize the other. They are said to be *exact duals* if each mesh equation of the one circuit is numerically identical with the corresponding nodal equation of the other; the current and voltage variables themselves cannot be identical, of course. *Duality* itself merely refers to any of the properties exhibited by dual circuits.

Let us interpret the definition and use it to construct an exact dual circuit

Fig. 4-20: A given circuit to which the definition of duality may be applied to determine the dual circuit.

Fig. 4-21: The exact dual of the circuit of Fig. 4-20.

by writing the two mesh equations for the circuit shown in Fig. 4-20. Two mesh currents i_1 and i_2 are assigned, and the mesh equations are

$$3i_1 + 4\frac{di_1}{dt} - 4\frac{di_2}{dt} = 2\cos 6t \tag{17}$$

$$-4\frac{di_1}{dt} + 4\frac{di_2}{dt} + \frac{1}{8}\int_0^t i_2\,dt + 5i_2 = -10 \tag{18}$$

It should be noted that the capacitor voltage v_C is assumed to be 10 V at $t = 0$.

We may now construct the two equations which describe the exact dual of the given circuit. We wish these to be nodal equations, and we thus begin by replacing the mesh currents i_1 and i_2 in Eqs. (17) and (18) by two node-to-reference voltages v_1 and v_2. We obtain

$$3v_1 + 4\frac{dv_1}{dt} - 4\frac{dv_2}{dt} = 2\cos 6t \tag{19}$$

$$-4\frac{dv_1}{dt} + 4\frac{dv_2}{dt} + \frac{1}{8}\int_0^t v_2\,dt + 5v_2 = -10 \tag{20}$$

and we now seek the circuit represented by these two nodal equations.

Let us first draw a line to represent the reference node, and then we may establish two nodes at which the positive references for v_1 and v_2 are located. Equation (19) indicates that a current source $2\cos 6t$ is connected between node 1 and the reference node, oriented to provide a current entering node 1. This equation also shows that a 3-℧ conductance appears between node 1 and the reference node. Turning to (20), we first consider the nonmutual terms, or those terms which do not appear in (19), and they instruct us to connect an 8-H inductor and a 5-℧ conductance (in parallel) between node 2 and the reference. The two similar terms in (19) and (20) represent a 4-F capacitor present mutually at nodes 1 and 2; the circuit is completed by connecting this capacitor between the two nodes. The constant term on the right side of (20) is the value of the inductor current at $t = 0$; thus, $i_L(0) = 10$ A. The dual circuit is shown in Fig. 4-21; since the two sets of equations are numerically identical, the circuits are exact duals.

Dual circuits may be obtained more readily than by the above method, for the equations need not be written. In order to construct the dual of a given circuit, we think of the circuit in terms of its mesh equations. With each mesh we must associate a nonreference node, and, in addition, we must supply the reference node. On a diagram of the given circuit we therefore place a node in the center of each mesh and supply the reference node as a line near the diagram or a loop enclosing the diagram. Each element which appears jointly in two meshes is a *mutual* element and gives rise to identical terms, except for sign, in the two corresponding mesh equations. It must be replaced by an element which supplies the dual term in the two corresponding nodal equations. This dual element must therefore be connected directly between the two nonreference nodes which are within the meshes in which the given mutual element appears. The nature of the dual element itself is easily determined; the mathematical form of the equations will be the same only if inductance is replaced by capacitance, capacitance by inductance, conductance by resistance, and resistance by conductance. Thus, the 4-H inductor which is common to meshes 1 and 2 in the circuit of Fig. 4-20 appears as a 4-F capacitor connected directly between nodes 1 and 2 in the dual circuit.

Elements which appear only in one mesh must have duals which appear between the corresponding node and the reference node. Referring again to Fig. 4-20, the voltage source $2 \cos 6t$ V appears only in mesh 1; its dual is a current source $2 \cos 6t$ A which is connected only to node 1 and the reference node. Since the voltage source is clockwise-sensed, the current source must be into-the-nonreference-node-sensed. Finally, provision must be made for the dual of the initial voltage present across the 8-F capacitor in the given circuit. The equations have shown us that the dual of this initial voltage across the capacitor is an initial current through the inductor in the dual circuit; the numerical values are the same, and the correct sign of the initial current may be determined most readily by considering both the initial voltage in the given circuit and the initial current in the dual circuit as sources. Thus, if v_C in the given circuit is treated as a source, it would appear as $-v_C$ on the right side of the mesh equation; in the dual circuit, treating the current i_L as a source would yield a term $-i_L$ on the right side of the nodal equation. Since each has the same sign when treated as a source, then, if $v_C(0) = 10$ V, $i_L(0)$ must be 10 A.

The circuit of Fig. 4-20 is repeated in Fig. 4-22, and its exact dual is constructed on the circuit diagram itself by merely drawing the dual of each

Fig. 4-22: The dual of the circuit of Fig. 4-20 is constructed directly from the circuit diagram.

Fig. 4-23: (a) The dual (in red) of a given circuit (in gray) is constructed on the given circuit. (b) The dual circuit is drawn in more conventional form.

(a)

(b)

given element between the two nodes which are centered in the two meshes which are common to the given element. A reference node which surrounds the given circuit may be helpful. After the dual circuit is redrawn in more standard form, it appears as shown in Fig. 4-21.

An additional example of the construction of a dual circuit is shown in Figs. 4-23a and b. Since no particular element values are specified, these two circuits are duals, but not necessarily exact duals. The original circuit may be recovered from the dual by placing a node in the center of each of the five meshes of Fig. 4-23b and proceeding as before.

The concept of duality may also be carried over into the language by which we describe circuit analysis or operation. One example of this was discussed previously in Sec. 4-4, and the duals of several words appeared there. Most of these pairs are obvious; whenever there is any question as to the dual of a word or phrase, the dual circuit may always be drawn or visualized and then described in similar language. For example, if we are given a voltage source in series with a capacitor, we might wish to make the important statement, "the voltage source causes a current to flow through the capacitor"; the dual statement is, "the current source causes a voltage to exist across the inductor." The dual of a less carefully worded statement, such as, "the current goes round and round the series circuit," often re-quires a little inventiveness.[3]

Practice in using dual language can be obtained by reading Thévenin's theorem in this sense; Norton's theorem should result.

We have spoken of dual elements, dual language, and dual circuits. What about a dual *network?* Consider a resistor R and an inductor L in series. The dual of this two-terminal network exists and is most readily obtained by connecting some ideal source to the given network. The dual circuit is then obtained as the dual source in parallel with a conductance G, $G = R$, and a capacitance C, $C = L$. We consider the dual network as the two-terminal network that is connected to the dual source; it is thus a pair of terminals between which G and C are connected in parallel.

Before leaving the definition of duality, it should be pointed out that dual-

---

[3] Someone has suggested, "the voltage is across all over the parallel circuit."

ity is defined on the basis of mesh and nodal equations. Since nonplanar circuits cannot be described by a system of mesh equations, a circuit which cannot be drawn in planar form does not possess a dual.

We shall use duality principally to reduce the work which we must do to analyze the simple standard circuits. After we have analyzed the series RL circuit, then the parallel RC circuit requires less attention, not because it is less important, but because the analysis of the dual network is already known. Since the analysis of some complicated circuit is not apt to be well known, duality will usually not provide us with any quick solution.

Drill Problem

4-7. Write the single mesh equation for the circuit of Fig. 4-24a and show, by direct substitution, that $i = 4e^{-2000t}$ A is a solution. Knowing this, refer to Fig. 4-24b and find: (a) i_1; (b) i_2.   *Ans:* $80e^{-2000t}$; $-48e^{-2000t}$ A

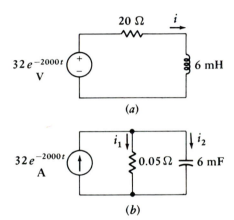

(a)

(b)

Fig. 4-24: See Drill Prob. 4-7.

4-7
Linearity and its consequences again

In the previous chapter we learned that the principle of superposition is a necessary consequence of the linear nature of the resistive circuits which we were analyzing. The resistive circuits are linear because the voltage-current relationship for the resistor is linear and Kirchhoff's laws are linear.

We now wish to show that the benefits of linearity apply to RLC circuits as well. In accordance with our previous definition of a linear circuit, these circuits are also linear because the voltage-current relationships for the inductor and capacitor are linear relationships. For the inductor, we have

$$v = L \frac{di}{dt}$$

and multiplication of the current by some constant K leads to a voltage which is also greater by a factor K. In the integral formulation,

$$i = \frac{1}{L} \int_{t_0}^{t} v \, dt + i_L (t_0)$$

it can be seen that, if each term is to increase by a factor of K, then the initial value of the current must also increase by this same factor. That is, the factor K applies not only to the current and voltage at time t but also to their past values.

A corresponding investigation of the capacitor shows that it too is linear. Thus, a circuit composed of independent sources, linear dependent sources, and linear resistors, inductors, and capacitors is a linear circuit.

In this linear circuit the response is again proportional to the forcing function. The proof of this statement is accomplished by first writing a general system of integrodifferential equations, say, in terms of loop currents. Let us place all the terms having the form of Ri, $L\,di/dt$, and $(1/C)\int i\,dt$ on the left side of each equation and keep the independent source voltages on the right side. As a simple example, one of the equations might have the form

$$Ri + L\frac{di}{dt} + \frac{1}{C}\int_{t_0}^{t} i\,dt + v_C(t_0) = v_s$$

If every independent source is now increased by a factor K, then the right side of each equation is greater by the factor K. Now each term on the left side is either a linear term involving some loop current or an initial capacitor voltage. In order to cause all the responses (loop currents) to increase by a factor K, it is apparent that we must also increase the initial capacitor voltages by a factor K. That is, we must treat the *initial capacitor voltage as an independent source voltage* and increase it also by a factor K. In a similar manner, initial inductor currents must be treated as independent source currents in nodal analysis.

The principle of proportionality between source and response is thus extensible to the general *RLC* circuit, and it follows that the principle of superposition is also applicable. It should be emphasized that initial inductor currents and capacitor voltages must be treated as independent sources in applying the superposition principle; each initial value must take its turn in being rendered inactive.

Before we can apply the superposition principle to *RLC* circuits, however, it is first necessary to develop methods of solving the equations describing these circuits when only one independent source is present. At this time we should feel convinced that a linear circuit will possess a response whose amplitude is proportional to the amplitude of the source. We should be prepared to apply superposition later, considering an inductor current or capacitor voltage specified at $t = t_0$ as a source which must be killed when its turn comes.

Thévenin's and Norton's theorems are based on the linearity of the initial circuit, the applicability of Kirchhoff's laws, and the superposition principle. The general *RLC* circuit conforms perfectly to these requirements, and it follows, therefore, that all linear circuits which contain any combinations of independent voltage and current sources, linear dependent voltage and current sources, and linear resistors, inductors, and capacitors may be analyzed with the use of these two theorems, if we wish. It is not necessary to repeat the theorems here, for they were previously stated in a manner that is equally applicable to the general *RLC* circuit.

Problems

1 Let v_L and i_L be assigned to a 20-mH inductor so as to satisfy the passive sign convention. (a) If $i_L = 12e^{-5t} \cos 10t$ A, find v_L at $t = 0.1$ s. (b) If $i_L = 8(e^{-6t} - e^{-2t})$ A, find $v_{L,\text{max}}$ and the value of time at which it occurs.

2 The current through a 2-mH inductor is $i_L = 0$ for $t < 0$, $500t$ A for $0 < t < 10$ ms, $10 - 500t$ A for $10 < t < 30$ ms, -5 A for $30 < t < 40$ ms, and $-5 \cos^2 [50\pi(t - 0.04)]$ A for $40 < t < 50$ ms. For the interval, $0 < t < 50$ ms: (a) sketch i_L vs. t; (b) sketch v_L vs. t, assuming the passive sign convention.

3 In Fig. 4-25, let $v_s(t) = 20t$ V for $0 < t < 3$ s, and $v_s = 0$ for $t < 0$ and $t > 3$. Find: (a) the value of i at $t = 4$ s; (b) the energy stored in the inductor at $t = 2$ s; (c) the power entering the inductor at $t = 2$ s.

4 The current in a 0.4-H inductor is zero for $t < 0$ and $3te^{-0.1t}$ A for $t > 0$. (a) At what instant is maximum power being delivered to the inductor? (b) At what instant is the energy stored in the inductor a maximum?

5 In the circuit of Fig. 4-26, let $v_s = 100 \cos 500t$ V for $t > -0.5$ s, and let $i_L(0) = -1$ A. (a) Find $i_s(t)$. (b) Find $w_L(t)$ at $t = 1$ ms.

Fig. 4-25: See Prob. 3.

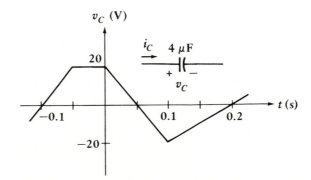

Fig. 4-26: See Prob. 5.

6 (a) If $v_C(t)$ is given by the waveform shown in Fig. 4-27, sketch $i_C(t)$ for $-0.1 < t < 0.2$ s. (b) Sketch the power entering the capacitor over the same time interval.

Fig. 4-27: See Prob. 6.

7 The current through a 0.2-μF capacitor is $60 \cos (10^4t + 36°)$ mA for all time. The average voltage across the capacitor is zero. (a) What is the maximum value of energy stored in the capacitor? (b) What is the first nonnegative value of t at which maximum energy is stored?

8 The energy stored in a 50-μF capacitor is given as $w_C(t) = 2e^{-50t}$ J for $t \geq 0$. Find the capacitor voltage, current, and absorbed power at $t = 30$ ms.

9 In the circuit of Fig. 4-10a, let $R = 1$ MΩ, $C = 1$ μF, $R_i = \infty$, and $R_o = 0$. Suppose that we wish the output to be $v_o(t) = e^{-10t} - 1$ V. Differentiate Eq. (16) to obtain the necessary $v_s(t)$ if: (a) $A = 1000$; (b) A is infinite.

10 Interchange the location of R and C in the circuit of Fig. 4-10a and assume that $R_i = \infty$, $R_o = 0$, and $A = \infty$ for the op-amp. (a) Find $v_o(t)$ as a function of $v_s(t)$. (b) Obtain an equation relating $v_o(t)$ and $v_s(t)$ if A is not assumed to be infinite.

11 In the circuit shown in Fig. 4-28, $v_C(t) = 4te^{-2t}$ V. At $t = 0.5$ s, find the value of: (a) the energy stored in the capacitor; (b) the energy stored in the inductor; (c) v_s.

Fig. 4-28: See Prob. 11.

12 (a) If each inductance in the network of Fig. 4-29 is 1 H, find the equivalent inductance at a-b. (b) Replace each inductor by a 1-F capacitor and find C_{eq}.

Fig. 4-29: See Prob. 12.

13 Find the equivalent inductance offered at terminals a-b in Fig. 4-30 if terminals x-x' are: (a) open-circuited; (b) short-circuited.

Fig. 4-30: See Prob. 13.   **Fig. 4-31:** See Prob. 14.

14 Each capacitor in Fig. 4-31 is 1 μF. Find C_{eq} at a-b if: (a) 1-2 and 1-3 are both open-circuited; (b) 1-2 and 1-3 are both short-circuited; (c) 1-2 is open-circuited and 1-3 is short-circuited; (d) 1-2 is short-circuited and 1-3 is open-circuited.

15 Given a bucketful of 1-nF capacitors, show how an equivalent capacitance of 0.7 nF might be obtained. Use as few capacitors as possible.

16 For the circuit of Fig. 4-32, find: (a) w_C; (b) w_L; (c) the current in each circuit element; (d) the voltage across each circuit element.

Fig. 4-32: See Prob. 16.

17 Let $i_s = 4(1 - e^{-3t})$ A for $t > 0$ and $v_C(0) = 20$ V in the circuit of Fig. 4-33. At $t = 0.5$ s, find the values of energy: (*a*) stored in the inductor; (*b*) stored in the capacitor; (*c*) dissipated by the resistor since $t = 0$.

Fig. 4-33: See Prob. 17.

18 Let $v(0) = 10$ V in the circuit shown in Fig. 4-34. (*a*) Find $i(t)$ for all t. (*b*) Find $v(t)$ for $t \geq 0$.

Fig. 4-34: See Prob. 18.

19 (*a*) Write nodal equations for the circuit of Fig. 4-35. (*b*) Write mesh equations for the same circuit.

Fig. 4-35: See Prob. 19.

$v_c(0) = 8$ V, $i_L(0) = 3$ A

20 In Fig. 4-36, let $i_s = 0.1e^{-400t}$ A with $i_2(0) = 0.03$ A. (*a*) Find $v(t)$ for all t. (*b*) Find $i_2(t)$ for $t \geq 0$. (*c*) Find $i_1(t)$ for $t \geq 0$.

Fig. 4-36: See Prob. 20.

21 Draw a tree for the circuit shown in Fig. 4-37 that satisfies not only the criteria enumerated in Secs. 3-7 and 3-8, but also places all capacitors in the tree and all inductors in the cotree. (a) Assign tree-branch voltages and write a set of nodal equations. Assume that there is no energy storage at $t = 0$. (b) Assign link currents and write a set of loop equations, again assuming no energy storage at $t = 0$.

Fig. 4-37: See Probs. 21 and 24.

22 In Fig. 4-38, let $v_{C1}(0) = 5$ V, $v_{C2}(0) = 6$ V, $v_{C3}(0) = 7$ V, $i_{L1}(0) = 8$ A, and $i_{L2}(0) = -4$ A. (a) Write a set of nodal equations in the variables v_{C1}, v_{C2}, and v_{C3}. (b) Write a set of mesh equations in the variables i_s, i_{L1}, and i_{L2}.

Fig. 4-38: See Probs. 22 and 23.

23 Construct the exact dual of the circuit shown in Fig. 4-38 and described in Prob. 22. Label all inductor currents and capacitor voltages and give their initial values.

24 Draw the exact dual of the circuit of Fig. 4-37.

25 If $i_L(0) = 10$ A and $v_C(0) = 4$ V in the circuit of Fig. 4-39, write a single nodal equation and (a) show that $v_C = -10 \sin 5000t + 4 \cos 5000t$ V satisfies your equation. (b) Obtain the exact dual circuit and find an expression for the capacitor voltage. (c) How do the original and the dual circuits compare topologically?

Fig. 4-39: See Prob. 25.

26 If all four sources in the circuit of Fig. 4-40 have been applied forever and ever, use the principle of superposition to find: (a) $i_C(t)$; (b) $i_L(t)$.

Fig. 4-40: See Prob. 26.

Chapter 5
Source-Free *RL* and *RC* Circuits

In the previous chapter we wrote equations governing the response of several circuits containing both inductance and capacitance, but we did not solve any of them. At this time we are ready to proceed with the solution for the simpler circuits. We shall restrict our attention to certain circuits which contain only resistors and inductors or only resistors and capacitors, and which contain no sources. We shall, however, allow the presence of energy which is stored in the inductors or capacitors, for without such energy every response would be zero.

Although the circuits which we are about to consider have a very elementary appearance, they are also of practical importance. They find use as coupling networks in electronic amplifiers, as compensating networks in automatic control systems and operational amplifiers, as equalizing networks in communications channels, and in many other ways. A familiarity with these simple circuits will enable us to predict the accuracy with which the output of an amplifier can follow an input which is changing rapidly with time or to predict how quickly the speed of a motor will change in response to a change in its field current. Our knowledge of the performance of the simple *RL* and *RC* circuits will also enable us to suggest modifications to the amplifier or motor in order to obtain a more desirable response.

The analysis of such circuits is dependent upon the formulation and solution of the integrodifferential equations which characterize the circuits. We shall call the special type of equation we obtain a homogeneous linear differential equation, which is simply a differential equation in which every term is of the first degree in the dependent variable or one of its derivatives. A *solution* is obtained when we have found an expression for the dependent variable, as a function of time, which satisfies the differential equation and also satisfies the prescribed energy distribution in the inductors or capacitors at a prescribed instant of time, usually $t = 0$.

The solution of the differential equation represents a response of the circuit, and it is known by many names. Since this response depends upon the general "nature" of the circuit (the types of elements, their sizes, the interconnection of the elements), it is often called a *natural response*. It is also obvious that any real circuit we construct cannot store energy forever; the resistances necessarily associated with inductors and capacitors will eventually convert all stored energy into heat. The response must eventually die out, and it is therefore referred to as the *transient response*. Finally, we must also be familiar with the mathematicians' contribution to the nomenclature; they call the solution of a homogeneous linear differential equation a *complementary function*. When we consider independent sources acting on a circuit, part of the response will partake of the nature of the particular source (forcing function) used; this part of the response, called the *particular solution* or the *forced response*, will be "complemented" by the complementary response produced in the source-free circuit, and the sum of the complementary function and the particular solution will be the complete response. That is, the complete response is the sum of the natural response, studied in this chapter, and the forced response which we shall examine in Chapter 6. The source-free response may be called the natural response, the transient response, the free response, or the complementary function, but because of its more descriptive nature, we shall most often call it the natural response.

We shall consider several different methods of solving these differential equations. This mathematics, however, is not circuit analysis. Our greatest interest lies in the solutions themselves, their meaning, and their interpretation, and we shall try to become sufficiently familiar with the form of the response that we are able to write down answers for new circuits by just plain thinking. Although complicated analytical methods are needed when simpler methods fail, an engineer must always remember that these complex techniques are only tools with which meaningful, informative answers may be obtained; they do not constitute engineering in themselves.

5-2
The simple *RL* circuit

We shall begin our study of transient analysis by considering the simple series *RL* circuit shown in Fig. 5-1. Let us designate the time-varying current as $i(t)$, and we shall let the value of $i(t)$ at $t = 0$ be prescribed as I_0. We therefore have

$$v_R + v_L = Ri + L\frac{di}{dt} = 0$$

or

$$\frac{di}{dt} + \frac{R}{L}i = 0 \qquad (1)$$

and we must determine an expression for $i(t)$ which satisfies this equation and also has the value I_0 at $t = 0$. The solution may be obtained by several different methods.

One very direct method of solving a differential equation consists of writ-

Fig. 5-1: A series *RL* circuit for which *i(t)* is to be determined, subject to the initial condition that $i(0) = I_0$.

ing the equation in such a way that the variables are separated and then integrating each side of the equation. The variables in (1) are *i* and *t*, and it is apparent that the equation may be multiplied by *dt*, divided by *i*, and arranged with the variables separated,

$$\frac{di}{i} = -\frac{R}{L}\,dt \tag{2}$$

Since the current is I_0 at $t = 0$ and $i(t)$ at time *t*, we may equate the two definite integrals which are obtained by integrating each side between the corresponding limits,

$$\int_{I_0}^{i(t)} \frac{di}{i} = \int_0^t -\frac{R}{L}\,dt$$

and therefore,

$$\ln i \,\Big|_{I_0}^{i} = -\frac{R}{L}\,t \,\Big|_0^t$$

or

$$\ln i - \ln I_0 = -\frac{R}{L}(t - 0)$$

Therefore,

$$i(t) = I_0 e^{-Rt/L} \tag{3}$$

We check our solution by first showing that substitution of (3) into (1) yields the identity $0 \equiv 0$ and then showing that substitution of $t = 0$ into (3) produces $i(0) = I_0$. Both steps are necessary; the solution must satisfy the differential equation which characterizes the circuit, and it must also satisfy the initial condition, or the response at zero time.

The solution may also be obtained by a slight variation of the method described above. After separating the variables, we may obtain the *indefinite* integral of each side of (2) if we also include a constant of integration. Thus,

$$\int \frac{di}{i} = -\int \frac{R}{L}\,dt + K$$

and integration gives us

$$\ln i = -\frac{R}{L}\,t + K \tag{4}$$

The constant *K* cannot be evaluated by substitution of (4) into the original differential equation (1); the identity $0 \equiv 0$ will result, because (4) is a solution of (1) for any value of *K*. The constant of integration must be selected to satisfy the initial condition $i(0) = I_0$. Thus, at $t = 0$, (4) becomes

$$\ln I_0 = K$$

and we use this value for K in (4) to obtain the desired response

$$\ln i = -\frac{R}{L}t + \ln I_0$$

or
$$i(t) = I_0 e^{-Rt/L}$$

as before.

Either of the methods above can be used when the variables can be separated, but this is only occasionally possible. In the remaining cases we shall rely on a very powerful method, the success of which will depend upon our intuition or experience. We simply guess or assume a form for the solution and then test our assumptions, first by substitution into the differential equation and then by applying the given initial conditions. Since we cannot be expected to guess the exact numerical expression for the solution, we shall assume a solution containing several unknown constants and select the values for these constants in order to satisfy the differential equation and the initial conditions. Many of the differential equations encountered in circuit analysis have a solution which may be represented by the exponential function or by the sum of several exponential functions. Let us assume a solution of (1) in exponential form,

$$i(t) = Ae^{s_1 t}$$

where A and s_1 are constants to be determined. After substituting this assumed solution into (1), we have

$$As_1 e^{s_1 t} + \frac{R}{L} Ae^{s_1 t} = 0$$

or
$$\left(s_1 + \frac{R}{L}\right) Ae^{s_1 t} = 0$$

In order to satisfy this equation for all values of time, it is necessary that either $A = 0$, or $s_1 = -\infty$, or $s_1 = -R/L$. But if $A = 0$ or $s_1 = -\infty$, then every response is zero; neither can be a solution to our problem. Therefore, we must choose

$$s_1 = -\frac{R}{L}$$

and our assumed solution takes on the form

$$i(t) = Ae^{-Rt/L}$$

The remaining constant must be evaluated by applying the initial condition $i = I_0$ at $t = 0$. Thus,

$$I_0 = A$$

and the final form of the assumed solution is

$$i(t) = I_0 e^{-Rt/L} \tag{5}$$

once again.

We shall not consider any other methods for solving (1), although a number of other techniques may be used. Watch for them in a study of differential equations.

Before we turn our attention to the interpretation of the response, let us check the power and energy relationships in this circuit. The power being dissipated in the resistor is

$$p_R = i^2 R = I_0^2 R e^{-2Rt/L}$$

and the total energy turned into heat in the resistor is found by integrating the instantaneous power from zero time to infinite time,

$$W_R = \int_0^\infty p_R dt = I_0^2 R \int_0^\infty e^{-2Rt/L} \, dt$$

$$= I_0^2 R \left(-\frac{L}{2R} \right) e^{-2Rt/L} \, \Big|_0^\infty = \frac{1}{2} L I_0^2$$

This is the result we expect, because the total energy stored initially in the inductor is $\frac{1}{2}LI_0^2$, and there is no energy stored in the inductor at infinite time. All the initial energy is accounted for by dissipation in the resistor.

Drill Problems

5-1. Each circuit shown in Fig. 5-2 has been in the condition shown for a long time. The switches in a and b are opened at $t = 0$. The switch in c is a

Fig. 5-2: See Drill Probs. 5-1, 5-2, and 5-5.

single-pole double-throw switch that is drawn to indicate that it closes one circuit before opening the other. It is called *make before break*. After reviewing the characteristics of an inductor at the end of Sec. 4-3, determine $i(0)$ in each circuit.   *Ans:* 600; 20; −300 mA

5-2. At the instant just *after* the switches are thrown in the circuit of Fig. 5-2, find v.   *Ans:* −12; 0.340; −24 V

5-3. Element values in Fig. 5-1 are $R = 0.8\ \Omega$, $L = 1.6$ H. Let $i = 20$ A at $t = -1$ s and find: (*a*) $i(0)$; (*b*) the power being absorbed by the inductor at $t = 1$ s; (*c*) the time at which the energy stored in the inductor is 100 J.   *Ans:* 12.13 A; −43.3 W; 0.1632 s

---

**5-3
Properties of the exponential response**

Let us now consider the nature of the response in the series RL circuit. We found that the current is represented by

$$i(t) = I_0 e^{-Rt/L} \tag{6}$$

At zero time, the current is the assumed value I_0, and as time increases, the current decreases and approaches zero. The shape of this decaying exponential is seen by a plot of $i(t)/I_0$ versus t, as shown in Fig. 5-3. Since the function we are plotting is $e^{-Rt/L}$, the curve will not change if R/L does not change. Thus, the same curve must be obtained for every series RL circuit having the same R/L or L/R ratio. Let us see how this ratio affects the shape of the curve.

If we double the ratio of L to R, then the exponent will be unchanged if t is also doubled. In other words, the original response will occur at a later time, and the new curve is obtained by moving each point on the original curve twice as far to the right. With this larger L/R ratio, the current takes longer to decay to any given fraction of its original value. We might have a tendency to say that the "width" of the curve is doubled, or that the "width" is proportional to L/R. However, we should have to define our term "width,"

Fig. 5-3: A plot of $e^{-Rt/L}$ versus t.

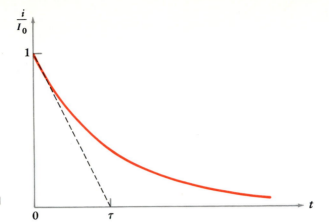

Fig. 5-4: The time constant τ is L/R for a series *RL* circuit. It is the time required for the response curve to drop to zero if it decays at a constant rate which is equal to its initial rate of decay.

because each curve extends from $t = 0$ to ∞. Instead, let us consider the time that would be required for the current to drop to zero *if it continued to drop at its initial rate.*

The initial rate of decay is found by evaluating the derivative at zero time,

$$\frac{d}{dt} \frac{i}{I_0} \bigg|_{t=0} = -\frac{R}{L} e^{-Rt/L} \bigg|_{t=0} = -\frac{R}{L}$$

We designate the value of time it takes for i/I_0 to drop from unity to zero, assuming a constant rate of decay, by the Greek letter τ *(tau)*. Thus,

$$\frac{R}{L} \tau = 1$$

or

$$\tau = \frac{L}{R} \tag{7}$$

The ratio L/R has the units of seconds since the exponent $-Rt/L$ must be dimensionless. This value of time τ is called the *time constant;* it is shown in Fig. 5-4. It is apparent that the time constant of a series *RL* circuit may easily be found graphically from the response curve; it is only necessary to draw the tangent to the curve at $t = 0$ and determine the intercept of this tangent line with the time axis. This is often a convenient way of approximating the time constant from the display on an oscilloscope.

An equally important interpretation of the time constant τ is obtained by determining the value of $i(t)/I_0$ at $t = \tau$. We have

$$\frac{i(\tau)}{I_0} = e^{-1} = 0.368 \qquad \text{or} \qquad i(\tau) = 0.368 I_0$$

Thus, in one time constant the response has dropped to 36.8 percent of its initial value; the value of τ may also be determined graphically from this fact,

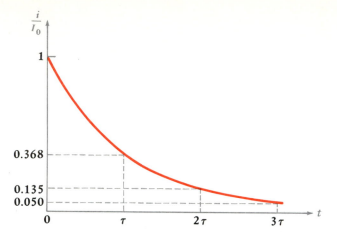

Fig. 5-5: The current in a series *RL* circuit is 36.8, 13.5, and 5 percent of its initial value at τ, 2τ, and 3τ, respectively.

as indicated by Fig. 5-5. It is convenient to measure the decay of the current at intervals of one time constant, and recourse to a hand calculator or a table of negative exponentials shows that $i(t)/I_0$ is 0.368 at $t = \tau$, 0.135 at $t = 2\tau$, 0.0498 at $t = 3\tau$, 0.0183 at $t = 4\tau$, and 0.0067 at $t = 5\tau$. At some point three to five time constants after zero time, most of us would agree that the current is a negligible fraction of its former self. So, if we are asked, "How long does it take for the current to decay to zero?" our answer might be "About five time constants." At the end of that time interval the current is less than one percent of its original value.

Why does a larger value of the time constant L/R produce a response curve which decays more slowly? Let us consider the effect of each element. An increase in L allows a greater energy storage for the same initial current, and this larger energy requires a longer time to be dissipated in the resistor. We may also increase L/R by reducing R. In this case, the power flowing into the resistor is less for the same initial current; again, a greater time is required to dissipate the stored energy.

In terms of the time constant τ, the response of the series *RL* circuit may be written simply as

$$i(t) = I_0 e^{-t/\tau}$$

Drill Problem

5-4. In a series *RL* circuit, find the ratio: (*a*) $i(2\tau)/i(0)$; (*b*) $i(4\tau)/i(2\tau)$; (*c*) t/τ if $i(t) = 0.1i(\tau)$; (*d*) t/τ if $i(t) - i(\tau) = 0.1i(0)$.

Ans: 0.1353; 0.1353; 3.30; 0.760

5-4
A more general *RL* circuit

It is not difficult to extend the results obtained for the series *RL* circuit to a circuit containing any number of resistors and one inductor. We fix our attention on the two terminals of the inductor and determine the equivalent resistance across these terminals. The circuit is thus reduced to the simple series case. As an example, conside the circuit shown in Fig. 5-6. The equivalent resistance which the inductor faces is

$$R_{eq} = R_3 + R_4 + \frac{R_1 R_2}{R_1 + R_2}$$

and the time constant is therefore

$$\tau = \frac{L}{R_{eq}}$$

The inductor current i_L is

$$i_L = i_L(0)e^{-t/\tau} \tag{8}$$

and (8) represents what we might call the basic solution to the problem. It is quite possible that some current or voltage other than i_L is needed, such as the current i_2 in R_2. We can always apply Kirchhoff's laws and Ohm's law to the resistive portion of the circuit without any difficulty, but current division provides the quickest answer in this circuit,

$$i_2 = -\frac{R_1}{R_1 + R_2} i_L(0)e^{-t/\tau}$$

It may also happen that we know the initial value of some current other than the inductor current. Since the current in a resistor may change instantaneously, we shall indicate the instant *after* any change that might have occurred at $t = 0$ by the use of the symbol 0^+; in more mathematical language, $i_1(0^+)$ is the limit from the right of $i_1(t)$ as t approaches zero. Thus, if we are given the initial value of i_1 as $i_1(0^+)$, then it is apparent that the initial value of i_2 is

$$i_2(0^+) = i_1(0^+) \frac{R_1}{R_2}$$

From these values, we obtain the necessary initial value of $i_L(0)$ [or $i_L(0^-)$ or $i_L(0^+)$],

$$i_L(0) = -[i_1(0^+) + i_2(0^+)] = -\frac{R_1 + R_2}{R_2} i_1(0^+)$$

and the expression for i_2 becomes

$$i_2 = i_1(0^+) \frac{R_1}{R_2} e^{-t/\tau}$$

Let us see if we can obtain this last expression more directly. Since the

Fig. 5-6: A source-free circuit containing one inductor and several resistors is analyzed by determining the time constant, $\tau = L/R_{eq}$.

inductor current decays exponentially as $e^{-t/\tau}$, then every current throughout the circuit must follow the same functional behavior. This is made clear by considering the inductor current as a source current which is being applied to a resistive network. Every current and voltage in the resistive network must have the same time dependence. Using these ideas, we therefore express i_2 as

$$i_2 = Ae^{-t/\tau}$$

where

$$\tau = \frac{L}{R_{eq}}$$

and A must be determined from a knowledge of the initial value of i_2. Since $i_1(0^+)$ is known, then the voltage across R_1 and R_2 is known, and

$$i_2(0^+) = i_1(0^+)\frac{R_1}{R_2}$$

Therefore,

$$i_2 = i_1(0^+)\frac{R_1}{R_2}e^{-t/\tau}$$

A similar sequence of steps will provide a rapid solution to a large number of problems. We first recognize the time dependence of the response as an exponential decay, determine the appropriate time constant by combining resistances, write the solution with an unknown amplitude, and then determine the amplitude from a given initial condition.

This same technique is also applicable to a circuit which contains one resistor and any number of inductors, as well as to those special circuits containing two or more inductors and also two or more resistors that may be simplified by resistance or inductance combination until the simplified circuits have only one inductance or one resistance. As an example of such a circuit we may consider Fig. 5-7. After $t = 0$, when the voltage source is disconnected, we easily calculate an equivalent inductance,

$$L_{eq} = \frac{2 \times 3}{2 + 3} + 1 = 2.2 \quad \text{mH}$$

an equivalent resistance,

$$R_{eq} = \frac{90(60 + 120)}{90 + 180} + 50 = 110 \quad \Omega$$

and the time constant,

$$\tau = \frac{L_{eq}}{R_{eq}} = \frac{2.2 \times 10^{-3}}{110} = 20 \quad \mu s$$

Thus, the form of the natural response is $Ae^{-50\,000t}$. With the independent

Fig. 5-7: After $t = 0$, this circuit simplifies to an equivalent resistance of 110 Ω in series with $L_{eq} = 2.2$ mH.

source connected ($t < 0$), i_L is $\frac{18}{50}$, or 0.36 A, while i_1 is $\frac{18}{90}$, or 0.2 A. At $t = 0^+$, i_L must still be 0.36 A, but i_1 will jump to a new value determined by $i_L(0^+)$. Thus,

$$i_1(0^+) = -i_L(0^+)\tfrac{180}{270} = -0.24 \quad \text{A}$$

Hence,

$$i_L = 0.36 \qquad\qquad (t < 0)$$
$$= 0.36e^{-50\ 000t} \qquad (t > 0)$$

and
$$i_1 = 0.2 \qquad\qquad (t < 0)$$
$$= -0.24e^{-50\ 000t} \qquad (t > 0)$$

   In idealized circuits in which a pure inductance loop is present, such as that through the 2- and 3-mH inductors of Fig. 5-7, a constant current may continue to circulate as $t \to \infty$. The current through either of these inductors is not necessarily of the form, $Ae^{-t/\tau}$, but takes the more general form, $A_1 + A_2e^{-t/\tau}$. This unimportant special case is illustrated by Prob. 14 at the end of this chapter.

   We have now considered the task of finding the natural response of any circuit which can be represented by an equivalent inductor in series with an equivalent resistor. A circuit containing several resistors and several inductors does not in general possess a form which allows either the resistors or inductors to be combined into single equivalent elements. There is no single negative exponential term or single time constant associated with the circuit. Rather, there will, in general, be several negative exponential terms, the number of terms being equal to the number of inductors that remain after all possible inductor combinations have been made. The natural response of these more complex circuits is obtained by using techniques that we shall study later. One of the methods will appear toward the end of Chap. 13; it is based on the concept of complex frequency. The most powerful methods rely on the use of Fourier or Laplace transforms and will arise in Chaps. 18 and 19.

Drill Problems

5-5. After $t = 0$, each of the circuits in Fig. 5-2 is source-free. Find values for i and v at $t = 50$ μs: (*a*) in Fig. 5-2*a*; (*b*) in Fig. 5-2*b*; (*c*) in Fig. 5-2*c*.
Ans: 283 mA, -5.67 V; 2.71 mA, 0.340 V; -182.0 mA, -14.56 V

5-6. Find values for these currents in the circuit of Fig. 5-8 at $t = 5$ ms: (a) i_L; (b) i_x; (c) i_y.   *Ans: 28.7; −22.9; 114.3 mA*

Fig. 5-8: See Drill Prob. 5-6.

5-5

The simple *RC* circuit

The series combination of a resistor and a capacitor has a greater practical importance than does the combination of a resistor and an inductor. When engineers have any freedom of choice between using a capacitor and using an inductor in the coupling network of an electronic amplifier, in the compensation networks of an automatic control system, or in the synthesis of an equalizing network, for example, they choose the *RC* network over the *RL* network whenever possible. The reasons for this choice are the smaller losses present in a physical capacitor, its lower cost, the better approximation which the mathematical model makes to the physical element it is intended to represent, and the smaller size and lighter weight as exemplified by capacitors in hybrid and integrated circuits.

Let us see how closely the analysis of the parallel (or is it series?) *RC* circuit corresponds to that of the *RL* circuit. The *RC* circuit is shown in Fig. 5-9. We shall assume an initial stored energy in the capacitor by selecting

$$v(0) = V_0$$

The total current leaving the node at the top of the circuit diagram must be zero, and, therefore,

$$C \frac{dv}{dt} + \frac{v}{R} = 0$$

Division by *C* gives us

$$\frac{dv}{dt} + \frac{v}{RC} = 0 \tag{9}$$

Equation (9) has a familiar form; comparison with (1),

$$\frac{di}{dt} + \frac{R}{L} i = 0 \tag{1}$$

Fig. 5-9: A parallel *RC* circuit for which $v(t)$ is to be determined, subject to the initial condition that $v(0) = V_0$.

shows that the replacement of *i* by *v* and *L/R* by *RC* produces the identical equation we considered previously. It should, for the *RC* circuit we are now analyzing is the dual of the *RL* circuit we considered first. This duality forces $v(t)$ for the *RC* circuit and $i(t)$ for the *RL* circuit to have identical expressions if the resistance of one circuit is equal to the reciprocal of the resistance of the other circuit and if L is numerically equal to C. Thus, the response of the *RL* circuit,

$$i(t) = i(0)e^{-Rt/L} = I_0 e^{-Rt/L}$$

enables us to write immediately

$$v(t) = v(0)e^{-t/RC} = V_0 e^{-t/RC} \tag{10}$$

for the *RC* circuit.

Now let us suppose that we had selected the current i as our variable in the *RC* circuit, rather than the voltage v. Applying Kirchhoff's voltage law,

$$\frac{1}{C} \int_{t_0}^{t} i \, dt - v(t_0) + Ri = 0$$

we obtain an integral equation and not a differential equation. However, if we take the time derivative of both sides of this equation,

$$\frac{i}{C} + R \frac{di}{dt} = 0 \tag{11}$$

and replace i by v/R,

$$\frac{v}{RC} + \frac{dv}{dt} = 0$$

we obtain (9) again. Equation (11) could have been used as our starting point, but duality would not have appeared as naturally.

Let us discuss the physical nature of the voltage response of the *RC* circuit as expressed by (10). At $t = 0$ we obtain the correct initial condition, and as t becomes infinite the voltage approaches zero. This latter result agrees with our thinking that if there were any voltage remaining across the capacitor, then energy would continue to flow into the resistor and be dissipated as heat. Thus, a final voltage of zero is necessary. The time constant of the *RC* circuit may be found by using the duality relationships on the expression for the time constant of the *RL* circuit, or it may be found by simply noting the time at which the response has dropped to 36.8 percent of its initial value,

$$\frac{\tau}{RC} = 1$$

and

$$\tau = RC \tag{12}$$

Our familiarity with the negative exponential and the significance of the time constant τ enables us to sketch the response curve readily, Fig. 5-10. Larger values of R or C provide larger time constants and slower dissipation of the stored energy. A larger resistance will dissipate a smaller power[1] with a given voltage across it, thus requiring a greater time to convert the stored

---

[1] "Greater resistance leads to less dissipation" might be the scholar's motto.

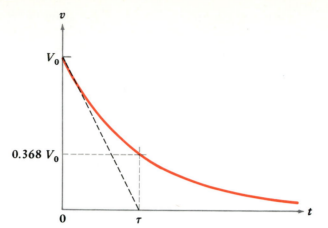

Fig. 5-10: The capacitor voltage $v(t)$ in the parallel RC circuit is plotted as a function of time. The initial value of $v(t)$ is assumed to be V_0.

energy into heat; a larger capacitance stores a larger energy with a given voltage across it, again requiring a greater time to lose this initial energy.

Drill Problems

5-7. Determine $v(0^+)$ for each of the circuits of Fig. 5-11. *Ans: 5; 1.5; 4 V*

5-8. Find $i(0^+)$ for each circuit in Fig. 5-11. *Ans: 0.5; 300 000; 20 mA*

Fig. 5-11: See Drill Probs. 5-7, 5-8, and 5-10.

5-9. Element values in Fig. 5-9 are $C = 0.02\ \mu\text{F}$ and $R = 500\ \Omega$. If $v = 20$ V at $t = 5\ \mu\text{s}$, find: (*a*) $v(10\ \mu\text{s})$; (*b*) the power being supplied by the capacitor at $t = 12\ \mu\text{s}$; (*c*) the time at which the energy stored in the capacitor is 1.5 μJ.   *Ans:* 12.13 V; 197.3 mW; 9.90 μs

---

5-6
A more general *RC* circuit

Many of the *RC* circuits for which we would like to find the natural response contain more than a single resistor and capacitor. Just as we did for the *RL* circuits, we first consider those cases in which the given circuit may be reduced to an equivalent circuit consisting of only one resistor and one capacitor.

Let us suppose first that we are faced with a circuit containing only one capacitor, but any number of resistors. It is possible to replace the two-terminal resistive network which is across the capacitor terminals by an equivalent resistance, and we may then write down the expression for the capacitor voltage immediately. For example, the circuit shown in Fig. 5-12*a* may be simplified to that of Fig. 5-12*b*, enabling us to write

$$v = V_0 e^{-t/R_{eq}C}$$

where

$$v(0) = V_0 \qquad \text{and} \qquad R_{eq} = R_2 + \frac{R_1 R_3}{R_1 + R_3}$$

Every current and voltage in the resistive portion of the network must have the form $Ae^{-t/R_{eq}C}$, where A is the initial value of that current or voltage. Thus, the current in R_1, for example, may be expressed as

$$i_1 = i_1(0^+)e^{-t/\tau}$$

where

$$\tau = \left(R_2 + \frac{R_1 R_3}{R_1 + R_3} \right) C$$

and $i_1(0^+)$ remains to be determined from some initial condition. Suppose that $v(0)$ is given. Since v cannot change instantaneously, we may think of the capacitor as being replaced by an independent dc source, $v(0)$. Thus,

$$i_1(0^+) = \frac{v(0)}{R_2 + R_1 R_3/(R_1 + R_3)} \frac{R_3}{R_1 + R_3}$$

The solution is obtained by collecting these results.

Another special case includes those circuits containing one resistor and

Fig. 5-12: (*a*) A given circuit containing one capacitor and several resistors. (*b*) The resistors have been replaced by a single equivalent resistor; the time constant is now obvious.

any number of capacitors. The resistor voltage is easily obtained by establishing the value of the equivalent capacitance and determining the time constant. Once again our mathematically perfect elements may lead to phenomena which would not exist in a physical circuit. Here, two capacitors in series may have equal and opposite voltages across each element and yet have zero voltage across the combination. Thus the general form of the voltage across either is $A_1 + A_2e^{-t/\tau}$, while the voltage across the series combination continues to be $Ae^{-t/\tau}$. An example of such a situation is provided by Prob. 27 at the end of the chapter.

Some circuits containing a number of both resistors and capacitors may be replaced by an equivalent circuit containing only one resistor and one capacitor; it is necessary that the original circuit be one which can be broken into two parts, one containing all resistors and the other containing all capacitors, such that the two parts are connected by only two ideal conductors. This is not possible in general.

More complicated circuits that cannot be reduced to simple series RC circuits will be considered later in Chaps. 13, 18, and 19.

Drill Problem

5-10. Find values for v and i in the circuit of: (a) Fig. 5-11a at $t = 0.03$ s; (b) Fig. 5-11b at $t = 0.1$ μs; (c) Fig. 5-11c at $t = 1$ ms.

Ans: 2.74 V, 0.274 mA; 0.0275 V, 5.49 A; 1.146 V, 5.73 mA

Problems

1 After being closed for a very long time, the switch in Fig. 5-13 is opened at $t = 0$. (a) Find $i_L(0^+)$ and $w_L(0^+)$. (b) Find $i_L(t)$. (c) Find $v_{10}(t)$.

Fig. 5-13: See Probs. 1 and 18.

2 In Fig. 5-14, let $i_s = 12$ A for $t < 0$ and zero for $t > 0$. (a) Find $i_L(t)$ and sketch it for $-0.01 < t < 0.05$ s. (b) Find $v_L(t)$ and sketch it for the same time interval.

3 (a) A 0.2-H inductor is in parallel with a 25-Ω resistor. If $i_L(0) = 5$ A, find $i_L(t)$ and the value of t, $t = t_1$, at which $i_L = 1$ A. (b) At this instant, a 5-Ω resistor is connected in parallel with the 25-Ω resistor. Write $i_L(t)$ for $t > t_1$ and find the value of t, $t = t_2$, at which $i_L = 1$ mA.

Fig. 5-14: See Prob. 2.

4 For the circuit of Fig. 5-15, find $i_L(t)$ and sketch it for $t > 0$.

5 Determine the energy stored in the inductor of Fig. 5-16 twenty microseconds after the switch is thrown.

6 (*a*) In Fig. 5-17, find the inductor current as a function of time, and sketch curves of v_1, v_2, and v_3 vs. *t*, $-1 < t < 1$ s. (*b*) Sketch $v_3(t)$ vs. *t*, $-1 < t < 1$ s, if another switch places a second 10-Ω resistor in parallel with the original 10-Ω resistor at $t = 0.25$ s.

7 A source-free *RL* circuit contains a 50-H inductor discharging through a 20-Ω resistor. The inductor current amplitude is 18 A at $t = 2$ s. (*a*) When was the energy stored in the inductor twice as great as it is at $t = 2$ s? (*b*) At what instant will the power dissipated in the resistor be 1 kW?

8 The switch in the circuit shown in Fig. 5-18 opens at $t = 0$. (*a*) Find $v(t)$ for $t > 0$. (*b*) Sketch $v(t)$ for $-1 < t < 1$ s.

9 The battery in Fig. 5-19 is disconnected at $t = 0$. (a) Find $i_L(0^-)$. (b) Find $i_L(0^+)$. (c) Find the (Thévenin) equivalent resistance seen by the inductor for $t > 0$. (d) Find τ. (e) Find $i_L(t)$, $t > 0$.

Fig. 5-19: See Prob. 9.

10 In Fig. 5-20, let $i_L(0) = 10$ A. Find $v_x(t)$ for $t > 0$ and sketch it over a three-time-constant interval.

Fig. 5-20: See Prob. 10.

11 Refer to Fig. 5-21 and find $i_x(t)$ for all time. Give its numerical value at $t = -2, 0^-$, 0^+, 2, and 4 ms.

Fig. 5-21: See Prob. 11.

12 The switch in Fig. 5-22 opens at $t = 0$ after being long-closed. Calculate v_x at $t = -10, 0^-, 0^+$, 10, and 20 ms.

Fig. 5-22: See Prob. 12.

Fig. 5-23: See Prob. 13.

13 Find $i_1(t)$ for $t > 0$ in Fig. 5-23 if $i_1(0) = 5$ A.
14 The circuit shown in Fig. 5-24 contains two inductors in parallel, thus providing the opportunity of a trapped current circulating around the inductive loop. Let $i_1(0^-) = 20$ A and $i_2(0^-) = 10$ A. (a) Find $i_1(0^+)$, $i_2(0^+)$, and $i(0^+)$. (b) Determine

the time constant τ for $i(t)$. (c) Find $i(t)$, $t > 0$. (d) Find $v(t)$. (e) Find $i_1(t)$ and $i_2(t)$ from $v(t)$ and the initial values. (f) Show that the stored energy at $t = 0$ is equal to the sum of the energy dissipated in the resistive network between $t = 0$ and $t = \infty$, plus the energy stored in the inductors at $t = \infty$.

Fig. 5-24: See Prob. 14.

15 Calculate i_L and i_1 in the circuit of Fig. 5-25 at $t = -10$, 0^-, 0^+, 10, and 20 ms.

Fig. 5-25: See Prob. 15.

16 The switch in Fig. 5-26 is opened at $t = 0$ after having been closed for a long time. Plot $v_C(t)$ and $v_x(t)$ on the same time axis, $-0.5 < t < 1$ ms.

Fig. 5-26: See Prob. 16.

17 Refer to the circuit shown in Fig. 5-27 and determine values at $t = 1$ s for: (a) v_C; (b) v_R; (c) v_{SW}.

Fig. 5-27: See Prob. 17.

18 Construct the exact dual of the circuit in Prob. 1, write the exact dual of the problem statement, and then solve your new problem.

19 At $t = 0$, the left-hand switch in the circuit shown in Fig. 5-28 is thrown down and

the right-hand switch is simultaneously opened. Find the total charge that leaves the lower terminal of the capacitor during the time interval, $0.3 < t < 1$ s.

Fig. 5-28: See Prob. 19.

20 In the circuit shown in Fig. 5-29, the switch moves from A to B at $t = 0$. Find expressions for and sketch $i(t)$ in the interval $-2 < t < 2$ ms.

Fig. 5-29: See Prob. 20.

21 At $t = 0$, the left switch in Fig. 5-30 is closed while the right one moves from A to B. At $t = -2$ μs and $+2$ μs, find: (a) i; (b) the rate at which energy is leaving the capacitor.

Fig. 5-30: See Prob. 21.

22 In the circuit shown in Fig. 5-31: (a) find $v_C(0)$; (b) calculate τ; (c) obtain an expression for $v_C(t)$.

Fig. 5-31: See Prob. 22.

23 Select values for R_1 and R_2 in Fig. 5-32 so that $v_C = 10$ V at $t = 0.5$ ms, and $v_C = 1$ V at $t = 2$ ms.

Fig. 5-32: See Prob. 23.

24 Plot v_R vs. t, $-1 < t < 4$ μs, for the circuit appearing in Fig. 5-33.

Fig. 5-33: See Prob. 24.

25 Derive an expression for $v_C(t)$ in Fig. 5-34 that is valid for: (*a*) $t < 0$; (*b*) $t > 0$.

Fig. 5-34: See Prob. 25.

26 The switch in Fig. 5-35 has been closed for a very long time. Sketch v_x vs. t for $-1 < t < 3$ s.

Fig. 5-35: See Prob. 26.

27 The switch in Fig. 5-36 is moved from A to B at $t = 0$ after being at A for a long time. This places the two capacitors in series, thus allowing the possibility of equal and opposite dc voltages to be trapped on the capacitors. (*a*) Determine $v_{C1}(0^-)$ and $v_{C2}(0^-)$. (*b*) Find $v_{C1}(0^+)$, $v_{C2}(0^+)$, and $v_R(0^+)$. (*c*) Determine the time constant of $v_R(t)$. (*d*) Find $v_R(t)$, $t > 0$. (*e*) Find $i(t)$. (*f*) Find $v_{C1}(t)$ and $v_{C2}(t)$.

from $i(t)$ and the initial values. (g) Show that the stored energy at $t = \infty$ plus the total energy dissipated in the 2.5-kΩ resistor is equal to the energy stored in the capacitors at $t = 0$.

Fig. 5-36: See Prob. 27.

28 The switch in Fig. 5-37 opens at $t = 0$ after having been closed for a long time. Find $v_{30}(t)$, $t > 0$.

Fig. 5-37: See Prob. 28.

29 The circuit shown in Fig. 5-38 has been operating for a long time in the configuration shown. At $t = 0$, a loose piece of copper wire falls on the circuit and short-circuits the upper terminals of four elements, as shown. Find $i_L(t)$, $t > 0$.

Fig. 5-38: See Prob. 29.

30 (a) Find $v_C(0^-)$ as a function of k for the circuit shown in Fig. 5-39. (b) Find $v_C(t)$ as a function of k for $t > 0$. (c) Evaluate your expressions for $k = 0$ and $k = 10^{-3}$.

Fig. 5-39: See Prob. 30.

31 In the circuit of Fig. 5-40, let $i_s = 50$ A for $t < 0$ and zero for $t > 0$. Find $v(t)$ for all t.

Fig. 5-40: See Prob. 31.

Chapter 6
The Application of the
Unit-Step Forcing Function

We have just spent a chapter's worth of our time studying the response of *RL* and *RC* circuits when no source or forcing function was present. We termed this response the natural response because its form depends only on the nature of the circuit. The reason that any response at all is obtained arises from the presence of an initial energy storage within the inductive or capacitive elements in the circuit. In many of the examples and problems, we were confronted with circuits containing sources and switches; we were informed that certain switching operations were performed at $t = 0$ in order to remove all the sources from the circuit, while leaving known amounts of energy stored here and there. In other words, we have been solving problems in which energy sources are suddenly *removed* from the circuit; now we must consider that type of response which results when energy sources are suddenly *applied* to a circuit.

We shall devote this chapter to a study of the response which occurs when the energy sources which are suddenly applied are dc sources. After we have studied sinusoidal and exponential sources, we may then consider the general problem of the sudden application of a more general source. Since every electrical device is intended to be energized at least once, and since most devices are turned on and off many times in the course of their lives, it should be evident that our study will be applicable to many practical cases. Even though we are now restricting ourselves to dc sources, there are still innumerable cases in which these simpler examples correspond to the operation of physical devices. For example, the first circuit we shall analyze may be considered to represent the build-up of the field current when a dc motor is started. The generation and use of the rectangular voltage pulses needed to represent a number or a command in a digital computer provide many examples in the field of electronic or transistor circuitry. Similar circuits are found in the synchronization and sweep circuits of television receivers, in communication systems using pulse modulation, and in radar systems, to name but a few examples. Furthermore, an important part of the analysis of most servomechanisms is the determination of their responses to suddenly applied constant inputs.

6-2
The unit-step
forcing function

We have been speaking of the "sudden application" of an energy source, and by this phrase we imply its application in zero time. The operation of a switch in series with a battery is thus equivalent to a forcing function which is zero up to the instant that the switch is closed and is equal to the battery voltage thereafter. The forcing function has a break or discontinuity at the instant the switch is closed. Certain special forcing functions which are discontinuous or have discontinuous derivatives are called *singularity functions*, the two most important of these singularity functions being the unit-step function and the unit-impulse function. The unit-step function is the subject of this chapter; the unit impulse is discussed in Chaps. 18 and 19.

We define the *unit-step forcing function* as a function of time which is zero for all values of its argument which are less than zero and which is unity for all positive values of its argument. If we let $(t - t_0)$ be the argument and represent the unit-step function by u, then $u(t - t_0)$ must be zero for all values of t less than t_0, and it must be unity for all values of t greater than t_0. At $t = t_0$, $u(t - t_0)$ changes abruptly from 0 to 1. Its value at $t = t_0$ is not defined, but its value is known for all instants of time that are arbitrarily close to $t = t_0$. We often indicate this by writing $u(t_{0-}) = 0$ and $u(t_{0+}) = 1$. The concise mathematical definition of the unit-step forcing function is

$$u(t - t_0) = \begin{cases} 0 & t < t_0 \\ 1 & t > t_0 \end{cases}$$

and it is shown graphically in Fig. 6-1. Note that a vertical line of unit length is shown at $t = t_0$. Although this "riser" is not strictly a part of the definition of the unit step, it is usually shown in each drawing.

We also note that the unit step need not be a *time* function, although our attention in this chapter will be directed only to time functions. For example, $u(x - x_0)$ might be used to denote a *unit-step function* which is not a unit-step *forcing* function because it is not a function of time t. Rather, it is a function of x, where x might be a distance in meters, for example, or a frequency as we shall see in Chap. 18.

Very often in circuit analysis a discontinuity or a switching action takes place at an instant that is defined as $t = 0$. In that case $t_0 = 0$, and we then

Fig. 6-1: The unit-step forcing function, $u(t - t_0)$.

represent the corresponding unit-step forcing function by $u(t - 0)$, or more simply $u(t)$. This is shown in Fig. 6-2. Thus

$$u(t) = \begin{cases} 0 & t < 0 \\ 1 & t > 0 \end{cases}$$

The unit-step forcing function is in itself dimensionless. If we wish it to represent a voltage, it is necessary to multiply $u(t - t_0)$ by some constant voltage, such as V. Thus, $v(t) = Vu(t - t_0)$ is an ideal voltage source which is zero before $t = t_0$ and a constant V after $t = t_0$. This forcing function is shown connected to a general network in Fig. 6-3a. We should now logically ask what physical source is the equivalent of this discontinuous forcing function. By equivalent, we mean simply that the voltage-current characteristics of the two networks are identical. For the step-voltage source of Fig. 6-3a, the voltage-current characteristic is quite simple; the voltage is zero prior to $t = t_0$, it is V after $t = t_0$, and the current may be any (finite) value in either time interval. Our first thoughts might produce the attempt at an equivalent shown in Fig. 6-3b, a dc source V in series with a switch which closes at $t = t_0$. This network is not equivalent for $t < t_0$, however, because the voltage across the battery and switch is completely unspecified in this time interval. The "equivalent" source is an open circuit, and the voltage across it may be anything. After $t = t_0$, the networks are equivalent, and if this is the only time interval in which we are interested, and if the initial currents which flow from the two networks are identical at $t = t_0$, then Fig. 6-3b becomes a useful equivalent of Fig. 6-3a.

In order to obtain an exact equivalent for the voltage-step forcing function, we may provide a single-pole double-throw switch. Before $t = t_0$, the switch serves to ensure zero voltage across the input terminals of the general network. After $t = t_0$, the switch is thrown to provide a constant input voltage V. At $t = t_0$, the voltage is indeterminate (as is the step forcing function), the battery is momentarily short-circuited, and it is fortunate that

Fig. 6-3: (a) A voltage-step forcing function is shown as the source driving a general network. (b) A simple circuit which, although not the exact equivalent of (a), may be used as its equivalent in many cases. (c) An exact equivalent of (a).

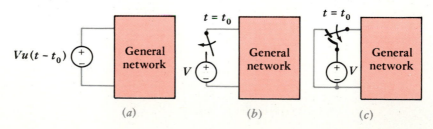

Fig. 6-4: (a) A current-step forcing function is applied to a general network. (b). A simple circuit which, although not the exact equivalent of (a), may be used as its equivalent in many cases.

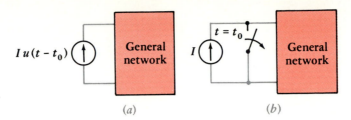

$I\,u(t - t_0)$

General network

$t = t_0$

I

General network

(a)   (b)

we are dealing with mathematical models. This exact equivalent of Fig. 6-3a is shown in Fig. 6-3c.

Before concluding our discussion of equivalence, it is enlightening to consider the exact equivalent of a battery and a switch. What is the voltage-step forcing function which is equivalent to Fig. 6-3b? We are searching for some arrangement which changes suddenly from an open circuit to a constant voltage; a change in resistance is involved, and this is the crux of our difficulty. The step function enables us to change a voltage discontinuously (or a current), but here we need a changing resistance as well. The equivalent therefore must contain a resistance or conductance step function, a passive element which is time-varying. Although we might construct such an element with the unit-step function, it should be apparent that the end product is a switch; a switch is merely a resistance which changes instantaneously from zero to infinite ohms, or vice versa. Thus, we conclude that the exact equivalent of a battery and switch in series must be a battery in series with some representation of a time-varying resistance; no arrangement of voltage- and current-step forcing functions is able to provide us with the exact equivalent.[1]

Figure 6-4a shows a current-step forcing function driving a general network. If we attempt to replace this circuit by a dc source in parallel with a switch (which *opens at* $t = t_0$), we must realize that the circuits are equivalent after $t = t_0$, but the responses are alike after $t = t_0$ only if the initial conditions are the same. Judiciously, then, we may often use the circuits of Figs. 6-4a and b interchangeably. The exact equivalent of Fig. 6-4a is the dual of the circuit of Fig. 6-3c; the exact equivalent of Fig. 6-4b cannot be constructed with current- and voltage-step forcing functions alone.[2]

Some very useful forcing functions may be obtained by manipulating the unit-step forcing function. Let us define a rectangular voltage pulse by the following conditions:

$$v(t) = \begin{cases} 0 & t < t_0 \\ V & t_0 < t < t_1 \\ 0 & t_1 < t \end{cases}$$

[1] An equivalent may always be determined *if some information about the general network is available* (the voltage across the switch for $t < t_0$); we assume no a priori knowledge about the general network.
[2] The equivalent can be drawn if the current through the switch prior to $t = t_0$ is known.

$v(t)$

Fig. 6-5: A useful forcing function, the rectangular voltage pulse.

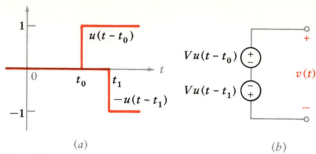

(a) (b)

Fig. 6-6: (a) The unit steps $u(t - t_0)$ and $u(t - t_1)$. (b) A source which yields the rectangular voltage pulse of Fig. 6-5.

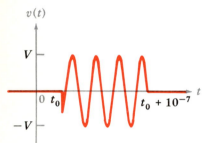

Fig. 6-7: A radio-frequency pulse, described by $v(t) = V[u(t - t_0) - u(t - t_0 - 10^{-7})]$ sin ωt. The sinusoidal frequency within the pulse shown is about 36 MHz, a value which is too low for radar, but about right for constructing legible drawings.

The pulse is drawn in Fig. 6-5. Can this pulse be represented in terms of the unit-step forcing function? Let us consider the difference of the two unit steps $u(t - t_0) - u(t - t_1)$. The two step functions are shown in Fig. 6-6a, and their difference is obviously a rectangular pulse. The source $Vu(t - t_0) - Vu(t - t_1)$ which will provide us with the desired voltage is indicated in Fig. 6-6b.

If we have a sinusoidal voltage source V sin ωt which is suddenly connected to a network at $t = t_0$, then an appropriate voltage forcing function would be $v(t) = Vu(t - t_0)$ sin ωt. If we wish to represent one burst of energy from a radar transmitter, we may turn the sinusoidal source off $\frac{1}{10}$ μs later by a second unit-step forcing function. The voltage pulse is thus

$$v(t) = V[u(t - t_0) - u(t - t_0 - 10^{-7})] \sin \omega t$$

This forcing function is sketched in Fig. 6-7.

As a final introductory remark, we should note that the unit-step forcing function must be considered only as the mathematical model of an actual switching operation. No physical resistor, inductor, or capacitor behaves entirely like its idealized circuit element; we also cannot perform a switching operation in zero time. However, switching times less than 1 ns are common in many circuits, and this time is often sufficiently short compared with the time constants in the rest of the circuit that it may be ignored.

Drill Problems

6-1. Evaluate each of the following at $t = 1.5$: (a) $2u(t - 2) + u(t + 1) + u(t - 1)$; (b) $[u(t - 1) - u(1 - t)]u(t + 1)$; (c) $u(\sin \pi t) - u(t + 4) + u(t)$.

Ans: 2; 1; 0

6-2. For the circuit of Fig. 6-8, find v_1 at $t = :$ (a) −5; (b) −1; (c) 1; (d) 3 s.

Ans: 50; 90; 34; 24 V

Fig. 6-8: See Drill Prob. 6-2.

6-3

A first look at the driven *RL* circuit

We are now ready to subject a simple network to the sudden application of a dc source. The circuit consists of a battery whose voltage is V in series with a switch, a resistor R, and an inductor L. The switch is closed at $t = 0$, as indicated on the circuit diagram of Fig. 6-9a. It is evident that the current $i(t)$ is zero before $t = 0$, and we are therefore able to replace the battery and switch by a voltage-step forcing function $Vu(t)$, which also produces no response prior to $t = 0$. After $t = 0$, the two circuits are obviously identical. Hence, we seek the current $i(t)$ either in the given circuit of Fig. 6-9a or in the equivalent circuit of Fig. 6-9b.

We shall find $i(t)$ at this time by writing the appropriate circuit equation and then solving it by separation of the variables and integration. After we obtain the answer and investigate the two parts of which it is composed, we shall next spend some time (the following section) in learning the general significance of these two terms. We can then construct the solution to this problem very easily; moreover, we shall be able to apply the general principles behind this simpler method to produce more rapid and more meaningful solutions to every problem involving the sudden application of any source. Let us now proceed with the more formal method of solution.

Applying Kirchhoff's voltage law to the circuit of Fig. 6-9b, we have

$$Ri + L\frac{di}{dt} = Vu(t)$$

Since the unit-step forcing function is discontinuous at $t = 0$, we shall first consider the solution for $t < 0$ and then for $t > 0$. It is evident that the application of zero voltage since $t = -\infty$ has not produced any response, and, therefore,

$$i(t) = 0 \qquad t < 0$$

Fig. 6-9: (a) The given circuit. (b) An equivalent circuit, possessing the same response $i(t)$ for all time.

(a) (b)

For positive time, however, $u(t)$ is unity and we must solve the equation

$$Ri + L\frac{di}{dt} = V \qquad t > 0$$

The variables may be separated in several simple algebraic steps, yielding

$$\frac{L\,di}{V - Ri} = dt$$

and each side may be integrated directly,

$$-\frac{L}{R}\ln\,(V - Ri) = t + k$$

In order to evaluate k, an initial condition must be invoked. Prior to $t = 0$, $i(t)$ is zero, and thus $i(0^-) = 0$; since the current in an inductor cannot change by a finite amount in zero time without being associated with an infinite voltage, we thus have $i(0^+) = 0$. Setting $i = 0$ at $t = 0$, we obtain

$$-\frac{L}{R}\ln\,V = k$$

and, hence,

$$-\frac{L}{R}\,[\ln\,(V - Ri) - \ln\,V] = t$$

Rearranging,

$$\frac{V - Ri}{V} = e^{-Rt/L}$$

or

$$i = \frac{V}{R} - \frac{V}{R}\,e^{-Rt/L} \qquad t > 0$$

Thus, an expression for the response valid for all t would be

$$i = \left(\frac{V}{R} - \frac{V}{R}\,e^{-Rt/L}\right)u(t) \tag{1}$$

This is the desired solution, but it has not been obtained in the simplest manner. In order to establish a more direct procedure, let us try to interpret the two terms appearing in (1). The exponential term has the functional form of the natural response of the RL circuit; it is a negative exponential, it approaches zero as time increases, and it is characterized by the time constant L/R. The *functional form* of this part of the response is thus identical with that which is obtained in the source-free circuit. However, the amplitude of this exponential term depends on V. We might generalize, then, that the response will be the sum of two terms, where one term has a functional form which is identical with that of the source-free response, but has an amplitude which depends on the forcing function. Now let us consider the nature of the second part of the response.

Equation (1) also contains a constant term V/R. Why is it present? The answer is simple: the natural response approaches zero as the energy is gradually dissipated, but the total response must not approach zero. Eventually the circuit behaves as a resistor and inductor in series with a battery, and a direct current V/R flows. This current is a part of the response which is directly attributable to the forcing function, and we call it the *forced response*. It is the response which is present a long time after the switch is closed.

The complete response is composed of two parts, the natural response and the forced response. The natural response is a characteristic of the circuit and not of the sources. Its form may be found by considering the source-free circuit, and it has an amplitude which depends on the initial amplitude of the source and the initial energy storage. The forced response has the characteristics of the forcing function; it is found by pretending that all switches have been thrown a long time ago. Since we are presently concerned only with switches and dc sources, the forced response is merely the solution of a simple dc circuit problem.

The reason for the two responses, forced and natural, may also be seen from physical arguments. We know that our circuit will eventually assume the forced response. However, at the instant the switches are thrown, the initial inductor currents (or the voltages across the capacitors in other circuits) will have values which depend only on the energy stored in these elements. These currents or voltages cannot be expected to be the same as the currents and voltages demanded by the forced response. Hence, there must be a transient period during which the currents and voltages change from their given initial values to their required final values. The portion of the response which provides the transition from initial to final values is the natural response (often called the transient response, as we found earlier). If we describe the response of the *source-free* simple RL circuit in these terms, then we should say that the forced response is zero and that the natural response serves to connect the initial response produced by stored energy with the zero value of the forced response. This description is appropriate only for those circuits in which the natural response eventually dies out. This always occurs in physical circuits where some resistance is associated with every element, but there are a number of pathologic circuits in which the natural response is nonvanishing as time becomes infinite; those circuits, in which trapped currents circulate around inductive loops or voltages are trapped in series strings of capacitors, are examples.

Now let us search out the mathematical basis for dividing the response into a natural response and a forced response.

Drill Problem

6-3. A 40-mH inductor, an 80-Ω resistor, and a voltage source, $v_s = 20 + 40u(t)$ V, are in series. Use superposition to find the inductor current magnitude: (*a*) at $t = 0^-$; (*b*) at $t = 0^+$; (*c*) as $t \to \infty$; (*d*) at $t = 1$ ms.

Ans: 0.250; 0.250; 0.750; 0.682 A

6-4
The natural and the forced response

There is also an excellent mathematical reason for considering the complete response to be composed of two parts, the forced response and the natural response. The reason is based on the fact that the solution of any linear differential equation may be expressed as the sum of two parts, the complementary solution (natural response) and the particular solution (forced response). Without delving into the general theory of differential equations, let us consider a general equation of the type met in the previous section,

$$\frac{di}{dt} + Pi = Q$$

or
$$di + Pi\,dt = Q\,dt \qquad (2)$$

We may identify Q as a forcing function and express it as $Q(t)$ to emphasize its general time dependence. In all our circuits, P will be a positive constant, but the remarks that follow about the solution of (2) are equally valid for the cases in which P is a general function of time. Let us simplify the discussion by assuming that P is a positive constant. Later, we shall also assume that Q is constant, thus restricting ourselves to dc forcing functions.

In any standard text on elementary differential equations, it is shown that if both sides of (2) are multiplied by a so-called integrating factor, then each side becomes an exact differential which can be integrated directly to obtain the solution. We are not separating the variables, but merely arranging them in such a way that integration is possible. For this equation, the integrating factor is $e^{\int P\,dt}$ or e^{Pt}, since P is a constant. We multiply each side of the equation by this integrating factor and obtain

$$e^{Pt}\,di + iPe^{Pt}\,dt = Qe^{Pt}\,dt$$

The form of the left side may now be improved when it is recognized as the exact differential of ie^{Pt},

$$d(ie^{Pt}) = e^{Pt}\,di + iPe^{Pt}\,dt$$

and, thus,

$$d(ie^{Pt}) = Qe^{Pt}\,dt$$

We may now integrate each side, finding

$$ie^{Pt} = \int Qe^{Pt}\,dt + A$$

where A is a constant of integration. Since this constant is explicitly shown, we should remember that no integration constant needs to be added to the remaining integral when it is evaluated. Multiplication by e^{-Pt} produces the solution for $i(t)$,

$$i = e^{-Pt} \int Qe^{Pt}\,dt + Ae^{-Pt} \qquad (3)$$

If $Q(t)$, the forcing function, is known, then it remains only to evaluate the integral to obtain the exact functional form for $i(t)$. We shall not evaluate such an integral for each problem, however; instead, we are interested in using (3) as an exemplary solution from which we shall draw several very general conclusions.

We should note first that, for a source-free circuit, Q must be zero, and the solution is the natural response

$$i_n = Ae^{-Pt} \tag{4}$$

We shall find that the constant P is never negative; its value depends only on the passive circuit elements[3] and their interconnection in the circuit. The natural response therefore approaches zero as time increases without limit. It must do so, of course, in the simple RL series circuit because the initial energy is gradually dissipated in the resistor. There are also idealized, non-physical circuits in which P is zero; in these circuits the natural response does not die out, but approaches a constant value, as exemplified by trapped currents or voltages.

We therefore find that one of the two terms making up the complete response has the form of the natural response; it has an amplitude which will depend on (but not usually be equal to) the initial value of the complete response and thus on the initial value of the forcing function also.

We next observe that the first term of (3) depends on the functional form of $Q(t)$, the forcing function. Whenever we have a circuit in which the natural response dies out as t becomes infinite, then this first term must describe the form of the response completely after the natural response has disappeared. This term we shall call the *forced response;* it is also called the steady-state response, the particular solution, or the particular integral.

For the present, we have elected to consider only those problems involving the sudden application of dc sources, and $Q(t)$ is therefore a constant for all values of time after the switch in Fig. 6-9a has been closed. If we wish, we can now evaluate the integral in (3), obtaining the forced response

$$i_f = \frac{Q}{P}$$

or the complete response

$$i(t) = \frac{Q}{P} + Ae^{-Pt}$$

For the RL series circuit, Q/P is the constant current V/R and $1/P$ is the time constant τ. We should see that the forced response might have been obtained without evaluating the integral, because it must be the complete response at infinite time; it is merely the source voltage divided by the series resistance. The forced response is thus obtained by inspection.

In the following section we shall attempt to find the complete response for

---

[3] If the circuit contains a dependent source or a negative resistance, it is possible for P to be negative.

several *RL* circuits by obtaining the natural and forced responses and then adding them.

6-4. Assume that the voltage source, $v_s = 10e^{-20t}u(t)$ V, is in series with a 25-Ω resistor and a 100-mH inductor. Substitute the applicable expressions into (3) and find the magnitude of the current at $t =$: (*a*) 0^+; (*b*) 3 ms; (*c*) 10 ms; (*d*) 50 ms.
Ans: 0; 204; 320; 159.9 mA

6-5
RL circuits

Let us use the simple *RL* series circuit to illustrate how to determine the complete response by the addition of the natural and forced responses. This circuit, shown in Fig. 6-10, has been analyzed earlier, but by a longer method. The desired response is the current $i(t)$, and we first express this current as the sum of the natural and the forced current,

$$i = i_n + i_f$$

The functional form of the natural response must be the same as that obtained without any sources. We therefore replace the step-voltage source by a short circuit and recognize the old *RL* series loop. Thus,

$$i_n = Ae^{-Rt/L}$$

where the amplitude A is yet to be determined.

We next consider the forced response, that part of the response which depends upon the nature of the forcing function itself. In this particular problem the forced response must be constant because the source is a constant V for all positive values of time. After the natural response has died out, therefore, there can be no voltage across the inductor; hence, a voltage V appears across R, and the forced response is simply

$$i_f = \frac{V}{R}$$

Note that the forced response is determined completely; there is no unknown amplitude. We next combine the two responses

$$i = Ae^{-Rt/L} + \frac{V}{R}$$

and apply the initial condition to evaluate A. The current is zero prior to $t = 0$, and it cannot change value instantaneously since it is the current flowing through an inductor. Thus, the current is zero immediately after $t = 0$, and

$$0 = A + \frac{V}{R}$$

and

$$i = \frac{V}{R}(1 - e^{-Rt/L}) \tag{5}$$

Note carefully that A is not the initial value of i, since $A = -V/R$, while $i(0) = 0$. In Chap. 5, where the circuits were source-free, A was indeed the

$i(t)$

$Vu(t)$ R L

Fig. 6-10: A series *RL* circuit which is used to illustrate the method by which the complete response is obtained as the sum of the natural and forced response.

Fig. 6-11: The current expressed by (5) is shown graphically. A line extending the initial slope meets the constant forced response at $t = \tau$.

initial value of the response. When forcing functions are present, however, we must first find the initial value of the response, and then substitute this into the equation for the complete response to find A.

This response is plotted in Fig. 6-11, and we can see the manner in which the current builds up from its initial value of zero to its final value of V/R. The transition is effectively accomplished in a time 3τ. If our circuit represents the field coil of a large dc motor, we might assign $L = 10$ H and $R = 20\ \Omega$, obtaining $\tau = 0.5$ s. The field current is thus established in about 1.5 s. In one time constant, the current has attained 63.2 percent of its final value.

Now let us apply this method to a more complicated circuit. In Fig. 6-12, a circuit is shown containing a dc voltage source as well as a step-voltage source. Let us determine $i(t)$ for all values of time. We might choose to replace everything to the left of the inductor by the Thévenin equivalent, but instead let us merely recognize the form of that equivalent as a resistor in series with some voltage source. The circuit contains only one energy-storage element, the inductor, and the natural response is therefore a negative exponential as before,

$$i = i_f + i_n$$

where

$$i_n = Ae^{-t/2} \qquad t > 0$$

since

$$\tau = \frac{L}{R_{eq}} = \frac{3}{1.5} = 2$$

The forced response must be that produced by a constant voltage of 100 V. The forced response is constant, and no voltage is present across the inductor; it behaves as a short circuit and, therefore,

$$i_f = \frac{100}{2} = 50$$

Fig. 6-12: A circuit used as an example.

Fig. 6-13: The response $i(t)$ of the circuit shown in Fig. 6-12 is sketched for values of time less and greater than zero.

Thus,

$$i = 50 + Ae^{-0.5t} \qquad t > 0$$

In order to evaluate A, we must establish the initial value of the inductor current. Prior to $t = 0$, this current is 25 A, and it cannot change instantaneously. Thus,

$$25 = 50 + A \qquad \text{or} \qquad A = -25$$

Hence,

$$i = 50 - 25e^{-0.5t} \qquad t > 0$$

We complete the solution by also stating

$$i = 25 \qquad t < 0$$

or by writing a single expression valid for all t,

$$i = 25 + 25(1 - e^{-0.5t})u(t) \qquad A$$

The complete response is sketched in Fig. 6-13. Note how the natural response serves to connect the response for $t < 0$ with the constant forced response.

As a final example of this method by which the complete response of any circuit subjected to a transient may be written down almost by inspection, let us apply a rectangular voltage pulse of amplitude V and duration t_0 to the simple RL series circuit. We represent the forcing function as the sum of two step-voltage sources $Vu(t)$ and $-Vu(t - t_0)$, as indicated in Figs. 6-14a and b, and plan to obtain the response by using the superposition principle. Suppose we designate that part of $i(t)$ whch is due to the upper source $Vu(t)$

Fig. 6-14: (a) A rectangular voltage pulse which is to be used as the forcing function in a simple series RL circuit. (b) The series RL circuit, showing the representation of the forcing function by the series combination of two independent voltage-step sources. The current $i(t)$ is desired.

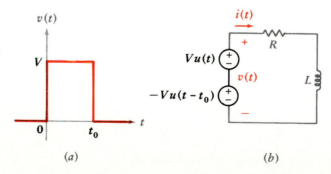

(a) (b)

acting alone by the symbol $i_1(t)$ and then let $i_2(t)$ represent that part due to $-Vu(t - t_0)$ acting alone. Then,

$$i(t) = i_1(t) + i_2(t)$$

Our object is now to write each of the partial responses i_1 and i_2 as the sum of a natural and a forced response. The response $i_1(t)$ is familiar; this problem was solved two or three pages back,

$$i_1(t) = \frac{V}{R} (1 - e^{-Rt/L}) \qquad t > 0$$

Note that the range of t, $t > 0$, in which this solution is valid, is indicated.

We now turn our attention to the lower source and its response $i_2(t)$. Only the polarity of the source and the time of its application are different. There is thus no need to determine the form of the natural response and the forced response; the solution for $i_1(t)$ enables us to write

$$i_2(t) = -\frac{V}{R} (1 - e^{-R(t - t_0)/L}) \qquad t > t_0$$

where the applicable range of t, $t > t_0$, must again be indicated.

We now add the two solutions, but do so carefully, since each is valid over a different interval of time. Thus,

$$i(t) = \frac{V}{R} (1 - e^{-Rt/L}) \qquad 0 < t < t_0$$

$$i(t) = \frac{V}{R} (1 - e^{-Rt/L}) - \frac{V}{R} (1 - e^{-R(t - t_0)/L}) \qquad t > t_0$$

or
$$i(t) = \frac{V}{R} e^{-Rt/L} (e^{Rt_0/L} - 1) \qquad t > t_0$$

The solution is completed by stating that $i(t)$ is zero for negative t and sketching the response as a function of time. The type of curve obtained depends upon the relative values of t_0 and the time constant τ. Two possible curves are shown in Fig. 6-15. The left curve is drawn for the case where the time constant is only one-half as large as the length of the applied pulse; the rising portion of the exponential has therefore almost reached V/R before the

Fig. 6-15: Two response curves shown for the ci Fig. 6-14b. In (a) lected as $t_0/2$, ar as $2t_0$.

(a)   (b)

decaying exponential begins. The opposite situation is shown to the right; there, the time constant is twice t_0 and the response never has a chance to reach the larger amplitudes.

The procedure we have been using to find the response of an *RL* circuit after dc sources have been switched on or off or in or out of the circuit at some instant of time, say $t = 0$, is summarized in the following. We assume that the circuit is reducible to a single equivalent resistance R_{eq} in series with a single equivalent inductance L_{eq} when all independent sources are set equal to zero. The response we seek is represented by $f(t)$.

1 With all independent sources killed, simplify the circuit to determine R_{eq}, L_{eq}, and the time constant $\tau = L_{eq}/R_{eq}$.
2 Viewing L_{eq} as a short circuit, use dc-analysis methods to find $i_L(0^-)$, the inductor current just prior to the discontinuity.
3 Again viewing L_{eq} as a short circuit, use dc-analysis methods to find the forced response. This is the value approached by $f(t)$ as $t \to \infty$; we represent it by $f(\infty)$.
4 Write the total response as the sum of the forced and natural responses: $f(t) = f(\infty) + Ae^{-t/\tau}$.
5 Find $f(0^+)$ by using the condition that $i_L(0^+) = i_L(0^-)$. If desired, L_{eq} may be replaced by a current source $i_L(0^+)$ [an open circuit if $i_L(0^+) = 0$] for this calculation. With the exception of inductor currents (and capacitor voltages), other currents and voltages in the circuit may change abruptly.
6 Then $f(0^+) = f(\infty) + A$, and $f(t) = f(\infty) + [f(0^+) - f(\infty)]e^{-t/\tau}$, or

total response = final value + (initial value − final value)$e^{-t/\tau}$.

Drill Problems

6-5. For the circuit shown in Fig. 6-16a, find: (a) $i_1(0^-)$, $i_2(0^-)$, and $v_2(0^-)$; (b) $i_1(0^+)$, $i_2(0^+)$, and $v_2(0^+)$; (c) $i_1(\infty)$, $i_2(\infty)$, and $v_2(\infty)$; (d) $i_1(0.15)$, $i_2(0.15)$, and $v_2(0.15)$.
Ans: 0, 0, 0; 1.2 A, 0, 120 V; 6 A, 6 A, 0; 4.93 A, 4.66 A, 26.8 V

Fig. 6-16: See Drill Probs. 6-5 and 6-6.

(a)   (b)

6-6. Find v_R for the circuit of Fig. 6-16b at $t =$: (a) −1 ms; (b) 0^-; (c) 0^+; (d) 1 ms.
Ans: 0; 0; 8; 11.46 V

6-6
RC circuits

The complete response of any *RC* circuit may also be obtained as the sum of the natural and the forced response. Figure 6-17 shows a friendly little circuit containing two batteries, four resistors, one capacitor, and a switch. The switch is assumed to have been in position *a* for a long time, or, in other words, the natural response which resulted from the original excitation of the circuit has decayed to a negligible amplitude, leaving only a forced response caused by the 120-V source. We are asked for $v_C(t)$, and we thus begin by finding this forced response prior to $t = 0$ when the switch is in position *a*. The voltages throughout the circuit are all constant, and there is thus no current through the capacitor. Simple voltage division gives us the initial voltage,

$$v_C(0) = \frac{50}{50 + 10} 120 = 100$$

Since the capacitor voltage cannot change instantaneously, this voltage is equally valid at $t = 0^-$ and $t = 0^+$.

The switch is now thrown to *b*, and the complete response is

$$v_C = v_{Cf} + v_{Cn}$$

The form of the natural response is obtained by replacing the 50-V source by a short circuit and evaluating the equivalent resistance,

$$v_{Cn} = Ae^{-t/R_{eq}C}$$

$$R_{eq} = \frac{1}{\frac{1}{50} + \frac{1}{200} + \frac{1}{60}} = 24$$

or

$$v_{Cn} = Ae^{-t/1.2}$$

In order to evaluate the forced response with the switch at *b*, we wait until all the voltages and currents have stopped changing, thus treating the capacitor as an open circuit, and use voltage division once more,

$$v_{Cf} = \frac{(50)(200)/(50 + 200)}{60 + (50)(200)/(50 + 200)} 50 = 20$$

Thus,

$$v_C = 20 + Ae^{-t/1.2}$$

Fig. 6-17: An *RC* circuit in which the complete responses v_C and *i* are obtained by adding a forced response and a natural response.

Fig. 6-18: The responses (a) v_C and (b) i are plotted as functions of time for the circuit of Fig. 6-17.

(a)

(b)

and from the initial condition already obtained,

$$100 = 20 + A$$

or

$$v_C = 20 + 80e^{-t/1.2} \qquad t > 0$$

This response is sketched in Fig. 6-18a; again the natural response is seen to form a transition from the initial to the final response.

Finally, let us calculate some response that need not remain constant during the instant of switching, such as $i(t)$ in Fig. 6-17. With the contact at a, it is evident that $i = \frac{50}{260} = 0.192$ A. When the switch moves to position b, the forced response for this current becomes

$$i_f = \frac{50}{60 + (50)(200)/(50 + 200)} \frac{50}{50 + 200} = 0.1$$

The form of the natural response is the same as that which we already determined for the capacitor voltage,

$$i_n = Ae^{-t/1.2}$$

Combining the forced and natural responses, we obtain

$$i = 0.1 + Ae^{-t/1.2}$$

To evaluate A, we need to know $i(0^+)$. This is found by fixing our attention on the energy-storage element, here the capacitor, for the fact that v_C must remain 100 V during the switching interval is the governing condition establishing the other currents and voltages at $t = 0^+$. Since $v_C(0^+) = 100$ V, and since the capacitor is in parallel with the 200-Ω resistor, we find $i(0^+) = 0.5$ A, $A = 0.4$, and thus,

$$i(t) = 0.192 \qquad t < 0$$

$$i(t) = 0.1 + 0.4e^{-t/1.2} \qquad t > 0$$

or

$$i(t) = 0.192 + (-0.092 + 0.4e^{-t/1.2})u(t) \qquad \text{A}$$

where the latter expression is correct for all t.

The complete response for all t may also be written concisely by using $u(-t)$, which is unity for $t < 0$ and 0 for $t > 0$. Thus,

$$i(t) = 0.192u(-t) + (0.1 + 0.4e^{-t/1.2})u(t) \qquad \text{A}$$

This response is sketched in Fig. 6-18*b*. Note that only four numbers are needed to write the functional form of the response for this single energy-storage-element circuit or to prepare the sketch: the constant value prior to switching (0.192 A), the instantaneous value just after switching (0.5 A), the constant forced response (0.1 A), and the time constant (1.2 s). The appropriate negative exponential function is then easily written or drawn.

We conclude by listing the duals of the statements given at the end of Sec. 6-5.

The procedure we have been using to find the response of an *RC* circuit after dc sources have been switched on or off or in or out of the circuit at some instant of time, say $t = 0$, is summarized in the following. We assume that the circuit is reducible to a single equivalent resistance R_{eq} in parallel with a single equivalent capacitance C_{eq} when all independent sources are set equal to zero. The response we seek is represented by $f(t)$.

---

1 With all independent sources killed, simplify the circuit to determine R_{eq}, C_{eq}, and the time constant $\tau = R_{eq}C_{eq}$.

2 Viewing C_{eq} as an open circuit, use dc-analysis methods to find $v_C(0^-)$, the capacitor voltage just prior to the discontinuity.

3 Again viewing C_{eq} as an open circuit, use dc-analysis methods to find the forced response. This is the value approached by $f(t)$ as $t \to \infty$; we represent it by $f(\infty)$.

4 Write the total response as the sum of the forced and natural responses: $f(t) = f(\infty) + Ae^{-t/\tau}$.

5 Find $f(0^+)$ by using the condition that $v_C(0^+) = v_C(0^-)$. If desired, C_{eq} may be replaced by a voltage source $v_C(0^+)$ [a short circuit if $v_C(0^+) = 0$] for this calculation. With the exception of capacitor voltages (and inductor currents), other voltages and currents in the circuit may change abruptly.

6 Then $f(0^+) = f(\infty) + A$, and $f(t) = f(\infty) + [f(0^+) - f(\infty)]e^{-t/\tau}$, or

total response = final value + (initial value − final value)$e^{-t/\tau}$.

Drill Problem

6-7. At $t = 1$ ms in the circuit of Fig. 6-19, find the value of: (*a*) v_C; (*b*) i_C; (*c*) v_x.　　　　　　　　　　　　　　　　　　　　*Ans:* 22.6 V; 0.294 A; 26.1 V

Fig. 6-19: See Drill Prob. 6-7.

Problems

1 At $t = 5$ s, determine the value of: (a) $u(t - 6)$; (b) $u(6 - t)$; (c) $2 - u(t^2 - 10t + 24)$; (d) $(t - 2)u(t - 2)$; (e) $2^{t-3}(-1)^{1+u(t-3)}$; (f) $-tu(-t) + tu(t)$; (g) i_x in Fig. 6-20.

Fig. 6-20: See Prob. 1.

5 u (t - 3) V
13 Ω   3 Ω
20 u (t) V   1 Ω   6 u (-t) A
i_x

2 Sketch the time function, $f(t) = [u(t + 5) - u(t - 5)][u(\cos \frac{1}{2}\pi t) + u(\sin \pi t)]$.

3 Find the value of each of the following time functions at $t = 1.5$ s: (a) $2u(t + 1) - 3u(t + 2) + 4u(t - 3)$; (b) $2u(1 - t) + 5[u(t)]^2$; (c) $4u(\cos 3t)$; (d) $4 \cos[3u(t)]$. (e) A voltage is 10 V for $t < 0$, $2e^{-0.4t}$ V for $0 < t < 5$ s, and zero for $t > 5$ s. Express it as a single function of time using unit-step forcing functions.

4 The "general network" in each part of Fig. 6-3 consists of a voltage source, $10u(2 - t)$ V, + reference connected to the upper terminal, in series with a 5-Ω resistor. If $V = 5$ V and $t_0 = -1$ s in the external circuits, sketch the current entering the network at the upper terminal in each case.

5 Find v_x in the network of Fig. 6-21 at 1-s intervals from $t = -0.5$ s to $t = 3.5$ s.

Fig. 6-21: See Prob. 5.

25 Ω   40 u (-t) V
2 u (3 - t) A v_x 100 Ω   75 Ω   u (t - 1) A
t = 2 s

6 The switch in Fig. 6-22 is at position A for $t < 0$. At $t = 0$ it moves to B, and then to C at $t = 4$ s and on to D at $t = 6$ s, where it remains. Sketch $v(t)$ as a function of time and express it as a sum of step forcing functions.

Fig. 6-22: See Prob. 6.

4 V
A
B
2 V   C D   100 Ω   v(t)
10 V   3 V

7 The excitation applied to the *RL* circuit in Fig. 6-9*b* may be divided into two parts as follows: $Vu(t) = V - Vu(-t)$. The first component is a dc source of *V* volts, while the second produces a source-free response after $t = 0$ of the type described in Chap. 5. Show that superposition of these two responses leads to a total current identical to Eq. (1) of Sec. 6-3.

8 For the circuit of Fig. 6-23, sketch $i_L(t)$ and $v_L(t)$ vs. *t*, $-3\tau < t < 3\tau$.

Fig. 6-23: See Prob. 8.

9 Replace the network to the left of the inductor in Fig. 6-24 with its Thévenin equivalent and then determine $i(t)$ for $t > 0$.

Fig. 6-24: See Prob. 9.

10 The switch in Fig. 6-25 has been open for a very long time. It closes at $t = 0$. (*a*) After replacing everything to the left of the inductor by its Thévenin equivalent, find $i_L(t)$ and sketch it for $t > 0$. (*b*) Write an expression for $i_L(t)$ that is correct for all *t*.

Fig. 6-25: See Prob. 10.

11 Equation (3) of Sec. 6-4 represents the general solution of the driven *RL* series circuit, where *Q* is a function of time and *A* and *P* are constants. Let $R = 200 \ \Omega$ and $L = 5$ H, and find $i(t)$ if the voltage forcing function $LQ(t)$ is: (*a*) $5u(t)$ V; (*b*) $5 + 5u(t)$ V; (*c*) $5u(t) \sin 40t$ V.

12 The switch in Fig. 6-26 has been closed for a long time. It is opened at $t = 0$. Find an expression for $i_L(t)$ for $t < 0$ and $t > 0$, and sketch it over the time interval, $-1 < t < 1$ s.

Fig. 6-26: See Prob. 12.

13 The switch in Fig. 6-27 is thrown down at $t = 0$. Find: (a) i at $t = 0.1$ s; (b) the maximum voltage magnitude that will appear across the inductor.

Fig. 6-27: See Prob. 13.

14 Determine an expression for i_L in Fig. 6-28 for all nonnegative t and then make an accurate sketch of i_L vs. t, showing scale values on both axes.

Fig. 6-28: See Prob. 14.

15 Find $i_L(t)$ for $t > 0$ in the circuit shown in Fig. 6-29.

Fig. 6-29: See Prob. 15.

16 If the switch in Fig. 6-30 is thrown down at $t = 0$, then: (*a*) at what value of t will i become zero, and (*b*) what will be the value of v when $i = 0$?

Fig. 6-30: See Prob. 16.

17 Calculate values for both i_L and v_R, 16 s before and 16 s after the switch in Fig. 6-31 closes.

Fig. 6-31: See Prob. 17.

18 Find i_L in the circuit of Fig. 6-32 at $t = :$ (*a*) -5 ms; (*b*) 5 ms; (*c*) 10 ms.

Fig. 6-32: See Prob. 18.

19 In Fig. 6-33, let $v_s = 2u(t)$ V and $i_s = 80u(t - 0.0002)$ mA. Find $i_L(t)$ for all t.

Fig. 6-33: See Prob. 19.

20 (a) Determine an expression for $v_C(t)$ in the circuit shown in Fig. 6-34. Graph it as a function of t for both negative and positive values of t. (b) Repeat for $i_5(t)$.

Fig. 6-34: See Prob. 20.

21 For the circuit shown in Fig. 6-35, the switch has been open for a long time. (a) Calculate $v_C(-2 \text{ ms})$ and $v_C(2 \text{ ms})$. (b) If the switch is now kept closed for a long time, find v_C two milliseconds after it suddenly opens.

Fig. 6-35: See Prob. 21.

22 Let $v_s = 9 - 12u(t)$ V for the circuit shown in Fig. 6-36. (a) Find $v_C(t)$ for all values of time. (b) Sketch v_C vs. t. (c) At what value of t is $v_C = 0$?

Fig. 6-36: See Prob. 22.

23 In the circuit of Fig. 6-37, obtain a single expression for $v_x(t)$ that is valid for all t.

Fig. 6-37: See Prob. 23.

24 The switch in the circuit of Fig. 6-38 has been open for a long time; it closes at $t = 0$. Find expressions valid for all t for: (a) $v_C(t)$; (b) $v_2(t)$.

Fig. 6-38: See Prob. 24.

25 The switch in Fig. 6-39 has been at A for a long time. At $t = 0$ it is moved to B. Find $v_C(t)$ for $t < 0$ and $t > 0$.

Fig. 6-39: See Probs. 25 and 26.

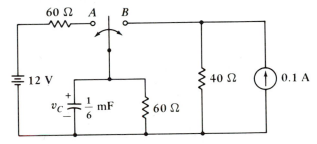

26 The switch in Fig. 6-39 has been at B for a long time. At $t = 0$ it is moved to A. Find $v_C(t)$ for $t < 0$ and $t > 0$.

27 The switch in Fig. 6-40 has been open for a long time. (a) Obtain an expression for $v_C(t)$ as a function of t, $t > 0$. (b) At what time is $v_C(t) = 0$?

Fig. 6-40: See Prob. 27.

28 The current source, $5u(t)$ A, the parallel combination of 20 Ω and 0.8 H, and the parallel combination of 20 Ω and C are all in series. Sketch the magnitude of the voltage across the current source as a function of time if $C = $: (a) 5 mF; (b) 2 mF.

29 Find the instant of time at which the capacitor voltage in the circuit of Fig. 6-41 is equal to zero.

Fig. 6-41: See Prob. 29.

30 The switch of Fig. 6-42 has been at A for a long time. It is moved to B at $t = 0$ and back to A at $t = 1$ s. Find R_1 and R_2 so that $v_C = 7.5$ V at $t = 1$ and 1 V at $t = 1.001$ s.

Fig. 6-42: See Prob. 30.

31 If the op-amp in Fig. 6-43 is assumed to be ideal, find $v_o(t)$ for all t.

Fig. 6-43: See Prob. 31.

32 The op-amp shown in Fig. 6-44 is assumed to be ideal. Find $v_x(t)$.

Fig. 6-44: See Prob. 32.

33 If the op-amp shown in Fig. 6-45 is ideal, find an expression for $v_o(t)$ valid for all t.

Fig. 6-45: See Prob. 33.

34 Assume that the op-amp of Fig. 6-46 is ideal and find v_{out} if $v_{in}(t) = u(t)$ V and $v_{out}(0) = 0$.

Fig. 6-46: See Prob. 34.

Chapter 7
The *RLC* Circuit

It would be very pleasant to learn that the detailed study we have just completed for the *RL* and *RC* circuits will make the analysis of the *RLC* circuit a simple task; unfortunately the analysis remains difficult. The presence of inductance and capacitance in the same circuit produces at least a *second-order system,* that is, one that is characterized by a linear differential equation including a second-order derivative, or by two simultaneous linear first-order differential equations. This increase in order will make it necessary to evaluate two arbitrary constants. Furthermore, it will be necessary to determine initial conditions for derivatives. And finally, we shall see that the presence of inductance and capacitance in the same circuit leads to a response which takes on different functional forms for circuits which have the same configuration but different element values. With this cheerful news, let us quickly review the methods and results we found useful for first-order systems, in order that we may extend this information as intelligently as possible to the second-order system.

We first considered the source-free first-order system. The response was termed the natural response, and it was determined completely by the types of passive elements in the network, by the manner in which they were interconnected, and by the initial conditions which were established by the stored energy. The natural response was invariably an exponentially decreasing function of time, and this response approached a constant value as time became infinite. The constant was usually zero, except in those circuits where paralleled inductors or series-connected capacitors allowed trapped currents or voltages to appear.

The addition of sources to the first-order system resulted in a two-part response, the familiar natural response and an additional term we called the forced response. This latter term was intimately related to the forcing function; its functional form was that of the forcing function itself, plus the integral and first derivative of the forcing function.[1] Since we treated only a

---

[1] Higher-order derivatives will appear in higher-order systems and, strictly speaking, we should say that all derivatives are present, although possibly with zero amplitude. Forcing functions which do not possess a finite number of different derivatives are exceptions which we shall not consider; the singularity functions are exceptions to the exceptions.

constant forcing function, we have not needed to devote much attention to the proper form of the forced response; this problem will not arise until sinusoidal forcing functions are encountered in the following chapter. To the known forced response, we added the correct expression for the natural response, complete except for a multiplicative constant. This constant was evaluated to make the total response fit the prescribed initial conditions.

We now turn to circuits which are characterized by linear second-order differential equations. Our first task is the determination of the natural response. This is most conveniently done by considering initially the source-free circuit. We may then include dc sources, switches, or step sources in the circuit, representing the total response once again as the sum of the natural response and the (usually constant) forced response. The second-order system that we are about to analyze is fundamentally the same as any lumped-constant mechanical second-order system. Our results, for example, will be of direct use to a mechanical engineer who is interested in the displacement of a spring-supported mass subjected to viscous damping, a system which can approximate the vertical motion of an automobile, with shock absorbers providing the damping; or to one who is excited by the motion of a simple pendulum or a torsional pendulum. Our results are still applicable, although less directly, to any distributed-parameter second-order system, such as a short-circuited transmission line, a diving board, a flute, or the ecology of the lemming.

7-2
The source-free parallel circuit

Our first goal is the determination of the natural response of a simple circuit formed by connecting R, L, and C in parallel; this modest goal will be reached after completing this and the next three sections. This particular combination of ideal elements is a suitable model for portions of many communications networks. It represents, for example, an important part of some of the electronic amplifiers found in every radio receiver, and it enables the amplifiers to produce a large voltage amplification over a narrow band of signal frequencies and nearly zero amplification outside this band. Frequency selectivity of this kind enables us to listen to the transmission of one station while rejecting the transmission of any other station. Other applications include the use of parallel RLC circuits in multiplexing filters, harmonic suppression filters, and so forth. But even a simple discussion of these principles requires an understanding of such terms as resonance, frequency response, and impedance which we have not yet discussed. Let it suffice to say, therefore, that an understanding of the natural behavior of the parallel RLC circuit is fundamentally important to future studies of communications networks and filter design.

When a physical inductor is connected in parallel with a capacitor, and the inductor has associated with it a nonzero ohmic resistance, the resulting network can be shown to have an equivalent circuit model like that shown in Fig. 7-1. Energy losses in the physical inductor are taken into account by the presence of the ideal resistor whose resistance R is dependent upon (but not equal to) the ohmic resistance of the inductor.

Fig. 7-1: The source-free parallel RLC circuit.

In the following analysis we shall assume that energy may be stored initially in both the inductor and the capacitor, and thus nonzero initial values of both inductor current and capacitor voltage may be present. With reference to the circuit of Fig. 7-1, we may then write the single nodal equation

$$\frac{v}{R} + \frac{1}{L}\int_{t_0}^{t} v \, dt - i(t_0) + C\frac{dv}{dt} = 0 \tag{1}$$

Note that the minus sign is a consequence of the assumed direction for i. We must solve (1) subject to the initial conditions

$$i(0^+) = I_0 \tag{2}$$

$$v(0^+) = V_0 \tag{3}$$

When both sides of (1) are differentiated once with respect to time, the result is the linear second-order homogeneous differential equation

$$C\frac{d^2v}{dt^2} + \frac{1}{R}\frac{dv}{dt} + \frac{1}{L}v = 0 \tag{4}$$

whose solution $v(t)$ is the desired natural response.

There are a number of interesting ways to solve (4). These methods we shall leave to a course in differential equations, selecting only the quickest and simplest method to use now. We shall assume a solution, relying upon our intuition and modest experience to select one of the several possible forms which are suitable. Our experience with the first-order equation should suggest that we at least try the exponential form once more. Moreover, the form of (4) indicates that this may work, because we must add three terms, the second derivative, first derivative, and the function itself, each multiplied by a constant factor, and achieve a sum of zero. A function whose derivatives have the same form as the function itself is obviously a sensible choice. With every hope for success, then, we assume

$$v = Ae^{st} \tag{5}$$

where we shall be as general as possible by allowing A and s to be complex numbers if necessary.[2] Substituting (5) into (4), we obtain

$$CAs^2e^{st} + \frac{1}{R}Ase^{st} + \frac{1}{L}Ae^{st} = 0$$

or

$$Ae^{st}\left(Cs^2 + \frac{1}{R}s + \frac{1}{L}\right) = 0$$

In order for this equation to be satisfied for all time, at least one of the three factors must be zero. If either of the first two factors is put equal to zero, then $v(t) = 0$. This is a trivial solution of the differential equation

---

[2] There is no cause for panic. Complex numbers will appear in this chapter only in an introductory way in the derivations. Their use as a tool will be necessary in Chap. 9.

which cannot satisfy our given initial conditions. We therefore equate the remaining factor to zero,

$$Cs^2 + \frac{s}{R} + \frac{1}{L} = 0 \tag{6}$$

This equation is usually called the *auxiliary equation* or the *characteristic equation* by mathematicians. If it can be satisfied, then our assumed solution is correct. Since (6) is a quadratic equation, there are two solutions, identified as s_1 and s_2,

$$s_1 = -\frac{1}{2RC} + \sqrt{\left(\frac{1}{2RC}\right)^2 - \frac{1}{LC}} \tag{7}$$

and

$$s_2 = -\frac{1}{2RC} - \sqrt{\left(\frac{1}{2RC}\right)^2 - \frac{1}{LC}} \tag{8}$$

If either of these two values is used for s in the assumed solution, then that solution satisfies the given differential equation; it thus becomes a valid solution of the differential equation.

Let us assume that we replace s by s_1 in (5), obtaining

$$v_1 = A_1 e^{s_1 t}$$

and, similarly,

$$v_2 = A_2 e^{s_2 t}$$

The former satisfies the differential equation

$$C \frac{d^2 v_1}{dt^2} + \frac{1}{R} \frac{dv_1}{dt} + \frac{1}{L} v_1 = 0$$

and the latter satisfies

$$C \frac{d^2 v_2}{dt^2} + \frac{1}{R} \frac{dv_2}{dt} + \frac{1}{L} v_2 = 0$$

Adding these two differential equations and combining similar terms,

$$C \frac{d^2(v_1 + v_2)}{dt^2} + \frac{1}{R} \frac{d(v_1 + v_2)}{dt} + \frac{1}{L} (v_1 + v_2) = 0$$

linearity triumphs, and it is seen that the sum of the two solutions is also a solution. We thus have the form of the natural response

$$v = A_1 e^{s_1 t} + A_2 e^{s_2 t} \tag{9}$$

where s_1 and s_2 are given by (7) and (8); A_1 and A_2 are two arbitrary constants which are to be selected to satisfy the two specified initial conditions.

The form of the natural response as given above can hardly be expected to bring forth any expressions of interested amazement, for, in its present form, it offers little insight into the nature of the curve we might obtain if $v(t)$ were plotted as a function of time. The relative amplitudes of A_1 and A_2, for example, will certainly be important in determining the shape of the response curve. Furthermore, the constants s_1 and s_2 can be real numbers or conjugate complex numbers, depending upon the values of R, L, and C in the given network. These two cases will produce fundamentally different response forms. Therefore it will be helpful to make some simplifying substitutions in (9) for the sake of conceptual clarity.

Since the exponents s_1t and s_2t must be dimensionless, s_1 and s_2 must have the unit of some dimensionless quantity "per second." From (7) and (8) it is apparent that the units of $1/2RC$ and $1/\sqrt{LC}$ must also be s^{-1}. Units of this type are called *frequencies*. Although we shall expand this concept in much more detail in Chap. 13, we shall introduce several of the terms now. Let us represent $1/\sqrt{LC}$ by ω_0 (omega),

$$\omega_0 = \frac{1}{\sqrt{LC}} \tag{10}$$

and reserve the term *resonant frequency*[3] for it. On the other hand, we shall call $1/2RC$ the *neper frequency* or the *exponential damping coefficient* and represent it by the symbol α (alpha),

$$\alpha = \frac{1}{2RC} \tag{11}$$

This latter descriptive expression is used because α is a measure of how rapidly the natural response decays or damps out to its steady final value (usually zero).[4] Finally, s, s_1, and s_2, which are quantities that will form the basis for some of our later work, will be called *complex frequencies*.

We should note that s_1, s_2, α, and ω_0 are merely symbols used to simplify the discussion of *RLC* circuits; they are not mysterious new properties of any kind. It is easier, for example, to say "alpha" than it is to say "the reciprocal of $2RC$."

Let us collect these results. The natural response of the parallel *RLC* circuit is

$$v(t) = A_1 e^{s_1 t} + A_2 e^{s_2 t} \tag{9}$$

where

$$s_1 = -\alpha + \sqrt{\alpha^2 - \omega_0^2} \tag{12}$$

$$s_2 = -\alpha - \sqrt{\alpha^2 - \omega_0^2} \tag{13}$$

$$\alpha = \frac{1}{2RC} \tag{11}$$

$$\omega_0 = \frac{1}{\sqrt{LC}} \tag{10}$$

and A_1 and A_2 must be found by applying the given initial conditions.

---

[3] More accurately, the resonant *radian* frequency.

[4] The ratio of α to ω_0 is called the *damping ratio* by control system engineers and is designated by ζ (zeta). We shall encounter it again in Chap. 14.

The response described by the equations above applies not only to the voltage $v(t)$, but also to the current flowing in each of the three circuit elements. The values of the constants, A_1 and A_2 for $v(t)$, will of course be different for the currents.

It is now apparent that the nature of the response depends upon the relative magnitudes of α and ω_0. The radical appearing in the expressions for s_1 and s_2 will be real when α is greater than ω_0, imaginary when α is less than ω_0, and zero when α and ω_0 are equal. Each of these cases will be considered separately in the following three sections.

Drill Problem	**7-1.** A parallel *RLC* circuit contains a 100-mH inductor and has the values $\omega_0 = 2000$ rad/s and $\alpha = 2500$ s^{-1}. Find: (*a*) C; (*b*) R; (*c*) s_1; (*d*) s_2. *Ans:* 2.5 μF; 80 Ω; -1000 s^{-1}; -4000 s^{-1}

7-3
The overdamped
parallel *RLC* circuit

A comparison of (10) and (11) in Sec. 7-2 shows that α will be greater than ω_0 if $LC > 4R^2C^2$. In this case the radical used in calculating s_1 and s_2 will be real, and both s_1 and s_2 will be real. Moreover, the following inequalities,

$$\sqrt{\alpha^2 - \omega_0{}^2} < \alpha$$

$$(-\alpha - \sqrt{\alpha^2 - \omega_0{}^2}) < (-\alpha + \sqrt{\alpha^2 - \omega_0{}^2}) < 0$$

may be applied to (12) and (13) to show that both s_1 and s_2 are *negative* real numbers. Thus, the response $v(t)$ can be expressed as the (algebraic) sum of two decreasing exponential terms, both of which approach zero as time increases without limit. In fact, since the absolute value of s_2 is larger than that of s_1, the term containing s_2 has the more rapid rate of decrease and, for large values of time, we may write the limiting expression

$$v(t) \rightarrow A_1 e^{s_1 t} \rightarrow 0 \qquad \text{as } t \rightarrow \infty$$

In order to discuss the method by which the arbitrary constants A_1 and A_2 are selected to conform with the initial conditions, and in order to provide a typical example of a response curve, let us turn to a numerical example. We shall select a parallel *RLC* circuit for which $R = 6\,\Omega$, $L = 7$ H, and, for ease of computation, the impractically large value $C = \frac{1}{42}$ F; the initial energy storage is specified by choosing an initial voltage across the circuit $v(0) = 0$ and an initial inductor current $i(0) = 10$ A, where v and i are defined in Fig. 7-2.

Fig. 7-2: A parallel *RLC* circuit used as a numerical example. The circuit is overdamped.

We may easily determine the values of the several parameters

$$\alpha = 3.5 \qquad \omega_0 = \sqrt{6}$$
$$s_1 = -1 \qquad s_2 = -6$$

(all s^{-1})

and immediately write the general form of the natural response

$$v(t) = A_1 e^{-t} + A_2 e^{-6t} \qquad (14)$$

Only the evaluation of the two constants A_1 and A_2 remains. If we knew the response $v(t)$ at two different values of time, these two pairs of values could be substituted in (14) and A_1 and A_2 easily found. However, we know only one instantaneous value of $v(t)$,

$$v(0) = 0$$

and, therefore,

$$0 = A_1 + A_2 \qquad (15)$$

We shall obtain a second equation relating A_1 and A_2 by taking the derivative of $v(t)$ with respect to time in (14), determining the initial value of this derivative through the use of the remaining initial condition $i(0) = 10$, and equating the results. Taking the derivative of both sides of (14),

$$\frac{dv}{dt} = -A_1 e^{-t} - 6A_2 e^{-6t}$$

evaluating the derivative at $t = 0$,

$$\frac{dv}{dt}\bigg|_{t=0} = -A_1 - 6A_2$$

we next pause to consider how the initial value of the derivative can be found numerically. This next step is always suggested by the derivative itself; dv/dt suggests capacitor current, for

$$i_C = C \frac{dv}{dt}$$

Thus,

$$\frac{dv}{dt}\bigg|_{t=0} = \frac{i_C(0)}{C} = \frac{i(0) + i_R(0)}{C} = \frac{i(0)}{C}$$
$$= 420 \quad \text{V/s}$$

since zero initial voltage across the resistor requires zero initial current through it. We thus have our second equation

$$420 = -A_1 - 6A_2 \qquad (16)$$

and simultaneous solution of (15) and (16) provides the two amplitudes $A_1 = 84$ and $A_2 = -84$. Therefore, the final numerical solution for the natural response is

$$v(t) = 84(e^{-t} - e^{-6t}) \qquad (17)$$

The evaluation of A_1 and A_2 for other conditions of initial energy storage, including initial energy storage in the capacitor, is considered in the first drill problem following this section.

Let us see what information we can glean from (17) without calculating unduly. We note that $v(t)$ is zero at $t = 0$, a comforting check on our original assumption. We may also interpret the first exponential term as having a time constant of 1 s and the other exponential, a time constant of $\frac{1}{6}$ s. Each starts with unity amplitude, but the latter decays more rapidly; $v(t)$ is thus never negative. As time becomes infinite, each term approaches zero, and the response itself dies out as it should. We therefore have a response curve which is zero at $t = 0$, zero at $t = \infty$, and is never negative; since it is not everywhere zero, it must possess at least one maximum, and this is not a difficult point to determine exactly. We differentiate the response

$$\frac{dv}{dt} = 84(-e^{-t} + 6e^{-6t})$$

set the derivative equal to zero to determine the time t_m at which the voltage becomes maximum,

$$0 = -e^{-t_m} + 6e^{-6t_m}$$

manipulate once,

$$e^{5t_m} = 6$$

and obtain

$$t_m = 0.358 \quad \text{s}$$

and

$$v(t_m) = 48.9 \quad \text{V}$$

A reasonable sketch of the response may be made by plotting the two exponential terms $84e^{-t}$ and $84e^{-6t}$ and then taking their difference. The usefulness of this technique is indicated by the curves of Fig. 7-3; the two exponentials are shown lightly, and their difference, the total response $v(t)$, is drawn as a colored line. The curves also verify our previous prediction

Fig. 7-3: The response $v(t) = 84(e^{-t} - e^{-6t})$ of the network which is shown in Fig. 7-2.

$i(0) = 10$ A

$v(0) = 0$

$\alpha = 3.5$

$\omega_0 = \sqrt{6}$

Overdamping

that the functional behavior of $v(t)$ for very large t is $84e^{-t}$, the exponential term containing the smaller magnitude of s_1 and s_2.

Another question that frequently arises in the consideration of network response is concerned with the length of time it takes for the transient part of the response to disappear (or damp out). In practice, it is often desirable to have this transient response approach zero as rapidly as possible, that is, to minimize the *settling time* t_s. Theoretically, of course, t_s is infinite, because $v(t)$ never settles to zero in a finite time. However, a negligible response is present after the magnitude of $v(t)$ has settled to values that remain less than 1 percent of its maximum absolute value $|v_m|$. The time which is required for this to occur we define as the settling time.[5] Since $|v_m| = v_m = 48.9$ V for our example, the settling time is the time required for the response to drop to 0.489 V. Substituting this value for $v(t)$ into (17) and neglecting the second exponential term, known to be negligible here, the settling time is found to be 5.15 s.

In comparison with the responses which we shall obtain in the following two sections, this is a comparatively large settling time; the damping time is overly long, and the response is called overdamped. We shall refer to the case for which α is greater than ω_0 as the *overdamped* case. Now let us see what happens as α is decreased.

Drill Problems

7-2. If the initial voltage across a parallel *RLC* circuit is not zero, then the resultant value of the initial resistor current must be considered in evaluating the initial value of dv/dt. For the numerical example considered in the foregoing section, let $v(0) = 30$ V and $i(0) = 10$ A. Find: (*a*) A_1; (*b*) A_2; (*c*) $v_{\max}$.
Ans: 78; −48; 50.1 V

7-3. After being open for more than 2 hours, the switch in Fig. 7-4 closes at $t = 0$. Find: (*a*) $v_C(0^+)$; (*b*) $i_L(0^+)$; (*c*) $i_R(0^+)$; (*d*) $i_C(0^+)$; (*e*) $v_C(10 \ \mu\text{s})$.
Ans: 60 V; −0.3 A; 0.3 A; 0; 45.8 V

Fig. 7-4: See Drill Prob. 7-3.

7-4
Critical damping

The overdamped case is characterized by

$$\alpha > \omega_0$$

or

$$LC > 4R^2C^2$$

---

[5] The 1 percent level is somewhat arbitrary. Some people prefer 2 percent or 5 percent.

and leads to negative real values for s_1 and s_2 and to a response expressed as the algebraic sum of two negative exponentials. Typical forms of the response $v(t)$ are obtained through the numerical example in the last section and in the drill problems following it.

Now let us adjust the element values until α and ω_0 are equal. This is a very special case which is termed _critical damping_. If we were to attempt to construct a parallel _RLC_ circuit that is critically damped, we would be attempting an impossible task; we could never make α exactly equal to ω_0. The result of such an attempt would result in an overdamped circuit, or an underdamped circuit, which we shall discuss in the next section. For completeness, however, we shall discuss the critically damped circuit here because it shows an interesting transition between overdamping and underdamping.

Critical damping is achieved when

$$\left. \begin{array}{c} \alpha = \omega_0 \\ LC = 4R^2C^2 \\ L = 4R^2C \end{array} \right\} \text{ critical damping}$$

or

or

It is obvious that we may produce critical damping by changing the value of any of the three elements in the numerical example discussed above. We shall select R, increasing its value until critical damping is obtained, and thus leave ω_0 unchanged. The necessary value of R is $7\sqrt{6}/2\ \Omega$; L is still 7 H, and C remains $\frac{1}{42}$ F. We thus find

$$\alpha = \omega_0 = \sqrt{6}$$
$$s_1 = s_2 = -\sqrt{6}$$

and we blithely construct the response as the sum of the two exponentials,

$$v(t) \overset{?}{=} A_1 e^{-\sqrt{6}t} + A_2 e^{-\sqrt{6}t}$$

which may be written as

$$v(t) \overset{?}{=} A_3 e^{-\sqrt{6}t}$$

At this point, some of us should feel we have lost our way. We have a response which contains only one arbitrary constant, but there are two initial conditions $v(0) = 0$ and $i(0) = 10$ which must be satisfied by this single constant. This is in general impossible. In our case, for example, the first initial condition requires A_3 to be zero, and it is then impossible to satisfy the second initial condition.

Our mathematics and our electricity have been unimpeachable; therefore, if a mistake has not led to our difficulties, we must have begun with an incorrect assumption, and only one assumption has been made. We originally hypothesized that the differential equation could be solved by assuming an exponential solution, and this turns out to be incorrect for this single special case of critical damping. When $\alpha = \omega_0$, the differential equation, (4), becomes

$$\frac{d^2v}{dt^2} + 2\,\alpha\,\frac{dv}{dt} + \alpha^2 v = 0$$

The solution of this equation is not a tremendously difficult process, but we shall avoid developing it here since the equation is a standard type found in the usual differential-equation texts. The solution is

$$v = e^{-\alpha t}(A_1 t + A_2) \tag{18}$$

It should be noted that the solution may be expressed as the sum of two terms, where one term is the familiar negative exponential but the second is t times a negative exponential. We should also note that the solution contains the *two* expected arbitrary constants.

Let us now complete our numerical example. After we substitute the known value of α into (18),

$$v = A_1 t e^{-\sqrt{6}t} + A_2 e^{-\sqrt{6}t}$$

we establish the values of A_1 and A_2 by first imposing the initial condition on $v(t)$ itself, $v(0) = 0$. Thus, $A_2 = 0$. This simple result occurs because the initial value of the response was selected as zero; the more general case, which leads to an equation determining A_2, may be expected to arise in the drill problems. The second initial condition must be applied to the derivative dv/dt just as in the overdamped case. We therefore differentiate, remembering that $A_2 = 0$,

$$\frac{dv}{dt} = A_1 t(-\sqrt{6})e^{-\sqrt{6}t} + A_1 e^{-\sqrt{6}t}$$

evaluate at $t = 0$,

$$\left.\frac{dv}{dt}\right|_{t=0} = A_1$$

express the derivative in terms of the initial capacitor current,

$$\left.\frac{dv}{dt}\right|_{t=0} = \frac{i_C(0)}{C} = \frac{i_R(0)}{C} + \frac{i(0)}{C}$$

where reference directions for i_C, i_R, and i are defined in Fig. 7-2. Thus,

$$A_1 = 420$$

The response is, therefore,

$$v(t) = 420t e^{-2.45t} \tag{19}$$

Before plotting this response in detail, let us again try to anticipate its form by qualitative reasoning. The specified initial value is zero, and (19) concurs. It is not immediately apparent that the response also approaches zero as t

becomes infinitely large because $te^{-2.45t}$ is an indeterminate form. However, this minor obstacle is easily overcome by use of L'Hôpital's rule. Thus,

$$\lim_{t \to \infty} v(t) = 420 \lim_{t \to \infty} \frac{t}{e^{2.45t}} = 420 \lim_{t \to \infty} \frac{1}{2.45 e^{2.45t}} = 0$$

and once again we have a response which begins and ends at zero and has positive values at all other times. A maximum value v_m again occurs at time t_m; for our example,

$$t_m = 0.408 \quad \text{s} \quad \text{and} \quad v_m = 63.1 \quad \text{V}$$

This maximum is larger than that obtained in the overdamped case and is a result of the smaller losses that occur in the larger resistor; the time of the maximum response is slightly later than it was with overdamping. The settling time may also be determined by solving

$$\frac{v_m}{100} = 420 t_s e^{-2.45 t_s}$$

for t_s (by trial-and-error methods or a calculator's SOLVE routine),

$$t_s = 3.12 \quad \text{s}$$

which is a considerably smaller value than arose in the overdamped case (5.15 s). As a matter of fact, it can be shown that, for given values of L and C, the selection of that value of R which provides critical damping will always give a shorter settling time than any choice of R which produces an overdamped response. However, a slight improvement (reduction) in settling time may be obtained by a further slight increase in resistance; a slightly underdamped response which will undershoot the zero axis before it dies out will yield the shortest settling time.

The response curve for critical damping is drawn in Fig. 7-5; it may be compared with the overdamped (and underdamped) case by reference to Fig. 7-8.

Fig. 7-5: The response $v(t) = 420\, te^{-2.45t}$ of the network shown in Fig. 7-2 with R changed to provide critical damping.

Drill Problem

7-4. The switch in Fig. 7-6 closes at $t = 0$ after being open for a long time. (a) Find L so that the circuit will be critically damped. Using this value of L, determine: (b) $i_L(0^+)$; (c) $v(0^+)$; (d) $i_L(1 \text{ ms})$.   *Ans:* 320 mH; 1 A; 0; 993 mA

Fig. 7-6: See Drill Prob. 7-4.

7-5
The underdamped parallel *RLC* circuit

Let us continue the process begun in the last section by increasing R once more. Thus, the damping coefficient α decreases while ω_0 remains constant, α^2 becomes smaller than ω_0^2, and the radicand appearing in the expressions for s_1 and s_2 becomes negative. This causes the response to take on quite a different character, but it is fortunately not necessary to return to the basic differential equation again. By using complex numbers, the exponential response turns into a damped sinusoidal response; this response is composed entirely of real quantities, the complex quantities being necessary only for the derivation.[6]

We therefore begin with the exponential form

$$v(t) = A_1 e^{s_1 t} + A_2 e^{s_2 t}$$

where

$$s_{1,2} = -\alpha \pm \sqrt{\alpha^2 - \omega_0^2}$$

and then let

$$\sqrt{\alpha^2 - \omega_0^2} = \sqrt{-1}\,\sqrt{\omega_0^2 - \alpha^2} = j\sqrt{\omega_0^2 - \alpha^2}$$

where

$$j = \sqrt{-1}$$

We now take the new radical, which is real for the underdamped case, and call it ω_d, the *natural resonant frequency*,

$$\omega_d = \sqrt{\omega_0^2 - \alpha^2}$$

Collecting, the response may now be written as

$$v(t) = e^{-\alpha t}(A_1 e^{j\omega_d t} + A_2 e^{-j\omega_d t})$$

---

[6] An introduction to the use of complex numbers appears in Chap. 9 and Appendix 4. At that time we shall emphasize the more general nature of complex quantities by identifying them with boldface type; no special symbolism need be adopted in these few pages.

or in the longer but equivalent form,

$$v(t) = e^{-\alpha t}\left\{(A_1 + A_2)\left[\frac{e^{j\omega_d t} + e^{-j\omega_d t}}{2}\right] + j(A_1 - A_2)\left[\frac{e^{j\omega_d t} - e^{-j\omega_d t}}{j2}\right]\right\}$$

Two of the most important identities in the field of complex numbers, identities which are later proved in Appendix 4, may now be readily applied. The first bracket in the above equation is identically equal to cos $\omega_d t$, and the second bracket is identically sin $\omega_d t$. Hence,

$$v(t) = e^{-\alpha t}[(A_1 + A_2)\cos \omega_d t + j(A_1 - A_2)\sin \omega_d t]$$

and the multiplying factors may be assigned new symbols,

$$v(t) = e^{-\alpha t}(B_1 \cos \omega_d t + B_2 \sin \omega_d t) \tag{20}$$

If we are dealing with the underdamped case, we have now left complex numbers behind. This is true since α, ω_d, and t are real quantities, $v(t)$ itself must be a real quantity (which might be presented on an oscilloscope, a voltmeter, or a sheet of graph paper), and thus B_1 and B_2 are real quantities. Equation (20) is the desired functional form for the underdamped response, and its validity may be checked by direct substitution into the original differential equation; this exercise is left to the doubters. The two real constants B_1 and B_2 are again selected to fit the given initial conditions.

Let us increase the resistance in our example from $7\sqrt{6}/2$ or 8.57 Ω to 10.5 Ω; L and C are unchanged. Thus,

$$\alpha = \frac{1}{2RC} = 2$$

$$\omega_0 = \frac{1}{\sqrt{LC}} = \sqrt{6}$$

and

$$\omega_d = \sqrt{\omega_0^2 - \alpha^2} = \sqrt{2} \quad \text{(rad/s)}$$

Except for the evaluation of the arbitrary constants, the response is now known,

$$v(t) = e^{-2t}(B_1 \cos \sqrt{2}t + B_2 \sin \sqrt{2}t)$$

The determination of the two constants proceeds as before. If we again assume that $v(0) = 0$ and $i(0) = 10$, then B_1 must be zero. Hence

$$v(t) = B_2 e^{-2t} \sin \sqrt{2}t$$

The derivative is

$$\frac{dv}{dt} = \sqrt{2}B_2 e^{-2t} \cos \sqrt{2}t - 2B_2 e^{-2t} \sin \sqrt{2}t$$

and at $t = 0$ it becomes

$$\left.\frac{dv}{dt}\right|_{t=0} = \sqrt{2}B_2 = \frac{i_C(0)}{C} = 420$$

where i_C is defined in Fig. 7-2. Therefore,

$$v(t) = 210\sqrt{2}\,e^{-2t} \sin \sqrt{2}t$$

Notice that, as before, this response function has an initial value of zero, because of the initial voltage condition we imposed, and a final value of zero, because the exponential term vanishes for large values of t. As t increases from zero through small positive values, $v(t)$ increases as $210\sqrt{2} \sin \sqrt{2}t$ because the exponential term remains essentially equal to unity. But at a time t_m, the exponential function begins to decrease more rapidly than $\sin \sqrt{2}t$ is increasing; so $v(t)$ reaches a maximum v_m and begins to decrease. We should note that t_m is not the value of t for which $\sin \sqrt{2}t$ is a maximum, but must occur somewhat before $\sin \sqrt{2}t$ reaches its maximum. When $t = \pi/\sqrt{2}$, $v(t)$ is zero; for the interval $\pi/\sqrt{2} < t < \sqrt{2}\pi$ the response is negative, becoming zero again at $t = \sqrt{2}\pi$. Thus $v(t)$ is an *oscillatory* function of time and crosses the time axis an infinite number of times at $t = n\pi/\sqrt{2}$, where n is any positive integer. In our example, however, the response is only slightly underdamped and the exponential term causes the function to die out so rapidly that most of the zero crossings will not be evident in a sketch.

The oscillatory nature of the response becomes more noticeable as α decreases. If α is zero, which corresponds to an infinitely large resistance, then $v(t)$ is an undamped sinusoid which oscillates with constant amplitude. There is never a time at which $v(t)$ drops below and stays below 1 percent of its maximum value; the settling time is therefore infinite. This is not perpetual motion; we have merely assumed an initial energy in the circuit and have not provided any means to dissipate this energy. It is transferred from its initial location in the inductor to the capacitor, then returns to the inductor, and so on, forever.

A noninfinite R in the parallel RLC circuit acts as a kind of electrical transfer agent. Every time energy is transferred from L to C or from C to L, the agent exacts its commission. Before long, the agent has taken all the energy, wantonly dissipating every last joule. The L and C are left without a joule of their own, sans voltage and sans current.

Fig. 7-7: The response $v(t) = 210\sqrt{2}\,e^{-2t} \sin \sqrt{2}\,t$ of the network shown in Fig. 7-2 with R increased to produce an underdamped response.

Actual parallel *RLC* circuits can be made to have effective values of *R* so large that a natural undamped sinusoidal response can be maintained for years without supplying any additional energy. We can also build active networks which introduce a sufficient amount of energy during each oscillation of $v(t)$ so that a sinusoidal response which is nearly perfect can be maintained for as long as we wish. This circuit is a sinusoidal oscillator, or signal generator, which is an important laboratory instrument. Section 7-8 develops an operational-amplifier version of such a sinusoidal oscillator.

Returning to our specific numerical problem, differentiation locates the first maximum of $v(t)$,

$$v_{m1} = 71.8 \quad \text{V} \qquad \text{at } t_{m1} = 0.435 \quad \text{s}$$

the succeeding minimum,

$$v_{m2} = -0.845 \quad \text{V} \qquad \text{at } t_{m2} = 2.657 \quad \text{s}$$

and so on. The response curve is shown in Fig. 7-7.

The settling time may be obtained by a trial-and-error solution, and it turns out to be 2.92 s, somewhat smaller than for critical damping. Note that t_s is *greater* than t_{m2} because the magnitude of v_{m2} is greater than 1 percent of the magnitude of v_{m1}. This suggests that a slight decrease in *R* would reduce the magnitude of the undershoot and permit t_s to be less than t_{m2}. Problem 30 at the end of this chapter is provided so that each student may satisfy the inevitable curiosity prompted by these remarks and determine the numerical value of the minimum possible settling time for this circuit, as well as the value of *R* that produces it.

The overdamped, critically damped, and underdamped responses for this network are shown on the same graph in Fig. 7-8. A comparison of the three curves makes these general conclusions plausible:

1 When the damping is changed by adjusting the size of the parallel resistance, the maximum magnitude of the response is greater with smaller damping.

2 The response becomes oscillatory when underdamping is present, and the minimum settling time is obtained for slight underdamping.

Fig. 7-8: Three response curves for a parallel *RLC* circuit for which $\omega_0 = \sqrt{6}$, $v(0) = 0$, $i(0) = 10$ A, and α is 3.5 (overdamped), 2.45 (critically damped), and 2 (underdamped).

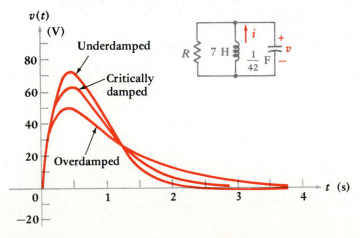

Drill Problem

7-5. The three sources in the circuit of Fig. 7-9 turn off at $t = 0$. Find: (a) $v_C(0)$; (b) $i_L(0)$; (c) $i_R(0^+)$; (d) $i_L(10 \text{ ms})$. *Ans:* -100 V; 7 A; -5 A; 2.64 A

Fig. 7-9: See Drill Prob. 7-5.

7-6
The source-free series *RLC* circuit

We now wish to determine the natural response of a circuit model composed of an ideal resistor, an ideal inductor, and an ideal capacitor connected in series. The ideal resistor may represent a physical resistor connected into a series *LC* or *RLC* circuit; it may represent the ohmic losses and the losses in the ferromagnetic core of the inductor; or it may be used to represent all these and other energy-absorbing devices. In a special case, the resistance of the ideal resistor may even be exactly equal to the measured resistance of the wire out of which the physical inductor is made.

The series *RLC* circuit is the dual of the parallel *RLC* circuit, and this single fact is sufficient to make its analysis a trivial affair. Figure 7-10*a* shows the series circuit. The fundamental integrodifferential equation is

$$L \frac{di}{dt} + Ri + \frac{1}{C} \int_{t_0}^{t} i \, dt - v_C(t_0) = 0$$

and should be compared with the analogous equation for the parallel *RLC* circuit, drawn again in Fig. 7-10*b*,

$$C \frac{dv}{dt} + \frac{1}{R} v + \frac{1}{L} \int_{t_0}^{t} v \, dt - i_L(t_0) = 0$$

The second-order equations obtained by differentiating each of these equations with respect to time are also duals,

$$L \frac{d^2 i}{dt^2} + R \frac{di}{dt} + \frac{i}{C} = 0 \qquad (21)$$

$$C \frac{d^2 v}{dt^2} + \frac{1}{R} \frac{dv}{dt} + \frac{v}{L} = 0 \qquad (22)$$

Fig. 7-10: (a) The series *RLC* circuit which is the dual of (b) the parallel *RLC* circuit. Element values are, of course, not identical in (a) and (b).

(a)　　　(b)

It is apparent that our complete discussion of the parallel RLC circuit is directly applicable to the series RLC circuit; the initial conditions on capacitor voltage and inductor current are equivalent to the initial conditions on inductor current and capacitor voltage; the voltage response becomes a current response. It is quite possible to reread the previous four sections (including the drill problems) using dual language and thereby obtain a complete description of the series RLC circuit.[7] This process, however, is apt to induce a mild neurosis after the first few paragraphs and does not really seem to be necessary.

A brief résumé of the series circuit response is easily collected. In terms of the circuit shown in Fig. 7-10a, the overdamped response is

$$i(t) = A_1 e^{s_1 t} + A_2 e^{s_2 t}$$

where

$$s_{1,2} = -\frac{R}{2L} \pm \sqrt{\left(\frac{R}{2L}\right)^2 - \frac{1}{LC}}$$
$$= -\alpha \pm \sqrt{\alpha^2 - \omega_0^2}$$

and thus

$$\alpha = \frac{R}{2L}$$

$$\omega_0 = \frac{1}{\sqrt{LC}}$$

The form of the critically damped response is

$$i(t) = e^{-\alpha t}(A_1 t + A_2)$$

and the underdamped case may be written

$$i(t) = e^{-\alpha t}(B_1 \cos \omega_d t + B_2 \sin \omega_d t)$$

where

$$\omega_d = \sqrt{\omega_0^2 - \alpha^2}$$

It is evident that if we work in terms of the parameters α, ω_0, and ω_d the mathematical forms of the responses for the dual situations are identical. An increase in α in either the series or parallel circuit, while keeping ω_0 constant, tends toward an overdamped response. The only caution that we need exert is in the computation of α, which is $1/2RC$ for the parallel circuit and

---

[7] In fact, during the writing of the first edition of this text, the authors wrote these first sections to describe the series RLC circuit. But, after deciding that it would be better to present the analysis of the more practical parallel RLC circuit first, it was easy to go back to the original writing and replace it with its dual. The numerical values of several of the elements were also scaled, a process described later in Chap. 14.

$R/2L$ for the series circuit; thus, α is increased by increasing the series resistance or decreasing the parallel resistance. Lest we forget,

$$\alpha = \frac{1}{2RC} \qquad \alpha = \frac{R}{2L}$$

$$\text{(parallel)} \qquad \text{(series)}$$

As a numerical example, let us consider a series RLC circuit in which $L = 1$ H, $R = 2$ kΩ, $C = \frac{1}{401}$ μF, $i(0) = 2$ mA, and $v_C(0) = 2$ V. We find that α is 1000 and ω_0 is 20 025, and thus an underdamped response is indicated; we therefore calculate the value of ω_d and obtain 20 000. Except for the evaluation of the two arbitrary constants, the response is now known:

$$i(t) = e^{-1000t}(B_1 \cos 20\,000t + B_2 \sin 20\,000t)$$

By applying the initial value of the current, we find

$$B_1 = 0.002$$

and thus

$$i(t) = e^{-1000t}(0.002 \cos 20\,000t + B_2 \sin 20\,000t)$$

The remaining initial condition must be applied to the derivative; thus,

$$\frac{di}{dt} = e^{-1000t}(-40 \sin 20\,000t + 20\,000B_2 \cos 20\,000t$$

$$- 2 \cos 20\,000t - 1000B_2 \sin 20\,000t)$$

$$\frac{di}{dt}\bigg|_{t=0} = 20\,000B_2 - 2 = \frac{v_L(0)}{L} = \frac{v_C(0) - Ri(0)}{L}$$

$$= \frac{2 - 2000(0.002)}{1} = -2$$

or
$$B_2 = 0$$

The desired response is, therefore,

$$i(t) = 2e^{-1000t} \cos 20\,000t \qquad \text{mA}$$

This response is more oscillatory, or shows less damping, than any we have considered up to this time, and the direct calculation of enough points to graph a smooth response curve is a tedious undertaking. A good sketch may be made by first drawing in the two portions of the exponential *envelope*, $0.002e^{-1000t}$ and $-0.002e^{-1000t}$, as shown by the broken lines in Fig. 7-11. The location of the quarter-cycle points of the sinusoidal wave at $20\,000t = 0$, $\pi/2$, π, etc., or $t = 0.078\,54k$ ms, $k = 0, 1, 2, \ldots$, by light marks on the time axis then permits the oscillatory curve to be sketched in quickly.

The settling time can be determined easily here by using the upper portion of the envelope. That is, $0.002e^{-1000t_s} = 0.01(0.002)$, and $t_s = 461$ μs is the approximate value that is usually used.

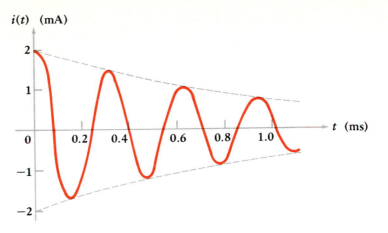

Fig. 7-11: The current response in an underdamped series *RLC* circuit for which $\alpha = 1000$ s^{-1}, $\omega_d = 20\,000$ s^{-1}, $i(0) = 2$ mA, $v_C(0) = 2$ V. The graphical construction is simplified by drawing in the envelope, shown as a pair of broken lines.

Drill Problems

7-6. Refer to the circuit of Fig. 7-12 and determine values for: (*a*) $v_C(0)$; (*b*) $i_L(0)$; (*c*) $v_x(0^-)$; (*d*) $v_x(0^+)$; (*e*) $v_C(20$ ms). *Ans:* 45 V; 3 A; 0; 15 V; 29.7 V

7-7. Change the 15-Ω resistor in Fig. 7-12 to 4.6 Ω and repeat Drill Prob. 7-6. *Ans:* 13.8 V; 3 A; 0; 15 V; -0.412 V

Fig. 7-12: See Drill Probs. 7-6 and 7-7.

7-7
The complete response of the *RLC* circuit

We must now consider those *RLC* circuits in which dc sources are switched into the network and produce forced responses that do not vanish as time becomes infinite. The general solution is obtained by the same procedure that was followed in *RL* and *RC* circuits: the forced response is determined completely; the natural response is obtained as a suitable functional form containing the appropriate number of arbitrary constants; the complete response is written as the sum of the forced and the natural responses; and the initial conditions are then determined and applied to the complete response to find the values of the constants. It is this last step which is quite frequently the most troublesome to students. Consequently, although the determination of the initial conditions is basically no different for a circuit containing dc sources than it is for the source-free circuits which we have already covered in some detail, this topic will receive particular emphasis in the example below.

Most of the confusion in determining and applying the initial conditions arises for the simple reason that we do not have laid down for us a rigorous set of rules to follow. At some point in each analysis there usually arises a situation in which some thinking is involved that is more or less unique to that particular problem. This originality and flexibility of thought, as simple as it is to achieve after several problems' worth of practice, is the source of the difficulty.

The complete response (arbitrarily assumed to be a voltage response) of a second-order system consists of a forced response, which is a constant for dc excitation,

$$v_f(t) = V_f$$

and a natural response

$$v_n(t) = Ae^{s_1t} + Be^{s_2t}$$

Thus,

$$v(t) = V_f + Ae^{s_1t} + Be^{s_2t}$$

We shall now assume that s_1, s_2, and V_f have already been determined from the circuit and the given forcing functions; A and B remain to be found. The last equation shows the functional interdependence of A, B, v, and t, and substitution of the known value of v at $t = 0^+$ thus provides us with a single equation relating A and B. This is the easy part. Another relationship between A and B is necessary, unfortunately, and this is normally obtained by taking the derivative of the response,

$$\frac{dv}{dt} = 0 + s_1Ae^{s_1t} + s_2Be^{s_2t}$$

and inserting in it the known value of dv/dt at $t = 0^+$. There is no reason that this process cannot be continued; a second derivative might be taken, and a third relationship between A and B will then result if the value of d^2v/dt^2 at $t = 0^+$ is used. This value is *not* usually known, however, in a second-order system; as a matter of fact, we are much more likely to use this method to *find* the initial value of the second derivative if we should need it. We thus have only two equations relating A and B, and these may be solved simultaneously to evaluate the two constants.

The only remaining problem is that of determining the values of v and dv/dt at $t = 0^+$. Let us suppose that v is a capacitor voltage, $v = v_C$. Since $i_C = C \, dv_C/dt$, we should recognize the relationship between the initial value of dv/dt and the initial value of some capacitor current. If we can establish a value for this initial capacitor current, then we shall automatically establish the value of dv/dt. Students are usually able to get $v(0^+)$ very easily, but are inclined to stumble a bit in finding the initial value of dv/dt. If we had selected an inductor current as our response, then the initial value of di_L/dt must be intimately related to the initial value of some inductor voltage. Variables other than capacitor voltages and inductor currents are determined by expressing their initial values and the initial values of their derivatives in terms of the corresponding values for v_C and i_L.

We shall illustrate the procedure and find all these values by the careful analysis of the circuit shown in Fig. 7-13. To simplify the analysis, an unreal-

Fig. 7-13: An *RLC* circuit which is used to illustrate several procedures by which the initial conditions may be obtained. The desired response is nominally taken to be $v_C(t)$.

istically large capacitance is used again. Our object is to find the value of each current and voltage at both $t = 0^-$ and $t = 0^+$; with these quantities known, the required derivatives may be easily calculated. We shall employ a logical step-by-step method first.

At $t = 0^-$, only the right-hand current source is active. Moreover, the circuit is assumed to have been in this state forever, and all currents and voltages are constant. In other words, a steady-state condition has been reached and the resultant forced response has the form of the forcing function, its integral, and its derivatives. The integral of the forcing function, a linearly increasing function of time, is not present in this circuit, for it can occur only when a constant current is forced through a capacitor or a constant voltage is maintained across an inductor. This situation should not normally be present because the capacitor voltage or inductor current would assume an unrealistic infinite value at $t = 0^-$. Continuing, then, a constant current through the inductor requires zero voltage across it,

$$v_L(0^-) = 0$$

and a constant voltage across the capacitor requires zero current through it,

$$i_C(0^-) = 0$$

We then apply Kirchhoff's current law to the right node to obtain

$$i_R(0^-) = -5 \quad \text{A}$$

which also yields

$$v_R(0^-) = -150 \quad \text{V}$$

We may now use Kirchhoff's voltage law around the central mesh, finding

$$v_C(0^-) = 150 \quad \text{V}$$

while KCL enables us to find the inductor current,

$$i_L(0^-) = 5 \quad \text{A}$$

Although the derivatives at $t = 0^-$ are of little interest to us, it is evident that they are all zero.

Now let time increase an incremental amount. During the interval from $t = 0^-$ to $t = 0^+$, the left-hand current source becomes active and most of the voltage and current values at $t = 0^-$ will change abruptly. However, we should *begin by focusing our attention on those quantities which cannot change, inductor current and capacitor voltage.* Both of these must remain constant during the switching interval. Thus,

$$i_L(0^+) = 5 \quad \text{A} \qquad \text{and} \qquad v_C(0^+) = 150 \quad \text{V}$$

Since two currents are now known at the left node, we next obtain[8]

$$i_R(0^+) = -1 \quad \text{A} \qquad \text{and} \qquad v_R(0^+) = -30 \quad \text{V}$$

Thus,

$$i_C(0^+) = 4 \quad \text{A} \qquad \text{and} \qquad v_L(0^+) = 120 \quad \text{V}$$

There remain six derivatives which might be evaluated, although not all are needed to evaluate the two arbitrary constants. The procedure must begin with the energy-storage elements by the direct application of their defining equations. For the inductor,

$$v_L = L \frac{di_L}{dt}$$

and, specifically,

$$v_L(0^+) = L \frac{di_L}{dt} \Big|_{t=0^+}$$

Thus,

$$\frac{di_L}{dt} \Big|_{t=0^+} = \frac{v_L(0^+)}{L} = 40 \quad \text{A/s}$$

Similarly,

$$\frac{dv_C}{dt} \Big|_{t=0^+} = \frac{i_C(0^+)}{C} = 108 \quad \text{V/s}$$

The other four derivatives may be determined by realizing that KCL and KVL are both satisfied by the derivatives also. For example, at the left node,

$$4 - i_L - i_R = 0 \qquad t > 0$$

and thus

$$0 - \frac{di_L}{dt} - \frac{di_R}{dt} = 0 \qquad t > 0$$

---

[8] This current is the only one of the four remaining quantities which can be obtained in one step. In more complicated circuits, it is quite possible that none of the remaining initial values can be obtained with a single step; either circuit equations must then be written or a simpler equivalent resistive circuit must be drawn which can be analyzed by writing simultaneous equations. This latter method will be described shortly.

Fig. 7-14: (a) A simple circuit which is the equivalent of the circuit of Fig. 7-13 for $t = 0^-$. (b) Another equivalent of the circuit of Fig. 7-13, valid during the switching interval, $t = 0^-$ to $t = 0^+$.

and therefore,

$$\left.\frac{di_R}{dt}\right|_{t=0^+} = -40 \quad \text{A/s}$$

The three remaining initial values of the derivatives are found to be

$$\left.\frac{dv_R}{dt}\right|_{t=0^+} = -1200 \quad \text{V/s} \qquad \left.\frac{dv_L}{dt}\right|_{t=0^+} = -1092 \quad \text{V/s} \qquad \left.\frac{di_C}{dt}\right|_{t=0^+} = -40 \quad \text{A/s}$$

Now let us turn to a slightly different method by which all these currents, voltages, and derivatives may be evaluated at $t = 0^-$ and $t = 0^+$. We shall construct two equivalent circuits, one which is valid for the steady-state condition reached at $t = 0^-$ and a second which is valid during the switching interval. The discussion which follows relies on some of the reasoning we did above and, for that reason, appears shorter than it would be if it were presented first.

Prior to the switching operation, only direct currents and voltages exist in the circuit, and the inductor may therefore be replaced by a short circuit, its dc equivalent, while the capacitor is replaced by an open circuit. Redrawn in this manner, the circuit of Fig. 7-13 appears as shown in Fig. 7-14a. The three voltages and three currents at $t = 0^-$ are now easily found by resistive circuit-analysis methods; the numerical values are the same as those found previously.

We now turn to the problem of drawing an equivalent circuit which will assist us in determining the several voltages and currents at $t = 0^+$. Each capacitor voltage and each inductor current must remain constant during the switching interval. These conditions may be ensured by replacing the inductor by a current source and the capacitor by a voltage source. Each source serves to maintain the necessary response constant during the discontinuity. The equivalent circuit of Fig. 7-14b results; it should be noted that this circuit is a true equivalent at $t = 0^-$ since it possesses the same currents and voltages as the simple equivalent of Fig. 7-14a. It is also a true equivalent at $t = 0^+$ since the step-current source appears as a function of time and not merely as 0 A or 4 A.

The voltages and currents at $t = 0^+$ are obtained by letting $4u(t) = 4$ A and solving the dc circuit which results. The solution is not difficult, but the relatively large number of sources present in the network does produce a somewhat strange sight. However, problems of this type were solved in Chap. 3, and nothing new is involved. The six responses at $t = 0^+$ must agree with those found by the previous method.

Before leaving this problem of the determination of the necessary initial values, it should be pointed out that at least one other powerful method of determining them has been omitted; we could have written general nodal or loop equations for the original circuit. Then, the substitution of the known zero values of inductor voltage and capacitor current at $t = 0^-$ would uncover several other response values at $t = 0^-$ and enable the remainder to be found easily. A similar analysis at $t = 0^+$ must then be made. This is an important method, and it becomes a necessary one in more complicated circuits which cannot be analyzed by the simpler step-by-step procedures we have followed. However, we must leave a few topics to be covered at the time operational methods of circuit analysis are introduced later.

Now let us briefly complete the determination of the response $v_C(t)$ for the original circuit of Fig. 7-13. With both sources dead, the circuit appears as a series RLC circuit and s_1 and s_2 are easily found to be -1 and -9, respectively. The forced response may be found by inspection or, if necessary, by drawing the dc equivalent, which is similar to Fig. 7-14a, with the addition of a 4-A current source. The forced response is 150 V. Thus,

$$v_C(t) = 150 + Ae^{-t} + Be^{-9t}$$

and

$$v_C(0^+) = 150 = 150 + A + B$$

Then,

$$\frac{dv_C}{dt} = -Ae^{-t} - 9Be^{-9t}$$

and

$$\left.\frac{dv_C}{dt}\right|_{t=0^+} = 108 = -A - 9B$$

Finally,

$$A = 13.5 \qquad B = -13.5$$

and

$$v_C(t) = 150 + 13.5(e^{-t} - e^{-9t})$$

In summary then, whenever we wish to determine the transient behavior of a simple three-element RLC circuit, we must first decide whether we are confronted with a series or a parallel circuit, so that we may use the correct relationship for α. The two equations are:

$$\alpha = \frac{1}{2RC} \qquad \text{(parallel } RLC)$$

$$\alpha = \frac{R}{2L} \qquad \text{(series } RLC)$$

Our second decision is made after comparing α with ω_0, which is given for either circuit by

$$\omega_0 = \frac{1}{\sqrt{LC}}$$

If $\alpha > \omega_0$, the circuit is overdamped, and the natural response has the form:

$$f_n(t) = A_1 e^{s_1 t} + A_2 e^{s_2 t}$$

where

$$s_{1,2} = -\alpha \pm \sqrt{\alpha^2 - \omega_0^2}$$

If $\alpha = \omega_0$, then the circuit is critically damped and

$$f_n(t) = e^{-\alpha t}(A_1 t + A_2)$$

And finally, if $\alpha < \omega_0$, then we are faced with the underdamped response,

$$f_n(t) = e^{-\alpha t}(A_1 \cos \omega_d t + A_2 \sin \omega_d t)$$

where

$$\omega_d = \sqrt{\omega_0^2 - \alpha^2}$$

Our last decision depends on the independent sources. If there are none acting in the circuit after the switching or discontinuity is completed, then the circuit is source-free and the natural response comprises the complete response. If independent sources are still present, then the circuit is driven and a forced response must be determined. The complete response is then the sum,

$$f(t) = f_f(t) + f_n(t)$$

This is applicable to any current or voltage in the circuit.

Drill Problems

7-8. For the circuit shown in Fig. 7-15, find values at $t = 0^-$ and 0^+ for: (a) i_1; (b) i_2; (c) i_3.   *Ans:* 0.6, 0.6; 0, −0.3; −0.6, −0.6 A

Fig. 7-15: See Drill Probs. 7-8 to 7-10.

7-9. For the circuit of Fig. 7-15, determine values at $t = 0^-$ and 0^+ for: (a) di_1/dt; (b) di_2/dt; (c) di_3/dt.   *Ans:* 0, −24; 0, 6; 0, 24 A/s

7-10. For the circuit of Fig. 7-15, find values at $t = 25$ ms for: (a) i_1; (b) i_2; (c) i_3.   *Ans:* 0.1278; −0.1820; −0.1278 A

**7-8
The lossless *LC*
circuit**

If the value of the resistance in a parallel *RLC* circuit becomes infinite, or that in a series *RLC* circuit becomes zero, we have a simple *LC* loop in which an oscillatory response can be maintained forever. Let us look briefly at an example of such a circuit, and then discuss another means of obtaining an identical response without the need of supplying any inductance.

Consider the circuit of Fig. 7-16 in which the large values, $L = 4$ H and $C = \frac{1}{36}$ F, are used so that the mathematics will be simple. We shall let $i(0) = -\frac{1}{6}$ A and $v(0) = 0$. The circuit is source-free, $\alpha = 0$, and $\omega_0^2 = 9$. Thus, $\omega_d = \omega_0 = 3$, and the voltage is

$$v = A \cos 3t + B \sin 3t$$

Since $v(0) = 0$, we see that $A = 0$. Next,

$$\frac{dv}{dt}\bigg|_{t=0} = 3B = -\frac{i(0)}{C}$$

But $i(0) = -\frac{1}{6}$, and therefore $dv/dt = 6$ V/s at $t = 0$ and $B = 2$ V. Thus

$$v = 2 \sin 3t \quad V$$

which is an undamped sinusoidal response.

Now let us see how we might obtain this voltage without using an *LC* circuit. Our intentions are to write the differential equation that v satisfies and then to develop a configuration of op-amps that will yield the solution of the equation. Although we are working with a specific example, the technique is a general one that can be used to solve any linear homogeneous differential equation.

For the *LC* circuit of Fig. 7-16, we select v as our variable and set the sum of the downward inductor and capacitor currents equal to zero,

$$\frac{1}{4} \int_0^t v \, dt - \frac{1}{6} + \frac{1}{36} \frac{dv}{dt} = 0$$

Differentiating once, we have

$$\frac{1}{4} v + \frac{1}{36} \frac{d^2v}{dt^2} = 0$$

or

$$\frac{d^2v}{dt^2} = -9v$$

In order to solve this equation, let us make use of the operational amplifier as an integrator twice. We assume that the highest-order derivative appear-

Fig. 7-16: This circuit is lossless and it provides the undamped response, $v = 2 \sin 3t$ V, if $v(0) = 0$ and $i(0) = -1/6$ A.

ing in the differential equation, here d^2v/dt^2, is available at some point in our configuration of op-amps, say at an arbitrary point A. We now make use of the integrator, with $RC = 1$, as discussed at the end of Sec. 4-4. The input is d^2v/dt^2, and the output must be $-dv/dt$, where we assume a sign change in the integrator. The initial value of dv/dt is 6 V/s, as we showed in our initial analysis of the circuit, and thus an initial value of -6 V must be set in the integrator. The negative of the first derivative now forms the input to a second integrator. Its output is therefore $v(t)$ and the initial value is $v(0) = 0$. Now it only remains to multiply v by -9 to obtain the second derivative we assumed at point A. This is amplification by 9 with a sign change, and it is easily accomplished by using the op-amp as an inverting amplifier.

Figure 7-17 shows the circuit of an inverting amplifier. For an ideal op-amp, both the input current and voltage are zero. Thus, the current going east through R_1 is v_s/R_1, while that traveling west through R_f is v_o/R_f. Since their sum is zero, we have

$$\frac{v_o}{v_s} = -\frac{R_f}{R_1}$$

Thus, we can achieve a gain of -9 by setting $R_f = 90$ kΩ and $R_1 = 10$ kΩ, for example.

If we let R be 1 MΩ and C be 1 μF in each of the integrators, then $v_o = -\int_0^t v_s \, dt + v_o(0)$ in each case. The output of the inverting amplifier now forms the assumed input at point A, leading to the configuration of op-amps shown in Fig. 7-18. If the left switch is closed at $t = 0$, while the two initial-condition switches are opened at the same time, the output of the second integrator will be the undamped sine wave, $v = 2 \sin 3t$ V.

Fig. 7-17: The inverting operational amplifier provides a gain, $v_o/v_s = -R_f/R_1$, with an ideal op-amp

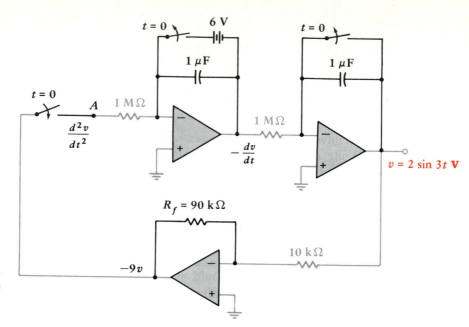

Fig. 7-18: Two integrators and an inverting amplifier are connected to provide the solution of the differential equation, $d^2v/dt^2 = -9v$.

Note that both the LC circuit of Fig. 7-16 and the op-amp circuit of Fig. 7-18 have the same output, but the op-amp circuit does not contain a single inductor. It simply *acts* as though it contained an inductor, providing the appropriate sinusoidal voltage between its output terminal and ground. This can be a considerable practical or economic advantage in circuit design.

Drill Problem

7-11. Give new values for the two initial voltages and for R_f in the circuit of Fig. 7-18 if the output represents the voltage $v(t)$ in the circuit of Fig. 7-19 for $t > 0$, and $R_1 = 10$ kΩ as before.   *Ans:* -10 V; -1 V; 250 kΩ

Fig. 7-19: See Drill Prob. 7-11.

Problems

1 Select element values for a parallel RLC circuit so that $s_1 = -200$ s^{-1}, $s_2 = -500$ s^{-1}, and the initial current through the resistor (in mA) is numerically equal to the initial voltage across the capacitor (in V).

2 The inductor current in the circuit shown in Fig. 7-20 is $i = 2e^{-5t} - 5e^{-10t}$ A. If $L = 0.2$ H, find: (a) $v(t)$; (b) $i_R(t)$; (c) $i_C(t)$.

Fig. 7-20: See Probs. 2, 3, 8, 10, and 11.

3 Element values in Fig. 7-20 are $R = 10\ \Omega$, $L = \frac{1}{32}$ H, and $C = 50\ \mu$F. If $i(0) = -2$ A and $v(0) = 40$ V, find $i(t)$ for $t > 0$.

4 Find $v_L(t)$ for the circuit shown in Fig. 7-21.

Fig. 7-21: See Prob. 4.

5 For the *RLC* circuit of Fig. 7-22: (a) find $i_R(t)$; (b) find the time at which $i_R = 250$ mA.

Fig. 7-22: See Prob. 5.

6 The switch in Fig. 7-23 has been closed for a long time. (a) Find $i_L(t)$ for all t. (b) Sketch $i_L(t)$ for the time interval, $-1 < t < 1$ ms.

Fig. 7-23: See Prob. 6.

7 After being open for an hour, the switch in Fig. 7-24 closes at $t = 0$. (a) Find $v_C(t)$ for $t > 0$. (b) Find $i_L(t)$ for $t > 0$. (c) Find the settling time t_s for v_C. (d) Find t_s for i_L.

Fig. 7-24: See Prob. 7.

8 Suppose the circuit of Fig. 7-20 is critically damped, with $v(0) = 100$ V and $i(0) = 4$ A. Then if $R = 10 \Omega$ and $\alpha = 100$ Np/s: (a) find L and C; (b) find $v(t)$ for $t > 0$ and sketch the response as a function of time. (c) At what time is $v = 0$? (d) Find the settling time t_s.

9 (a) In the circuit shown in Fig. 7-25, what value of R will cause critical damping? (b) Using this value of R, find $i_C(t)$ for $t > 0$. (c) What is t_s for i_C?

Fig. 7-25: See Prob. 9.

10 Let $v(0) = 0$ and $i(0) = 5$ A in the circuit shown in Fig. 7-20. If $L = 0.1$ H, $\omega_0 = 100$ rad/s, and the circuit is critically damped: (a) find R and C; (b) find $v(t)$ for $t > 0$. (c) Find $|v|_{\max}$ and the time t_m at which it occurs. (d) Find the settling time t_s.

11 Let $i = 10e^{-4t} \sin 28t$ A in the circuit shown in Fig. 7-20. If $R = 200 \Omega$, find: (a) $v(t)$; (b) $i_C(t)$; (c) $i_R(t)$.

12 The switch appearing in Fig. 7-26 has been closed for a long time. Find $i(t)$ after it opens at $t = 0$.

Fig. 7-26: See Probs. 12 and 27.

13 The switch in the circuit of Fig. 7-27 has been open for several hours before it is closed at $t = 0$. (a) Find $v_C(t)$ for $t > 0$. (b) Find i_{SW} for $t > 0$.

Fig. 7-27: See Prob. 13.

14 (a) Select the value of R that will make $\omega_d = 800$ krad/s in the circuit of Fig. 7-28 and then find i_L for $t > 0$. (b) Sketch i_L vs. t, $-1 \le t \le 7$ μs.

Fig. 7-28: See Prob. 14.

15 After being open for a long time, the switch in the circuit shown in Fig. 7-29 closes at $t = 0$. (a) Find $v_C(t)$ for $t > 0$. (b) Find the maximum and minimum values of $v_C(t)$.

Fig. 7-29: See Prob. 15.

16 The switch in the circuit of Fig. 7-30 has been closed for a very long time. It opens at $t = 0$ just as the source turns off. (a) Find $v_R(t)$ for $t > 0$. (b) Find $v_1(t)$ for $t > 0$.

Fig. 7-30: See Prob. 16.

17 If the switch in the circuit of Fig. 7-31 is closed at $t = -10^6$ s and opened at $t = 0$, find $v(t)$ for $t > 0$.

Fig. 7-31: See Prob. 17.

18 (a) Find $v(t)$ for $t > 0$ in the circuit of Fig. 7-32. (b) What is the settling time t_s for $v(t)$?

Fig. 7-32: See Prob. 18.

Fig. 7-33: See Prob. 19.

19 (a) Find $i_L(t)$ in the circuit of Fig. 7-33. (b) What is the value of $d^2 i_L/dt^2$ at $t = 0^+$?

20 Find $v_C(t)$ for the circuit shown in Fig. 7-34 and sketch the response.

Fig. 7-34: See Probs. 20 and 25.

21 For the circuit of Fig. 7-35: (a) find $i_L(t)$ for all t; (b) find $v_C(t)$ for $t > 0$.

Fig. 7-35: See Probs. 21 and 24.

22 Sketch v_C vs. t, $0 < t < 3$ s, for a series *RLC* circuit in which $\alpha = 1$ Np/s, $\omega_d = 25$ rad/s, $v_C(0^+) = 0$, and $v_C'(0^+) = 100$ V/s.

23 (*a*) Find $i_L(t)$ for $t > 0$ in the circuit shown in Fig. 7-36, and (*b*) determine its minimum and maximum values.

Fig. 7-36: See Prob. 23.

24 Change $4u(-t)$ A in Fig. 7-35 to $4u(t)$ A and rework Prob. 21.

25 The source in the circuit shown in Fig. 7-34 is changed from $8u(-t)$ to $4u(t) + 4$ A. (*a*) Find $v_C(t)$ for $t > 0$. (*b*) Find $v_{C,max}$.

26 At $t = 0$, the strength of the current source in Fig. 7-37 is suddenly increased from 10 to 20 A. Find v_s for $t > 0$.

Fig. 7-37: See Prob. 26.

27 Assume that the switch in Fig. 7-26 has been open for a long time and is closed at $t = 0$. Find $i(t)$ for $t > 0$.

28 (*a*) Refer to the circuit of Fig. 7-38, and determine $i(t)$ for $t > 0$. (*b*) Sketch the current as a function of time for $-1 < t < 4$ ms.

Fig. 7-38: See Prob. 28.

29 (a) Determine a correct expression for $v_C(t)$ in the circuit of Fig. 7-39. (b) Write the differential equation for $v_C(t)$ and use it to help determine values for d^2v_C/dt^2 and d^3v_C/dt^3 at $t = 0^+$.

Fig. 7-39: See Prob. 29.

30 Determine the value of R for the underdamped circuit of Sec. 7-5 ($L = 7$ H, $C = \frac{1}{42}$ F, $i(0) = 10$ A, $v(0) = 0$) that will lead to a minimum value of the settling time t_s. What is t_s?

31 Both switches in Fig. 7-40 have been in the position shown for a long time. At $t = 0$, S_A is closed. (a) Assuming $v_C(0) = 0$, find and sketch $v_C(t)$, $0 < t < 5$ s. (b) At $t = 5$ s, S_B is thrown to the right. Find and sketch $v_C(t)$, $5 < t < 8$ s.

Fig. 7-40: See Prob. 31.

32 Refer to the simple parallel RC circuit of Fig. 5-9 with $R = 10$ kΩ, $C = 5$ μF, and $v(0) = 10$ V. Write the differential equation for v, $t > 0$, and design a circuit using an op-amp integrator to provide $v(t)$ as the output. First use an integrator having $R_1 = 1$ MΩ, $C_f = 1$ μF, and then repeat for $R_1 = 10$ kΩ, $C_f = 5$ μF.

33 After being in the position shown for a long time, the switch in the circuit of Fig. 7-41 is moved down at $t = 0$. Construct an op-amp circuit whose output is $v(t)$ for $t > 0$.

Fig. 7-41: See Prob. 33.

PART THREE
SINUSOIDAL ANALYSIS

Chapter 8
The Sinusoidal Forcing Function

The complete response of a linear electric circuit is composed of two parts, the natural response and the forced response. The first part of our study was devoted to the resistive circuit, in which only the forced response was required or present. For simplicity, we usually restricted our forcing functions to dc sources, and we therefore became exceedingly familiar with the various techniques useful in finding the dc forced response. We then passed on to the next part and considered the natural response of a number of different circuits containing one or two energy-storage elements. Without undue strain we were then able to determine the complete response of these circuits by adding the natural response, which is characteristic of the circuit and not of the forcing function, to the forced response produced by dc forcing functions, the only forced response with which we are familiar. We are therefore now in a position where our mastery of the natural response is greater than our knowledge of the forced response.

In this third part of our study we shall extend our knowledge of the forced response by considering the sinusoidal forcing function.

Why should we select the sinusoidal forcing function as the second functional form to study? Why not the linear function, the exponential function, or a modified Bessel function of the second kind? There are many reasons for the choice of the sinusoid, and any one of them would probably be sufficient to lead us in this direction.

One of these reasons is apparent from the results of the preceding chapter; the natural response of an underdamped second-order system is a damped sinusoid, and if no losses are present it is a pure sinusoid. The sinusoid thus appears naturally (as does the negative exponential). Indeed, Nature in general seems to have a decidedly sinusoidal character; the motion of a pendulum, the bouncing of a ball, the vibration of a guitar string, the political atmosphere in any country, and the ripples on the surface of a stein of chocolate milk will always display a reasonably sinusoidal character.

Perhaps it was observations of these natural phenomena that led the great French mathematician Fourier to his discovery of the important analytical method embodied in the Fourier theorem. In Chap. 17 we shall see that this theorem enables us to represent most of the useful mathematical functions of time which repeat themselves f_0 times per second by the sum of an infinite number of sinusoidal time functions with frequencies that are integral multiples of f_0; the given periodic function $f(t)$ can also be approximated as closely as we wish by the sum of a finite number of such terms, even though the graph of $f(t)$ might look very nonsinusoidal. The accuracy of such an approximation is illustrated by Drill Prob. 8-1.

This decomposition of a periodic forcing function into a number of appropriately chosen sinusoidal forcing functions is a very powerful analytical method, for it enables us to superimpose the partial responses produced in any linear circuit by each sinusoidal component and thereby obtain the desired response caused by the given periodic forcing function. Thus, another reason for studying the response to a sinusoidal forcing function is found in the dependence of other forcing functions on sinusoidal analysis.

A third reason is found in an important mathematical property of the sinusoidal function. Its derivatives and integrals are also all sinusoids.[1] Since the forced response takes on the form of the forcing function, its integral, and its derivatives, the sinusoidal forcing function will produce a sinusoidal forced response throughout a linear circuit. The sinusoidal forcing function thus allows a much easier mathematical analysis than does almost every other forcing function.

Finally, the sinusoidal forcing function has important practical applications. It is an easy function to generate and is the waveform used predominantly throughout the electric power industry; and every electrical laboratory contains a number of sinusoidal generators which operate throughout a tremendous range of useful frequencies.

[1] We are using the term "sinusoid" collectively here to include cosinusoidal functions of time also. After all, a cosine function can be written as a sine function if the angle is increased by 90°.

Fig. 8-1: See Drill Prob. 8-1.

Drill Problem

8-1. The Fourier theorem, which we shall study in Chap. 17, shows that the periodic triangular waveform $v_1(t)$, shown in Fig. 8-1a, and the infinite sum of sine terms,

$$v_1(t) = \frac{8}{\pi^2}\left(\sin \pi t - \frac{1}{3^2}\sin 3\pi t + \frac{1}{5^2}\sin 5\pi t - \frac{1}{7^2}\sin 7\pi t + \cdots\right)$$

are equal. Let us suppose that the sum of the first three terms of the series is called $v_2(t)$ and used as an approximation to $v_1(t)$. Figures 8-1b and c show $v_1(t)$ and $v_2(t)$ as sources in identical resistive circuits. Calculate $i_1(t)$ and $i_2(t)$ at $t =$: (a) 0.3; (b) 1.5 s.   *Ans:* 12, 11.91; −20, −18.66 mA

**8-2
Characteristics of
sinusoids**

In this section we shall define the trigonometric nomenclature which is used to describe sinusoidal (or cosinusoidal) functions. The definitions should be familiar to most of us, and if we remember a little trigonometry, the section can be read very rapidly.

Let us consider a sinusoidally varying voltage

$$v(t) = V_m \sin \omega t$$

shown graphically in Figs. 8-2a and b. The *amplitude* of the sine wave is V_m, and the *argument* is ωt. The *radian frequency* or *angular frequency* is ω. In Fig. 8-2a, $V_m \sin \omega t$ is plotted as a function of the argument ωt, and the periodic nature of the sine wave is evident. The function repeats itself every 2π radians, and its *period* is therefore 2π radians. In Fig. 8-2b, $V_m \sin \omega t$ is

Fig. 8-2: The sinusoidal function $v(t) = V_m \sin \omega t$ is plotted versus ωt in (a) and versus t in (b).

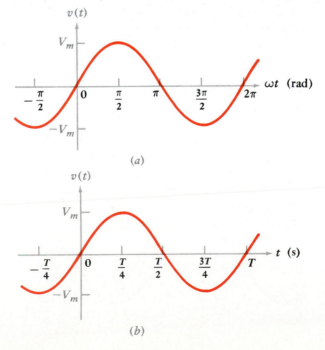

plotted as a function of t and the *period* is now T. The period may also be expressed in degrees, or occasionally in other units such as centimeters or inches. A sine wave having a period T must execute $1/T$ periods each second; its *frequency* f is $1/T$ hertz, abbreviated Hz. Thus, one hertz is identical to one cycle per second, a term whose use is now discouraged because so many people incorrectly used "cycle" for "cycle per second." Thus,

$$f = \frac{1}{T}$$

and since

$$\omega T = 2\pi$$

we obtain the common relationship between frequency and radian frequency,

$$\omega = 2\pi f$$

A more general form of the sinusoid

$$v(t) = V_m \sin(\omega t + \theta) \tag{1}$$

includes a *phase angle* θ in its argument $(\omega t + \theta)$. Equation (1) is plotted in Fig. 8-3 as a function of ωt, and the phase angle appears as the number of radians by which the original sine wave, shown as a broken line in the sketch, is shifted to the left, or earlier in time. Since corresponding points on the sinusoid $V_m \sin(\omega t + \theta)$ occur θ rad, or θ/ω s, earlier, we say that $V_m \sin(\omega t + \theta)$ *leads* $V_m \sin \omega t$ by θ rad. Conversely, it is correct to describe $\sin \omega t$ as *lagging* $\sin(\omega t + \theta)$ by θ rad, as leading $\sin(\omega t + \theta)$ by $-\theta$ rad, or as leading $\sin(\omega t - \theta)$ by θ rad.

In either case, leading or lagging, we say that the sinusoids are *out of phase;* if the phase angles are equal, they are said to be *in phase*.

In electrical engineering, the phase angle is commonly given in degrees, rather than radians, and no confusion will arise if the degree symbol is always used. Thus, instead of writing

$$v = 100 \sin\left(2\pi 1000 t - \frac{\pi}{6}\right)$$

Fig. 8-3: The sine wave $V_m \sin(\omega t + \theta)$ leads $V_m \sin \omega t$ by θ rad.

we customarily use

$$v = 100 \sin (2\pi 1000t - 30°)$$

In evaluating this expression at a specific instant of time, say $t = 10^{-4}$ s, $2\pi 1000t$ becomes 0.2π *radians,* and this should be expressed as 36° before 30° is subtracted from it. Don't confuse your apples with your oranges.

Two sinusoidal waves that are to be compared in phase must both be written as sine waves, or both as cosine waves; both waves must be written with positive amplitudes; and each must have the same frequency. It is also evident that multiples of 360° may be added to or subtracted from the argument of any sinusoidal function without changing the value of the function. Hence, we may say that

$$v_1 = V_{m1} \sin (5t - 30°)$$

lags

$$
\begin{aligned}
v_2 &= V_{m2} \cos (5t + 10°) \\
&= V_{m2} \sin (5t + 90° + 10°) \\
&= V_{m2} \sin (5t + 100°)
\end{aligned}
$$

by 130°, or it is also correct to say that v_1 leads v_2 by 230°, since v_2 may be written as

$$v_2 = V_{m2} \sin (5t - 260°)$$

V_{m1} and V_{m2} are each assumed to be positive quantities. Normally, the difference in phase between two sinusoids is expressed by that angle which is less than or equal to 180° in magnitude.

The concept of a leading or lagging relationship between two sinusoids will be used extensively, and the relationship should be recognizable both mathematically and graphically.

Drill Problems

8-2. Find the angle by which i_2 leads i_1 if $i_1 = 120 \cos (100\pi t + 30°)$ and $i_2 = :$ (*a*) $-0.1 \cos (100\pi t + 100°)$; (*b*) $20 \sin (100\pi t - 50°)$; (*c*) $-18 \sin (100\pi t + 40°)$.   *Ans:* $-110°$; $-170°$; $100°$

8-3. Find *A, B, C* and ϕ if $-50 \cos (120\pi t + 30°) + 20 \sin (120\pi t - 40°) = A \cos 120\pi t + B \sin 120\pi t = C \cos (120\pi t + \phi)$.   *Ans:* -56.2; 40.3; 69.1; $-144.3°$

8-3
Forced response to sinusoidal forcing functions

Now that we are familiar with the mathematical characteristics of sinusoids and can describe and compare them intelligently, we are ready to apply a sinusoidal forcing function to a simple circuit and to obtain the forced response. We shall first write the differential equation which applies to the given circuit. The complete solution of this equation is composed of two parts, the complementary solution (which we call the natural response) and the particular integral (or forced response). The natural response is independent of the mathematical form of the forcing function and depends only upon the type of circuit, the element values, and the initial conditions. We find it by setting all the forcing functions equal to zero, thus reducing the equation to the simpler linear *homogeneous* differential equation. We have already determined the natural response of many *RL*, *RC*, and *RLC* circuits.

The forced response has the mathematical form of the forcing function, plus all its derivatives and its first integral. From this knowledge, it is apparent that one of the methods by which the forced response may be found is to assume a solution composed of a sum of such functions, where each function has an unknown amplitude to be determined by direct substitution into the differential equation. This is a lengthy method, but it is the one which we shall use in this chapter to introduce sinusoidal analysis because it involves a minimum of new concepts. However, if the simpler method to be described in the following chapters were not available, circuit analysis would be an impractical, useless art.

The term *steady-state* response is used synonymously with forced response, and the circuits we are about to analyze are commonly said to be in the "sinusoidal steady state." Unfortunately, steady state carries the connotation of "not changing with time" in the minds of many students. This is true for dc forcing functions, but the sinusoidal steady-state response is definitely changing with time. The steady state simply refers to the condition which is reached after the transient or natural response has died out.

Now let us consider the series *RL* circuit shown in Fig. 8-4. The sinusoidal source voltage $v_s = V_m \cos \omega t$ has been switched into the circuit at some remote time in the past, and the natural response has died out completely. We seek the forced response, or steady-state response, and it must satisfy the differential equation

$$L \frac{di}{dt} + Ri = V_m \cos \omega t$$

The functional form of the forced response is next obtained by integration and repeated differentiation of the forcing function. Only two different forms are obtained, $\sin \omega t$ and $\cos \omega t$. The forced response must therefore have the general form

$$i(t) = I_1 \cos \omega t + I_2 \sin \omega t$$

where I_1 and I_2 are real constants whose values depend upon V_m, R, L, and ω. No constant or exponential function can be present. Substituting the assumed form for the solution into the differential equation yields

$$L(-I_1\omega \sin \omega t + I_2\omega \cos \omega t) + R(I_1 \cos \omega t + I_2 \sin \omega t) = V_m \cos \omega t$$

If we collect the cosine and sine terms, we obtain

$$(-LI_1\omega + RI_2) \sin \omega t + (LI_2\omega + RI_1 - V_m) \cos \omega t = 0$$

This equation must be true for all values of t, and this can be achieved only if the factors multiplying $\cos \omega t$ and $\sin \omega t$ are each zero. Thus,

$$-\omega LI_1 + RI_2 = 0 \qquad \omega LI_2 + RI_1 - V_m = 0$$

$v_s(t) =$
$V_m \cos \omega t$

Fig. 8-4: A series *RL* circuit for which the forced response is desired.

and simultaneous solution for I_1 and I_2 leads to

$$I_1 = \frac{RV_m}{R^2 + \omega^2 L^2} \qquad I_2 = \frac{\omega L V_m}{R^2 + \omega^2 L^2}$$

Thus, the forced response is obtained,

$$i(t) = \frac{RV_m}{R^2 + \omega^2 L^2} \cos \omega t + \frac{\omega L V_m}{R^2 + \omega^2 L^2} \sin \omega t \tag{2}$$

This expression is slightly cumbersome, however, and a clearer picture of the response can be obtained by expressing the response as a single sinusoid or cosinusoid with a phase angle. Let us select the cosine function in anticipation of the method presented in the following chapter,

$$i(t) = A \cos (\omega t - \theta) \tag{3}$$

At least two methods of obtaining the values of A and θ should suggest themselves. We might substitute (3) directly into the original differential equation, or we could simply equate the two solutions (2) and (3). Let us select the latter method, since the former makes an excellent problem for the end of the chapter, and equate (2) and (3) after expanding the function $\cos (\omega t - \theta)$,

$$A \cos \theta \cos \omega t + A \sin \theta \sin \omega t = \frac{RV_m}{R^2 + \omega^2 L^2} \cos \omega t + \frac{\omega L V_m}{R^2 + \omega^2 L^2} \sin \omega t$$

Thus, again collecting and setting the coefficients of $\cos \omega t$ and $\sin \omega t$ equal to zero, we find

$$A \cos \theta = \frac{RV_m}{R^2 + \omega^2 L^2} \qquad \text{and} \qquad A \sin \theta = \frac{\omega L V_m}{R^2 + \omega^2 L^2}$$

To find A and θ, we divide one equation by the other,

$$\frac{A \sin \theta}{A \cos \theta} = \tan \theta = \frac{\omega L}{R}$$

and also square both equations and add the results,

$$A^2 \cos^2 \theta + A^2 \sin^2 \theta = A^2 = \frac{R^2 V_m^2}{(R^2 + \omega^2 L^2)^2} + \frac{\omega^2 L^2 V_m^2}{(R^2 + \omega^2 L^2)^2} = \frac{V_m^2}{R^2 + \omega^2 L^2}$$

Hence,

$$\theta = \tan^{-1} \frac{\omega L}{R}$$

and

$$A = \frac{V_m}{\sqrt{R^2 + \omega^2 L^2}}$$

The alternative form of the forced response therefore becomes

$$i(t) = \frac{V_m}{\sqrt{R^2 + \omega^2 L^2}} \cos \left(\omega t - \tan^{-1} \frac{\omega L}{R} \right) \tag{4}$$

The electrical characteristics of the response $i(t)$ should now be considered. The amplitude of the response is proportional to the amplitude of the forcing function; if it were not, the linearity concept would have to be discarded. The amplitude of the response also decreases as R, L, or ω is increased, but not proportionately. This is confirmed by the differential equation, for an increase in R, L, or di/dt requires a decrease in current amplitude if the source-voltage amplitude is not changed. The current is seen to lag the applied voltage by $\tan^{-1}(\omega L/R)$, an angle between 0 and 90°. When $\omega = 0$ or $L = 0$, the current must be in phase with the voltage; since the former situation is direct current and the latter provides a resistive circuit, the result is expected. If $R = 0$, the current lags the voltage by 90°; then $v_s = L(di/dt)$, and the derivative-integral relationship between the sine and cosine indicates the validity of the 90° phase difference. In an inductor alone, then, the current lags the voltage by exactly 90°, if the passive sign convention is satisfied. In a similar manner we can show that the current through a capacitor *leads* the voltage across it by 90°.

The applied voltage and the resultant current are both plotted on the same ωt axis in Fig. 8-5, but arbitrary current and voltage ordinates are assumed. The fact that the current lags the voltage in this simple RL circuit is now visually apparent. We shall later be able to show easily that this phase relationship is found at the input of every circuit composed only of inductors and resistors.[2]

The phase difference between the current and voltage depends upon the ratio of the quantity ωL to R. We call ωL the *inductive reactance* of the inductor; it is measured in ohms, and it is a measure of the opposition which is offered by the inductor to the passage of a sinusoidal current. Much more will be said about reactance in the following chapter.

The method by which we have found the sinusoidal steady-state response for this simple series RL circuit has not been a trivial problem. We might think of the analytical complications as arising through the presence of the inductor; if both the passive elements had been resistors, the analysis would have been ridiculously easy, even with the sinusoidal forcing function present. The reason the analysis would be so easy results from the simple voltage-current relationship specified by Ohm's law. The voltage-current relationship for an inductor is not as simple, however; instead of solving an algebraic equation, we were faced with a nonhomogeneous differential equation. It would be quite impractical to analyze every circuit by the method described above, and in the following chapter we shall therefore take steps to simplify the analysis. Our result will be an algebraic relationship between sinusoidal current and sinusoidal voltage for inductors and capacitors as well as resistors, and we shall be able to produce a set of *algebraic* equations for a circuit of any complexity. The constants and the variables in the algebraic

---

[2] Once upon a time the symbol E (for electromotive force) was used to designate a voltage. Then every student learned the phrase "ELI the ICE man" as a reminder that voltage leads current in an inductive circuit, while current leads voltage in a capacitive circuit, and they all lived happily ever after.

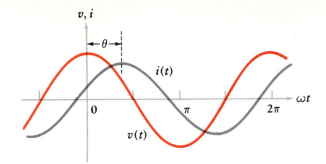

Fig. 8-5: The applied sinusoidal forcing function (colored) and the resultant sinusoidal current response (black) of the series *RL* circuit shown in Fig. 8-4.

equations will be complex numbers rather than real numbers, but the analysis of any circuit in the sinusoidal steady state becomes almost as easy as the analysis of a similar resistive circuit.

Drill Problems

8-4. Let R be 200 Ω and L be 8 mH in the circuit shown in Fig. 8-4. If $v_s = 40 \cos 10^4 t$ V, find: (a) $v_R(t)$, the voltage across R, + reference at the left; (b) $v_L(t)$, the voltage across L, + reference at the top; (c) $p_s(t)$, the power being supplied by the source.
Ans: 37.1 cos $(10^4 t - 21.8°)$ V; 14.86 cos $(10^4 t + 68.2°)$ V;
$$7.43 \cos 10^4 t \cos (10^4 t - 21.8°) \text{ W}$$

8-5. Use Thévenin's theorem to simplify the circuit of Fig. 8-6, and then find values at $t = 0$ for: (a) i_L; (b) v_6; (c) v_1. *Ans:* −5.76 mA; 11.52 V; −3.84 V

Fig. 8-6: See Drill Prob. 8-5.

Problems

1 If $y(t) = 26 \cos 190t - 18 \sin 190t$, determine values for the following characteristics of $y(t)$: (a) f; (b) T; (c) ω; (d) the amplitude; (e) the angle by which $y(t)$ leads 5 cos 190t; (f) the angle by which $y(t)$ lags 8 sin 190t; (g) A and θ if $A \cos (190t + \theta) = y(t) + 20 \cos (190t + 160°)$.

2 Express each of these voltages in the form, $v(t) = A \cos (\omega t + \phi)$, by giving values for A and ϕ: (a) 6 sin 20t; (b) 4 sin $(20t - 20°)$; (c) 8 cos 5t + 3 sin 5t. (d) What is the smallest positive value of t for which 5 cos $(100t + 40°) = 0$? (e) What is the smallest positive value of t for which $-3 \cos 20t + 2 \sin 20t$ reaches its maximum value, and what is the maximum value?

3 (a) A 100-Hz sinusoid has a maximum positive value of 50 at $t = 0.1234$ s. Express it in the form, $A \cos \omega t + B \sin \omega t$. (b) The sinusoid shown in Fig. 8-2b passes through zero every 2 ms and has a value of −12 at $t = 2.4$ ms. Express it in the form $C \cos (\omega t + \phi)$.

4 Carry out the exercise threatened in the text by substituting the assumed current response (3), $i(t) = A \cos (\omega t - \theta)$, directly into the differential equation,

$L(di/dt) + Ri = V_m \cos \omega t$, to show that values for A and θ are obtained which agree with (4).

5 In Fig. 8-4, let $v_s = 20 \cos (500t + 100°)$ V and $v_R = K \cos (500t + 45°)$ V (+ reference on left). If $L = 80$ mH, find R, K, $i(t)$, and $v_L(t)$ (+ reference on top).

6 A 0.2-H inductor is in series with a 10-Ω resistor, a 30-Ω resistor, and a voltage source, $v_s = 100 \cos 300t$ V. Determine the amplitude of the voltage across the series combination of the inductor and the 10-Ω resistor.

7 A voltage source, $v_s = 10 \cos (800t - 30°)$ V, a 20-Ω resistor, and a 50-mH inductor are in series. Determine the amplitude of the maximum instantaneous power: (*a*) generated by the source; (*b*) dissipated by the resistor; (*c*) absorbed by the inductor.

8 Use Thévenin's theorem on the circuit of Fig. 8-7 to help find: (*a*) i_L; (*b*) v_{60}.

Fig. 8-7: See Probs. 8 and 10.

9 Use repeated source transformations on the circuit shown in Fig. 8-8 and find $i_L(t)$.

Fig. 8-8: See Probs. 9 and 11.

10 Replace the 5-Ω resistor in Fig. 8-7 by a dependent voltage source, + reference at the top, and labeled $0.1v_{60}$. Now rework Prob. 8*a*.

11 Change the right current source in Fig. 8-8 to $5 \cos 200t$ A, and rework Prob. 9.

12 Assume that the op-amp in Fig. 8-9 is ideal ($R_i = \infty$, $R_o = 0$, and $A = \infty$). Note also that the integrator input has two signals applied to it, $-V_m \cos \omega t$ and v_{out}. If the

Fig. 8-9: See Prob. 12.

product $R_1 C_1$ is set equal to the ratio L/R in the series RL circuit of Fig. 8-4, show that v_{out} equals the voltage across R (+ reference on the left) in that circuit.

13 A voltage source $V_m \cos \omega t$, a resistor R, and a capacitor C are all in series. (a) Write an integrodifferential equation in terms of the loop current i and then differentiate it to obtain the differential equation for this circuit. (b) Assume a suitable general form for the forced response $i(t)$, substitute it into the differential equation, and determine the exact form of the forced response.

Chapter 9
The Phasor Concept

9-1
Introduction

Throughout the earlier portions of our study of circuit analysis, we devoted our entire attention to the resistive circuit. However, we might remember that we were often promised that those methods which we were applying to resistive circuits would later prove applicable to circuits containing inductors and capacitors as well. In this chapter we shall lay the descriptive groundwork which will make this prediction come true. We shall develop a method for representing a sinusoidal forcing function or a sinusoidal response by a complex-number symbolism called a phasor transform, or simply a phasor. This is nothing more than a number which, by specifying both the amplitude and phase angle of a sinusoid, characterizes that sinusoid just as completely as if it were expressed as an analytical function of time. By working with phasors, rather than with derivatives and integrals of sinusoids as we did in the preceding chapter, we shall effect a truly remarkable simplification in the steady-state sinusoidal analysis of general *RLC* circuits. The simplification should become apparent toward the end of this chapter.

The use of a mathematical transformation to simplify a problem should not be a new idea to us. For example, we are all familiar with the procedure of using logarithms to simplify arithmetic multiplication and division. In order to multiply several numbers together, we first determined the logarithm of each of the numbers, or "transformed" the numbers into an alternative mathematical description. We might now describe that operation as obtaining the "logarithmic transform." We then added all the logarithms to obtain the logarithm of the desired product. Then, finally, we found the antilogarithm, a process which might be termed an "inverse transformation"; the antilogarithm was our desired answer. Our solution carried us from the domain of everyday numbers to the logarithmic domain, and back again.

Other familiar examples of transform operations may be found in the alternative representations of a circle as a mathematical equation, as a geometric figure on a rectangular-coordinate plane, or merely as a set of three numbers, where it is understood that the first is the *x*-coordinate value of the center, the second is the *y*-coordinate value, and the third is the magnitude of the radius. Each of the three representations contains exactly the same information, and once the rules of the transformations are laid down in analytic geometry, we find no difficulty in passing from the algebraic domain to the geometric domain or to the ''domain of the ordered triplet.''

Few other transforms with which we are familiar provide the simplification that can be achieved with the phasor concept.

9-2
The complex forcing function

We are now ready to think about applying a complex forcing function (that is, one that has both a real and an imaginary part) to an electrical network.[1] It may seem strange, but we shall find that the use of complex quantities in sinusoidal steady-state analysis leads to methods which are much simpler than those involving purely real quantities. We should expect a complex forcing function to produce a complex response; we might even suspect, and suspect correctly, that the real part of the forcing function will produce the real part of the response, while the imaginary portion of the forcing function will result in the imaginary portion of the response. Our goal in this section is to prove, or at least demonstrate, that these suspicions are correct.

Let us first discuss the problem in rather general terms, thus indicating the method by which we might prove our assertions if we were to construct a general network and analyze it by means of a set of simultaneous equations. In Fig. 9-1, a sinusoidal source

$$V_m \cos (\omega t + \theta) \tag{1}$$

is connected to a general network, which we shall arbitrarily assume to be passive in order to avoid complicating our use of the superposition principle later. A current response in some other branch of the network is to be determined. The parameters appearing in (1) are all real quantities.

The discussion in Chap. 8 of the method whereby the response to a sinusoidal forcing function may be determined, through the assumption of a sinusoidal form with arbitrary amplitude and arbitrary phase angle, shows that the response may be represented by

$$I_m \cos (\omega t + \phi) \tag{2}$$

A sinusoidal forcing function always produces a sinusoidal forced response in a linear circuit.

Now let us change our time reference by shifting the phase of the forcing

---

[1] Appendix 4 defines the complex number and related terms, describes the arithmetic operations on complex numbers, and develops Euler's identity and the exponential and polar forms.

Fig. 9-1: The sinusoidal forcing function $V_m \cos (\omega t + \theta)$ produces the steady-state sinusoidal response $I_m \cos (\omega t + \phi)$.

function by 90°, or changing the instant that we call $t = 0$. Thus, the forcing function

$$V_m \cos (\omega t + \theta - 90°) = V_m \sin (\omega t + \theta) \qquad (3)$$

when applied to the same network will produce a corresponding response

$$I_m \cos (\omega t + \phi - 90°) = I_m \sin (\omega t + \phi) \qquad (4)$$

We must next depart from physical reality by applying an imaginary forcing function, one which cannot be applied in the laboratory but can be applied mathematically.

We construct an imaginary source very simply; it is only necessary to multiply the source, as it appears in (3), by j, the imaginary operator. We thus apply

$$jV_m \sin (\omega t + \theta) \qquad (5)$$

What is the response? If we had doubled the source, then the principle of linearity would require that we double the response; multiplication of the forcing function by a constant k would result in the multiplication of the response by the same constant k. The fact that this constant is the imaginary operator j does not destroy this relationship, even though our earlier definition and discussion of linearity did not specifically include complex constants. It is now more realistic to conclude that it did not specifically *exclude* them, for the entire discussion is equally applicable if all the constants in the equations are complex. The response to the imaginary source of (5) is thus

$$jI_m \sin (\omega t + \phi) \qquad (6)$$

The imaginary source and response are indicated in Fig. 9-2.

We have applied a real source and obtained a real response; we have also applied an imaginary source and obtained an imaginary response. Now we may use the superposition theorem to find the response to that complex forcing function which is the sum of the real and imaginary forcing functions. The applicability of superposition, of course, is guaranteed by the linearity of the circuit and does not depend on the form of the forcing functions. Thus, the sum of the forcing functions of (1) and (5),

$$V_m \cos (\omega t + \theta) + jV_m \sin (\omega t + \theta) \qquad (7)$$

must therefore produce a response which is the sum of (2) and (6),

$$I_m \cos (\omega t + \phi) + jI_m \sin (\omega t + \phi) \qquad (8)$$

Fig. 9-2: The imaginary sinusoidal forcing function $jV_m \sin (\omega t + \theta)$ produces the imaginary sinusoidal response $jI_m \sin (\omega t + \phi)$ in the network of Fig. 9-1.

Fig. 9-3: The complex forcing function $V_m e^{j(\omega t + \theta)}$ produces the complex response $I_m e^{j(\omega t + \phi)}$ in the network of Fig. 9-1.

$$V_m e^{j(\omega t + \theta)} \qquad N \qquad I_m e^{j(\omega t + \phi)}$$

The complex source and response may be represented more simply by applying Euler's identity. The source of (7) thus becomes

$$V_m e^{j(\omega t + \theta)} \qquad\qquad (9)$$

and the response of (8) is

$$I_m e^{j(\omega t + \phi)} \qquad\qquad (10)$$

The complex source and response are illustrated in Fig. 9-3.

There are several important conclusions to be drawn from this general example. A real, an imaginary, or a complex forcing function will produce a real, an imaginary, or a complex response, respectively. Moreover, a complex forcing function may be considered, by the use of Euler's identity and the superposition theorem, as the sum of a real and an imaginary forcing function; thus the real part of the complex response is produced by the real part of the complex forcing function, while the imaginary part of the response is caused by the imaginary part of the complex forcing function.

Instead of applying a real forcing function to obtain the desired real response, we apply a complex forcing function whose real part is the given real forcing function; we obtain a complex response whose real part is the desired real response. Through this procedure, the integrodifferential equations describing the steady-state response of a circuit will become simple algebraic equations.

Let us try out this idea on the simple *RL* series circuit shown in Fig. 9-4. The real source $V_m \cos \omega t$ is applied; the real response $i(t)$ is desired.

We first construct the complex forcing function which, upon the application of Euler's identity, yields the given real forcing function. Since

$$\cos \omega t = \text{Re } e^{j\omega t}$$

then the necessary complex source is

$$V_m e^{j\omega t}$$

Fig. 9-4: A simple circuit in the sinusoidal steady state is to be analyzed by the application of a complex forcing function.

$$v_s = V_m \cos \omega t$$

The complex response which results is expressed in terms of an unknown amplitude I_m and an unknown phase angle ϕ,

$$I_m e^{j(\omega t + \phi)}$$

Writing the differential equation for this particular circuit,

$$Ri + L\frac{di}{dt} = v_s$$

we insert our complex expressions for v_s and i,

$$RI_m e^{j(\omega t + \phi)} + L\frac{d}{dt}(I_m e^{j(\omega t + \phi)}) = V_m e^{j\omega t}$$

take the indicated derivative,

$$RI_m e^{j(\omega t + \phi)} + j\omega L I_m e^{j(\omega t + \phi)} = V_m e^{j\omega t}$$

and obtain a complex *algebraic* equation. In order to determine the value of I_m and ϕ, we divide throughout by the common factor $e^{j\omega t}$,

$$RI_m e^{j\phi} + j\omega L I_m e^{j\phi} = V_m \tag{11}$$

factor the left side,

$$I_m e^{j\phi}(R + j\omega L) = V_m$$

rearrange,

$$I_m e^{j\phi} = \frac{V_m}{R + j\omega L}$$

and identify I_m and ϕ by expressing the right side of the equation in exponential or polar form,

$$I_m e^{j\phi} = \frac{V_m}{\sqrt{R^2 + \omega^2 L^2}}\, e^{j[-\tan^{-1}(\omega L/R)]} \tag{12}$$

Thus,

$$I_m = \frac{V_m}{\sqrt{R^2 + \omega^2 L^2}}$$

and

$$\phi = -\tan^{-1}\frac{\omega L}{R}$$

The complex response is given by (12). Since I_m and ϕ are readily identified, we can write the expression for $i(t)$ immediately. Using a more rigorous approach, we may obtain the real response $i(t)$ by reinserting the $e^{j\omega t}$ factor on both sides of (12) and taking the real part, obtained by applying Euler's omnipotent formula. Thus,

$$i(t) = I_m \cos(\omega t + \phi)$$

$$= \frac{V_m}{\sqrt{R^2 + \omega^2 L^2}}\cos\left(\omega t - \tan^{-1}\frac{\omega L}{R}\right)$$

which agrees with the response obtained for this same circuit in the previous chapter, Eq. (8-4).

Although we have successfully worked a sinusoidal steady-state problem by applying a complex forcing function and obtaining a complex response, we have not taken advantage of the full power of the complex representation. In order to do so, we must carry the concept of the complex source or response one additional step and define the quantity called a "phasor."

Drill Problems	**9-1.** (If one has trouble working this Drill Problem, one should study Appendix 4.) Evaluate and express the answer in rectangular form: (a) $8\underline{/35°} - 5e^{-j100°}$; (b) $[4/(3 - j2)] + 1/(j3)$. Evaluate and express the result in polar form: (c) $2.5\underline{/-120°} - j8$; (d) $(4 + j3)e^{j(200t+50°)}$ at $t = 4$ ms. *Ans:* $7.42 + j9.51$; $0.923 + j0.282$; $10.24\underline{/-97.0°}$; $5.00\underline{/132.7°}$

9-2. Assuming use of the passive sign convention, find the total complex: (a) current that results when the complex voltage, $60e^{j(500t-50°)}$ V, is applied to the parallel combination of 20 Ω and 0.1 mF; (b) voltage that results when the complex current $5e^{j(500t+90°)}$ A is applied to the series combination of 3 Ω and 2 mH. *Ans:* $4.24e^{j(500t-5°)}$ A; $15.81e^{j(500t+108.4°)}$ V

9-3
The phasor

A sinusoidal current or voltage *at a given frequency* is characterized by only two parameters, an amplitude and a phase angle. The complex representation of the voltage or current is also characterized by these same two parameters. For example, the assumed sinusoidal form of the current response in the example above was

$$I_m \cos (\omega t + \phi)$$

and the corresponding representation of this current in complex form is

$$I_m e^{j(\omega t + \phi)}$$

Once I_m and ϕ are specified, the current is exactly defined. Throughout any linear circuit operating in the sinusoidal steady state at a single frequency ω, every current or voltage may be characterized completely by a knowledge of its amplitude and phase angle. Moreover, the complex representation of every voltage and current will contain the same factor $e^{j\omega t}$. The factor is superfluous; since it is the same for every quantity, it contains no useful information. Of course, the value of the frequency may be recognized by inspecting one of these factors, but it is a lot simpler to write down the value of the frequency near the circuit diagram once and for all and avoid carrying redundant information throughout the solution. Thus, we could simplify the voltage source and the current response of the example by representing them concisely as

$$V_m \quad \text{or} \quad V_m e^{j0°} \quad \text{and} \quad I_m e^{j\phi}$$

These complex quantities are usually written in polar form rather than exponential form in order to achieve a slight additional saving of time and effort.

Thus, the source voltage

$$v(t) = V_m \cos \omega t$$

we now express in complex form as

$$V_m \underline{/0°}$$

and the current response

$$i(t) = I_m \cos (\omega t + \phi)$$

becomes

$$I_m \underline{/\phi}$$

This abbreviated complex representation is called a *phasor*.[2] Let us review the steps by which a real sinusoidal voltage or current is transformed into a phasor, and then we shall be able to define a phasor more meaningfully and to assign a symbol to represent it.

A real sinusoidal current

$$i(t) = I_m \cos (\omega t + \phi)$$

is expressed as the real part of a complex quantity by an equivalent form of Euler's identity

$$i(t) = \text{Re} \, (I_m e^{j(\omega t + \phi)})$$

We then represent the current as a complex quantity by dropping the instruction Re, thus adding an imaginary component to the current without affecting the real component; further simplification is achieved by suppressing the factor $e^{j\omega t}$,

$$\mathbf{I} = I_m e^{j\phi}$$

and writing the result in polar form,

$$\mathbf{I} = I_m \underline{/\phi}$$

This abbreviated complex representation is the *phasor representation*; phasors are complex quantities and hence are printed in boldface type. Capital letters are used for the phasor representation of an electrical quantity because the phasor is not an instantaneous function of time; it contains only amplitude and phase information. We recognize this difference in viewpoint by referring to $i(t)$ as a *time-domain representation* and terming the phasor **I** a *frequency-domain representation*. It should be noted that the frequency-domain expression of a current or voltage does not explicitly include the frequency; however, we might think of the frequency as being so fundamental in the frequency domain that it is emphasized by its omission.[3]

The process by which we change $i(t)$ into **I** is called a phasor transforma-

---

[2] Contrary to popular opinion, the phasor was not invented by Captain Kirk.
[3] Very little local mail in this country includes "U.S.A." in its address.

tion from the time domain to the frequency domain. The mathematical steps of the time-domain to frequency-domain transformation are as follows:

1 Given the sinusoidal function $i(t)$ in the time domain, write $i(t)$ as a cosine wave with a phase angle. For example, $\sin \omega t$ should be written as $\cos(\omega t - 90°)$.
2 Express the cosine wave as the real part of a complex quantity by using Euler's identity.
3 Drop Re.
4 Suppress $e^{j\omega t}$.

In practice, we find it very easy to jump directly from the first step to the answer by extracting the amplitude and phase angle of the cosine wave from its time-domain expression. The four steps above are listed for technical fulfillment only.

As an example, let us transform the time-domain voltage

$$v(t) = 100 \cos(400t - 30°)$$

into the frequency domain. The time-domain expression is already in the form of a cosine wave with a phase angle, and the rigorous time-domain to frequency-domain transformation is accomplished by taking the real part of the complex representation,

$$v(t) = \text{Re}\,(100e^{j(400t - 30°)})$$

and dropping Re and suppressing $e^{j\omega t}$,

$$\mathbf{V} = 100\underline{/-30°}$$

However, it is much simpler to identify the 100 and $-30°$ in the time-domain cosine representation, and to write $\mathbf{V} = 100\underline{/-30°}$ directly.

In a similar fashion, the time-domain current

$$i(t) = 5 \sin(377t + 150°)$$

transforms into the phasor

$$\mathbf{I} = 5\underline{/60°}$$

Before we consider the analysis of circuits in the sinusoidal steady state through the use of phasors, it is necessary to learn how to shift our transformation smoothly into reverse to return to the time domain from the frequency domain. The process is exactly the reverse of the sequence given above. Thus, the detailed steps in the frequency-domain to time-domain transformation are as follows:

1 Given the phasor current $\mathbf{I}$ in polar form in the frequency domain, write the complex expression in exponential form.
2 Reinsert (multiply by) the factor $e^{j\omega t}$.
3 Replace the real-part operator Re.
4 Obtain the time-domain representation by applying Euler's identity. The resultant cosine-wave expression may be changed to a sine wave, if desired, by increasing the argument by 90°.

Once again, we should be able to bypass the mathematics and write the desired time-domain expression by utilizing the amplitude and phase angle of the polar form. Thus, given the phasor voltage

$$\mathbf{V} = 115\underline{/-45°}$$

we would write the time-domain equivalent directly,

$$v(t) = 115 \cos (\omega t - 45°)$$

As a sine wave, $v(t)$ could be written

$$v(t) = 115 \sin (\omega t + 45°)$$

In anticipation of the methods of applying phasors in the analysis of the sinusoidal steady-state circuit, we may treat ourselves to a quick preview by returning to the example of the *RL* series circuit. A number of steps after writing the applicable differential equation, we arrived at (11), rewritten below,

$$RI_m e^{j\phi} + j\omega L I_m e^{j\phi} = V_m$$

If we substitute phasors for the current,

$$\mathbf{I} = I_m\underline{/\phi}$$

and the voltage

$$\mathbf{V} = V_m\underline{/0°}$$

we obtain

$$R\mathbf{I} + j\omega L\mathbf{I} = \mathbf{V}$$

or
$$(R + j\omega L)\mathbf{I} = \mathbf{V} \qquad (13)$$

a complex algebraic equation in which the current and voltage are expressed in phasor form. This equation is only slightly more complicated than Ohm's law for a single resistor. The next time we analyze this circuit, we shall begin with (13).

Drill Problems

9-3. Express each of these currents as a phasor: (*a*) 8 sin ($\omega t - 20°$) A; (*b*) 6 sin $\omega t - 2 \cos \omega t$ A; (*c*) 4 cos ($\omega t - 80°$) $- 3 \sin (\omega t + 30°)$.

Ans: $8\underline{/-110°}$; $6.32\underline{/-108.4°}$; $1.564\underline{/-121.0°}$ A

9-4. If $f = 60$ Hz, find the instantaneous value at $t = 1$ ms of the voltage represented by the phasor: (*a*) $120\underline{/0°}$ V; (*b*) $80 + j75$ V; (*c*) $50 + 80\underline{/70°}$ V.

Ans: 111.6; 46.8; 44.3 V

9-4
Phasor relationships for *R*, L, and C

Now that we are able to transform into and out of the frequency domain, we can proceed to our simplification of sinusoidal steady-state analysis by establishing the relationship between the phasor voltage and phasor current for each of the three passive elements. We shall begin with the defining equation for each of the elements, a time-domain relationship, and then let both the current and the voltage become complex quantities. After suppressing $e^{j\omega t}$

Fig. 9-5: A resistor and its associated voltage and current in: (a) the time domain, $v = Ri$; (b) the frequency domain, $\mathbf{V} = R\mathbf{I}$.

throughout the equation, the desired relationship between the phasor voltage and phasor current will become apparent.

The resistor provides the simplest case. In the time domain, as indicated by Fig. 9-5a, the defining equation is

$$v(t) = Ri(t) \tag{14}$$

Now let us apply the complex voltage

$$V_m e^{j(\omega t + \theta)} = V_m \cos(\omega t + \theta) + jV_m \sin(\omega t + \theta) \tag{15}$$

and assume the complex current response

$$I_m e^{j(\omega t + \phi)} = I_m \cos(\omega t + \phi) + jI_m \sin(\omega t + \phi) \tag{16}$$

and obtain

$$V_m e^{j(\omega t + \theta)} = RI_m e^{j(\omega t + \phi)}$$

By dividing throughout by $e^{j\omega t}$ (or suppressing $e^{j\omega t}$ on both sides of the equation), we find

$$V_m e^{j\theta} = RI_m e^{j\phi}$$

or, in polar form,

$$V_m\underline{/\theta} = RI_m\underline{/\phi}$$

But $V_m\underline{/\theta}$ and $I_m\underline{/\phi}$ merely represent the general voltage and current phasors $\mathbf{V}$ and $\mathbf{I}$. Thus,

$$\mathbf{V} = R\mathbf{I} \tag{17}$$

The voltage-current relationship in phasor form for a resistor has the same form as the relationship between the time-domain voltage and current. The defining equation in phasor form is illustrated in Fig. 9-5b. The equality of the angles θ and ϕ is apparent, and the current and voltage are thus in phase.

As an example of the use of both the time-domain and frequency-domain relationships, let us assume a voltage of $8 \cos(100t - 50°)$ V across a 4-Ω resistor. Working in the time domain, we find that the current must be

$$i(t) = \frac{v(t)}{R} = 2 \cos(100t - 50°)$$

The phasor form of the same voltage is $8\underline{/-50°}$ V, and therefore

$$\mathbf{I} = \frac{\mathbf{V}}{R} = 2\underline{/-50°} \quad \text{A}$$

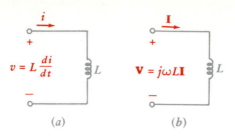

Fig. 9-6: An inductor and its associated voltage and current in: (a) the time domain, v = L di/dt; (b) the frquency domain, **V** = jωL**I**.

If we transform this answer back to the time domain, it is evident that the same expression for the current is obtained.

It is apparent that there is no saving in time or effort when a resistive circuit is analyzed in the frequency domain. As a matter of fact, if it is necessary to transform a given time-domain source to the frequency domain and then translate the desired response back to the time domain, we should be much better off working completely in the time domain. This definitely does not apply to any circuit containing resistance and either inductance or capacitance, unless the complexity of the problem calls forth the digital computer, a machine that is quite indifferent to tedious calculations.

Let us now turn to the inductor. The time-domain network is shown in Fig. 9-6a, and the defining equation, a time-domain expression, is

$$v(t) = L\frac{di(t)}{dt} \tag{18}$$

After substituting the complex voltage equation (15) and complex current equation (16) into (18), we have

$$V_m e^{j(\omega t + \theta)} = L\frac{d}{dt}(I_m e^{j(\omega t + \phi)})$$

Taking the indicated derivative

$$V_m e^{j(\omega t + \theta)} = j\omega L I_m e^{j(\omega t + \phi)}$$

and suppressing $e^{j\omega t}$,

$$V_m e^{j\theta} = j\omega L I_m e^{j\phi}$$

we obtain the desired phasor relationship

$$\mathbf{V} = j\omega L\mathbf{I} \tag{19}$$

The time-domain differential equation (18) has become an algebraic equation in the frequency domain. The phasor relationship is indicated in Fig. 9-6b. Note that the angle of the factor $j\omega L$ is exactly +90° and that **I** must therefore lag **V** by 90° in an inductor.

As an illustration of the phasor relationship, let us apply the voltage $8\underline{/-50°}$ V at a frequency $\omega = 100$ rad/s to a 4-H inductor. From (19), the phasor current is

$$I = \frac{V}{j\omega L} = \frac{8\underline{/-50°}}{j100(4)} = -j0.02\underline{/-50°}$$

or

$$I = 0.02\underline{/-140°} \quad A$$

If we express this current in the time domain, it becomes

$$i(t) = 0.02 \cos(100t - 140°) \quad A$$

This response is also easily obtained by working entirely in the time domain; it is not so easily obtained if resistance or capacitance is combined with the inductance.

The final element we must consider is the capacitor. The definition of capacitance, a familiar time-domain expression, is

$$i(t) = C \frac{dv(t)}{dt} \tag{20}$$

The equivalent expression in the frequency domain is obtained once more by letting $v(t)$ and $i(t)$ be the complex quantities of (15) and (16), taking the indicated derivative, suppressing $e^{j\omega t}$, and recognizing the phasors **V** and **I**. It is

$$I = j\omega C V \tag{21}$$

Thus, **I** leads **V** by 90° in a capacitor. This, of course, does not mean that a current response is present one-quarter of a period earlier than the voltage that caused it! We are studying steady-state response, and we find that the current maximum is caused by the increasing voltage that occurs 90° earlier than the voltage maximum.

If the phasor voltage $8\underline{/-50°}$ V is applied to a 4-F capacitor at $\omega = 100$ rad/s, the phasor current is

$$I = j100(4)(8\underline{/-50°}) = 3200\underline{/40°} \quad A$$

The amplitude of the current is tremendous, but the assumed size of the capacitor is also unrealistic. If a 4-F capacitor were constructed of two flat plates separated 1 mm in air, each plate would have the area of about 85 000 football fields.[4] The time-domain and frequency-domain representations are compared in Figs. 9-7a and b.

[4] Including both end zones.

Fig. 9-7: (a) The time-domain and (b) the frequency-domain relationships between capacitor current and voltage.

(a) (b)

We have now obtained the **V-I** relationships for the three passive elements. These results are summarized in Table 9-1 where the time-domain v-i expressions and the frequency-domain **V-I** relationships are shown in adjacent columns for the three elements. All the phasor equations are algebraic. Each is also linear, and the equations relating to inductance and capacitance bear a great similarity to Ohm's law. We shall use them as we use Ohm's law.

Table 9-1: A comparison and summary of the relationships between v and i in the time domain and **V** and **I** in the frequency domain for R, L, and C

Time domain		Frequency domain	
	$v = Ri$	$\mathbf{V} = \mathbf{RI}$	
	$v = L\dfrac{di}{dt}$	$\mathbf{V} = j\omega L\mathbf{I}$	
	$v = \dfrac{1}{C}\int i\,dt$	$\mathbf{V} = \dfrac{1}{j\omega C}\mathbf{I}$	

Before we do so, we must show that phasors satisfy Kirchhoff's two laws. Kirchhoff's voltage law in the time domain is

$$v_1(t) + v_2(t) + \cdots + v_N(t) = 0$$

We now use Euler's identity to replace each real voltage by the complex voltage having the same real part, suppress $e^{j\omega t}$ throughout, and obtain

$$\mathbf{V}_1 + \mathbf{V}_2 + \cdots + \mathbf{V}_n = 0$$

Kirchhoff's current law is shown to hold for phasor currents by a similar argument.

Now let us look briefly at the series RL circuit that we have considered several times before. The circuit is shown in Fig. 9-8, and a phasor current and several phasor voltages are indicated. We may obtain the desired response, a time-domain current, by first finding the phasor current. The method is similar to that used in analyzing our first single-loop resistive circuit. From Kirchhoff's voltage law,

$$\mathbf{V}_R + \mathbf{V}_L = \mathbf{V}_s$$

and the recently obtained **V-I** relationships for the elements give

$$R\mathbf{I} + j\omega L\mathbf{I} = \mathbf{V}_s$$

The phasor current is then found in terms of the source voltage $\mathbf{V}_s$,

$$\mathbf{I} = \frac{\mathbf{V}_s}{R + j\omega L}$$

Fig. 9-8: The series RL circuit with a phasor voltage applied.

Let us select a source-voltage amplitude of V_m and phase angle of $0°$; the latter merely represents the simplest possible choice of a reference. Thus,

$$\mathbf{I} = \frac{V_m \underline{/0°}}{R + j\omega L}$$

The current may be transformed to the time domain by first writing it in polar form,

$$\mathbf{I} = \frac{V_m}{\sqrt{R^2 + \omega^2 L^2}} \left/ -\tan^{-1} \frac{\omega L}{R} \right.$$

and then following the familiar sequence of steps to obtain in a very simple manner the same result we obtained the "hard way" as Eq. (4) of Section 8-3.

Drill Problem

9-5. In Fig. 9-9, let $\omega = 800$ rad/s, $\mathbf{V}_L = 66\underline{/57°}$ V, and $\mathbf{V}_C = 50\underline{/-85°}$ V, and find: (a) $\mathbf{V}_s$; (b) $\mathbf{I}_s$; (c) $i_x(t)$.

$Ans:$ $119.0\underline{/0.7°}$ V; $4.27\underline{/41.9°}$ A; $7.85 \cos(800t + 24.1°)$ A

Fig. 9-9: See Drill Prob. 9-5.

9-5
Impedance

The current-voltage relationships for the three passive elements in the frequency domain are (assuming the passive sign convention is satisfied)

$$\mathbf{V} = R\mathbf{I} \qquad \mathbf{V} = j\omega L\mathbf{I} \qquad \mathbf{V} = \frac{\mathbf{I}}{j\omega C}$$

If these equations are written as phasor-voltage phasor-current ratios

$$\frac{\mathbf{V}}{\mathbf{I}} = R \qquad \frac{\mathbf{V}}{\mathbf{I}} = j\omega L \qquad \frac{\mathbf{V}}{\mathbf{I}} = \frac{1}{j\omega C}$$

we find that these ratios are simple functions of the element values, and frequency also, in the case of inductance and capacitance. We treat these ratios in the same manner we treat resistances, with the exception that they are complex quantities and all algebraic manipulations must be those appropriate for complex numbers.

Let us define the ratio of the phasor voltage to the phasor current as *impedance*, symbolized by the letter **Z**. The impedance is a complex quantity having the dimensions of ohms. Impedance is not a phasor and cannot be transformed to the time domain by multiplying by $e^{j\omega t}$ and taking the real part. Instead, we think of an inductor as being represented in the time domain by its inductance L and in the frequency domain by its impedance $j\omega L$. A capacitor in the time domain has a capacitance C and an impedance $1/j\omega C$ in the frequency domain. Impedance is a part of the frequency domain and not a concept which is a part of the time domain.

The validity of Kirchhoff's two laws in the frequency domain enables it to be easily demonstrated that impedances may be combined in series and parallel by the same rules we have already established for resistances. For example, at $\omega = 10^4$ rad/s, a 5-mH inductor in series with a 100-μF capacitor may be replaced by the single impedance which is the sum of the individual impedances. The impedance of the inductor is

$$\mathbf{Z}_L = j\omega L = j50 \quad \Omega$$

the impedance of the capacitor is

$$\mathbf{Z}_C = \frac{1}{j\omega C} = -j1 \quad \Omega$$

and the impedance of the series combination is therefore

$$\mathbf{Z}_{eq} = j50 - j1 = j49 \quad \Omega$$

The impedance of inductors and capacitors is a function of frequency, and this equivalent impedance is thus applicable only at the single frequency at which it was calculated, $\omega = 10\,000$. At $\omega = 5000$, $\mathbf{Z}_{eq} = j23 \ \Omega$.

The parallel combination of these same two elements at $\omega = 10^4$ yields an impedance which is the product divided by the sum,

$$\mathbf{Z}_{eq} = \frac{(j50)(-j1)}{j50 - j1} = \frac{50}{j49} = -j1.02 \quad \Omega$$

At $\omega = 5000$, the parallel equivalent is $-j2.17 \ \Omega$.

The complex number or quantity representing impedance may be expressed in either polar or rectangular form. In polar form, an impedance, such as $100\underline{/-60°}$, is described as having an impedance magnitude of 100 Ω and a phase angle of $-60°$. The same impedance in rectangular form, $50 - j86.6$, is said to have a *resistive component*, or *resistance*, of 50 Ω and a *reactive component*, or *reactance*, of $-86.6 \ \Omega$. The resistive component is the real part of the impedance, and the reactive component is the imaginary component of the impedance, including sign, but of course excluding the imaginary operator.

It is important to note that the resistive component of the impedance is not necessarily equal to the resistance of the resistor which is present in the network. For example, a 20-Ω resistor in series with a 5-H inductor at $\omega = 4$ has an equivalent impedance $\mathbf{Z} = 20 + j20 \ \Omega$, or in polar form, $28.3\underline{/45°} \ \Omega$. In this case, the resistive component of the impedance is equal to the resistance of the series resistor because the network is a simple series network. However, if these same two elements are placed in parallel, the equivalent

Fig. 9-10: (a) An *RLC* circuit for which the sinusoidal forced response *i(t)* is desired. (b) The frequency-domain equivalent of the given circuit at ω = 3000 rad/s.

impedance is $20(j20)/(20 + j20)$, or $10 + j10\ \Omega$. The resistive component of the impedance is now $10\ \Omega$.

No special symbol is assigned for impedance magnitude or phase angle. A general form for an impedance in polar form might be

$$\mathbf{Z} = |\mathbf{Z}|\underline{/\theta}$$

In rectangular form, the resistive component is represented by R and the reactive component by X. Thus,

$$\mathbf{Z} = R + jX$$

Let us now use the impedance concept to analyze an *RLC* circuit, shown in Fig. 9-10*a*. The circuit is shown in the time domain, and the time-domain current response $i(t)$ is required. However, the analysis should be carried out in the frequency domain. We therefore begin by drawing a frequency-domain circuit; the source is transformed to the frequency domain, becoming $40\underline{/-90°}$ V; the response is transformed to the frequency domain, being represented as **I**; and the impedances of the inductor and capacitor, determined at $\omega = 3000$, are $j1$ and $-j2$ kΩ, respectively. The frequency-domain circuit is shown in Fig. 9-10*b*.

The equivalent impedance offered to the source is now calculated:

$$\mathbf{Z}_{eq} = 1.5 + \frac{(j1)(1 - j2)}{j1 + 1 - j2} = 1.5 + \frac{2 + j1}{1 - j1}$$

$$= 1.5 + \frac{2 + j1}{1 - j1}\frac{1 + j1}{1 + j1} = 1.5 + \frac{1 + j3}{2}$$

$$= 2 + j1.5 = 2.5\underline{/36.9°}\quad \text{k}\Omega$$

The phasor current is thus

$$\mathbf{I} = \frac{\mathbf{V}_s}{\mathbf{Z}_{eq}} = \frac{40\underline{/-90°}}{2.5\underline{/36.9°}} = 16\underline{/-126.9°}\quad \text{mA}$$

Upon transforming the current to the time domain, the desired response is obtained,

$$i(t) = 16\cos(3000t - 126.9°)\quad \text{mA}$$

If the capacitor current is desired, current division should be applied in the frequency domain.

Before we begin to write great numbers of equations in the time domain or in the frequency domain, it is most important that we shun the construction of equations which are partly in the time domain, partly in the frequency domain, and wholly incorrect. One clue that a faux pas of this type has been committed is the sight of both a complex number and a *t* in the same equation, except when the factor $e^{j\omega t}$ is also present. And, since $e^{j\omega t}$ is much bigger in the derivations than it is in applications, it is pretty safe to say that students who find they have just created an equation containing *j and t* or ∟ *and t* have created a monster that they and the world would be better off without.

For example, two equations back, we saw:

$$\mathbf{I} = \frac{\mathbf{V}_s}{\mathbf{Z}_{eq}} = \frac{40\underline{/-90°}}{2.5\underline{/36.9°}} = 16\underline{/-126.9°} \quad \text{mA}$$

Please do not try anything like the following:

$$i(t) = \frac{40 \sin 3000t}{2.5\underline{/36.9°}} \qquad \text{(No! No! No!)}$$

or

$$i(t) = \frac{40 \sin 3000t}{2 + j1.5} \qquad \text{(also, No!)}$$

Drill Problems

9-6. Find the input impedance that would be measured in the circuit of Fig. 9-11 between points: (*a*) *a* − *x*; (*b*) *b* − *x*; (*c*) *a* − *b*.

Ans: $10 - j20$; $10 + j0$; $40 - j20$ Ω

Fig. 9-11: See Drill Prob. 9-6.

9-7. The positions of the inductor and capacitor in the circuit of Fig. 9-10 are interchanged. Find the amplitudes of the voltages across each of the four circuit elements.

Ans: 17.14; 22.9; 16.16; 16.16 V

**9-6
Admittance**

Just as conductance, the reciprocal of resistance, proved to be a useful quantity in the analysis of resistive circuits, so does the reciprocal of impedance offer some convenience in the sinusoidal steady-state analysis of a general *RLC* circuit. We define the *admittance* **Y** of a circuit element as the ratio of phasor current to phasor voltage (assuming the passive sign convention is satisfied):

$$\mathbf{Y} = \frac{\mathbf{I}}{\mathbf{V}}$$

and thus

$$\mathbf{Y} = \frac{1}{\mathbf{Z}}$$

The real part of the admittance is the *conductance G*, and the imaginary part of the admittance is the *susceptance B*. Thus,

$$\mathbf{Y} = G + jB = \frac{1}{\mathbf{Z}} = \frac{1}{R + jX} \tag{22}$$

Equation (22) should be scrutinized carefully; it does *not* state that the real part of the admittance is equal to the reciprocal of the real part of the impedance or that the imaginary part of the admittance is equal to the reciprocal of the imaginary part of the impedance. Admittance, conductance, and susceptance are all measured in mhos.

An impedance

$$\mathbf{Z} = 1 - j2 \quad \Omega$$

which might be represented, for example, by a 1-ohm resistor in series with a 0.1-μF capacitor at $\omega = 5$ Mrad/s, possesses an admittance

$$\mathbf{Y} = \frac{1}{\mathbf{Z}} = \frac{1}{1 - j2} = \frac{1}{1 - j2}\frac{1 + j2}{1 + j2}$$
$$= 0.2 + j0.4 \quad \mho$$

Without stopping to inspect a formal proof, it should be apparent that the equivalent admittance of a network consisting of a number of parallel branches is the sum of the admittances of the individual branches. Thus, the numerical value of the admittance above might be obtained from a conductance of 0.2 $\mho$ in parallel with a positive susceptance of 0.4 $\mho$. The former could be represented by a 5-Ω resistor and the latter by a 0.08-μF capacitor at $\omega = 5$ Mrad/s, since the admittance of a capacitor is evidently $j\omega C$.

As a check on our analysis, let us compute the impedance of this latest network, a 5-Ω resistor in parallel with a 0.08-μF capacitor at $\omega = 5$ Mrad/s. The equivalent impedance is

$$\mathbf{Z} = \frac{5(1/j\omega C)}{5 + 1/j\omega C} = \frac{5(-j2.5)}{5 - j2.5} = 1 - j2 \quad \Omega$$

as before. These two networks represent only two of an infinite number of different networks which possess this same impedance and admittance at this frequency. They do, however, represent the only two two-element networks, and thus might be considered to be the two simplest networks having an impedance of $1 - j2$ Ω and an admittance of $0.2 + j0.4$ $\mho$ at $\omega = 5 \times 10^6$ rad/s.

The term *immittance,* a combination of the words "impedance" and "admittance," is sometimes used as a general term for both impedance and admittance. For example, it is evident that a knowledge of the phasor voltage across a known immittance enables the current through the immittance to be calculated.

Drill Problems

9-8. Find in rectangular form the admittance of: (*a*) a network whose impedance is $100 - j160 \ \Omega$; (*b*) the series combination of 50 Ω, 20 mH, and 2 μF, if $\omega = 4$ krad/s; (*c*) the parallel combination of 50 Ω, 20 mH, and 2 μF at 4 krad/s.   *Ans: 2.81 + j4.49; 11.05 + j9.94; 20 − j4.5 m℧*

9-9. A 15-mH inductor is in series with the parallel combination of a resistor R and 20 μF at $\omega = 1000$ rad/s. (*a*) Find the admittance of the network if $R = 80 \ \Omega$. (*b*) If $\mathbf{Y} = G - j0.02$ ℧, find R.   *Ans: 23.8 + j22.2 m℧; 21.0 Ω*

Problems

1 Find in polar form: (*a*) $5\underline{/27°} - 3\underline{/112°}$; (*b*) $4\underline{/30°}/(1 + 2\underline{/-70°})$; (*c*) $(-4 + j5)/(1 + j0.4)$. Find in rectangular form: (*d*) $(2\underline{/30°} + 1\underline{/60°})^2$; (*e*) $8e^{j0.5}$.

2 Evaluate the following in polar form: (*a*) $(36\underline{/21°}) \left(\dfrac{1}{j}\right) + 19\underline{/-108°}$; (*b*) $11j + \sqrt{80\underline{/83°}}$. Evaluate in rectangular form: (*c*) $\dfrac{1}{\dfrac{1}{7 - j3.1} + \dfrac{1}{4 + j9.2}}$;

(*d*) Re[$6.11\underline{/26°}$] + Im[$4.18\underline{/-99°}$] + [$5.50\underline{/412°}$]*.

3 In Fig. 9-12, let $i(t)$ be the complex current, $0.8e^{j20t}$ A, while the source voltage $v_s(t)$ is $(30 - j40)e^{j20t}$ V, and find the complex input voltage $v_{in}(t)$.

Fig. 9-12: See Prob. 3.

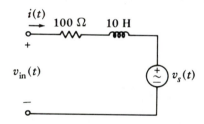

4 Let $i_s(t) = j2e^{j50t}$ A and $v(t) = 240e^{j50t}$ V in the network of Fig. 9-13. Find $i_{in}(t)$.

Fig. 9-13: See Prob. 4.

5 In a linear network, similar to that shown in Fig. 9-1, a sinusoidal voltage source, $v_s = 40 \cos 1000t$ V, produces a current response, $i_o = 2.5 \cos (1000t - 24°)$ A. Find $i_o(t)$ if $v_s(t) = :$ (*a*) $20 \sin 1000t$ V; (*b*) $20 \cos (1000t - 40°)$ V; (*c*) $20e^{j27°}e^{j1000t}$ V; (*d*) $(10 - j6)e^{j1000t}$ V.

6 Express as a phasor: (*a*) $v(t) = 165 \cos (120\pi t + 30°)$ V; (*b*) $i(t) = 2 \cos 10^6 t + 3 \sin 10^6 t$ mA. Find the instantaneous value at $t = 1$ ms of: (*c*) $\mathbf{V} = 40 - j70$ V, $\omega = 400$ rad/s; (*d*) $\mathbf{I} = -8.1 - j2.4$ A, $f = 60$ Hz.

7 Let $\mathbf{I} = 10\underline{/-130°}$ A and find i at $t = 1$ ms if $\omega = :$ (*a*) 1200 rad/s; (*b*) 600 rad/s. (*c*) Let $i_1 = 5 \cos (100t + 50°)$, $i_2 = 4 \cos 100t$, and $i_3 = 3 \sin (100t + 70°)$ A, and $i_x = i_1 + i_2 + i_3$. By transforming i_1, i_2, and i_3 into phasors, find $\mathbf{I}_x$ and $i_x(t)$.

8 If $\omega = 200$ rad/s, find the instantaneous value at $t = 1$ ms of: (*a*) $\mathbf{I} = 2\underline{/70°}$ A; (*b*) $\mathbf{I} = 4 - j1$ A. Find the phasor voltage corresponding to: (*c*) $v(t) = 6 \sin (500t -$

50°) V; (d) $v(t) = -3 \cos 50t + 6 \sin 50t$ V; (e) $8 \cos (100t - 100°) - 6 \cos (100t - 40°)$ V.

9 In Fig. 9-14, let $\mathbf{I}_1 = 6 + j1$ A, $\mathbf{I}_2 = 2 - j5$ A, and $\omega = 1$ krad/s. At $t = 2.5$ ms, find the instantaneous value of: (a) v_A; (b) i_A; (c) the power being absorbed by element A.

Fig. 9-14: See Prob. 9.

10 A purple- and yellow-striped box contains a voltage source $\mathbf{V}_{s1}$ and a current source $\mathbf{I}_{s2}$. The voltage between an available pair of terminals is labeled $\mathbf{V}_{AB}$. If $\mathbf{V}_{s1} = 10\underline{/20°}$ V and $\mathbf{I}_{s2} = 0.2\underline{/-30°}$ A, then $\mathbf{V}_{AB} = -50 + j80$ V. However, if $\mathbf{V}_{s1} = 20\underline{/-80°}$ V and $\mathbf{I}_{s2} = 0.5\underline{/30°}$ A, then $\mathbf{V}_{AB} = 40 + j60$ V. Find $\mathbf{V}_{AB}$ if $\mathbf{V}_{s1} = 10 - j30$ V and $\mathbf{I}_{s2} = 0.3 - j0.1$ A.

11 A resistor R, an inductor L, and an ideal source $v_s = 100 \cos \omega t$ V are in series. If $\omega = 200$ rad/s, the phasor current is $5 - j2$ A. What will the phasor current be if $\omega = 400$ rad/s?

12 Let $\omega = 500$ rad/s for the network of Fig. 9-15 and find the phasor voltage $\mathbf{V}$.

Fig. 9-15: See Probs. 12 and 16.

13 Find $v_1(t)$, $v_2(t)$, and $v_3(t)$ in the circuit shown in Fig. 9-16 if $\omega = 5$ krad/s.

Fig. 9-16: See Prob. 13.

14 Find $\mathbf{V}$ in Fig. 9-17 if the box contains: (a) 8 Ω in series with 5 mH; (b) 8 Ω in series with 50 μF; (c) 8 Ω, 5 mH, and 50 μF in series; (d) 8 Ω, 5 mH, and 50 μF in series, but $\omega = 4$ krad/s.

Fig. 9-17: See Prob. 14.

15 A 200-Ω resistor, a 25-μF capacitor, and an inductor L are in parallel. The phasor voltage across the combination is $100\underline{/0°}$ V. (a) Find L if the current entering the plus-marked terminal is $\mathbf{I} = 0.5 - j2$ A and $\omega = 1$ krad/s. (b) Find L if $\omega = 100$ rad/s and $|\mathbf{I}| = 1$ A.

16 Find the input impedance of the network shown in Fig. 9-15 if $\omega = 500$ rad/s.

17 A 200-Ω resistor, a 25-μF capacitor, and an inductor L are in parallel. (a) Find the impedance of the parallel combination at $\omega = 100$ rad/s if $L = 10$ H. (b) If the magnitude of the impedance is 80 Ω at $\omega = 100$ rad/s, find L. (c) At what values of ω is the impedance magnitude equal to 150 Ω if $L = 10$ H?

18 A 2-kΩ resistor is in parallel with a 4-μF capacitor. If the impedance of the parallel combination is $\mathbf{Z}_{in}$, at what frequency is: (a) $|\mathbf{Z}_{in}| = 1500$ Ω? (b) the angle of $\mathbf{Z}_{in} = -30°$? (c) Re$[\mathbf{Z}_{in}] = 200$ Ω? (d) $|X_{in}| = 400$ Ω?

19 A certain two-terminal network has an input impedance of $5 - j8$ Ω. At a frequency of 4 krad/s, what inductance should be placed in parallel with the network to cause the input impedance to: (a) have zero reactance? (b) have a magnitude of 4 Ω?

20 Find $\mathbf{Z}_{in}$ at $\omega = 5$ rad/s for the network of Fig. 9-18 if terminals $a–b$ are: (a) open-circuited; (b) short-circuited.

Fig. 9-18: See Prob. 20.

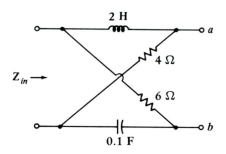

21 A 10-Ω resistor, a 2-mH inductor, and a 0.2-μF capacitor are in series. The impedance of the series combination is $\mathbf{Z}_{in}$. (a) At what value of ω is $|\mathbf{Z}_{in}|$ a minimum? (b) At what values of ω is $|\mathbf{Z}_{in}| = 2|\mathbf{Z}_{in}|_{min}$?

22 A 2-Ω resistor and a 0.5-F capacitor are in series, and a 3-H inductor is in parallel with the series combination. If the impedance of the parallel network is designated $\mathbf{Z}_{in}(\omega) = R_{in}(\omega) + jX_{in}(\omega)$, find $R_{in}(\omega)$ and $X_{in}(\omega)$.

23 A 100-Ω resistor is in series with a 10-μF capacitor. Let the admittance of the series combination be $\mathbf{Y}_{in} = G_{in} + jB_{in}$. At what frequency is the magnitude of: (a) $\mathbf{Y}_{in}$ equal to 4 m℧? (b) G_{in} equal to 4 m℧? (c) B_{in} equal to 4 m℧?

24 Let $\omega = 400$ rad/s for the network of Fig. 9-19 and find: (a) the impedance seen at terminals $A–B$ if terminals $C–D$ are open-circuited; (b) the admittance seen at terminals $C–D$ if terminals $A–B$ are short-circuited.

Fig. 9-19: See Prob. 24.

Fig. 9-20: See Prob. 25.

25 For the network shown in Fig. 9-20, at what frequency is: (*a*) $R_{in} = 200 \, \Omega$?
(*b*) $X_{in} = -200 \, \Omega$? (*c*) $G_{in} = 1/200 \, \mho$? (*d*) $B_{in} = 1/240 \, \mho$?

26 Three admittances, $\mathbf{Y}_1 = 2 - j4 \, \mu\mho$, $\mathbf{Y}_2 = 4 - j1 \, \mu\mho$, and $\mathbf{Y}_3 = 1 + j3 \, \mu\mho$, are in series, and the current $\mathbf{I} = 20\underline{/40°} \, \mu$A flows through the series combination. Find the amplitude of the voltage across each admittance and also across the series combination.

27 The admittance of the parallel combination of $100 \, \Omega$ and 0.2 H at $\omega = 1500$ rad/s is the same as the admittance of R_1 and L_1 in series at that frequency. (*a*) Find R_1 and L_1. (*b*) Repeat for $\omega = 500$ rad/s.

28 A network is composed of 5 kΩ, 0.1 H, and 0.1 μF in parallel. (*a*) At what frequency does the input admittance of the network have its minimum magnitude? (*b*) At what frequency is the phase angle of the input admittance 45°? (*c*) −45°?

Chapter 10
The Sinusoidal Steady-State Response

10-1
Introduction

In Chap. 2 and, particularly, Chap. 3 we learned several methods which are useful in analyzing resistive circuits. No matter how complex the resistive circuit, we are able to determine any desired response by using nodal, mesh, or loop analysis, superposition, source transformations, or Thévenin's or Norton's theorems. Sometimes one method is sufficient, but more often we find it convenient to combine several methods to obtain the response in the most direct manner. We now wish to extend these techniques to the analysis of circuits in the sinusoidal steady state, and we have already seen that impedances combine in the same manner as do resistances. The extension of the techniques of resistive circuit analysis has been promised several times, and we must now find out why the extension is justified and practice its use.

10-2
Nodal, mesh, and loop analysis

Let us first review the arguments by which we accepted nodal analysis for a purely resistive N-node circuit. After designating a reference node and assigning voltage variables between each of the $N - 1$ remaining nodes and the reference, we applied Kirchhoff's current law to each of these $N - 1$ nodes. The application of Ohm's law to all the resistors then led to $N - 1$ equations in $N - 1$ unknowns if no voltage sources or dependent sources were present; if they were, additional equations were written in accordance with the definitions of the types of sources involved.

Now, is a similar procedure valid in terms of phasors and impedances for the sinusoidal steady state? We already know that both of Kirchhoff's laws are valid for phasors; also, we have an Ohm-like law for the passive elements, $\mathbf{V} = \mathbf{ZI}.$ In other words, the laws upon which nodal analysis rests are true for phasors, and we may proceed, therefore, to analyze circuits by nodal techniques in the sinusoidal steady state. It is also evident that mesh- and loop-analysis methods are valid as well.

As an example of nodal analysis, consider the frequency-domain circuit

Fig. 10-1: A frequency-domain circuit for which node voltages V_1 and V_2 are identified.

shown in Fig. 10-1. Each passive element is specified by its impedance, although the analysis might be simplified slightly by using admittance values. Two current sources are given as phasors, and phasor node voltages V_1 and V_2 are indicated. At the left node we apply KCL and $I = V/Z$,

$$\frac{V_1}{5} + \frac{V_1}{-j10} + \frac{V_1 - V_2}{-j5} + \frac{V_1 - V_2}{j10} = 1 + j0$$

At the right node,

$$\frac{V_2 - V_1}{-j5} + \frac{V_2 - V_1}{j10} + \frac{V_2}{j5} + \frac{V_2}{10} = -(-j0.5)$$

Combining terms, we have

$$(0.2 + j0.2)V_1 - j0.1V_2 = 1 \tag{1}$$

and

$$-j0.1V_1 + (0.1 - j0.1)V_2 = j0.5 \tag{2}$$

Using determinants to solve (1) and (2), we obtain

$$V_1 = \frac{\begin{vmatrix} 1 & -j0.1 \\ j0.5 & (0.1 - j0.1) \end{vmatrix}}{\begin{vmatrix} (0.2 + j0.2) & -j0.1 \\ -j0.1 & (0.1 - j0.1) \end{vmatrix}} = \frac{0.1 - j0.1 - 0.05}{0.02 - j0.02 + j0.02 + 0.02 + 0.01}$$

$$= \frac{0.05 - j0.1}{0.05} = 1 - j2 \quad V$$

$$V_2 = \frac{\begin{vmatrix} (0.2 + j0.2) & 1 \\ -j0.1 & j0.5 \end{vmatrix}}{0.05} = \frac{-0.1 + j0.1 + j0.1}{0.05} = -2 + j4 \quad V$$

The time-domain solutions are therefore obtained by expressing V_1 and V_2 in polar form,

$$V_1 = 2.24\underline{/-63.4°} \qquad V_2 = 4.47\underline{/116.6°}$$

and passing to the time domain:

$$v_1(t) = 2.24 \cos(\omega t - 63.4°) \qquad v_2(t) = 4.47 \cos(\omega t + 116.6°) \quad V$$

Note that the value of ω would have to be known in order to compute the impedance values given on the circuit diagram. Also, both sources are assumed to operate at the same frequency.

(a)   (b)

Fig. 10-2: (a) A time-domain circuit containing a dependent source. (b) The corresponding frequency-domain circuit on which mesh currents I_1 and I_2 are indicated.

As an example of loop or mesh analysis, we consider the circuit given in Fig. 10-2a. Noting from the left source that $\omega = 10^3$ rad/s, we draw the frequency-domain circuit of Fig. 10-2b and assign mesh currents I_1 and I_2. Around mesh 1,

$$3I_1 + j4(I_1 - I_2) = 10\underline{/0°}$$

or

$$(3 + j4)I_1 - j4I_2 = 10$$

while mesh 2 leads to

$$j4(I_2 - I_1) - j2I_2 + 2I_1 = 0$$

or

$$(2 - j4)I_1 + j2I_2 = 0$$

Solving,

$$I_1 = \frac{14 + j8}{13} = 1.24\underline{/29.7°} \quad A$$

$$I_2 = \frac{20 + j30}{13} = 2.77\underline{/56.3°} \quad A$$

or

$$i_1(t) = 1.24 \cos (10^3 t + 29.7°) \quad A$$

$$i_2(t) = 2.77 \cos (10^3 t + 56.3°) \quad A$$

The solutions for either of the problems above could be checked by working entirely in the time domain, but it would be quite an undertaking and one which is typical of those we may safely ignore because the phasor method is providentially available.

Drill Problems

10-1. Assuming that both sources in Fig. 10-3 operate at the same frequency, find: (a) V_1; (b) V_2; (c) V_3.

Ans: $33.9\underline{/81.9°}$; $15.62\underline{/-13.32°}$; $36.9\underline{/-65.7°}$ V

Fig. 10-3: See Drill Prob. 10-1.

Fig. 10-4: See Drill Prob. 10-2.

10-2. If $i_s = 3 \cos 500t$ A, transform the circuit of Fig. 10-4 to the frequency domain and determine the amplitude of the current in the: (*a*) inductor; (*b*) capacitor; (*c*) resistor. *Ans:* 1.843; 1.106; 0.553 A

**10-3
Superposition,
source
transformations,
and Thévenin's
theorem**

After inductors and capacitors were introduced in Chap. 4, we found that circuits containing these elements were still linear, and that the benefits of linearity were again available. Included among these were the superposition principle and Thévenin's and Norton's theorems; source transformations we now recognize as a simple case of the latter two theorems. Thus, we know that these methods may be used on the circuits we are now considering; the fact that we happen to be applying sinusoidal sources and are seeking only the forced response is immaterial. The fact that we are analyzing the circuits in terms of phasors is also immaterial; they are still linear circuits. We might also remember that linearity and superposition were invoked when we combined real and imaginary sources to obtain a complex source.

We therefore shall consider several examples in which answers are obtained more readily through applying superposition, source transformations, or Thévenin's or Norton's theorems.

First we look again at the circuit of Fig. 10-1, redrawn as Fig. 10-5 with each pair of parallel impedances replaced by a single equivalent impedance. That is, 5 and $-j10$ in parallel yield $4 - j2$ Ω, $j10$ in parallel with $-j5$ gives $-j10$ Ω, and 10 in parallel with $j5$ provides $2 + j4$ Ω. To find $\mathbf{V}_1$, we first activate only the left source and find the partial response

$$\mathbf{V}_{1L} = 1\underline{/0°}\ \frac{(4 - j2)(-j10 + 2 + j4)}{4 - j2 - j10 + 2 + j4} = \frac{-4 - j28}{6 - j8} = 2 - j2$$

With only the right source active, current division helps us to obtain

$$\mathbf{V}_{1R} = (-0.5\underline{/-90°})\left(\frac{2 + j4}{4 - j2 - j10 + 2 + j4}\right)(4 - j2) = \frac{-6 + j8}{6 - j8} = -1$$

Summing, then

$$\mathbf{V}_1 = 2 - j2 - 1 = 1 - j2 \quad \text{V}$$

which agrees with our previous result.

Fig. 10-5: $\mathbf{V}_1$ and $\mathbf{V}_2$ may be found by using superposition of the separate phasor responses.

We might also see whether or not Thévenin's theorem can help the analysis of this circuit (Fig. 10-5). Suppose we determine the Thévenin equivalent faced by the $-j10$-Ω impedance; the open-circuit voltage (+ reference to left) is

$$\mathbf{V}_{oc} = (1\underline{/0^\circ})(4 - j2) + (0.5\underline{/-90^\circ})(2 + j4)$$

$$= 4 - j2 + 2 - j1 = 6 - j3$$

The impedance of the inactive circuit, as viewed from the load terminals, is simply the sum of the two remaining impedances. Hence,

$$\mathbf{Z}_{th} = 6 + j2$$

Thus, when we reconnect the circuit, the current directed from node 1 toward node 2 through the $-j10$-Ω load is

$$\mathbf{I}_{12} = \frac{6 - j3}{6 + j2 - j10} = 0.6 + j0.3$$

Subtracting this from the left source current, the downward current through the $4 - j2$-Ω branch is found:

$$\mathbf{I}_1 = 1 - 0.6 - j0.3 = 0.4 - j0.3$$

and, thus,

$$\mathbf{V}_1 = (0.4 - j0.3)(4 - j2) = 1 - j2 \quad \text{V}$$

We might have been cleverer and used Norton's theorem on the three right elements, assuming that our chief interest is in $\mathbf{V}_1$. Source transformations can also be used repeatedly to simplify the circuit. Thus, all the short-cuts and tricks that arose in Chaps. 2 and 3 are available for circuit analysis in the frequency domain. The slight additional complexity that is apparent now arises from the necessity of using complex numbers and not from any more involved theoretical considerations.

Finally, we should also be pleased to hear that these same techniques will be applicable to the forced response of circuits driven by damped sinusoidal forcing functions, exponential forcing functions, and forcing functions having a *complex frequency* in general. Thus, we shall meet these same techniques again in Chap. 13.

Drill Problems

10-3. If superposition is used to find the frequency-domain response $\mathbf{I}$ in the circuit of Fig. 10-6*a*, what is the partial response produced by: (*a*) i_{s1}? (*b*) i_{s2}? (*c*) v_{s3}?
Ans: $j2$; -2; $-4 + j4$ A

10-4. Find the Thévenin and Norton equivalents of the network shown in Fig. 10-6*b*. Give values for the: (*a*) open-circuit voltage $\mathbf{V}_{ab}$; (*b*) short-circuit current $\mathbf{I}_{ab}$; (*c*) Thévenin impedance $\mathbf{Z}_{th}$.
Ans: $89.4\underline{/63.4^\circ}$ V; $20\underline{/0^\circ}$ A; $2 + j4$ Ω

$v_{s3} = \sqrt{32} \cos (10^5 t + 45°)$ V

20 μH   10 μF

$i_{s1} =$ 2 sin $10^5 t$ A   I   2 Ω $i_{s2} =$ 2 cos $10^5 t$ A

(a)

I_{ab}
0.1$\mathbf{V}_{ab}$ 5 Ω $j5$ Ω $\mathbf{V}_{ab}$
100 $\underline{/0°}$ V

(b)

Fig. 10-6: (a) See Drill Prob. 10-3. (b) See Drill Prob. 10-4.

10-4
Phasor diagrams

The phasor diagram is a name given to a sketch in the complex plane of the phasor voltages and phasor currents throughout a specific circuit. It also provides a graphical method for solving certain problems which may be used to check more exact analytical methods. It proves to be of considerable help in simplifying the analytical work in certain symmetrical polyphase problems by enabling the symmetry to be recognized and helpfully applied. In the following chapter we shall encounter similar diagrams which display the complex power relationships in the sinusoidal steady state. The use of other complex planes will appear in connection with complex frequency in Chap. 13.

We are already familiar with the use of the complex plane in the graphical identification of complex numbers and in their addition and subtraction. Since phasor voltages and currents are complex numbers, they may also be identified as points in a complex plane. For example, the phasor voltage $\mathbf{V}_1$ = $6 + j8$ = $10\underline{/53.1°}$ is identified on the complex voltage plane shown in Fig. 10-7. The axes are the real voltage axis and the imaginary voltage axis; the voltage $\mathbf{V}_1$ is located by an arrow drawn from the origin. Since addition and subtraction are particularly easy to perform and display on a complex plane, it is apparent that phasors may be easily added and subtracted in a phasor

Fig. 10-7: A simple phasor diagram shows the single voltage phasor $\mathbf{V}_1$ = $6 + j8$ = $10\underline{/53.1°}$ V.

Imaginary axis (V)

$j8$

10

$\mathbf{V}_1$

53.1°

6   Real axis (V)

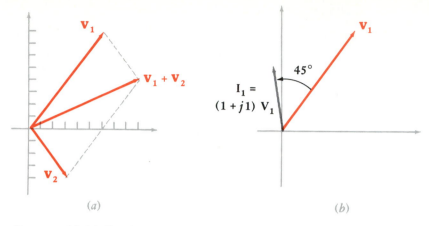

Fig. 10-8: (a) A phasor diagram showing the sum of $V_1 = 6 + j8$ and $V_2 = 3 - j4$, $V_1 + V_2 = 9 + j4 = 9.85\underline{/24.0°}$. (b) The phasor diagram shows V_1 and I_1, where $I_1 = YV_1$ and $Y = 1 + j1\,\mho$.

diagram. Multiplication and division result in the addition and subtraction of *angles* and a change of amplitude; the latter is less clearly shown, since the amplitude change depends on the amplitude of each phasor and on the scale of the diagram. Figure 10-8a shows the sum of V_1 and a second phasor voltage $V_2 = 3 - j4 = 5\underline{/-53.1°}$, and Fig. 10-8$b$ shows the current I_1, which is the product of V_1 and the admittance $Y = 1 + j1$.

This last phasor diagram shows both current and voltage phasors on the same complex plane; it is understood that each will have its own amplitude scale, but a common angle scale. For example, a phasor voltage 1 cm long might represent 100 V, while a phasor current 1 cm long could indicate 3 mA.

The phasor diagram also offers an interesting interpretation of the time-domain to frequency-domain transformation, since the diagram may be interpreted from either the time- or frequency-domain viewpoint. Up to this time, it is obvious that we have been using the frequency-domain interpretation, because we have been showing phasors directly on the phasor diagram. However, let us proceed to a time-domain viewpoint by first showing the phasor $V = V_m\underline{/\alpha}$, as sketched in Fig. 10-9a. In order to transform V to the time domain, the next necessary step is the multiplication of the phasor by $e^{j\omega t}$; thus we now have the complex voltage $V_m e^{j\alpha} e^{j\omega t} = V_m\underline{/\omega t + \alpha}$. This voltage may also be interpreted as a phasor, one which possesses a phase

Fig. 10-9: (a) The phasor voltage $V_m\underline{/\alpha}$. (b) The complex voltage $V_m\underline{/\omega t + \alpha}$ is shown as a phasor at a particular instant of time. This phasor leads $V_m\underline{/\alpha}$ by ωt rad.

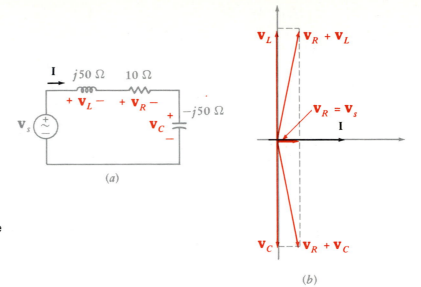

(a)

(b)

Fig. 10-10: (a) A series *RLC* circuit shown in the frequency domain. (b) The associated phasor diagram, drawn with the single mesh current as the reference phasor.

angle which increases linearly with time. On a phasor diagram it therefore represents a rotating line segment, the instantaneous position being ωt rad ahead (counterclockwise) of $V_m/\underline{\alpha}$. Both $V_m/\underline{\alpha}$ and $V_m/\underline{\omega t + \alpha}$ are shown on the phasor diagram of Fig. 10-9*b*.

The passage to the time domain is now completed by taking the real part of $V_m/\underline{\omega t + \alpha}$. The real part of this complex quantity, however, is merely the projection of $V_m/\underline{\omega t + \alpha}$ on the real axis. In summary, then, the frequency-domain phasor appears on the phasor diagram, and the transformation to the time domain is accomplished by allowing the phasor to rotate in a counter-clockwise direction at an angular velocity of ω rad/s and then visualizing the projection on the real axis. It is helpful to think of the arrow representing the phasor **V** on the phasor diagram as the snapshot, taken at $\omega t = 0$, of a rotating arrow whose projection on the real axis is the instantaneous voltage $v(t)$.

Let us now construct the phasor diagrams for several simple circuits. The series *RLC* circuit shown in Fig. 10-10*a* has several different voltages associated with it, but only a single current. The phasor diagram is constructed most easily by employing the single current as the reference phasor. Let us arbitrarily select $\mathbf{I} = I_m/\underline{0°}$ and place it along the real axis of the phasor diagram, Fig. 10-10*b*. The resistor, capacitor, and inductor voltages may next be calculated and placed on the diagram, where the 90° phase relationships stand out clearly. The sum of these three voltages is the source voltage, and for this circuit, which is in the resonant condition[1] where $\mathbf{Z}_C = -\mathbf{Z}_L$, the source voltage and resistor voltage are equal. The total voltage across the resistor and inductor or resistor and capacitor is easily obtained from the phasor diagram.

[1] Resonance will be defined in Chap. 14.

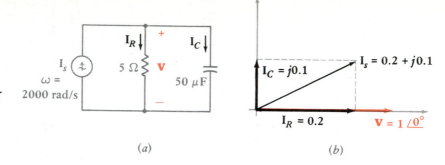

Fig. 10-11: (a) A parallel
RC circuit. (b) The phasor
diagram for this circuit;
the node voltage **V** is
used as a convenient
reference phasor.

Figure 10-11*a* shows a simple parallel circuit in which it is logical to use the single voltage between the two nodes as a reference phasor. Suppose that $V = 1\underline{/0°}$ V. The resistor current is in phase with this voltage, $I_R = 0.2\underline{/0°}$ A, and the capacitor current leads the reference voltage by 90°, $I_C = j0.1$ A. After these two currents are added to the phasor diagram, shown as Fig. 10-11*b*, they may be summed to obtain the source current. The result is $I_s = 0.2 + j0.1$ A.

If the source current were specified initially as, for example, $1\underline{/0°}$ A, and the node voltage is not initially known, it is still convenient to begin construction of the phasor diagram by assuming a node voltage, say $V = 1\underline{/0°}$ V once again, and using it as the reference phasor. The diagram is then completed as before, and the source current which flows as a result of the assumed node voltage is again found to be $0.2 + j0.1$ A. The true source current is $1\underline{/0°}$ A, however, and thus the true node voltage is greater by the factor $1\underline{/0°}/(0.2 + j0.1)$; the true node voltage is therefore $4 - j2$ V. The assumed voltage leads to a phasor diagram which differs from the true phasor diagram by a change of scale (the assumed diagram is smaller by a factor of $1/\sqrt{20}$) and an angular rotation (the assumed diagram is rotated counterclockwise through 26.6°).

Phasor diagrams are usually very simple to construct, and most sinusoidal steady-state analyses will be more meaningful if such diagrams are included. Additional examples of the use of phasor diagrams will appear frequently throughout the remainder of our study.

Drill Problems

10-5. Assume a convenient value for **V** in Fig. 10-12*a* and construct a phasor diagram showing I_R, I_L, and I_C. By combining these currents, determine the angle by which I_s leads: (a) I_R; (b) I_C; (c) I_x.   *Ans:* 63°; −27°; −13°

Fig. 10-12: (a) See Drill
Prob. 10-5. (b) See Drill
Prob. 10-6.

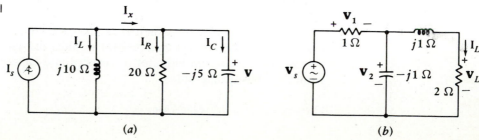

10-6. Select a reference value for I_L in Fig. 10-12b, draw a phasor diagram showing V_L, V_2, V_1, and V_s, and measure the ratio of the lengths of: (a) V_s to V_1; (b) V_s to V_2; (c) V_s to V_L.   *Ans: 1.80; 1.61; 1.80*

**10-5
Response as a
function of ω**

We will now consider methods of obtaining and presenting the response of a circuit with sinusoidal excitation as a function of the radian frequency ω. With the possible exception of the 60-Hz power area in which frequency is a constant and the load is the variable, sinusoidal frequency response is extremely important in almost every branch of electrical engineering as well as in related areas, such as the theory of mechanical vibrations.

Let us suppose that we have a circuit which is excited by a single source $V_s = V_s \underline{/\theta}$. This phasor voltage may also be transformed into the time-domain source voltage $V_s \cos(\omega t + \theta)$. Somewhere in the circuit exists the desired response, say, a current I. As we know, this phasor response is a complex number, and its value cannot be specified in general without the use of two quantities: either a real part and an imaginary part, or an amplitude and a phase angle. The latter pair of quantities is more useful and more easily determined experimentally, and is the information which we shall obtain analytically as a function of frequency. The data may be presented as two curves, the magnitude of the response as a function of ω and the phase angle of the response as a function of ω. We often normalize the curves by plotting the magnitude of the current-voltage ratio and the phase angle of the current-voltage ratio versus ω. It is evident that an alternative description of the resultant curves is the magnitude and phase angle of an admittance as a function of frequency. The admittance might be an input admittance or, if the current and voltage are measured at different locations in a circuit, a *transfer* admittance. A normalized voltage response to a current source may be similarly presented as the magnitude and phase angle of an input or transfer impedance versus ω. Other possibilities are voltage-voltage ratios (voltage gains) or current-current ratios (current gains). Let us consider the details of this process by thoroughly discussing two examples.

For the first example, we select the series *RL* circuit. The phasor voltage V_s is therefore applied to this simple circuit, and the phasor current I (leaving the positively marked end of the voltage source) is selected as the desired response. We are dealing with the forced response only, and the familiar phasor methods enable the current to be obtained:

$$I = \frac{V_s}{R + j\omega L}$$

Let us immediately express this result in normalized form as a ratio of current to voltage, that is, as an input admittance:

$$Y = \frac{I}{V_s}$$

or

$$Y = \frac{1}{R + j\omega L} \tag{3}$$

If we like, we may consider the admittance as the current produced by a source voltage $1\underline{/0°}$ V. The magnitude of the response is

$$|\mathbf{Y}| = \frac{1}{\sqrt{R^2 + \omega^2 L^2}} \tag{4}$$

while the angle of the response is found to be

$$\text{ang } \mathbf{Y} = -\tan^{-1}\frac{\omega L}{R} \tag{5}$$

Equations (4) and (5) are the analytical expressions for the magnitude and phase angle of the response as a function of ω; we now desire to present this same information graphically.

First consider the magnitude curve. It is important to note that we are plotting the absolute value of some quantity versus ω, and the entire curve must therefore lie *above* the ω axis. The response curve is constructed by noting that the value of the response at zero frequency is $1/R$, the initial slope is zero, and the response approaches zero as frequency approaches infinity; the graph of the magnitude of the response as a function of ω is shown in Fig. 10-13a.

For the sake of generality and completeness, the response is shown for

Fig. 10-13: (a) The magnitude of $\mathbf{Y} = \mathbf{I}/\mathbf{V}_s$ and (b) the angle of $\mathbf{Y}$ are plotted as functions of ω for a series *RL* circuit with sinusoidal excitation.

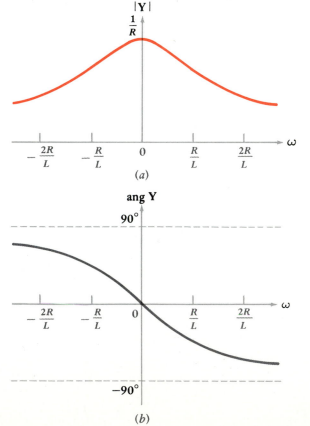

both positive and negative values of frequency; the symmetry results from the fact that (4) indicates that $|\mathbf{Y}|$ has the same value when ω is replaced by $(-\omega)$. The physical interpretation of a negative radian frequency, such as $\omega = -100$ rad/s, depends on the time-domain function, and it may always be obtained by inspection of the time-domain expression. Suppose, for example, that we consider the voltage $v(t) = 50 \cos (\omega t + 30°)$. If $\omega = 100$, the voltage is $v(t) = 50 \cos (100t + 30°)$, but if $\omega = -100$, $v(t) = 50 \cos (-100t + 30°)$ or $50 \cos (100t - 30°)$. These voltages have different values at $t = 1$ ms, for example. Any sinusoidal response may be treated in a similar manner.

The second part of the response, the phase angle of $\mathbf{Y}$ versus ω, is an inverse tangent function. The tangent function itself is quite familiar, and we should have no difficulty turning that curve on its side; asymptotes of $+90°$ and $-90°$ are helpful. The response curve is shown in Fig. 10-13b. The points at which $\omega = \pm R/L$ are marked on both the magnitude and phase curves. At these frequencies the magnitude is 0.707 times the maximum magnitude at zero frequency and the phase angle has a magnitude of $45°$. At the frequency at which the admittance magnitude is 0.707 times its maximum value, the current magnitude is 0.707 times its maximum value, and the average power supplied by the source is 0.707^2 or 0.5 times its maximum value. It is not very strange that $\omega = R/L$ is identified as a *half-power frequency*.

As a second example, let us select a parallel LC circuit driven by a sinusoidal current source, as illustrated in Fig. 10-14a. The voltage response $\mathbf{V}$ is easily obtained:

$$\mathbf{V} = \mathbf{I}_s \frac{(j\omega L)(1/j\omega C)}{j\omega L - j(1/\omega C)}$$

Fig. 10-14: (*a*) A sinusoidally excited parallel LC circuit. (*b*) The magnitude of the input impedance, $\mathbf{Z} = \mathbf{V}/\mathbf{I}_s$, and (*c*) the angle of the input impedance are plotted as functions of ω.

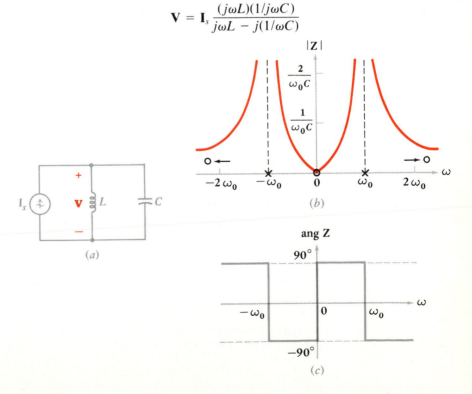

and it may be expressed as an input impedance

$$\mathbf{Z} = \frac{\mathbf{V}}{\mathbf{I}_s} = \frac{L/C}{j(\omega L - 1/\omega C)}$$

or

$$\mathbf{Z} = -j\frac{1}{C}\frac{\omega}{\omega^2 - 1/LC} \tag{6}$$

By letting

$$\omega_0 = \frac{1}{\sqrt{LC}}$$

and factoring the expression for the input impedance, the magnitude of the impedance may be written in a form which enables those frequencies to be identified at which the response is zero or infinite,

$$|\mathbf{Z}| = \frac{1}{C}\frac{|\omega|}{|(\omega - \omega_0)(\omega + \omega_0)|} \tag{7}$$

Such frequencies are termed *critical frequencies,* and their early identification simplifies the construction of the response curve. We note first that the response has zero amplitude at $\omega = 0$; when this happens, we say that the response has a *zero* at $\omega = 0$, and we also describe the frequency at which it occurs as a zero. Response of infinite amplitude is noted at $\omega = \omega_0$ and $\omega = -\omega_0$; these frequencies are called *poles,* and the response is said to have a pole at each of these frequencies. Finally, we note that the response approaches zero as $\omega \to \infty$, and thus $\omega = \pm\infty$ is also a zero.[2]

The locations of the critical frequencies should be marked on the ω axis, by using small circles for the zeros and crosses for the poles. Poles or zeros at infinite frequency should be indicated by an arrow near the axis, as shown in Fig. 10-14b. The actual drawing of the graph is made easier by adding broken vertical lines as asymptotes at each pole location. The completed graph of magnitude versus ω is shown in Fig. 10-14b; the slope at the origin is *not* zero.

An inspection of (6) shows that the phase angle of the input impedance must be either $+90$ or $-90°$; no other values are possible, as must apparently be the case for any circuit composed entirely of inductors and capacitors. An analytical expression for ang $\mathbf{Z}$ would therefore consist of a series of statements that the angle is $+90$ or $-90°$ in certain frequency ranges. It is simpler to present the information graphically, as shown in Fig. 10-14c. Although this curve is only a collection of horizontal straight line segments, errors are often made in its construction, and it is a good idea to make certain that it can be drawn directly from an inspection of (6).

---

[2] It is customary to consider plus infinity and minus infinity as being the same point. The phase angle of the response at very large positive and negative values of ω need not be the same, however.

Drill Problems

10-7. For the circuit of Fig. 10-15, sketch, as a function of ω: (a) $|\mathbf{V}_a|$; (b) ang $\mathbf{V}_a$; (c) $|\mathbf{I}_c|$.

Ans: $|\mathbf{V}_a(j3)| = 25.7$ V; ang $\mathbf{V}_a(-j10) = -26.6°$; $|\mathbf{I}_c(j5)| = 7.91$ A

10-8. Replace both 10-Ω resistors in the circuit of Fig. 10-15 with 100-μF capacitors and let $\mathbf{I}_c$ be the desired response. Sketch the magnitude and phase angle of the response as functions of ω and determine all critical frequencies of the response.

Ans: $\mathbf{I}_c(j80) = 1.03\underline{/-90°}$ A; 0, ±70.7, ±100 rad/s

Fig. 10-15. See Drill Probs. 10-7 and 10-8.

Problems

1 Use nodal analysis on the circuit shown in Fig. 10-16 to find the phasor voltage $\mathbf{V}_{RL}$.

Fig. 10-16: See Prob. 1.

2 (a) Find $\mathbf{V}_1$ in Fig. 10-17. (b) To what identical values should the capacitive impedances be changed so that $\mathbf{V}_1$ is 180° out of phase with the source voltage?

Fig. 10-17: See Prob. 2.

3 If $i_s(t) = 10^{-3} \cos 10^4 t$ A in the circuit shown in Fig. 10-18, find $v_1(t)$.

Fig. 10-18: See Prob. 3.

Fig. 10-19: See Probs. 4
and 5.

4 Find $v_3(t)$ in the circuit of Fig. 10-19 by using nodal analysis.

5 Write three mesh equations and solve them to determine $i_3(t)$ for the circuit shown in Fig. 10-19.

6 If $\omega = 100$ rad/s for the source in Fig. 10-20, find $\mathbf{I}_1$.

Fig. 10-20: See Prob. 6.

7 (a) Draw a tree for the circuit shown in Fig. 10-21 so that i_x is a link current, assign a complete set of link currents, and find $i_x(t)$. (b) Construct another tree in which v_x is a tree-branch voltage, assign a complete set of tree-branch voltages, and find $v_x(t)$.

Fig. 10-21: See Prob. 7.

8 The op-amp shown in Fig. 10-22 has an infinite input impedance, zero output impedance, and a large but finite (positive, real) gain, $\mathbf{A} = -\mathbf{V}_o/\mathbf{V}_i$. (a) Construct

Fig. 10-22: See Prob. 8.

a basic differentiator by letting $\mathbf{Z}_f = R_f$, find $\mathbf{V}_o/\mathbf{V}_s$, and then show that $\mathbf{V}_o/\mathbf{V}_s \rightarrow -j\omega C_1 R_f$ as $\mathbf{A} \rightarrow \infty$. (b) Let $\mathbf{Z}_f$ represent C_f and R_f in parallel, find $\mathbf{V}_o/\mathbf{V}_s$, and then show that $\mathbf{V}_o/\mathbf{V}_s \rightarrow -j\omega C_1 R_f/(1 + j\omega C_f R_f)$ as $\mathbf{A} \rightarrow \infty$.

9 Use superposition to find $\mathbf{I}_x$ and $i_x(t)$ in the circuit of Fig. 10-23.

Fig. 10-23: See Prob. 9.

10 Find the Thévenin equivalent of the network shown in: (a) Fig. 10-24a; (b) Fig. 10-24b.

Fig. 10-24: See Prob. 10.

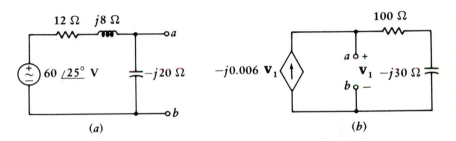

(a)   (b)

11 Let $\omega = 1$ rad/s in Fig. 10-25, and find the Thévenin equivalent as seen at terminals a–b. Draw the equivalent circuit as a voltage source $\mathbf{V}_{th}$ in series with a resistance R_{th} and either an inductance L_{th} or a capacitance C_{th}.

Fig. 10-25: See Prob. 11.

12 Find the Thévenin equivalent of the network shown in Fig. 10-26 at: (a) $\omega = 10$ rad/s; (b) $\omega = 20$ rad/s.

Fig. 10-26: See Prob. 12.

13 Find $\mathbf{V}_x$ in the circuit of Fig. 10-27 by using: (*a*) superposition; (*b*) repeated source transformations.

Fig. 10-27: See Prob. 13.

14 Use superposition to help find $i_x(t)$ in the circuit of Fig. 10-28 if $v_s(t) = $: (*a*) $100 \cos 10^4 t$ V; (*b*) $100 \sin 10^4 t$ V; (*c*) $100 \cos 5000t$ V.

Fig. 10-28: See Prob. 14.

15 Let $\omega = 50$ rad/s for the source of Fig. 10-29. (*a*) Construct a phasor diagram showing the three currents. Use a scale of 1 A/in. (*b*) Construct another phasor diagram showing the three voltages, using a scale of 1 V/in.

Fig. 10-29: See Prob. 15.

16 In Fig. 10-30, let $\mathbf{I}_s$ be a current source with an amplitude of 3 A at 1000 rad/s. Using this current as the reference phasor, find the four indicated voltages and show them on an accurate phasor diagram. Identify the equality, $\mathbf{V}_1 + \mathbf{V}_2 = \mathbf{V}_3 + \mathbf{V}_4$.

Fig. 10-30: See Prob. 16.

17 For the circuit shown in Fig. 10-31, let $\mathbf{V}_s = 100\underline{/0°}$ V, $|\mathbf{V}_1| = 80$ V, and $|\mathbf{V}_2| = 50$ V. (*a*) Use a phasor diagram to determine $\mathbf{V}_1$ and $\mathbf{V}_2$. There are two possible answers; give both. (*b*) If $\mathbf{Z}_1$ is an inductive impedance and $\mathbf{Z}_2$ is a capacitive impedance, find $\mathbf{V}_1$ and $\mathbf{V}_2$.

Fig. 10-31: See Prob. 17.

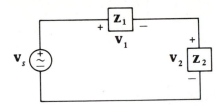

18 In the circuit of Fig. 10-32, it is known that $|\mathbf{V}_1| = 50$ V, $|\mathbf{V}_2| = 125$ V, and $\mathbf{V}_3 = 200\underline{/-45°}$ V. Use graphical methods (straightedge, ruler, protractor, compass, and all that fun stuff) to find the angle of $\mathbf{V}_1$. Although there may seem to be two answers to this problem, one is right, and the other is wrong.

Fig. 10-32: See Prob. 18.

19 A 4-Ω resistor is in parallel with a 125-μF capacitor. The parallel combination is in series with a 2-mH inductor and a sinusoidal voltage source $\mathbf{V}_s$. Calculate a few points, and then plot a curve of the magnitude of the ratio of the capacitor voltage to the source voltage. Cover the frequency interval, $0 \leq \omega \leq 5$ krad/s.

20 For the circuit shown in Fig. 10-33: (*a*) locate all the critical frequencies of the response $|\mathbf{V}_C/\mathbf{I}_s|$; (*b*) calculate the value of the response at one point between each pair of critical frequencies; (*c*) sketch the response over the range, $-10 < \omega < 10$ rad/s.

Fig. 10-33: See Prob. 20.

21 Plot a curve of $|\mathbf{V}_{out}/\mathbf{V}_{in}|$ vs. ω, $-5000 < \omega < 5000$ rad/s, for the circuit of Fig. 10-34.

Fig. 10-34: See Prob. 21.

Fig. 10-35: See Prob. 22.

22 Refer to Fig. 10-35 and determine all the critical frequencies; then prepare magnitude and phase plots of the response: (*a*) $\mathbf{V}_{in}/\mathbf{I}_s$; (*b*) $\mathbf{I}_C/\mathbf{I}_s$.

23 A two-terminal network consists of an 80-mH inductor in series with the parallel combination of 100 mH and 10 μF. Prepare plots of the magnitude and angle of $\mathbf{Z}_{in}$ vs. ω, $-4000 < \omega < 4000$ rad/s.

24 Construct plots of $|\mathbf{Y}_{in}|$ and ang $\mathbf{Y}_{in}$ vs. ω for the network of Fig. 10-36. Use an abscissa on which ω ranges from 0 to 30 rad/s.

Fig. 10-36: See Prob. 24.

25 Prepare sketches of $|\mathbf{Z}_{in}|$ and ang $\mathbf{Z}_{in}$ vs. ω for the two-terminal network shown in Fig. 10-37.

Fig. 10-37: See Prob. 25.

Chapter 11
Average Power and
RMS Values

11-1
Introduction

Nearly all problems in circuit analysis are concerned with applying one or more sources of electrical energy to a circuit and then quantitatively determining the consequent response throughout the circuit. The response may be a current or a voltage, but we are also interested in the amount of energy supplied from the sources, in the amount of energy dissipated or stored within the circuit, and in the manner in which energy is delivered to the points at which the responses are determined. Primarily, however, we are concerned with the *rate* at which energy is being generated and absorbed; our attention must now be directed to *power*.

We shall begin by considering instantaneous power, the product of the time-domain voltage and time-domain current associated with the element or network in which we are interested. The instantaneous power is sometimes quite useful in its own right, because its maximum value might have to be limited in order to avoid exceeding the safe or useful operating range of a physical device. For example, transistor and vacuum-tube power amplifiers both produce a distorted output and speakers give a distorted sound when the peak power exceeds a certain limiting value. However, we are mainly interested in instantaneous power for the simple reason that it provides us with the means to calculate a more important quantity, the average power. In a similar way, the progress of a cross-country automobile trip is best described by the average velocity; our interest in the instantaneous velocity is limited to the avoidance of maximum velocities which will endanger our safety or arouse the highway patrol.

In practical problems we shall deal with values of average power which range from the small fraction of a picowatt available in a telemetry signal from outer space, to the few watts of audio power supplied to the speakers in a high-fidelity stereo system, the several hundred watts required to invigorate the morning coffeepot, or the 10 billion watts generated at the Grand Coulee dam.

Our discussion will not be concerned entirely with the average power delivered by a sinusoidal current or voltage; we shall therefore define a quantity called the effective value, a mathematical measure of the effectiveness of other waveforms in delivering power. Our study of power will be completed by considering the descriptive quantities, power factor and complex power, two concepts which introduce the practical and economic aspects associated with the distribution of electrical power.

11-2 Instantaneous power

The power delivered to any device as a function of time is given by the product of the instantaneous voltage across the device and the instantaneous current through it, as we well know; the passive sign convention is assumed. Thus,

$$p = vi \tag{1}$$

A knowledge of both the current and the voltage is presumed. If the device in question is a resistor of resistance R, then the power may be expressed solely in terms of either the current or the voltage,

$$p = vi = i^2 R = \frac{v^2}{R} \tag{2}$$

If the voltage and current are associated with a device which is entirely inductive, then

$$p = vi = Li\frac{di}{dt} = \frac{1}{L} v \int_{-\infty}^{t} v \, dt \tag{3}$$

where we have arbitrarily assumed that the voltage is zero at $t = -\infty$. In the case of a capacitor,

$$p = vi = Cv\frac{dv}{dt} = \frac{1}{C} i \int_{-\infty}^{t} i \, dt \tag{4}$$

where a like assumption about the current is made. This listing of equations for power in terms of only a current or voltage soon becomes unwieldy, however, as we begin to consider more general networks. The listing is also quite unnecessary, for we need only find *both* the current and voltage at the network terminals. As an example, we may consider the series RL circuit, as shown in Fig. 11-1, excited by a step-voltage source. The familiar current response is

$$i(t) = \frac{V_0}{R} (1 - e^{-Rt/L})u(t)$$

Fig. 11-1: The instantaneous power that is delivered to R is $p_R = i^2 R = (V_0{}^2/R)(1 - e^{-Rt/L})^2 u(t)$.

and thus the total power delivered by the source or absorbed by the passive network is

$$p = vi = \frac{V_0^2}{R}(1 - e^{-Rt/L})u(t)$$

since the square of the unit-step function is obviously the unit-step function itself.

The power delivered to the resistor is

$$p_R = i^2R = \frac{V_0^2}{R}(1 - e^{-Rt/L})^2u(t)$$

In order to determine the power absorbed by the inductor, we first obtain the inductor voltage

$$v_L = L\frac{di}{dt}$$

$$= V_0e^{-Rt/L}u(t) + \frac{LV_0}{R}(1 - e^{-Rt/L})\frac{du(t)}{dt}$$

$$= V_0e^{-Rt/L}u(t)$$

since $du(t)/dt$ is zero for $t > 0$ and $(1 - e^{-Rt/L})$ is zero at $t = 0$. The power absorbed by the inductor is thus

$$p_L = v_Li = \frac{V_0^2}{R}e^{-Rt/L}(1 - e^{-Rt/L})u(t)$$

Only a few algebraic manipulations are required to show that

$$p = p_R + p_L$$

which serves to check our work.

The majority of the problems which involve power calculations are perhaps those which deal with circuits excited by sinusoidal forcing functions in the steady state; as we have been told previously, even when periodic non-sinusoidal forcing functions are employed, it is possible to resolve the problem into a number of subproblems in which the forcing functions *are* sinusoidal. The special case of the sinusoid therefore deserves special attention.

Let us change the voltage source in the circuit of Fig. 11-1 to the sinusoidal source $V_m \cos \omega t$. The familiar time-domain response is

$$i(t) = I_m \cos(\omega t + \phi)$$

where

$$I_m = \frac{V_m}{\sqrt{R^2 + \omega^2L^2}} \quad \text{and} \quad \phi = -\tan^{-1}\frac{\omega L}{R}$$

The instantaneous power delivered to the entire circuit in the sinusoidal steady state is, therefore,

$$p = vi = V_mI_m \cos(\omega t + \phi) \cos \omega t$$

which we shall find convenient to rewrite in a form obtained by using the trigonometric identity for the product of two cosine functions. Thus,

$$p = \frac{V_m I_m}{2} \left[\cos (2\omega t + \phi) + \cos \phi \right]$$

$$= \frac{V_m I_m}{2} \cos \phi + \frac{V_m I_m}{2} \cos (2\omega t + \phi)$$

The last equation possesses several characteristics which are true in general for circuits in the sinusoidal steady state. One term, the first, is not a function of time; and a second term is included which has a cyclic variation at *twice* the applied frequency. Since this term is a cosine wave, and since sine waves and cosine waves have average values which are zero (when averaged over an integral number of periods), this introductory example may serve to indicate that the *average* power is $\frac{1}{2}V_m I_m \cos \phi$. This is true, and we shall now establish this relationship in more general terms.

Drill Problems

11-1. A current source, $10 + 20u(t)$ A, a 0.1-H inductor, and a 5-Ω resistor are in parallel. At $t = 15$ ms, find the power being absorbed by the: (*a*) inductor; (*b*) resistor; (*c*) $20u(t)$-A portion of the source.

Ans: 971; 446; -945 W

11-2. A current source, $10 + 5 \cos 200t$ A, a 16-Ω resistor, and a 250-μF capacitor have been connected in parallel for a long time. At $t = 10$ ms, find the power being absorbed by the: (*a*) resistor; (*b*) capacitor; (*c*) sinusoidal source; (*d*) dc source. *Ans:* 1918; -531; 365; -1752 W

11-3
Average power

When we speak of an average value for the instantaneous power, the time interval over which the averaging process takes place must be clearly defined. Let us first select a general interval of time from t_1 to t_2. We may then obtain the average value by integrating $p(t)$ from t_1 to t_2 and dividing the result by the time interval $t_2 - t_1$. Thus,

$$P = \frac{1}{t_2 - t_1} \int_{t_1}^{t_2} p(t) \, dt \tag{5}$$

The average value is denoted by the capital letter P since it is not a function of time, and it usually appears without any specific subscripts that identify it as an average value. Although P is not a function of time, it is a function of t_1 and t_2, the two instants of time which define the interval of integration. This dependence of P on a specific time interval may be expressed in a simpler manner if $p(t)$ is a periodic function. We shall consider this important case first.

Let us assume that our forcing function and the circuit responses are all periodic; a steady-state condition has been reached, although not necessarily the sinusoidal steady state. We may define a *periodic* function $f(t)$ mathematically by requiring that

$$f(t) = f(t + T) \tag{6}$$

Fig. 11-2: The average value P of a periodic power function $p(t)$ is the same over any period T.

where T is the period. We now show that the average value of the instantaneous power as expressed by (5) may be computed over an interval of one period having an arbitrary beginning.

A general periodic waveform is shown in Fig. 11-2 and identified as $p(t)$. We first compute the average power by integrating from t_1 to a time t_2 which is one period later, $t_2 = t_1 + T$,

$$P_1 = \frac{1}{T} \int_{t_1}^{t_1+T} p(t)\, dt$$

and then by integrating from some other time t_x to $t_x + T$,

$$P_x = \frac{1}{T} \int_{t_x}^{t_x+T} p(t)\, dt$$

The equality of P_1 and P_x should be evident from the graphical interpretation of the integrals; the area which represents the integral to be evaluated in determining P_x is smaller by the area from t_1 to t_x, but greater by the area from $t_1 + T$ to $t_x + T$, and the periodic nature of the curve requires these two areas to be equal. Thus, the average power may be computed by integrating the instantaneous power over any interval which is one period in length and then dividing by the period

$$P = \frac{1}{T} \int_{t_x}^{t_x+T} p\, dt \tag{7}$$

It is important to note that we might also integrate over any integral number of periods, provided that we divide by this same integral number of periods. Thus,

$$P = \frac{1}{nT} \int_{t_x}^{t_x+nT} p\, dt \qquad n = 1,2,3, \ldots \tag{8}$$

If we carry this concept to the extreme by integrating over all time, another useful result is obtained. We first provide ourselves with symmetrical limits on the integral

$$P = \frac{1}{nT} \int_{-nT/2}^{nT/2} p\, dt$$

and then take the limit as n becomes infinite,

$$P = \lim_{n\to\infty} \frac{1}{nT} \int_{-nT/2}^{nT/2} p\, dt$$

If $p(t)$ is a mathematically well-behaved function, as all *physical* forcing functions and responses are, it is apparent that if a large integer n is replaced by a slightly larger number which is not an integer, then the value of the integral and of P is changed by a negligible amount; moreover, the error decreases as n increases. Without justifying this step rigorously, we therefore replace the discrete variable nT by the continuous variable τ,

$$P = \lim_{\tau \to \infty} \frac{1}{\tau} \int_{-\tau/2}^{\tau/2} p\ dt \tag{9}$$

We shall find it convenient on several occasions to integrate periodic functions over this "infinite period." Examples of the use of (7), (8), and (9) are given below.

Let us illustrate the calculation of the average power of a periodic wave by finding the average power delivered to an R-ohm resistor by the (periodic) sawtooth current waveform shown in Fig. 11-3a. We have

$$i(t) = \frac{I_m}{T}\ t \qquad 0 < t \le T$$

$$i(t) = \frac{I_m}{T}\ (t - T) \qquad T < t \le 2T$$

etc.

and
$$p(t) = \frac{1}{T^2}\ I_m^2 R t^2 \qquad 0 < t \le T$$

$$p(t) = \frac{1}{T^2}\ I_m^2 R (t - T)^2 \qquad T < t \le 2T$$

etc.

as sketched in Fig. 11-3b. Integrating over the simplest range of one period, from $t = 0$ to $t = T$, we have

$$P = \frac{1}{T} \int_0^T \frac{I_m^2 R}{T^2}\ t^2\ dt = \tfrac{1}{3} I_m^2 R$$

The selection of other ranges of one period, such as from $t = 0.1T$ to $t = 1.1T$, would produce the same answer. Integration from 0 to $2T$ and division

Fig. 11-3: (a) A sawtooth current waveform and (b) the instantaneous power waveform it produces in a resistor R.

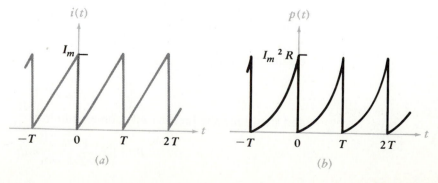

(a) $\qquad\qquad\qquad$ (b)

by $2T$, that is, the application of (8) with $n = 2$ and $t_x = 0$, would also provide the same answer.

Now let us obtain the general result for the sinusoidal steady state. We shall assume the general sinusoidal voltage

$$v(t) = V_m \cos (\omega t + \theta)$$

and current

$$i(t) = I_m \cos (\omega t + \phi)$$

associated with the device in question. The instantaneous power is

$$p(t) = V_m I_m \cos (\omega t + \theta) \cos (\omega t + \phi)$$

Again expressing the product of two cosine functions as one-half the sum of the cosine of the difference angle and the cosine of the sum angle,

$$p(t) = \tfrac{1}{2} V_m I_m \cos (\theta - \phi) + \tfrac{1}{2} V_m I_m \cos (2\omega t + \theta + \phi) \qquad (10)$$

we may save ourselves some integration by an inspection of the result. The first term is a constant, independent of t. The remaining term is a cosine function; $p(t)$ is therefore periodic, and its period is $\tfrac{1}{2} T$. Note that the period T is associated with the given current and voltage, and not with the power; the power function has a period $\tfrac{1}{2} T$. However, we may integrate over an interval of T to determine the average value if we wish; it is only necessary to divide also by T. Our familiarity with cosine and sine waves, however, shows that the average value of either over a period is zero. There is thus no need to integrate (10) formally; by inspection, the average value of the second term is zero over a period T (or $\tfrac{1}{2} T$), and the average value of the first term, a constant, must be that constant itself. Thus,

$$P = \tfrac{1}{2} V_m I_m \cos (\theta - \phi) \qquad (11)$$

This important result, introduced in the previous section for a specific circuit, is therefore quite general for the sinusoidal steady state. The average power is one-half the product of the crest amplitude of the voltage, the crest amplitude of the current, and the cosine of the phase-angle difference between the current and the voltage; the sense of the difference is immaterial.

As a numerical illustration, let us assume that a voltage

$$v(t) = 4 \cos \frac{\pi t}{6}$$

or

$$\mathbf{V} = 4\underline{/0^\circ}$$

is applied across an impedance $\mathbf{Z} = 2\underline{/60^\circ}\ \Omega$. The phasor current is therefore $2\underline{/-60^\circ}$ A, and the average power is

$$P = \tfrac{1}{2}(4)(2) \cos 60^\circ = 2 \quad \text{W}$$

The time-domain voltage

$$v(t) = 4 \cos \frac{\pi t}{6}$$

Fig. 11-4: Curves of $v(t)$, $i(t)$, and $p(t)$ are plotted as functions of time for a simple circuit in which the phasor voltage $\mathbf{V} = 4\underline{/0°}$ V is applied to the imped-ance $\mathbf{Z} = 2\underline{/60°}$ Ω at $\omega = \pi/6$ rad/s.

time-domain current

$$i(t) = 2 \cos\left(\frac{\pi t}{6} - 60°\right)$$

and instantaneous power

$$p(t) = 8 \cos\frac{\pi t}{6} \cos\left(\frac{\pi t}{6} - 60°\right)$$

$$= 2 + 4 \cos\left(\frac{\pi t}{3} - 60°\right)$$

are all sketched on the same time axis in Fig. 11-4. Both the 2-W average value of the power and its period of 6 s, one-half the period of either the current or the voltage, are evident. The zero value of the instantaneous power at each instant when either the voltage or current is zero is also apparent.

Two special cases are worth isolating for consideration, the average power delivered to an ideal resistor and that to an ideal reactor (any combi-nation of only capacitors and inductors).

The phase-angle difference between the current through and the voltage across a pure resistor is zero, and therefore

$$P_R = \tfrac{1}{2} V_m I_m$$

or

$$P_R = \tfrac{1}{2} I_m^2 R \tag{12}$$

or

$$P_R = \frac{V_m^2}{2R} \tag{13}$$

The last two formulas, enabling us to determine the average power delivered to a pure resistance from a knowledge of either the sinusoidal current or voltage, are simple and important. They are often misused. The most com-mon error is made in trying to apply them in cases where, say, the voltage included in (13) is *not the voltage across the resistor*. If care is taken to use the current through the resistor in (12) and the voltage across the resistor in (13), satisfactory operation is guaranteed. Also, do not forget the factor of $\tfrac{1}{2}$!

The average power delivered to any device which is purely reactive must be zero. This is evident from the 90° phase difference which must exist between current and voltage; hence, $\cos(\theta - \phi) = 0$ and

$$P_x = 0$$

The *average* power delivered to any network composed entirely of ideal inductors and capacitors is zero; the *instantaneous* power is zero only at specific instants. Thus, power flows into the network for a part of the cycle and out of the network during another portion of the cycle.

As an example, the average power delivered to an impedance, $\mathbf{Z}_L = 8 - j11\Omega$, by a current, $\mathbf{I} = 5\underline{/20°}$ A, is, therefore, found rapidly by (12),

$$P = \tfrac{1}{2}(5)^2 8 = 100 \quad \text{W}$$

since no average power can be absorbed by the $-j11\ \Omega$. Note also that if a current is given in rectangular form, say $\mathbf{I} = 2 + j5$ A, then the magnitude squared is $2^2 + 5^2$, and the average power delivered to $\mathbf{Z}_L = 8 - j11\ \Omega$ would be

$$P = \tfrac{1}{2}(2^2 + 5^2)8 = 116 \quad \text{W}$$

A further example illustrating these power relationships is afforded by the circuit shown in Fig. 11-5. The values of $\mathbf{I}_1$ and $\mathbf{I}_2$ are found by any of several methods, such as mesh analysis, nodal analysis, or superposition:

$$\mathbf{I}_1 = 5 - j10 = 11.18\underline{/-63.45°}$$

$$\mathbf{I}_2 = 5 - j5 = 7.07\underline{/-45°}$$

The current through the 2-Ω resistor is

$$\mathbf{I}_1 - \mathbf{I}_2 = -j5 = 5\underline{/-90°}$$

so $I_m = 5$, and the average power absorbed by the resistor is found most easily by (12),

$$P_R = \tfrac{1}{2}I_m^2 R = \tfrac{1}{2}(5)^2 2 = 25 \quad \text{W}$$

This result may be checked by using (11) or (13). Turning to the left source, the voltage $20\underline{/0°}$ and current $11.18\underline{/-63.45°}$ satisfy the *active* sign convention, and thus the power *delivered by* this source is

$$P_{\text{left}} = \tfrac{1}{2}(20)(11.18) \cos (0° + 63.45°) = 50 \quad \text{W}$$

In a similar manner, we find the power *absorbed* by the right source,

$$P_{\text{right}} = \tfrac{1}{2}(10)(7.07) \cos (0° + 45°) = 25 \quad \text{W}$$

The power delivered to each of the two reactive elements is zero; the power relations check.

Fig. 11-5: The average power delivered to each reactive element is zero in the sinusoidal steady state.

In Sec. 5 of Chap. 3 we considered the maximum-power-transfer theorem as it applied to resistive loads and resistive source impedances. For a Thévenin source $\mathbf{V}_s$ and impedance $\mathbf{Z}_{th} = R_{th} + jX_{th}$ connected to a load $\mathbf{Z}_L = R_L + jX_L$, it may be shown readily (Prob. 10) that the average power delivered to the load is a maximum when $R_L = R_{th}$ and $X_L = -X_{th}$; that is, $\mathbf{Z}_L = \mathbf{Z}_{th}^*$. This result is often dignified by calling it the *maximum-power-transfer theorem for the sinusoidal steady state:*

> An independent voltage source in series with an impedance $\mathbf{Z}_{th}$ or an independent current source in parallel with an impedance $\mathbf{Z}_{th}$ delivers a maximum average power to that load impedance $\mathbf{Z}_L$ which is the conjugate of $\mathbf{Z}_{th}$, or $\mathbf{Z}_L = \mathbf{Z}_{th}^*$.

It is apparent that the resistive condition considered in Chapter 3 is merely a special case.

We must now pay some attention to *nonperiodic* functions. One practical example of a nonperiodic power function for which an average power value is desired is the power output of a radio telescope which is directed toward a ''radio star.'' Another is the sum of a number of periodic functions, each function having a different period, such that no greater common period can be found for the combination. For example, the current

$$i(t) = \sin t + \sin \pi t \tag{14}$$

is nonperiodic because the ratio of the periods of the two sine waves is an irrational number. At $t = 0$, both terms are zero and increasing. But the first term is zero and increasing only when $t = 2\pi n$, where n is an integer, and thus periodicity demands that πt or $\pi(2\pi n)$ must equal $2\pi m$, where m is also an integer. No solution (integral values for both m and n) for this equation is possible. It may be illuminating to compare the nonperiodic expression (14) with the *periodic* function

$$i(t) = \sin t + \sin 3.14t \tag{15}$$

where 3.14 is an exact decimal expression and is not intended to be interpreted as $3.141\,592\,\ldots$. With a little effort, it can be shown that the period of this current wave is 100π s.[1]

The average value of the power delivered to a 1-Ω resistor by either the periodic current (15) or the nonperiodic current (14) may be found by inte-

---

[1] $T_1 = 2\pi$ and $T_2 = 2\pi/3.14$. Therefore, we seek integral values of m and n such that $2\pi n = 2\pi m/3.14$, or $3.14n = m$, or $(\frac{314}{100}) n = m$, or $157n = 50m$. Thus, the smallest integral values for n and m are $n = 50$ and $m = 157$. The period is therefore $T = 2\pi n = 100\pi$, or $T = 2\pi(157)/3.14 = 100\pi$ s.

grating over an infinite interval; much of the actual integration can be avoided because of our thorough knowledge of the average values of simple functions. We therefore obtain the instantaneous power delivered by the current in (14) by applying (9),

$$P = \lim_{\tau \to \infty} \frac{1}{\tau} \int_{-\tau/2}^{\tau/2} (\sin^2 t + \sin^2 \pi t + 2 \sin t \sin \pi t) \, dt$$

We now consider P as the sum of three average values. The average value of $\sin^2 t$ over an infinite interval is found by replacing $\sin^2 t$ by $(\frac{1}{2} - \frac{1}{2} \cos 2t)$; it is obviously $\frac{1}{2}$. Similarly, the average value of $\sin^2 \pi t$ is also $\frac{1}{2}$. And the last term can be expressed as the sum of two cosine functions, each of which must certainly have an average value of zero. Thus,

$$P = \tfrac{1}{2} + \tfrac{1}{2} = 1 \quad \text{W}$$

An identical result is obtained for the periodic current (15).

Applying this same method to a current function which is the sum of several sinusoids of *different periods* and arbitrary amplitudes,

$$i(t) = I_{m1} \cos \omega_1 t + I_{m2} \cos \omega_2 t + \cdots + I_{mN} \cos \omega_N t \qquad (16)$$

we find the average power delivered to a resistance R,

$$P = \tfrac{1}{2}(I_{m1}^2 + I_{m2}^2 + \cdots + I_{mN}^2)R \qquad (17)$$

The result is unchanged if an arbitrary phase angle is assigned to each component of the current. This important result is surprisingly simple when we think of the steps required for its derivation: squaring the current function, integrating, and taking the limit. The result is also just plain surprising, because it shows that, *in this special case of a current such as (16), superposition is applicable to power.* Superposition is *not* applicable for a current which is the sum of two direct currents, nor is it applicable for a current which is the sum of two sinusoids of the same frequency. That is, the current, $i_1 = 2 \cos 10t - 3 \cos 20t$ A, delivers 26 W to a 4-Ω resistor, whereas $i_2 = 2 \cos 10t - 3 \cos 10t$ A delivers only 2-W average power to that same resistor.

Drill Problems

11-3. Determine the average power delivered to each of the three boxed networks in the circuit of Fig. 11-6.　　　　　　　　*Ans:* 60; 20; 120 W

Fig. 11-6: See Drill Prob. 11-3.

(a)

(b)

(c)

Fig. 11-7: See Drill Probs. 11-4 and 11-6.

11-4. Find the average power delivered to a 6-Ω resistor by each of the periodic waveforms shown in Fig. 11-7. *Ans: 12.5; 12.0; 12.18 W*

11-5. A current source i_s is connected to a 50-Ω resistor. Find the average power delivered to the resistor if $i_s =$: (a) 4 cos 100t A; (b) 4 cos 100t + 2 cos 120t A; (c) 4 cos 100t + 3 sin 100t + 2 cos 120t A; (d) 4 cos 100t + 3 sin 100t + 2 cos 120t − 1 A; (e) 4 cos 100t + 3 sin 100t + 4 cos(100t + 120°) + 2 cos 120t − 1 A. *Ans: 400; 500; 725; 775; 255 W*

11-4
Effective values of current and voltage

Most of us are aware that the voltage available at the power outlets in our homes is a sinusoidal voltage having a frequency of 60 Hz and a "voltage" of 115 V. But what is meant by "115 volts"? This is certainly not the instantaneous value of the voltage, for the voltage is not a constant. The value of 115 V is also not the amplitude which we have been symbolizing as V_m; if we displayed the voltage waveform on a calibrated cathode-ray oscilloscope, we should find that the amplitude of this voltage at one of our ac outlets is $115\sqrt{2}$ or 162.6 V. We also cannot fit the concept of an average value to the 115 V because the average value of the sine wave is zero. We might come a little closer by trying the magnitude of the average over a positive or negative half cycle; by using a rectifier-type voltmeter at the outlet, we should measure 103.5 V. As it turns out, however, the 115 V is the *effective value* of this sinusoidal voltage. This value is a measure of the effectiveness of a voltage source in delivering power to a resistive load.

Let us now proceed to define the effective value of any periodic waveform representing either a current or voltage. We shall consider the sinusoidal

Fig. 11-8: If the resistor receives the same average power in (a) and (b), then the effective value of i(t) is equal to I_{eff}, and the effective value of v(t) is equal to V_{eff}.

waveform as only a special, albeit practically important, case. Let us arbitrarily define effective value in terms of a current waveform, although a voltage could equally well be selected. The *effective value* of any periodic current is equal to the value of the direct current which, flowing through an *R*-ohm resistor, delivers the same power to the resistor as does the periodic current. In other words, we allow the given periodic current to flow through the resistor, determine the instantaneous power i^2R, and then find the average value of i^2R over a period; this is the average power. We then cause a direct current to flow through this same resistor and adjust the value of the direct current until the same value of average power is obtained. The magnitude of the direct current is equal to the effective value of the given periodic current. These ideas are illustrated in Fig. 11-8.

The general mathematical expression for the effective value of i(t) is now easily obtained. The average power delivered to the resistor by the periodic current i(t) is

$$P = \frac{1}{T}\int_0^T i^2R \; dt = \frac{R}{T}\int_0^T i^2 \; dt$$

where the period of i(t) is T. The power delivered by the direct current is

$$P = I_{eff}^2 R$$

Equating the power expressions and solving for I_{eff},

$$I_{eff} = \sqrt{\frac{1}{T}\int_0^T i^2 \; dt} \qquad (18)$$

The result is independent of the resistance *R*, as it must be to provide us with a worthwhile concept. A similar expression is obtained for the effective value of a periodic voltage by replacing i and I_{eff} by v and V_{eff}, respectively.

Notice that the effective value is obtained by first squaring the time function, then taking the average value of the squared function over a period, and finally taking the square root of the average of the squared function. In abbreviated language, the operation involved in finding an effective value is the (square) *root* of the *mean* of the *square*; for this reason, the effective value is often called the *root-mean-square* value, or simply the rms value.

The most important special case is that of the sinusoidal waveform. Let us select the sinusoidal current

$$i(t) = I_m \cos (\omega t + \phi)$$

which has a period

$$T = \frac{2\pi}{\omega}$$

and substitute into (18) to obtain the effective value

$$
\begin{aligned}
I_{eff} &= \sqrt{\frac{1}{T} \int_0^T I_m^2 \cos^2 (\omega t + \phi)\, dt} \\
&= I_m \sqrt{\frac{\omega}{2\pi} \int_0^{2\pi/\omega} [\tfrac{1}{2} + \tfrac{1}{2} \cos (2\omega t + 2\phi)]\, dt} \\
&= I_m \sqrt{\frac{\omega}{4\pi} [t]_0^{2\pi/\omega}} = \frac{I_m}{\sqrt{2}}
\end{aligned}
$$

Thus the effective value of a sinusoidal current is a real quantity which is independent of the phase angle and numerically equal to 0.707 times its amplitude. A current $\sqrt{2} \cos (\omega t + \phi)$, therefore, has an effective value of 1 A and will deliver the same power to any resistor as will a direct current of 1 A.

It should be noted carefully that the $\sqrt{2}$ factor that we obtained as the ratio of the amplitude of the periodic current to the effective value is applicable only when the periodic function is sinusoidal. For the sawtooth waveform of Fig. 11-3, for example, the effective value is equal to the maximum value divided by $\sqrt{3}$. The factor by which the maximum value must be divided to obtain the effective value depends on the mathematical form of the given periodic function; it may be either rational or irrational, depending on the nature of the function.

The use of the effective value also simplifies slightly the expression for the average power delivered by a sinusoidal current or voltage by avoiding use of the factor $\tfrac{1}{2}$. For example, the average power delivered to an R-ohm resistor by a sinusoidal current is

$$P = \tfrac{1}{2} I_m^2 R$$

Since $I_{eff} = I_m/\sqrt{2}$, the average power may be written

$$P = I_{eff}^2 R \tag{19}$$

The other familiar power expressions may also be written in terms of effective values:

$$P = V_{eff} I_{eff} \cos (\theta - \phi) \tag{20}$$

and

$$P = \frac{V_{eff}^2}{R} \tag{21}$$

The fact that the effective value is defined in terms of an equivalent dc quantity provides us with average power formulas for resistive circuits which are identical with those used in dc analysis.

Although we have succeeded in eliminating the factor $\tfrac{1}{2}$ from our average-power relationships, we must now take care to determine whether a sinusoidal quantity is expressed in terms of its amplitude or its effective value. In practice, the effective value is usually used in the fields of power transmission or distribution and of rotating machinery; in the areas of electronics and communications, the amplitude is more often used. We shall assume that the amplitude is specified unless the term rms is explicitly used.

In the sinusoidal steady state, phasor voltages and currents may be given as either effective values or amplitudes; the two expressions differ only by a factor $\sqrt{2}$. The voltage $50\underline{/30°}$ V is expressed in terms of an amplitude; as an rms voltage, we should write $35.36\underline{/30°}$ V rms.

In order to determine the effective value of a periodic or nonperiodic waveform which is composed of the sum of a number of sinusoids of different frequencies, we may use the appropriate average-power relationship (17) developed in the previous section, rewritten in terms of the effective values of the several components,

$$P = (I_{1eff}^2 + I_{2eff}^2 + \cdots + I_{Neff}^2)R \tag{22}$$

These results indicate that if a sinusoidal current of 5 A rms at 60 Hz flows through a 2-Ω resistor, an average power of 50 W is absorbed by the resistor; if a second current, say 3 A rms at 120 Hz, is also present, the absorbed power is 68 W; however, if the second current is also at 60 Hz, then the absorbed power may have any value between 8 and 128 W, depending on the relative phase of the two current components.

We therefore have found the effective value of a current which is composed of any number of sinusoidal currents of *different* frequencies,

$$I_{eff} = \sqrt{I_{1eff}^2 + I_{2eff}^2 + \cdots + I_{Neff}^2} \tag{23}$$

The total current may or may not be periodic; the result is the same. The effective value of the sum of the 60- and 120-Hz currents in the example above is 5.83 A; the effective value of the sum of the two 60-Hz currents may have any value between 2 and 8 A.

Drill Problems

11-6. Use (18), written for voltage, to calculate the effective values of the three periodic voltage waveforms shown in Fig. 11-7.

Ans: 8.66; 8.49; 8.55 V

11-7. Determine the effective value of each of these periodic currents: (*a*) $\sin 5t$; (*b*) $\sin 5t + \sin 5t$; (*c*) $\sin 5t + \cos 5t$; (*d*) $\sin 5t + \cos(5t + 60°)$; (*e*) $\sin 5t + \sin 8t$; (*f*) $\sin 5t + 1$.

Ans: 0.707; 1.414; 1.000; 0.366; 1.000; 1.225 A

**11-5
Apparent power
and power factor**

Historically, the introduction of the concepts of apparent power and power factor can be traced to the electric-power industry, where large amounts of electrical energy must be transferred from one point to another; the efficiency with which this transfer is effected is related directly to the cost of the electrical energy, which is eventually paid by the consumer. A customer who provides a load which results in a relatively poor transmission efficiency must pay a greater price for each kilowatthour (kWh) of electrical energy he actually receives and uses. In a similar way, a customer who requires a costlier investment in transmission and distribution equipment by the power company will also pay more for each kilowatthour unless the company is benevolent and enjoys losing money.

Let us first define apparent power and power factor and then show briefly how these terms are related to the economic situations mentioned above. We shall assume that the sinusoidal voltage

$$v = V_m \cos (\omega t + \theta)$$

is applied to a network, and the resultant sinusoidal current is

$$i = I_m \cos (\omega t + \phi)$$

The phase angle by which the voltage leads the current is therefore $(\theta - \phi)$. The average power delivered to the network, assuming a passive sign convention at its input terminals, may be expressed either in terms of the maximum values,

$$P = \tfrac{1}{2} V_m I_m \cos (\theta - \phi)$$

or in terms of the effective values,

$$P = V_{eff} I_{eff} \cos (\theta - \phi)$$

If our applied voltage and current responses had been dc quantities, the average power delivered to the network would have been given simply by the product of the voltage and the current. Applying this dc technique to the sinusoidal problem, we should obtain a value for the absorbed power which is "apparently" given by the product $V_{eff} I_{eff}$. This product of the effective values of the voltage and current is not the average power; we define it as the *apparent power*. Dimensionally, apparent power must be measured in the same units as real power, since $\cos (\theta - \phi)$ is dimensionless, but in order to avoid confusion the term voltamperes, or VA, is applied to apparent power. Since $\cos (\theta - \phi)$ cannot have a magnitude greater than unity, it is evident that the magnitude of the real power can never be greater than the magnitude of the apparent power.

Apparent power is not a concept which is limited to sinusoidal forcing functions and responses. It may be determined for any current and voltage waveshapes by simply taking the product of the effective values of the current and voltage.

The ratio of the real or average power to the apparent power is called the *power factor*, symbolized by PF. Hence,

$$\text{PF} = \frac{\text{average power}}{\text{apparent power}} = \frac{P}{V_{eff} I_{eff}}$$

In the sinusoidal case, the power factor is simply $\cos (\theta - \phi)$, where $(\theta - \phi)$ is the angle by which the voltage leads the current. This relationship is the reason why the angle $(\theta - \phi)$ is often referred to as the *PF angle*.

For a purely resistive load, the voltage and current are in phase, $(\theta - \phi)$ is zero, and the PF is unity. The apparent power and the average power are equal. Unity PF, however, may also be achieved for loads which contain both inductance and capacitance if the element values and the operating frequency are selected to provide an input impedance having a zero phase angle.

A purely reactive load, that is, one containing no resistance, will cause a phase difference between the voltage and current of either plus or minus 90°, and the PF is therefore zero.

Between these two extreme cases there are the general networks for which the PF can range from zero to unity. A PF of 0.5, for example, indicates a load having an input impedance with a phase angle of either 60° or −60°; the former describes an inductive load, since the voltage leads the current by 60°, while the latter refers to a capacitive load. The ambiguity in the exact nature of the load is resolved by referring to a leading PF or a lagging PF, the term leading or lagging referring to the *phase of the current with respect to the voltage*. Thus, an inductive load will have a lagging PF and a capacitive load a leading PF.

Before considering the practical consequences of these ideas, we can practice the necessary techniques by analyzing the circuit illustrated in Fig. 11-9. Values for the average power delivered to each of the two loads, the apparent power supplied by the source, and the power factor of the combined loads are desired.

The source current is

$$\mathbf{I}_s = \frac{60}{2 - j1 + 1 + j5} = 12\underline{/-53.1°} \quad \text{A rms}$$

Therefore, the source supplies an apparent power of $60(12) = 720$ VA and an average power of $720 \cos (0 - 53.1°) = 432$ W. The upper load receives an average power of $12^2(2) = 288$ W. For the right-hand load, we find an average power of $12^2(1) = 144$ W.

Thus, the source provides 432 W, of which 288 W is dissipated in the upper load and 144 W in the one to the right. The power balance is correct.

Finally, the power factor of the combined loads is found by considering the voltage and current of the combined loads, which is, of course, identical to that for the source. Thus,

$$\text{PF} = \frac{P}{V_{eff}I_{eff}} = \frac{432}{60(12)} = 0.6$$

We might also combine the two loads in series, obtain $3 + j4$ or $5\underline{/53.1°}$ Ω,

Fig. 11-9: A circuit in which we seek the average power delivered to each element, the apparent power supplied by the source, and the power factor of the combined load.

identify 53.1° as the PF angle, and thus we have a PF of cos 53.1° = 0.6. We also note that the combined load is inductive, and the PF is therefore 0.6 lagging.

The practical importance of these new terms is shown by the several examples which follow. Let us first assume that we have a sinusoidal ac generator, which is a rotating machine driven by some other device whose output is a mechanical torque such as a steam turbine, an electric motor, or an internal-combustion engine. We shall let our generator produce an output voltage of 200 V rms at 60 Hz. Suppose now that an additional rating of the generator is stated as a maximum power output of 1 kW. The generator would therefore be capable of delivering an rms current of 5 A to a resistive load. If, however, a load requiring 1 kW at a lagging power factor of 0.5 is connected to the generator, then an rms current of 10 A is necessary. As the PF decreases, greater and greater currents must be delivered to the load if operation at 200 V and 1 kW is maintained. If our generator were correctly and economically designed to furnish safely a maximum current of 5 A, then these greater currents would cause unsatisfactory operation, such as causing the insulation to overheat and begin smoking, which could be injurious to its health.

The rating of the generator is more informatively given in terms of apparent power in voltamperes. Thus a 1000-VA rating at 200 V indicates that the generator can deliver a maximum current of 5 A at rated voltage; the power it delivers depends on the load, and in an extreme case might be zero. An apparent power rating is equivalent to a current rating when operation is at a constant voltage.

When electric power is being supplied to large industrial consumers by a power company, the company will frequently include a PF clause in its rate schedules. Under this clause, an additional charge is made to the consumer whenever the PF drops below a certain specified value, usually about 0.85 lagging. Very little industrial power is consumed at leading PFs because of the nature of the typical industrial loads. There are several reasons that force the power company to make this additional charge for low PFs. In the first place, it is apparent that larger current-carrying capacity must be built into its generators in order to provide the larger currents that go with lower PF operation at constant power and constant voltage. Another reason is found in the increased losses in its transmission and distribution system.

As an example, let us suppose that a certain consumer is using an average power of 11 kW at unity PF and 220 V rms. We also assume a total resistance of 0.2 Ω in the transmission lines through which the power is delivered to the consumer. An rms current of 50 A therefore flows in the load and in the lines, producing a line loss of 500 W. In order to supply 11 kW to the consumer, the power company must generate 11.5 kW (at the higher voltage, 230 V). Since the energy is necessarily metered at the location of each consumer, this consumer would be billed for 95.6 percent of the energy which the power company actually produced.

Now let us hypothesize another consumer, also requiring 11 kW, but at a PF angle of 60° lagging. This consumer forces the power company to push

100 A through the load and (of particular interest to the company) through the line resistance. The line losses are now found to be 2 kW, and the customer's meter indicates only 84.6 percent of the actual energy generated. This figure departs from 100 percent by more than the power company will tolerate; this costs it money. Of course, the transmission losses might be reduced by using heavier transmission lines which have lower resistance, but this costs more money too. The power company's solution to this problem is to encourage operation at PFs which exceed 0.9 lagging by offering slightly reduced rates and to discourage operation at PFs which are less than 0.85 lagging by invoking increased rates.

The power drawn by most homes is used at reasonably high PFs (and reasonably small power levels); no charge is customarily made for low PF operation.

Besides paying for the actual energy consumed and for operation at excessively low PFs, industrial consumers are also billed for inordinate *demand*.[2] An energy of 100 kWh is delivered much more economically as 5 kW for 20 h than it is as 20 kW for 5 h, particularly if everyone else demands large amounts of power at the same time.

Drill Problem

11-8. A 230-V rms source supplies power to a load $\mathbf{Z}_L$ through a transmission line having a total resistance of 2 Ω. If $\mathbf{Z}_L = 11 + j4$ Ω, find: (*a*) the average and apparent power supplied to the load; (*b*) the average and apparent power lost in the transmission line; (*c*) the average and apparent power supplied by the source; (*d*) the power factor at which the source operates.

Ans: 3145 W, 3350 VA; 572 W, 572 VA; 3717 W, 3890 VA; 0.956 lagging

**11-6
Complex power**

Some simplification in power calculations is achieved if power is considered to be a complex quantity. The magnitude of the complex power will be found to be the apparent power, and the real part of the complex power will be shown to be the (real) average power. The new quantity, the imaginary part of the complex power, we shall call *reactive power*.

We define complex power with reference to a general sinusoidal voltage $\mathbf{V}_{eff} = V_{eff}\underline{/\theta}$ across a pair of terminals and a general sinusoidal current $\mathbf{I}_{eff} = I_{eff}\underline{/\phi}$ flowing into one of the terminals in such a way as to satisfy the passive sign convention. The average power P absorbed by the two-terminal network is thus

$$P = V_{eff}I_{eff}\cos(\theta - \phi)$$

Complex nomenclature is next introduced by making use of Euler's formula in the same way as we did in introducing phasors. We express P as

$$P = V_{eff}I_{eff}\,\text{Re}[e^{j(\theta - \phi)}]$$

or

$$P = \text{Re}[V_{eff}\,e^{j\theta}I_{eff}\,e^{-j\phi}]$$

---

[2] Several electrical utilities have made experimental installations of demand kilowatthour meters on residential customer premises.

The phasor voltage may now be recognized as the first two factors within the brackets in the equation above, but the second two factors do not quite correspond to the phasor current because the angle includes a minus sign which is not present in the expression for the phasor current. That is, the phasor current is

$$\mathbf{I}_{eff} = I_{eff}\, e^{j\phi}$$

and we therefore must make use of conjugate notation,

$$\mathbf{I}^{*}_{eff} = I_{eff}\, e^{-j\phi}$$

Hence

$$P = \mathrm{Re}[\mathbf{V}_{eff}\mathbf{I}^{*}_{eff}]$$

and we may now let power become complex by defining the complex power **S** as

$$\mathbf{S} = \mathbf{V}_{eff}\mathbf{I}^{*}_{eff} \tag{24}$$

If we first inspect the polar or exponential form of the complex power

$$\mathbf{S} = V_{eff}I_{eff}\, e^{j(\theta-\phi)}$$

it is evident that the magnitude of **S** is the apparent power, and the angle of **S** is the PF angle, that is, the angle by which the voltage leads the current. In rectangular form,

$$\mathbf{S} = P + jQ \tag{25}$$

where P is the real average power, as before. The imaginary part of the complex power is symbolized as Q and is termed the *reactive power*. The dimensions of Q are obviously the same as those of the real power P, the complex power **S**, and the apparent power $|\mathbf{S}|$. In order to avoid confusion with these other quantities, the unit of Q is defined as the *var* (abbreviated VAR), standing for voltampere reactive. From (24), it is seen that

$$Q = V_{eff}I_{eff}\sin(\theta - \phi)$$

Another interpretation of the reactive power may be seen by constructing a

Fig. 11-10: The current phasor $\mathbf{I}_{eff}$ is resolved into two components, one in phase with the voltage phasor $\mathbf{V}_{eff}$ and the other 90° out of phase with the voltage phasor. This latter component is called a quadrature component.

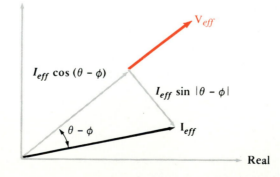

Imaginary

$I_{eff}\cos(\theta - \phi)$

V_{eff}

$I_{eff}\sin|\theta - \phi|$

$\theta - \phi$

$\mathbf{I}_{eff}$

Real

phasor diagram containing $\mathbf{V}_{eff}$ and $\mathbf{I}_{eff}$, as shown in Fig. 11-10. If the phasor current is resolved into two components, one in phase with the voltage, having a magnitude $I_{eff}\cos(\theta - \phi)$, and one 90° out of phase with the voltage, with magnitude $I_{eff}\sin|\theta - \phi|$, then it is clear that the real power is given by the product of the magnitude of the voltage phasor and the component of the phasor current which is in phase with the voltage. Moreover, the product of the magnitude of the voltage phasor and the component of the phasor current which is 90° out of phase with the voltage is the reactive power Q. It is common to speak of the component of a phasor which is 90° out of phase with some other phasor as a *quadrature component*. Thus Q is simply V_{eff} times the quadrature component of $\mathbf{I}_{eff}$; Q is also known as the quadrature power.

The sign of the reactive power characterizes the nature of a passive load at which $\mathbf{V}_{eff}$ and $\mathbf{I}_{eff}$ are specified. If the load is inductive, then $(\theta - \phi)$ is an angle between 0 and 90°, the sine of this angle is positive, and the reactive power is positive. A capacitive load results in a negative reactive power.

Just as a wattmeter reads the average real power drawn by a load, a varmeter will read the average reactive power Q drawn by the load. Both quantities may be metered simultaneously. In addition, watthourmeters and varhourmeters may be used simultaneously to record real and reactive energy used by any consumer during any desired time interval. From these records the average PF may be determined and the consumer's bill may be adjusted accordingly.

It is easy to show that the complex power delivered to several interconnected loads is the sum of the complex powers delivered to each of the individual loads, no matter how the loads are interconnected. For example, consider the two loads shown connected in parallel in Fig. 11-11. If rms values are assumed, the complex power drawn by the combined load is

$$\mathbf{S} = \mathbf{V}\mathbf{I}^* = \mathbf{V}(\mathbf{I}_1 + \mathbf{I}_2)^* = \mathbf{V}(\mathbf{I}_1^* + \mathbf{I}_2^*)$$

and thus

$$\mathbf{S} = \mathbf{V}\mathbf{I}_1^* + \mathbf{V}\mathbf{I}_2^*$$

as stated.

These new ideas can be clarified by a practical numerical example. Let us suppose that an industrial consumer is operating a 50-kW (67.1-hp) induction motor at a lagging PF of 0.8. In order to obtain lower electrical rates, the customer wishes to raise the PF to 0.95 lagging. Although the PF might be raised by increasing the real power and maintaining the reactive power con-

Fig. 11-11: A circuit used to show that the complex power drawn by two parallel loads is the sum of the complex powers drawn by the individual loads.

stant, this would not result in a lower bill; this cure therefore does not interest the consumer. A purely reactive load must be added to the system, and it is clear that it must be added in parallel, since the supply voltage to the inductor motor must not change. The circuit of Fig. 11-11 is thus applicable if we interpret S_1 as the induction motor's complex power and S_2 as the complex power drawn by the corrective device. Let us assume a voltage of $230\underline{/0°}$ V rms.

The complex power supplied to the induction motor must have a real part of 50 kW and an angle of $\cos^{-1}(0.8)$. Hence,

$$S_1 = \frac{50\underline{/\cos^{-1}(0.8)}}{0.8} = 50 + j37.5 \quad \text{kVA}$$

In order to achieve a PF of 0.95, the total complex power must become

$$S = \frac{50}{0.95} \underline{/\cos^{-1}(0.95)} = 50 + j16.43 \quad \text{kVA}$$

Thus, the complex power drawn by the corrective load is

$$S_2 = -j21.07 \quad \text{kVA}$$

The necessary load impedance Z_2 may be found in several simple steps. The current drawn by Z_2 is

$$I_2^* = \frac{S_2}{V} = \frac{-j21\,070}{230} = -j91.6 \quad \text{A}$$

or

$$I_2 = j91.6 \quad \text{A}$$

and, therefore,

$$Z_2 = \frac{V}{I_2} = \frac{230}{j91.6} = -j2.51 \quad \Omega$$

If the operating frequency is 60 Hz, this load can be provided by a 1056-μF capacitor. The load may also be simulated by a synchronous capacitor, a type of rotating machine. This is usually the more economical procedure for small capacitive reactances. Whatever device is selected, its initial costs, maintenance, and depreciation must be covered by the reduction in the electric bill.

Drill Problem

Fig. 11-12: See Drill Prob. 11-9.

11-9. For the circuit of Fig. 11-12, find the complex power absorbed by the: (*a*) dependent source; (*b*) inductor; (*c*) independent source; (*d*) resistor; (*e*) capacitor.   *Ans:* $-42 - j10.5$; $j25.5$; $-3 + j15$; 45; $-j30$ VA

Problems

1 (*a*) Find the instantaneous power being generated by the dependent voltage source in the circuit of Fig. 11-13. (*b*) Evaluate that power at $t = 0.03$ s.

Fig. 11-13: See Prob. 1.

4 sin 20*t* A

v_x

10 Ω   20 Ω

100 cos 20*t* V   50 Ω   0.6v_x

2 For the circuit shown in Fig. 11-14: (*a*) find the power being absorbed by the inductor as a function of time; (*b*) sketch the power as a function of time, $0 < t < 2$ ms; (*c*) and calculate the value of the power at $t = 1$ ms.

Fig. 11-14: See Prob. 2.

300 Ω

0.1 H   200 Ω   500 Ω   8$u(t)$ A

3 A voltage source, 10 cos 1000*t* V, a 0.4-Ω resistor, a 1-mH inductor, and a 1.25-mF capacitor are all in series. Find the power being absorbed by each of the four elements at $t = 1$ ms.

4 A 100-V battery, an open switch, a 10-Ω resistor, and an uncharged 1-mF capacitor are in series. Let the switch close at $t = 0$. (*a*) Obtain an expression for the instantaneous power being absorbed by the capacitor as a function of time. (*b*) Sketch it. (*c*) Find the time at which this power is a maximum and also the value of this maximum power.

5 Using the time interval, $10 < t < 12$ ms, find the average value of: (*a*) 40 cos 2000πt; (*b*) 40 cos² 2000πt; (*c*) the periodic waveform illustrated in Fig. 11-15; (*d*) the square of that waveform.

$f(t)$

30

etc

t (ms)

−1   0   1   2   3

Fig. 11-15: See Prob. 5.

6 (*a*) In Fig. 11-16, how much average power is delivered to the 8-Ω load? (*b*) How much of this power comes from the 20/0°-V source?

Fig. 11-16: See Prob. 6.

1.5 kΩ   500 Ω

+ $\mathbf{V}_1$ −

20 /0° V   0.5$\mathbf{V}_1$ 3 kΩ   8Ω

Source   Amplifier   Load

7 Find the average power absorbed by each of the five elements in the circuit of Fig. 11-17.

Fig. 11-17: See Prob. 7.

8 A square wave of voltage is 100 V for $0 < t < 1$ ms, 0 for $1 < t < 2$ ms, and continues in this pattern. As some help to a busy student, this leads to a current i in the RC circuit of Fig. 11-18 that is $73.11e^{-1000t}$ mA for $0 < t < 1$ ms, $-73.11e^{-1000(t-0.001)}$ mA for $1 < t < 2$ ms, and repeats. Find the average value of the: (a) voltage square wave; (b) current; (c) power delivered to the resistor; (d) power supplied by the source.

Fig. 11-18: See Prob. 8.

9 A sinusoidal voltage source, 100 cos ωt V, a 1-kΩ resistor, and a 1-μF capacitor are in series. Find the average power delivered to the resistor if $\omega =$: (a) 400 rad/s; (b) 3000 rad/s.

10 A frequency-domain Thevénin equivalent circuit consists of a sinusoidal source $\mathbf{V}_{th}$ in series with an impedance $\mathbf{Z}_{th} = R_{th} + jX_{th}$. Specify the conditions on a load, $\mathbf{Z}_L = R_L + jX_L$, if it is to receive a maximum average power subject to the constraint that: (a) $X_{th} = 0$; (b) R_L and X_L may be selected independently; (c) R_L is fixed (not equal to R_{th}); (d) X_L is fixed (independent of X_{th}); (e) $X_L = 0$.

11 Find the average power dissipated in the 5-Ω resistor of Fig. 11-19 for the following source voltages: (a) $A = C = 0, B = D = 10$ V; (b) $A = C = 10$ V, $B = D = 0$; (c) $A = B = C = 10$ V, $D = 0$; (d) $A = B = C = D = 10$ V.

Fig. 11-19: See Prob. 11.

Fig. 11-20: See Prob. 12.

12 In Fig. 11-20, find the value of $\mathbf{Z}_L$ to which a maximum average power can be delivered, and the value of $P_{L,av,max}$.

13 Select values of R and C in Fig. 11-21 so that a maximum average power is delivered to R, and then calculate the value of that power.

Fig. 11-21: See Prob. 13.

14 For the circuit of Fig. 11-22, make a sketch of the average power delivered to R_L vs. the value of R_L, $0 \le R_L \le 40\ \Omega$.

Fig. 11-22: See Prob. 14.

15 With reference to the circuit of Fig. 11-23: (a) what load impedance should be connected between x and y to absorb a maximum average power? (b) What is $P_{L,av,max}$?

Fig. 11-23: See Prob. 15.

16 For the periodic voltage waveform shown in Fig. 11-24, find: (a) the average value of the voltage; (b) the effective value of the voltage; (c) the average power the voltage would deliver to a 6-Ω resistor.

Fig. 11-24: See Prob. 16.

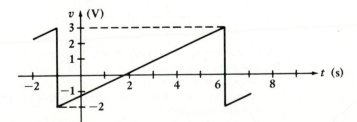

17 A periodic current waveform is $3e^{-t/4}$ A for $0 < t < 3$ s, -3 A for $3 < t < 5$ s, and zero for $5 < t < 7$ s. The period is 7 s. Find: (a) the average value of the current; (b) the effective value of the current.

18 A periodic current $i_1(t)$ is 1 A for $0 < t < 3$ s, 5 A for $3 < t < 5$ s, -2 A for $5 < t < 7$ s, and zero for $7 < t < 8$ s. If the period is 8 s, find: (a) $I_{1,av}$; (b) $|i_1|_{av}$; (c) $I_{1,eff}$.

19 The three waveforms shown in Fig. 11-25 all have $T = 10$ ms. Find the effective value of each.

Fig. 11-25: See Prob. 19.

(a)

(b)

(c)

20 Find the effective value of: (a) $4 - 5 \cos 100t$; (b) $4 \cos 100t + 5 \cos 200t$; (c) $4 \cos 100t + 5 \cos (100t + 150°)$; (d) a periodic waveform which is 6 from $t = 0$ to $t = 2$ ms decreases linearly to zero between 2 and 4 ms, remains at zero for 2 ms, and has a period of 6 ms.

21 (a) Find the effective value of the periodic semicircular waveform shown in Fig. 11-26. (b) What average power would this voltage deliver to a 15-Ω resistor?

Fig. 11-26: See Prob. 21.

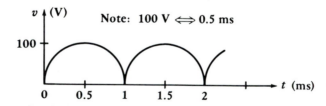

22 A voltage source, $\mathbf{V}_s = 88 + j66$ V rms, in series with 3 Ω of resistance, is supplying two loads in parallel: $\mathbf{Z}_A = 80 + j60$ Ω and $\mathbf{Z}_B = 50 - j100$ Ω. Find: (a) the average power delivered to $\mathbf{Z}_A$; (b) the average power delivered to $\mathbf{Z}_B$; (c) the average power supplied by the ideal source; (d) the power factor at which the source is operating.

23 (a) Find the apparent power and the real power being supplied by the source in the circuit of Fig. 11-27. (b) Find the power factor of the entire load. (c) To what value should the inductance be changed to cause the power factor to be 0.92 lagging?

Fig. 11-27: See Prob. 23.

24 (a) Find the power factor of the load to the right of a-b in the circuit shown in Fig. 11-28. (b) What value of capacitance should be connected across a-b to achieve a lagging PF of 0.95?

Fig. 11-28: See Prob. 24.

25 A source, $v_s(t) = 200 \cos 100t$ V, is in series with 8 Ω, 100 mH, and 2.5 mF. Find: (a) the PF at which the source is operating; (b) the value of capacitance that would have to be connected directly across the source so that it operates with a power factor of 0.9 leading.

26 A certain two-terminal network has across it a steady-state sinusoidal voltage whose frequency is 60 Hz and whose effective voltage is 120 V. The impedance of the network is $4 + j5$ Ω. (a) At what power factor is it operating? (b) What average power is it absorbing? (c) What value of capacitance should be placed in parallel with it so that the power factor becomes 0.9 leading?

27 The following three electrical devices are all connected in parallel: 3 kVA at 0.707 PF leading, 7 kVA at 0.8 PF lagging, and a series RL network with $R = 2$ Ω and $L = 10.61$ mH. This composite load is supplied with power from an ideal 100-V rms, 60-Hz, sinusoidal voltage source. Find: (a) the power factor at which the source is operating; (b) the effective value of the source current; (c) the capacitance that must be placed in parallel with the composite load so that the power factor will be 0.95 lagging.

28 A two-terminal network has a voltage, $v_1(t) = 4 \cos (10t - 30°)$ V, across its terminals, and a current, $i_1(t) = 2 \cos (10t + 20°)$ A, entering the terminal at which the positive voltage reference is located. Find: (a) $V_{1,eff}$; (b) $\mathbf{I}_{1,rms}$; (c) the average power absorbed by the network; (d) the power factor at which the network is operating; (e) the apparent power drawn by the network; (f) the complex power drawn by the network.

29 (a) Find the complex power absorbed by each of the five circuit elements in Fig. 11-29. (b) At what PF is the source operating?

Fig. 11-29: See Prob. 29.

30 A source of $500 \underline{/0°}$ V is in series with an impedance of $-j\,10\ \Omega$ and a load $\mathbf{Z}_L$. If the source is generating a complex power of $4 + j1.5$ kVA: (a) what complex power is delivered to $\mathbf{Z}_L$, and (b) what is the load PF?

31 Let $C = 0$ in the circuit shown in Fig. 11-30, and find the PF of: (a) the network to the right of C; (b) the entire load being supplied by the source. (c) Select the value of C so that the PF of the network to the right of the 0.3-Ω resistor is 0.95 lagging.

Fig. 11-30: See Prob. 31.

32 A 230-V rms source has two loads connected to it in parallel. One draws 8 kVA at a lagging power factor of 0.9, and the other draws 10 kW at a lagging power factor of 0.8. (a) Find the rms source current amplitude. (b) At what PF is the source operating?

Chapter 12
Polyphase Circuits

12-1
Introduction

One of the reasons for studying the sinusoidal steady state is that most household and industrial electric power is utilized as alternating current. The sinusoidal waveform may characterize a special mathematical function, but it represents a very common and very useful forcing function. In the same way, a *polyphase* source is even more specialized, but we consider it because almost the entire output of the electric power industry in this country is generated and distributed as polyphase power at a 60-Hz frequency. Before defining our terms carefully, let us look briefly at the most common polyphase system, a balanced three-phase system. The source has perhaps three terminals, and voltmeter measurements will show that sinusoidal voltages of equal amplitude are present between any two terminals. However, these voltages are not in phase; it will be easily shown later that each of the three voltages is 120° out of phase with each of the other two, the sign of the phase angle depending on the sense of the voltages. A balanced load draws power equally from the three phases, but when one of the voltages is instantaneously zero, the phase relationship shows that the other two must each be at half amplitude. At no instant does the instantaneous power drawn by the total load reach zero; as a matter of fact, this total instantaneous power is constant. This is an advantage in rotating machinery, for it keeps the torque on the rotor much more constant than it would be if a single-phase source were used. There is less vibration.

There are also advantages in using rotating machinery to generate three-phase power rather than single-phase power, and there are economical advantages in favor of the transmission of power in a three-phase system.

The use of a higher number of phases, such as 6- and 12-phase systems, is limited almost entirely to the supply of power to large rectifiers. Here, the rectifiers convert the alternating current to direct current, which is required for certain processes such as electrolysis. The rectifier output is a direct current plus a smaller pulsating component, or ripple, which decreases as the number of phases increases.

Almost without exception, polyphase systems in practice contain sources which may be closely approximated by ideal voltage sources or by ideal

Fig. 12-1: (a) The definition of the voltage V_{ab}. (b) $V_{ad} = V_{ab} + V_{bc} + V_{cd} = V_{ab} + V_{cd}$.

(a)   (b)

voltage sources in series with small internal impedances. Three-phase current sources are extremely rare.

It is convenient to describe polyphase voltages and currents using a double-subscript notation. With this notation, a voltage or current, such as $\mathbf{V}_{ab}$ or $\mathbf{I}_{aA}$, has more meaning than if it were indicated simply as $\mathbf{V}_3$ or $\mathbf{I}_x$. By definition, let the voltage of point a with respect to point b be $\mathbf{V}_{ab}$. Thus, the plus sign is located at a, as indicated in Fig. 12-1a. We therefore consider the double subscripts to be *equivalent to* a plus-minus-sign pair. The use of both would be redundant. With reference to Fig. 12-1b, it is now obvious that $\mathbf{V}_{ad} = \mathbf{V}_{ab} + \mathbf{V}_{cd}$. The advantage of the double-subscript notation lies in the fact that Kirchhoff's voltage law requires the voltage between two points to be the same, regardless of the path chosen between the points, and thus $\mathbf{V}_{ad} = \mathbf{V}_{ab} + \mathbf{V}_{bd} = \mathbf{V}_{ac} + \mathbf{V}_{cd} = \mathbf{V}_{ab} + \mathbf{V}_{bc} + \mathbf{V}_{cd}$, etc. It is apparent that KVL may be satisfied without reference to the circuit diagram; correct equations may be written even though a point, or subscript letter, is included which is not marked on the diagram. For example, we might have written above, $\mathbf{V}_{ad} = \mathbf{V}_{ax} + \mathbf{V}_{xd}$, where x identifies the location of any interesting point of our choice.

One possible representation of a three-phase system of voltages[1] will be found to be that of Fig. 12-2. Let us assume that the voltages $\mathbf{V}_{an}$, $\mathbf{V}_{bn}$, and $\mathbf{V}_{cn}$ are known,

$$\mathbf{V}_{an} = 100\underline{/0°} \quad \text{V rms}$$

$$\mathbf{V}_{bn} = 100\underline{/-120°}$$

$$\mathbf{V}_{cn} = 100\underline{/-240°}$$

and thus the voltage $\mathbf{V}_{ab}$ may be found, with an eye on the subscripts,

$$\mathbf{V}_{ab} = \mathbf{V}_{an} + \mathbf{V}_{nb} = \mathbf{V}_{an} - \mathbf{V}_{bn}$$

$$= 100\underline{/0°} - 100\underline{/-120°}$$

$$= 100 - (-50 - j86.6)$$

$$= 173.2\underline{/30°}$$

---

[1] Rms values of currents and voltages will be used throughout this chapter.

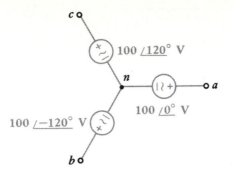

Fig. 12-2: A network used as a numerical example of double-subscript voltage notation.

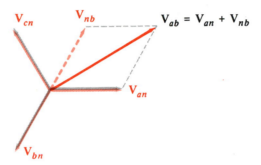

Fig. 12-3: This phasor diagram illustrates the graphical use of the double-subscript voltage convention to obtain $\mathbf{V}_{ab}$ for the network of Fig. 12-2.

The three given voltages and the construction of the phasor $\mathbf{V}_{ab}$ are shown on the phasor diagram of Fig. 12-3.

A double-subscript notation may also be applied to currents. We define the current $\mathbf{I}_{ab}$ as the current flowing from a to b *by the direct path.* In every complete circuit we consider, there must of course be at least two possible paths between the points a and b, and we agree that we shall not use double-subscript notation unless it is obvious that one path is much shorter, or much more direct. Usually this path is through a single element. Thus, the current $\mathbf{I}_{ab}$ is correctly indicated in Fig. 12-4. In fact, we do not even need the direction arrow when talking about this current; the subscripts *tell* us the direction. However, the mere identification of a current as $\mathbf{I}_{cd}$ would cause confusion.

Before considering polyphase systems, we shall make use of double-subscript notation to help with the analysis of a special single-phase system.

Fig. 12-4: An illustration of the use and misuse of the double-subscript convention for current notation.

Drill Problems

12-1. If $\mathbf{V}_{13} = 20\underline{/30°}$ V, $\mathbf{V}_{24} = 15\underline{/-100°}$ V, and $\mathbf{V}_{14} = 30\underline{/80°}$ V, find: (a) $\mathbf{V}_{12}$; (b) $\mathbf{V}_{23}$; (c) $\mathbf{V}_{34}$. *Ans:* $45.0\underline{/80.0°}$; $35.6\underline{/-74.5°}$; $23.0\underline{/121.8°}$ V

12-2. In the circuit of Fig. 12-5, let $I_{ad} = 3$ A, $I_{hi} = 2$ A, and $I_{ie} = -6$ A. Find: (a) I_{dh}; (b) I_{fj}; (c) I_{lk}. *Ans:* 0; -1; 7 A

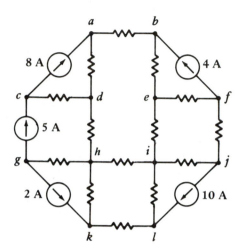

Fig. 12-5: See Drill Prob. 12-2.

12-2
Single-phase
three-wire systems

A single-phase three-wire source is defined as a source having three output terminals, such as a, n, and b in Fig. 12-6a, at which the phasor voltages $\mathbf{V}_{an}$ and $\mathbf{V}_{nb}$ are equal. The source may therefore be represented by the combination of two identical voltage sources; in Fig. 12-6b, $\mathbf{V}_{an} = \mathbf{V}_{nb} = \mathbf{V}_1$. It is apparent that $\mathbf{V}_{ab} = 2\mathbf{V}_{an} = 2\mathbf{V}_{nb}$, and we therefore have a source to which loads operating at either of two voltages may be connected. The normal household system is single-phase three-wire, permitting the operation of both 115-V and 230-V appliances. The higher-voltage appliances are normally those drawing larger amounts of power, and thus they cause currents in the lines which are only half those which operation at the same power and half the voltage would produce. Smaller-diameter wire may consequently be used safely in the appliance, the household distribution system, and the distribution system of the utility company.

Fig. 12-6: (a) A single-phase three-wire source. (b) The representation of a single-phase three-wire source by two identical voltage sources.

(a)   (b)

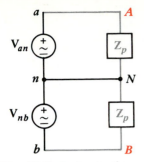

Fig. 12-7: A simple single-phase three-wire system. The two loads are identical and the neutral current is zero.

The name *single phase* arises because the voltages $\mathbf{V}_{an}$ and $\mathbf{V}_{nb}$, being equal, must have the same phase angle. From another viewpoint, however, the voltages between the outer wires and the central wire, which is usually referred to as the *neutral,* are exactly 180° out of phase. That is, $\mathbf{V}_{an} = -\mathbf{V}_{bn}$ and $\mathbf{V}_{an} + \mathbf{V}_{bn} = 0$. In a following section we shall see that balanced polyphase systems are characterized by a set of voltages of equal *amplitude* whose (phasor) sum is zero. From this viewpoint, then, the single-phase three-wire system is really a balanced two-phase system. ''Two phase,'' however, is a term that is traditionally reserved for a relatively unimportant system utilizing two voltage sources 90° out of phase; we shall not discuss it further.

Let us now consider a single-phase three-wire system which contains identical loads $\mathbf{Z}_p$ between each outer wire and the neutral (Fig. 12-7). We shall first assume that the wires connecting the source to the load are perfect conductors. Since

$$\mathbf{V}_{an} = \mathbf{V}_{nb}$$

then,

$$\mathbf{I}_{aA} = \frac{\mathbf{V}_{an}}{\mathbf{Z}_p} = \mathbf{I}_{Bb} = \frac{\mathbf{V}_{nb}}{\mathbf{Z}_p}$$

and, therefore

$$\mathbf{I}_{nN} = \mathbf{I}_{Bb} + \mathbf{I}_{Aa} = \mathbf{I}_{Bb} - \mathbf{I}_{aA} = 0$$

Thus there is no current in the neutral wire, and it could be removed without changing any current or voltage in the system. This result is achieved through the equality of the two loads and of the two sources.

We next consider the effect of a finite impedance in each of the wires. If lines aA and bB each have the same impedance, this impedance may be added to $\mathbf{Z}_p$, resulting in two equal loads once more and zero neutral current. Now let us allow the neutral wire to possess some impedance $\mathbf{Z}_n$. Without carrying out any detailed analysis, superposition should show us that the symmetry of the circuit will still cause zero neutral current. Moreover, the addition of any impedance connected directly from one of the outer lines to the other outer line also yields a symmetrical circuit and zero neutral current. Thus, zero neutral current is a consequence of a balanced, or symmetrical, load; any impedance in the neutral wire does not destroy the symmetry.

The most general single-phase three-wire system will contain unequal loads between each outside line and the neutral and another load directly between the two outer lines; the impedances of the two outer lines may be expected to be approximately equal, but the neutral impedance may be slightly larger. An example of such a system is shown in Fig. 12-8. The analysis of the circuit may be achieved by assigning mesh currents and writing the appropriate equations. The results of this labor are

Fig. 12-8: A typical single-phase three-wire system.

$$\mathbf{I}_1 = 11.24\underline{/-19.83°} \quad \text{A rms}$$

$$\mathbf{I}_2 = 9.39\underline{/-24.47°}$$

$$\mathbf{I}_3 = 10.37\underline{/-21.80°}$$

The currents in the outer lines are thus

$$\mathbf{I}_{aA} = \mathbf{I}_1 = 11.24\underline{/-19.83°} \quad \text{A rms}$$

$$\mathbf{I}_{bB} = -\mathbf{I}_3 = 10.37\underline{/158.20°}$$

and the smaller neutral current is

$$\mathbf{I}_{nN} = \mathbf{I}_3 - \mathbf{I}_1 = 0.946\underline{/-177.7°} \quad \text{A rms}$$

The power drawn by each load may be determined,

$$P_{50} = |\mathbf{I}_1 - \mathbf{I}_2|^2(50) = 206 \quad \text{W}$$

which could represent two 100-W lamps in parallel. Also,

$$P_{100} = |\mathbf{I}_3 - \mathbf{I}_2|^2(100) = 117 \quad \text{W}$$

which might represent one 100-W lamp. Finally,

$$P_{20+j10} = |\mathbf{I}_2|^2(20) = 1763 \quad \text{W}$$

which we may think of as a 2-hp induction motor. The total load power is 2086 W. The loss in each of the wires is next found:

$$P_{aA} = |\mathbf{I}_1|^2(1) = 126 \quad \text{W}$$

$$P_{bB} = |\mathbf{I}_3|^2(1) = 108 \quad \text{W}$$

$$P_{nN} = |\mathbf{I}_{nN}|^2(3) = 3 \quad \text{W}$$

or a total line loss of 237 W. The wires are evidently quite long; otherwise, the relatively high power loss in the two outer lines would cause a dangerous temperature rise. The total generated power must therefore be 2323 W, and this may be checked by finding the power delivered by each voltage source:

Fig. 12-9: The source voltages and three of the currents in the circuit of Fig. 12-8 are shown on a phasor diagram. Note that $I_{aA} + I_{bB} + I_{nN} = 0$.

$$P_{an} = 115(11.24) \cos 19.83° = 1216 \quad W$$

$$P_{bn} = 115(10.37) \cos 21.80° = 1107 \quad W$$

or a total of 2323 W. The transmission efficiency for this system is

$$\text{Eff.} = \frac{2086}{2086 + 237} = 89.8\%$$

This value would be unbelievable for a steam engine or an internal-combustion engine, but it is too low for a well-designed distribution system. Larger-diameter wires should be used if the source and the load cannot be placed closer to each other.

A phasor diagram showing the two source voltages, the currents in the outer lines, and the current in the neutral is constructed in Fig. 12-9. The fact that $I_{aA} + I_{bB} + I_{nN} = 0$ is indicated on the diagram.

Drill Problem

12-3. In Fig. 12-8, add 2 Ω to each line and neutral resistance, and then find new values for the average power delivered to each load.
Ans: 142.2; 88.4; 1272.0 W

12-3 Three-phase Y-Y connection

Three-phase sources have three terminals, called the *line* terminals, and they may or may not have a fourth terminal, the *neutral* connection. We shall begin by discussing a three-phase source which does have a neutral connection. It may be represented by three ideal voltage sources connected in a Y, as shown in Fig. 12-10; terminals a, b, c, and n are available. We shall

Fig. 12-10: A Y-connected three-phase, four-wire source.

Fig. 12-11: (a) Positive, or abc, phase sequence. (b) Negative, or cba, phase sequence.

consider only balanced three-phase sources which may be defined as having

$$|\mathbf{V}_{an}| = |\mathbf{V}_{bn}| = |\mathbf{V}_{cn}|$$

and

$$\mathbf{V}_{an} + \mathbf{V}_{bn} + \mathbf{V}_{cn} = 0$$

These three voltages, each existing between one line and the neutral, are called *phase voltages*. If we arbitrarily choose $\mathbf{V}_{an}$ as the reference,

$$\mathbf{V}_{an} = V_p\underline{/0°}$$

where we shall consistently use V_p to represent the rms *amplitude* of any of the phase voltages, then the definition of the three-phase source indicates that either

$$\mathbf{V}_{bn} = V_p\underline{/-120°} \qquad \mathbf{V}_{cn} = V_p\underline{/-240°}$$

or

$$\mathbf{V}_{bn} = V_p\underline{/120°} \qquad \mathbf{V}_{cn} = V_p\underline{/240°}$$

The former is called *positive phase sequence,* or *abc* phase sequence, and is shown in Fig. 12-11a; the latter is termed *negative phase sequence,* or *cba* phase sequence, and is indicated by the phasor diagram of Fig. 12-11b. It is apparent that the phase sequence of a physical three-phase source depends on the arbitrary choice of the three terminals to be lettered *a, b,* and *c.* They may always be chosen to provide positive phase sequence, and we shall assume that this has been done in most of the systems we consider.

Let us next find the line-to-line voltages (or simply "line" voltages) which are present when the phase voltages are those of Fig. 12-11a. It is easiest to do this with the help of a phasor diagram, since the angles are all multiples of 30°. The necessary construction is shown in Fig. 12-12; the results are

$$\mathbf{V}_{ab} = \sqrt{3}\,V_p\underline{/30°} \qquad \mathbf{V}_{bc} = \sqrt{3}\,V_p\underline{/-90°} \qquad \mathbf{V}_{ca} = \sqrt{3}\,V_p\underline{/-210°}$$

Kirchhoff's voltage law requires the sum of these three voltages to be zero, and it is zero.

Denoting the amplitude of any of the line voltages by V_L, then one of the important characteristics of the Y-connected three-phase source may be expressed as

$$V_L = \sqrt{3}\,V_p$$

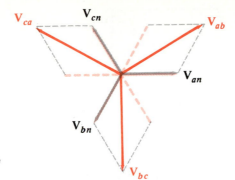

Fig. 12-12: A phasor diagram which is used to determine the line voltages from the given phase voltages.

Note that with positive phase sequence $\mathbf{V}_{an}$ leads $\mathbf{V}_{bn}$ and $\mathbf{V}_{bn}$ leads $\mathbf{V}_{cn}$, in each case by 120°, and also that $\mathbf{V}_{ab}$ leads $\mathbf{V}_{bc}$ and $\mathbf{V}_{bc}$ leads $\mathbf{V}_{ca}$, again by 120°. The statement is true for negative sequence if "lags" is substituted for "leads."

Now let us connect a balanced Y-connected three-phase load to our source, using three lines and a neutral, as drawn in Fig. 12-13. The load is represented by an impedance $\mathbf{Z}_p$ between each line and the neutral. The three line currents are found very easily, since we really have three single-phase circuits which possess one common lead:

$$\mathbf{I}_{aA} = \frac{\mathbf{V}_{an}}{\mathbf{Z}_p}$$

$$\mathbf{I}_{bB} = \frac{\mathbf{V}_{bn}}{\mathbf{Z}_p} = \frac{\mathbf{V}_{an}/-120°}{\mathbf{Z}_p} = \mathbf{I}_{aA}/-120°$$

$$\mathbf{I}_{cC} = \mathbf{I}_{aA}/-240°$$

and therefore

$$\mathbf{I}_{Nn} = \mathbf{I}_{aA} + \mathbf{I}_{bB} + \mathbf{I}_{cC} = 0$$

Thus, the neutral carries no current if the source and load are both balanced and if the four wires have zero impedance. How will this change if an impedance $\mathbf{Z}_L$ is inserted in series with each of the three lines and an impedance $\mathbf{Z}_n$ is inserted in the neutral? Evidently, the line impedances may be

Fig. 12-13: A balanced three-phase system, connected Y-Y and including a neutral.

Fig. 12-14: A balanced three-phase three-wire Y-Y connected system.

combined with the three load impedances; this effective load is still balanced, and a perfectly conducting neutral wire could be removed. Thus, if no change is produced in the system with a short circuit or an open circuit between n and N, any impedance may be inserted in the neutral and the neutral current will remain zero.

It follows that, if we have balanced sources, balanced loads, and balanced line impedances, a neutral wire of any impedance may be replaced by any other impedance, including a short circuit and an open circuit. It is often helpful to *visualize* a short circuit between the two neutral points, whether a neutral wire is actually present or not; the problem is then reduced to three single-phase problems, all identical except for the consistent difference in phase angle. We say that we thus work the problem on a "per-phase" basis.

Let us work several problems involving a balanced three-phase system having a Y-Y connection. A straightforward problem is suggested by the circuit of Fig. 12-14; we are asked to find the several currents and voltages throughout the circuit and to find the total power.

Since one of the source phase voltages is given, and since positive phase sequence is assumed, the three phase voltages are

$$\mathbf{V}_{an} = 200\underline{/0°} \qquad \mathbf{V}_{bn} = 200\underline{/-120°} \qquad \mathbf{V}_{cn} = 200\underline{/-240°}$$

The line voltage is $200\sqrt{3}$, or 346 V rms; the phase angle of each line voltage can be determined by constructing a phasor diagram, as before. As a matter of fact, the phasor diagram of Fig. 12-12 is applicable, and $\mathbf{V}_{ab}$ is $346\underline{/30°}$ V.

Let us work with phase A. The line current is

$$\mathbf{I}_{aA} = \frac{\mathbf{V}_{an}}{\mathbf{Z}_p} = \frac{200\underline{/0°}}{100\underline{/60°}} = 2\underline{/-60°} \quad \text{A rms}$$

and the power absorbed by this phase is, therefore,

$$P_{AN} = 200(2) \cos(0° + 60°) = 200 \quad \text{W}$$

Thus, the total power drawn by the three-phase load is 600 W. The problem is completed by drawing a phasor diagram and reading from it the appropriate phase angles which apply to the other line voltages and currents. The completed diagram is shown in Fig. 12-15.

We may also use per-phase methods to work problems in what might be called the backward direction. Suppose that we have a balanced three-phase

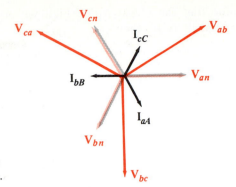

Fig. 12-15: The phasor diagram that is applicable to the circuit of Fig. 12-14.

system with a line voltage of 300 V rms, and we know that it is supplying a balanced Y-connected load with 1200 W at a leading PF of 0.8. What is the line current and the per-phase load impedance? It is evident that the phase voltage is $300/\sqrt{3}$ V rms and the per-phase power is 400 W. Thus the line current may be found from the power relationship

$$400 = \frac{300}{\sqrt{3}} (I_L)(0.8)$$

and the line current is therefore 2.89 A rms. The phase impedance is given by

$$|\mathbf{Z}_p| = \frac{V_p}{I_L} = \frac{300/\sqrt{3}}{2.89} = 60 \quad \Omega$$

Since the PF is 0.8, leading, the impedance phase angle is $-36.9°$, and $\mathbf{Z}_p = 60\underline{/-36.9°}\ \Omega$.

More complicated loads can be handled easily, since the problems reduce to simpler single-phase problems. Suppose that a balanced 600-W lighting load is added (in parallel) to the system above. A suitable per-phase circuit is first sketched, as shown in Fig. 12-16.

The amplitude of the lighting current is determined by

$$200 = \frac{300}{\sqrt{3}} |\mathbf{I}_1| \cos 0°$$

and $$|\mathbf{I}_1| = 1.155$$

In a similar way, the amplitude of the capacitive load current is found to be unchanged from its previous value,

$$|\mathbf{I}_2| = 2.89$$

Fig. 12-16: The per-phase circuit that is used to analyze a balanced three-phase example.

If we assume that the phase with which we are working has a phase voltage with an angle of 0°, then

$$\mathbf{I}_1 = 1.155\underline{/0°} \qquad \mathbf{I}_2 = 2.89\underline{/+36.9°}$$

and the line current is

$$\mathbf{I}_L = \mathbf{I}_1 + \mathbf{I}_2 = 3.87\underline{/+26.6°} \quad \text{A rms}$$

The power generated by this phase of the source is, therefore,

$$P_p = \frac{300}{\sqrt{3}} \, 3.87 \cos{(+26.6°)} = 600 \quad \text{W}$$

which checks with the original hypothesis.

If an *unbalanced* Y-connected load is present in an otherwise balanced three-phase system, the circuit may still be analyzed on a per-phase basis *if* the neutral wire is present and *if* it has zero impedance. If either of these conditions is not met, other methods must be used, such as mesh or nodal analysis. However, engineers who spend most of their time with unbalanced three-phase systems will find the use of *symmetrical components* a great timesaver. We shall not discuss this method here.

Drill Problems

12-4. A balanced three-phase three-wire system has a Y-connected load. Each phase contains a 50-μF capacitor in parallel with the series combination of 50 Ω and 50 mH. Using positive phase sequence with $\mathbf{V}_{an} = 240\underline{/-90°}$ V rms and $\omega = 500$ rad/s, find: (a) $\mathbf{V}_{bn}$; (b) $\mathbf{V}_{bc}$; (c) $\mathbf{I}_{aA}$; (d) the total power drawn by the load.
  Ans: 240$\underline{/150°}$ V rms; 416$\underline{/180°}$ V rms; 5.60$\underline{/-43.3°}$ A rms; 2.76 kW

12-5. A balanced three-phase three-wire system has a line voltage of 440 V rms and feeds two balanced Y-connected loads. One is an inductive load with $7 + j2$ Ω per phase, and the other is a capacitive load with $5 - j2$ Ω per phase. Find: (a) the rms line current provided by the source; (b) the average power delivered to the inductive load; (c) the average power generated by the source.
  Ans: 77.8 A rms; 25.6 kW; 58.9 kW

12-6. Three balanced Y-connected loads are present on a balanced three-phase three-wire system. Load A draws 21 kW at PF = 0.8 lag, load B requires 30 kVA at PF = 0.9 lead, and load C is a resistive load dissipating 12 kW. If the line voltage is 220 V rms, find: (a) the current in load A; (b) the current in load B; (c) the current in load C; (d) the line current.
  Ans: 68.9; 78.7; 31.5; 157.6 A rms

**12-4
The delta (Δ)
connection**

A three-phase load is more apt to be found Δ-connected than Y-connected. One reason for this, at least for the case of an unbalanced load, is the flexibility with which loads may be added or removed on a single phase. This is difficult (or impossible) to do with a Y-connected three-wire load.

Let us consider a balanced Δ-connected load which consists of an impedance $\mathbf{Z}_p$ inserted between each pair of lines. We shall assume a three-wire

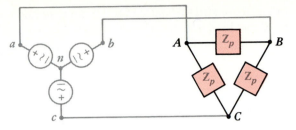

Fig. 12-17: A balanced Δ-connected load is present on a three-wire three-phase system. The source happens to be Y-connected.

system for obvious reasons. With reference to Fig. 12-17, let us assume known line voltages

$$V_L = |\mathbf{V}_{ab}| = |\mathbf{V}_{bc}| = |\mathbf{V}_{ca}|$$

or known phase voltages

$$V_p = |\mathbf{V}_{an}| = |\mathbf{V}_{bn}| = |\mathbf{V}_{cn}|$$

where

$$V_L = \sqrt{3}\, V_p \qquad \text{and} \qquad \mathbf{V}_{ab} = \sqrt{3}\, \mathbf{V}_{an}\underline{/30°}$$

and so forth, as before. Since the voltage across each branch of the Δ is known, the *phase currents* are found,

$$\mathbf{I}_{AB} = \frac{\mathbf{V}_{ab}}{\mathbf{Z}_p} \qquad \mathbf{I}_{BC} = \frac{\mathbf{V}_{bc}}{\mathbf{Z}_p} \qquad \mathbf{I}_{CA} = \frac{\mathbf{V}_{ca}}{\mathbf{Z}_p}$$

and their differences provide us with the line currents, such as

$$\mathbf{I}_{aA} = \mathbf{I}_{AB} - \mathbf{I}_{CA}$$

The three phase currents are of equal amplitude,

$$I_p = |\mathbf{I}_{AB}| = |\mathbf{I}_{BC}| = |\mathbf{I}_{CA}|$$

The line currents are also equal in amplitude, because the phase currents are equal in amplitude and 120° out of phase. The symmetry is apparent from the phasor diagram of Fig. 12-18. We thus have

$$I_L = |\mathbf{I}_{aA}| = |\mathbf{I}_{bB}| = |\mathbf{I}_{cC}|$$

and

$$I_L = \sqrt{3}\, I_p$$

Fig. 12-18: A phasor diagram which could apply to the circuit of Fig. 12-17 if $\mathbf{Z}_p$ were an inductive impedance.

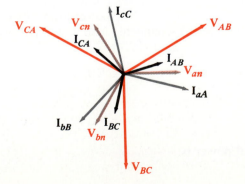

Let us disregard the source for the moment and consider only the balanced load. If the load is Δ-connected, then the phase voltage and the line voltage are indistinguishable, but the line current is larger than the phase current by a factor of $\sqrt{3}$; with a Y-connected load, however, the phase current and the line current refer to the same current, and the line voltage is greater than the phase voltage by a factor of $\sqrt{3}$.

The solution of three-phase problems will be speedily accomplished if the $\sqrt{3}$'s are used properly. Let us consider a typical numerical example. We are to determine the amplitude of the line current in a 300-V rms three-phase system which supplies 1200 W to a Δ-connected load at a lagging PF of 0.8. Let us again consider a single phase. It draws 400 W, 0.8 lagging PF, at a 300-V rms line voltage. Thus,

$$400 = 300(I_p)(0.8)$$

and

$$I_p = 1.667 \quad \text{A rms}$$

and the relationship between phase currents and line currents yields

$$I_L = \sqrt{3}(1.667) = 2.89 \quad \text{A rms}$$

Furthermore, the impedance in each phase must be

$$\mathbf{Z}_p = \frac{300}{1.667} \underline{/\cos^{-1} 0.8} = 180\underline{/36.9°} \quad \Omega$$

Now let us change the statement of the problem: the load is Y-connected instead of Δ-connected. On a per-phase basis, we now have a phase voltage of $300/\sqrt{3}$ V rms, a power of 400 W, and a lagging PF of 0.8. Thus,

$$400 = \frac{300}{\sqrt{3}}(I_p)(0.8)$$

and

$$I_p = 2.89 \quad \text{or} \quad I_L = 2.89 \quad \text{A rms}$$

The impedance in each phase of the Y is

$$\mathbf{Z}_p = \frac{300/\sqrt{3}}{2.89} \underline{/\cos^{-1} 0.8} = 60\underline{/36.9°} \quad \Omega$$

The $\sqrt{3}$ factor not only relates phase and line quantities but also appears in a useful expression for the total power drawn by any balanced three-phase load. If we assume a Y-connected load with a power-factor angle θ, then the power taken by any phase is

$$P_p = V_p I_p \cos \theta$$
$$= V_p I_L \cos \theta$$
$$= \frac{V_L}{\sqrt{3}} I_L \cos \theta$$

and the total power is

$$P = 3P_p$$
$$= \sqrt{3}V_L I_L \cos\theta$$

In a similar way, the power delivered to each phase of a Δ-connected load is

$$P_p = V_p I_p \cos\theta$$
$$= V_L I_p \cos\theta$$
$$= V_L \frac{I_L}{\sqrt{3}} \cos\theta$$

or a total power

$$P = 3P_p$$

or
$$P = \sqrt{3}V_L I_L \cos\theta \qquad (1)$$

Thus (1) enables us to calculate the total power delivered to a balanced load from a knowledge of the magnitude of the line voltage, of the line current, and of the phase angle of the load impedance (or admittance), regardless of whether the load is Y-connected or Δ-connected. The numerical example above can be worked in one line:

$$1200 = \sqrt{3}(300)(I_L)(0.8)$$

Therefore

$$I_L = \frac{5}{\sqrt{3}} = 2.89 \quad \text{A rms}$$

The source may also be connected in Δ. This is not typical, however, for a slight unbalance in the source phases can lead to large currents circulating in the Δ loop. As an example, let us call the three single-phase sources $\mathbf{V}_{ab}$, $\mathbf{V}_{bc}$, and $\mathbf{V}_{cd}$. Before closing the Δ by connecting d to a, let us determine the unbalance by measuring the sum $\mathbf{V}_{ab} + \mathbf{V}_{bc} + \mathbf{V}_{cd}$. Suppose that the amplitude of the resultant is only 1 percent of the line voltage. The circulating current is thus approximately $\frac{1}{3}$ percent of the line voltage divided by the internal impedance of any source. How large is this impedance apt to be? It must depend on the current that the source is expected to deliver with a negligible drop in terminal voltage. If we assume that this maximum current causes a 1 percent drop in the terminal voltage, then it is seen that the circulating current is one-third of the maximum current. This reduces the useful current capacity of the source and also increases the losses in the system.

We should also note that balanced three-phase sources may be transformed from Y to Δ, or vice versa, without affecting the load currents or voltages. The necessary relationships between the line and phase voltages are shown in Fig. 12-12 for the case where $\mathbf{V}_{an}$ has a reference phase angle of $0°$. This transformation enables us to use whichever source connection we prefer, and all the load relationships will be correct. Of course, we cannot specify any currents or voltages within the source until we know how it is actually connected.

Drill Problems

12-7. Each phase of a Δ-connected load contains a 50-mH inductor in series with the parallel combination of 50 Ω and 50 μF. The supply is a three-wire system with $(+)$ phase sequence and $\mathbf{V}_{ab} = 550\underline{/0°}$ V rms at $\omega = 800$ rad/s. Find: (a) $\mathbf{I}_{BC}$; (b) $\mathbf{I}_{bB}$; (c) the total power drawn by the load.

Ans: $24.6\underline{/-183.4°}$ A rms; $42.6\underline{/146.6°}$ A rms; 18.15 kW

12-8. A balanced three-phase three-wire system is terminated with two Δ-connected loads in parallel. Load 1 draws 30 kW at a lagging power factor of 0.8, while load 2 uses 24 kVA at PF = 0.9 leading. The line voltage is 660 V rms *at the loads*, and each line has a resistance of 0.6 Ω. Find: (a) the impedance per phase for each load; (b) the total power drawn by the combined loads; (c) the total power lost in the lines.

Ans: $27.9 + j20.9$ and $49.0 - j23.7$ Ω; 51.6 kW; 3.87 kW

Problems

1 Some current values for the circuit of Fig. 12-19 are $\mathbf{I}_{AD} = 11$ A, $\mathbf{I}_{BE} = -12$ A, $\mathbf{I}_{EF} = 7$ A, $\mathbf{I}_{CF} = 13$ A, $\mathbf{I}_{DG} = 16$ A, $\mathbf{I}_{EH} = -49$ A, $\mathbf{I}_{AE} = -4$ A, and $\mathbf{I}_{HI} = -27$ A. Find $\mathbf{I}_{BC}$, $\mathbf{I}_{EG}$, and $\mathbf{I}_{IE}$.

Fig. 12-19: See Prob. 1.

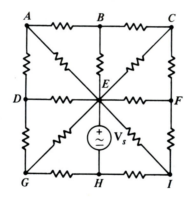

2 Given the voltages, $\mathbf{V}_{ax} = 161.8\underline{/144°}$ V, $\mathbf{V}_{ay} = 161.8\underline{/180°}$ V, $\mathbf{V}_{wz} = 161.8\underline{/-108°}$ V, and $\mathbf{V}_{yw} = 161.8\underline{/36°}$ V, draw a phasor diagram showing $\mathbf{V}_{wx}$, $\mathbf{V}_{xy}$, $\mathbf{V}_{yz}$, $\mathbf{V}_{za}$, and $\mathbf{V}_{aw}$.

3 The 110/220 V rms, 60 Hz, single-phase, three-wire system shown in Fig. 12-20

Fig. 12-20: See Prob. 3.

supplies power to three loads: load AN draws 1100 W at unity PF, load NB requires 2200 VA at PF = 0.707 lagging, and load AB uses a complex power of $5500\underline{/-30°}$ VA. If $\mathbf{V}_{an} = 110\underline{/0°}$ V, find: (*a*) $\mathbf{I}_{aA}$; (*b*) $\mathbf{I}_{nN}$.

4 A single-phase, three-wire source provides 220 and 440 V rms to the following loads: $10 + j0\ \Omega$ in one 220-V phase, $3 + j4\ \Omega$ in the other 220-V phase, and $10\underline{/36.87°}\ \Omega$ at 440 V. Find: (*a*) the effective value of each line current; (*b*) the effective value of the neutral current; (*c*) the total power absorbed by the composite load.

5 A single-phase, three-wire power source is supplying a load through lines that have line resistances of 1 Ω each, and a neutral wire resistance of 2 Ω. Voltages *at the load* are 120 and 240 V rms. The loads (as in Fig. 12-20) are $\mathbf{Z}_{AN} = 3 + j4\ \Omega$, $\mathbf{Z}_{NB} = 12 - j5\ \Omega$, and $\mathbf{Z}_{AB} = 8 + j0\ \Omega$. Find the rms amplitudes of: (*a*) $\mathbf{I}_{nN}$; (*b*) $\mathbf{I}_{aA}$; (*c*) the source voltage $\mathbf{V}_{an}$.

6 (*a*) If $\mathbf{Z}_{AN} = 20 + j10\ \Omega$, find the power supplied by each source in the circuit of Fig. 12-21. (*b*) Let $\mathbf{Z}_{AN} = 10 + j20\ \Omega$ and find the power supplied by the upper source.

Fig. 12-21: See Prob. 6.

7 In the three-wire single-phase system of Fig. 12-22, let $V_1 = 120$ V rms at 60 Hz. (*a*) What size should C be to provide a unity power-factor load? (*b*) How many kVA does C handle?

Fig. 12-22: See Prob. 7.

Fig. 12-23: See Probs. 8 to 11.

8 The balanced three-phase, three-wire circuit of Fig. 12-23 operates with $\mathbf{V}_{BC} = 100\underline{/20°}$ V rms with positive phase sequence. The total load draws 3 kW at PF = 0.8 lagging. If $R_w = 0.6\ \Omega$, find: (a) the total power lost in line resistances; (b) $\mathbf{V}_{ab}$.

9 Refer to the balanced Y-Y circuit of Fig. 12-23 and let $\mathbf{V}_{an} = 4600\underline{/0°}$ V rms with (+) phase sequence. Let the source supply $\mathbf{S} = 240 + j60$ kVA with $R_w = 3.2\ \Omega$. Find: (a) $\mathbf{V}_{AN}$; (b) $\mathbf{I}_{aA}$; (c) $\mathbf{Z}_p$; (d) the transmission efficiency.

10 In the system shown in Fig. 12-23, $\mathbf{Z}_p = 20 + j8\ \Omega$, and (+) phase sequence is assumed. If $\mathbf{I}_{aA} = 20\underline{/-46°}$ A rms, and the source is operating with PF = 0.94 lagging, find: (a) R_w; (b) the complex power supplied by the source; (c) $\mathbf{V}_{AB}$; (d) $\mathbf{V}_{ab}$.

11 In Fig. 12-23, let $\mathbf{Z}_p$ represent a capacitive impedance of $80\underline{/-24°}\ \Omega$ in parallel with 0.25 H. If $\mathbf{V}_{an} = 230\underline{/0°}$ V rms at 60 Hz, and $R_w = 2.5\ \Omega$, determine: (a) $\mathbf{I}_{aA}$; (b) the total power delivered to the load; (c) the total power lost in the wire resistance; (d) the PF at which the source operates.

12 An unbalanced three-phase, four-wire system is shown in Fig. 12-24. Assume positive phase sequence with $\mathbf{V}_{an} = 100\underline{/0°}$ V rms, and let the loads be $\mathbf{Z}_{AN} = 10 - j5\ \Omega$, $\mathbf{Z}_{BN} = 8 + j5\ \Omega$, and $\mathbf{Z}_{CN} = 12 + j0\ \Omega$. Find $\mathbf{I}_{nN}$.

Fig. 12-24: See Prob. 12.

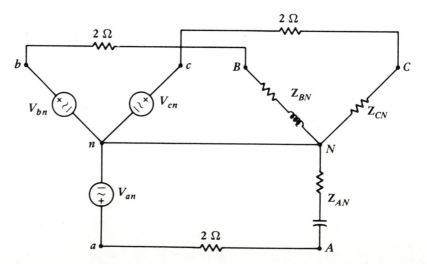

13 A Y-connected load has an impedance of $25 + j12\ \Omega$ in each phase. It is desired to raise the load power factor to 0.95 lagging. (a) What size capacitor should be

placed in parallel with each phase impedance if $f = 60$ Hz? (b) Let the line voltage at the load have an amplitude of 500 V rms. How many kVAR (capacitive) are required at the load to obtain PF = 0.95 lagging?

14 The balanced three-wire, three-phase circuit of Fig. 12-25 has $R_w = 0$ and $\mathbf{V}_{an} = 288.7\underline{/-30°}$ V rms. Each phase of the load absorbs 2.4 kW at 0.8 PF leading. Assume abc sequence. Find: (a) $\mathbf{V}_{ab}$; (b) $\mathbf{Z}_p$; (c) $\mathbf{I}_{cC}$; (d) the total load power.

Fig. 12-25: See Probs. 14 to 16.

15 The balanced three-phase load shown in Fig. 12-25 requires a total power of 6 kW at PF = 0.83 lagging. If $R_w = 0.8$ Ω and $\mathbf{V}_{CA} = 100\underline{/20°}$ V rms with (+) phase sequence, find: (a) $\mathbf{V}_{ab}$; (b) the total complex power being supplied by the source.

16 Let $R_w = 4$ Ω in the balanced three-phase circuit of Fig. 12-25. The load draws a total complex power, $\mathbf{S} = 2400 + j2100$ VA, and an additional 300 W is dissipated in the lines. Determine the rms amplitude of: (a) $\mathbf{I}_{aA}$; (b) $\mathbf{I}_{AB}$; (c) $\mathbf{V}_{an}$.

17 The source in Fig. 12-26 is balanced and has positive phase sequence with $f = 60$ Hz. Find $\mathbf{I}_{aA}$, $\mathbf{I}_{bB}$, $\mathbf{I}_{cC}$, and the power furnished by each phase of the source.

Fig. 12-26: See Probs. 17 and 19.

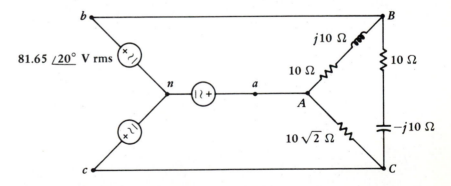

18 A balanced, three-phase, Y-connected voltage source has $\mathbf{V}_{an} = 110\underline{/60°}$ V rms with (+) phase sequence, and it supplies a balanced, Δ-connected, three-phase load. The total power drawn by the load is 12 kW at 0.75 lagging power factor. Find: (a) the line current $\mathbf{I}_{bB}$; (b) the load phase current $\mathbf{I}_{CA}$; (c) the line voltage $\mathbf{V}_{bc}$.

19 Add 1 Ω of resistance in each line of Fig. 12-26 and again work Prob. 17.

PART FOUR
COMPLEX FREQUENCY

Chapter 13
Complex Frequency

13-1
Introduction

We are now about to begin the fourth major portion of our study of circuit analysis, a discussion of the concept of complex frequency. This, we shall see, is a remarkably unifying concept which will enable us to tie together all our previously developed analytical techniques into one neat package. Resistive circuit analysis, steady-state sinusoidal analysis, transient analysis, the forced response, the complete response, and the analysis of circuits excited by exponential forcing functions and exponentially damped sinusoidal forcing functions will all become special cases of the general techniques of circuit analysis which are associated with the complex-frequency concept.

We introduce complex frequency by considering an exponentially damped sinusoidal function, such as the voltage

$$v(t) = V_m e^{\sigma t} \cos (\omega t + \theta) \tag{1}$$

where σ (sigma) is a real quantity and is usually negative. Although we refer to this function as being "damped," it is possible that the sinusoidal amplitude may increase if σ is positive; the more practical case is that of the damped function. Our work with the natural response of the *RLC* circuit also indicates that σ is the negative of the exponential damping coefficient.

We may first construct a constant voltage from (1) by letting both σ and ω be zero,

$$v(t) = V_m \cos \theta = V_0 \tag{2}$$

If we set only σ equal to zero, then we obtain a general sinusoidal voltage

$$v(t) = V_m \cos (\omega t + \theta) \tag{3}$$

And if $\omega = 0$, we have the exponential voltage

$$v(t) = V_m \cos \theta \, e^{\sigma t} = V_0 e^{\sigma t} \tag{4}$$

Thus, the damped sinusoid (1) includes as special cases the dc (2), sinusoidal (3), and exponential (4) functions.

Some additional insight into the significance of σ can be obtained by comparing the exponential function (4) with the complex representation of a sinusoidal function with a zero-degree phase angle,

$$v(t) = V_0 e^{j\omega t} \tag{5}$$

It is apparent that the two functions, (4) and (5), have much in common. The only difference is that the exponent in (4) is real, and that in (5) is imaginary. The similarity between the two functions is emphasized by describing σ as a "frequency." This choice of terminology will be discussed in detail in the following sections, but for now we need merely note that σ is specifically termed the real part of the complex frequency. It should not be called the "real frequency," however, for this is a term which is more suitable for f (or, loosely, for ω). We shall also refer to σ as the *neper frequency,* the name arising from the dimensionless unit of the exponent of e. Thus, given e^{7t}, the dimensions of $7t$ are nepers (Np), and 7 is the neper frequency in nepers per second. The neper itself was named after Napier and his napierian logarithm system; the spelling of his name is historically uncertain.

Let us now think in terms of a forcing function or source that has a damped sinusoidal or exponential form. When $\sigma < 0$, the amplitude of the forcing function can have extremely large values at times in the distant past. That is, we have considered the forced response as that produced by a forcing function applied since $t = -\infty$; its application at some finite time gives rise to a transient response in addition to the forced response. Since the infinite amplitude of the forcing function at $t = -\infty$ may make us somewhat uncomfortable, we might note that initial conditions can be established in any circuit so that the application of a specified forcing function at a specified instant of time produces thereafter a response identical to the forced response without any transient response. Examples of this will appear later. From a practical point of view, we know that, although it may be impossible in the lab to generate damped sinusoidal or exponential forcing functions which are accurate for all time, we can produce approximations that are satisfactory for circuits whose transient responses do not last very long.

The forced response of a network to a general forcing function of the form of (1) is found very simply by using a method almost identical with that used for the sinusoidal forcing function; we shall discuss the method in Sec. 13-3. When we are able to find the forced response to this damped sinusoid, we should realize that we shall also have found the forced response to a dc voltage, an exponential voltage and a sinusoidal voltage. Now let us see how we may consider σ and ω as the real and imaginary parts of a complex frequency.

**13-2
Complex frequency**

Let us first provide ourselves with a purely mathematical definition of complex frequency and then gradually develop a physical interpretation as the chapter progresses. We say that any function which may be written in the form

$$\mathbf{f}(t) = \mathbf{K}e^{st} \tag{6}$$

where $\mathbf{K}$ and $\mathbf{s}$ are complex constants (independent of time), is characterized by the *complex frequency* $\mathbf{s}$. The complex frequency $\mathbf{s}$ is therefore simply the factor which multiplies t in this complex exponential representation. Until we are able to determine the complex frequency of a given function by inspection, it is necessary to write the function in the form of (6).

We may apply this definition first to the more familiar forcing functions. For example, a constant voltage

$$v(t) = V_0$$

may be written in the form

$$v(t) = V_0 e^{(0)t}$$

The complex frequency of a dc voltage or current is thus zero; $\mathbf{s} = 0$. The next simple case is the exponential function

$$v(t) = V_0 e^{\sigma t}$$

which is already in the required form; the complex frequency of this voltage is therefore σ; $\mathbf{s} = \sigma + j0$. Finally, let us consider a sinusoidal voltage, one which may provide a slight surprise. Given

$$v(t) = V_m \cos(\omega t + \theta)$$

it is necessary to find an equivalent expression in terms of the complex exponential. From our past experience, we therefore use the formula we derived from Euler's identity,

$$\cos(\omega t + \theta) = \tfrac{1}{2}(e^{j(\omega t + \theta)} + e^{-j(\omega t + \theta)})$$

and obtain

$$v(t) = \tfrac{1}{2}V_m(e^{j(\omega t + \theta)} + e^{-j(\omega t + \theta)})$$
$$= (\tfrac{1}{2}V_m e^{j\theta})e^{j\omega t} + (\tfrac{1}{2}V_m e^{-j\theta})e^{-j\omega t}$$

or

$$v(t) = \mathbf{K}_1 e^{\mathbf{s}_1 t} + \mathbf{K}_2 e^{\mathbf{s}_2 t}$$

We have the *sum* of two complex exponentials, and *two* complex frequencies are therefore present, one for each term. The complex frequency of the first term is $\mathbf{s} = \mathbf{s}_1 = j\omega$ and that of the second term is $\mathbf{s} = \mathbf{s}_2 = -j\omega$. These two values of $\mathbf{s}$ are conjugates, or $\mathbf{s}_2 = \mathbf{s}_1^*$, and the two values of $\mathbf{K}$ are also conjugates, $\mathbf{K}_1 = \tfrac{1}{2}V_m e^{j\theta}$ and $\mathbf{K}_2 = \mathbf{K}_1^* = \tfrac{1}{2}V_m e^{-j\theta}$. The entire first term and the entire second term are therefore conjugates, which we should expect inasmuch as their sum must be a real quantity.

Now let us determine the complex frequency or frequencies associated with the exponentially damped sinusoidal function (1). We again use Euler's formula to obtain a complex exponential representation:

$$v(t) = V_m e^{\sigma t} \cos(\omega t + \theta)$$
$$= \tfrac{1}{2}V_m e^{\sigma t}(e^{j(\omega t + \theta)} + e^{-j(\omega t + \theta)})$$

and thus

$$v(t) = \tfrac{1}{2}V_m e^{j\theta} e^{(\sigma + j\omega)t} + \tfrac{1}{2}V_m e^{-j\theta} e^{(\sigma - j\omega)t}$$

We therefore find once again that a conjugate complex pair of frequencies is required to describe the exponentially damped sinusoid, $s_1 = \sigma + j\omega$ and $s_2 = s_1^* = \sigma - j\omega$. In general, neither σ nor ω is zero, and we see that the exponentially varying sinusoidal waveform is the general case; the constant, sinusoidal, and exponential waveforms are special cases.

As numerical illustrations, we should now recognize at sight the complex frequencies associated with these voltages:

$$v(t) = 100 \qquad\qquad\qquad\qquad \mathbf{s} = 0$$

$$v(t) = 5e^{-2t} \qquad\qquad\qquad\quad \mathbf{s} = -2 + j0$$

$$v(t) = 2 \sin 500t \qquad\qquad \begin{cases} \mathbf{s}_1 = j500 \\ \mathbf{s}_2 = \mathbf{s}_1^* = -j500 \end{cases}$$

$$v(t) = 4e^{-3t} \sin (6t + 10°) \qquad \begin{cases} \mathbf{s}_1 = -3 + j6 \\ \mathbf{s}_2 = \mathbf{s}_1^* = -3 - j6 \end{cases}$$

The reverse type of example is also worth consideration. Given a complex frequency or a pair of conjugate complex frequencies, we must be able to identify the nature of the function with which they are associated. The most special case, $\mathbf{s} = 0$, defines a constant or dc function. With reference to the defining functional form (6), it is apparent that the constant $\mathbf{K}$ must be real if the function is to be real.

Let us next consider real values of $\mathbf{s}$. A positive real value, such as $\mathbf{s} = 5 + j0$, identifies an exponentially increasing function $\mathbf{K}e^{5t}$, where again $\mathbf{K}$ must be real if the function is to be a physical one. A negative real value for $\mathbf{s}$, such as $\mathbf{s} = -5 + j0$, refers to an exponentially decreasing function $\mathbf{K}e^{-5t}$.

A purely imaginary value of $\mathbf{s}$, for example, $j10$, can never be associated with a real quantity; the functional form is $\mathbf{K}e^{j10t}$, which can also be written as $\mathbf{K}(\cos 10t + j \sin 10t)$, and obviously possesses both real and imaginary parts. Each part is sinusoidal. In order to construct a real function, it is necessary to consider conjugate values of $\mathbf{s}$, such as $\mathbf{s}_{1,2} = \pm j10$, with which must be associated conjugate values of $\mathbf{K}$. Loosely speaking, however, we may identify either of the complex frequencies $\mathbf{s}_1 = j10$ or $\mathbf{s}_2 = -j10$ with a sinusoidal voltage at the radian frequency of 10 rad/s. The presence of the conjugate complex frequency is understood. The amplitude and phase angle of the sinusoidal voltage will depend on the choice of $\mathbf{K}$ for each of the two frequencies. Thus, selecting $\mathbf{s}_1 = j10$ and $\mathbf{K}_1 = 6 - j8$, where $v(t) = \mathbf{K}_1 e^{\mathbf{s}_1 t} + \mathbf{K}_2 e^{\mathbf{s}_2 t}$, $\mathbf{s}_2 = \mathbf{s}_1^*$ and $\mathbf{K}_2 = \mathbf{K}_1^*$, we obtain the real sinusoid $20 \cos (10t - 53.1°)$.

In a similar manner, a general value for $\mathbf{s}$, such as $3 - j5$, can be associated with a real quantity only if it is accompanied by its conjugate $3 + j5$. Speaking loosely again, we may think of either of these two conjugate frequencies as describing an exponentially increasing sinusoidal function $e^{3t} \cos 5t$; the specific amplitude and phase angle will again depend on the specific values of the conjugate complex $\mathbf{K}$'s.

By now we should have achieved some appreciation of the physical nature of the complex frequency $\mathbf{s}$; in general, it describes an exponentially varying sinusoid. The real part of $\mathbf{s}$ is associated with the exponential variation; if it is negative, the function decays as *t* increases; if positive, the function increases; and if it is zero, the sinusoidal amplitude is constant. The larger the *magnitude* of the real part of **s**, the greater is the rate of exponential increase or decrease. The imaginary part of **s** describes the sinusoidal variation; it is specifically the radian frequency. A large magnitude for the imaginary part of **s** indicates a more rapidly changing function of time. Thus, larger magnitudes for the real part of **s**, the imaginary part of **s**, or the magnitude of **s** indicate a more rapidly varying function.

It is customary to call the real part of **s**, σ, and the imaginary part ω (*not $j\omega$*):

$$\mathbf{s} = \sigma + j\omega \tag{7}$$

The radian frequency is sometimes referred to as the "real frequency," but this terminology can be very confusing when we find that we must then say that "the real frequency is the imaginary part of the complex frequency"! When we need to be specific, we shall call **s** the complex frequency, σ the neper frequency, ω the radian frequency, and *f* the cyclic frequency; when no confusion seems likely, it is permissible to use "frequency" to refer to any of these four quantities. The neper frequency is measured in nepers per second, radian frequency is measured in radians per second, and complex frequency **s** is measured in units which are variously termed complex nepers per second or complex radians per second.

Drill Problems

13-1. Specify all the complex frequencies present in the time function: (*a*) $(2 - 3e^{-7t}) \cos 5t$; (*b*) $2 - 3e^{-7t} \cos 5t$; (*c*) $2 \cos 5t - 5 \sin 10t \cos 2t$; (*d*) $3 + 2 \cos (7t - 30°)$.
Ans: $j5, -j5, -7 + j5, -7 - j5; 0, -7 + j5, -7 - j5; j5, -j5, j8, -j8, j12, -j12; 0, j7, -j7$ s^{-1}

13-2. Write the general form of a real voltage having components at the complex frequencies: (*a*) $-7, 5$ s^{-1}; (*b*) $0, -4, -2 + j7$ s^{-1}; (*c*) $-3, -3 + j3, 3 - j3$ s^{-1}.
Ans: $Ae^{-7t} + Be^{5t}; A + Be^{-4t} + Ce^{-2t} \cos (7t + \phi); Ae^{-3t} + Be^{-3t} \cos (3t + \phi) + Ce^{3t} \cos (3t + \theta)$ V

13-3
The damped sinusoidal forcing function

We have devoted enough time to the definition and introductory interpretation of complex frequency; it is now time to put this concept to work and become familiar with it by seeing what it will do and how it is used.

The general exponentially varying sinusoid, which we may represent as a voltage for the moment,

$$v(t) = V_m e^{\sigma t} \cos (\omega t + \theta) \tag{8}$$

is expressible in terms of the complex frequency **s** by making use of Euler's identity as before:

$$v(t) = \text{Re} \ (V_m e^{\sigma t} e^{j(\omega t + \theta)}) \tag{9}$$

or

$$v(t) = \text{Re} \ (V_m e^{\sigma t} e^{j(-\omega t - \theta)}) \tag{10}$$

Either representation is suitable, and the two expressions should remind us that a pair of conjugate complex frequencies is associated with a sinusoid or an exponentially damped sinusoid. Equation (9) is more directly related to the given damped sinusoid, and we shall concern ourselves principally with it. Collecting factors,

$$v(t) = \text{Re } (V_m e^{j\theta} e^{(\sigma + j\omega)t})$$

we now substitute $\mathbf{s} = \sigma + j\omega$, and obtain

$$v(t) = \text{Re } (V_m e^{j\theta} e^{\mathbf{s}t}) \tag{11}$$

Before we apply a forcing function of this form to any circuit, we should note the resemblance of this last representation of the damped sinusoid to the corresponding representation of the *undamped* sinusoid that we studied back in Sec. 9-3,

$$\text{Re } (V_m e^{j\theta} e^{j\omega t})$$

The only difference is that we now have **s** where we previously had $j\omega$. Instead of restricting ourselves to sinusoidal forcing functions and their radian frequencies, we have now extended our notation to include the damped sinusoidal forcing function at a complex frequency. It should be no surprise at all to see later in this section and the following one that we shall develop a *frequency-domain* description of the exponentially damped sinusoid in exactly the same way that we did for the sinusoid; we shall simply omit the Re notation and suppress $e^{\mathbf{s}t}$.

We are now ready to apply the exponentially damped sinusoid, as given by (8), (9), (10), or (11), to an electrical network. The forced response, say a current in some branch of the network, is the desired response. Since the forced response has the form of the forcing function, its integral, and its derivatives, the response may be assumed to be

$$i(t) = I_m e^{\sigma t} \cos (\omega t + \phi)$$

or
$$i(t) = \text{Re } (I_m e^{j\phi} e^{\mathbf{s}t})$$

where the complex frequency of the source and the response must be identical.

If we now recall that the real part of a complex forcing function produces the real part of the response, while the imaginary part of the complex forcing function causes the imaginary part of the response, then we are again led to the application of a *complex* forcing function to our network. We shall obtain a complex response whose real part is the desired real response. Actually, we shall work with the Re notation omitted, but we should realize that it may be reinserted at any time and that it *must* be reinserted whenever we desire the time-domain response. Thus, given the real forcing function

$$v(t) = \text{Re } (V_m e^{j\theta} e^{\mathbf{s}t})$$

we apply the complex forcing function $V_m e^{j\theta} e^{\mathbf{s}t}$; the resultant forced response $I_m e^{j\phi} e^{st}$ is complex and it must have as its real part the desired time-domain forced response

$$i(t) = \text{Re } (I_m e^{j\phi} e^{st})$$

The solution of our circuit-analysis problem consists of the determination of the unknown response amplitude I_m and phase angle ϕ.

Before we actually carry out the details of an analysis problem and see how exactly the procedure follows that which was used in the sinusoidal analysis, it is worthwhile to outline the steps of the basic method. We first characterize the circuit with a set of loop or nodal integrodifferential equations. The given forcing functions, in complex form, and the assumed forced responses, also in complex form, are then substituted into the equations and the indicated integrations and differentiations performed. Each term in every equation will then contain the same factor e^{st}. We therefore divide throughout by this factor, or "suppress e^{st}," understanding that it must be reinserted if a time-domain description of any response function is desired. With the Re symbol and the e^{st} factor gone, we have converted all the voltages and currents from the time domain to the frequency domain. The integrodifferential equations have become algebraic equations, and their solution is obtained just as easily as they were in the sinusoidal steady state. Let us illustrate the basic method by a numerical example.

We shall apply the forcing function

$$v(t) = 60e^{-2t} \cos (4t + 10°)$$

to the series *RLC* circuit shown in Fig. 13-1, and we desire the forced response

$$i(t) = I_m e^{-2t} \cos (4t + \phi)$$

We first express the forcing function in Re notation,

$$v(t) = 60e^{-2t} \cos (4t + 10°)$$
$$= \text{Re } (60e^{-2t} e^{j(4t+10°)})$$
$$= \text{Re } (60e^{j10°} e^{(-2+j4)t})$$

or $\qquad\qquad v(t) = \text{Re } (\mathbf{V}e^{st})$

where

$$\mathbf{V} = 60\underline{/10°} \qquad \text{and} \qquad \mathbf{s} = -2 + j4$$

Fig 13-1: A series *RLC* circuit to which a damped sinusoidal forcing function is applied. A frequency-domain solution for *i(t)* is desired.

After dropping Re, we are left with the complex forcing function

$$60\underline{/10°}e^{st}$$

In a similar manner, we represent the unknown response by the complex quantity

$$\mathbf{I}e^{st}$$

where

$$\mathbf{I} = I_m\underline{/\phi}$$

Since we have not as yet extended the impedance concept to the exponentially damped sinusoid and complex frequency, our next step must be the integrodifferential equation for our circuit. From Kirchhoff's voltage law we obtain

$$v(t) = Ri + L\frac{di}{dt} + \frac{1}{C}\int i\,dt$$

$$= 2i + 3\frac{di}{dt} + 10\int i\,dt$$

and we substitute the given complex forcing function and the assumed complex forced response into this equation,

$$60\underline{/10°}e^{st} = 2\mathbf{I}e^{st} + 3\mathbf{s}\mathbf{I}e^{st} + \frac{10}{\mathbf{s}}\mathbf{I}e^{st}$$

The common factor e^{st} is next suppressed:

$$60\underline{/10°} = 2\mathbf{I} + 3\mathbf{s}\mathbf{I} + \frac{10}{\mathbf{s}}\mathbf{I}$$

and

$$\mathbf{I} = \frac{60\underline{/10°}}{2 + 3\mathbf{s} + 10/\mathbf{s}}$$

Before we evaluate this complex current, it is worth examining the form of this equation. The left side of the equation is a current, the numerator of the right side is a voltage, and the denominator must therefore have the dimensions of ohms; it is apparent that we shall soon interpret it as an impedance. Moreover, we might make some pretty astute guesses as to the impedance of each of the three passive elements at a complex frequency $\mathbf{s}$. The impedance of the 2-Ω resistor is simply 2 Ω; the impedance of the 3-H inductor is $3\mathbf{s}$ or $\mathbf{s}L$; and the impedance of the 0.1-F capacitor is $10/\mathbf{s}$ or $1/\mathbf{s}C$. These statements will be proved in the following section.

Going back to our numerical example, we now let $\mathbf{s} = -2 + j4$ in the last equation and solve for the complex current $\mathbf{I}$:

$$\mathbf{I} = \frac{60\underline{/10°}}{2 + 3(-2 + j4) + 10/(-2 + j4)}$$

After manipulating the complex numbers for a few minutes, we find

$$\mathbf{I} = 5.37\underline{/-106.6°}$$

Thus, I_m is 5.37 A, ϕ is $-106.6°$, and the desired forced response is

$$i(t) = 5.37e^{-2t} \cos(4t - 106.6°)$$

Now we must show that the impedance concept may be extended to complex frequencies. Except for the presence of complex numbers, the actual solution of the circuit equations will follow the same procedure used for purely resistive circuits.

Drill Problems

13-3. Specify the phasor voltage that corresponds to the time-domain expression: (a) $5 \cos(7t + 20°)$ V; (b) $5e^{-4t} \cos(7t + 20°)$ V; (c) $5e^{-4t} \cos 20°$ $\cos 7t$ V. If $\mathbf{I} = 5\underline{/50°}$ A, find $i(t)$ if $\mathbf{s} = :$ (d) 0; (e) -7 s^{-1}; (f) $-7 - j4$ s^{-1}.
Ans: $5\underline{/20°}$ V; $5\underline{/20°}$ V; $4.70\underline{/0°}$ V; 3.21 A; $3.21e^{-7t}$ A; $5e^{-7t} \cos(-4t + 50°)$ A

13-4. Give the numerical value of the voltage $v(t)$ at $t = 10$ ms if: (a) $\mathbf{V} = 20\underline{/90°}$ V at $\mathbf{s} = -80 + j200$ s^{-1}; (b) $\mathbf{V} = 15\underline{/120°}$ V at $\mathbf{s} = -60 + j120$ s^{-1}; (c) $\mathbf{V} = 15\underline{/-120°}$ V at $\mathbf{s} = -60 - j120$ s^{-1}.   *Ans: -8.17; -8.14; -8.14 V*

13-4
Z(s) and Y(s)

In order to apply Kirchhoff's laws directly to the complex forcing functions and complex forced responses, it is necessary to know the constant of proportionality between the complex voltage across an element and the complex current through it. This proportionality constant is the impedance or admittance of the element; it is easily determined for the resistor, inductor, and capacitor.

Let us consider the inductor carefully and then merely present the results for the other elements without proof. Suppose that a voltage source

$$v(t) = V_m e^{\sigma t} \cos(\omega t + \theta)$$

is applied to an inductor L; the current response must have the form

$$i(t) = I_m e^{\sigma t} \cos(\omega t + \phi)$$

The passive sign convention is employed, as indicated in Fig. 13-2a, the time-domain circuit. If we represent the voltage as

$$v(t) = \text{Re}(V_m e^{j\theta} e^{st}) = \text{Re}(\mathbf{V} e^{st})$$

Fig. 13-2: (a) The time-domain inductor voltage and current are related by $v = L\, di/dt$. (b) The (complex) frequency-domain inductor voltage and current are related by $\mathbf{V} = s L \mathbf{I}$.

$$i(t) = I_m e^{\sigma t} \cos(\omega t + \phi)$$

$$v(t) = V_m e^{\sigma t} \cos(\omega t + \theta)$$

(a)

(b)

and the current as

$$i(t) = \mathrm{Re}\,(I_m e^{j\phi} e^{st}) = Re\,(\mathbf{I}e^{st})$$

then the substitution of these expressions into the defining equation of an inductor,

$$v(t) = L\,\frac{di(t)}{dt}$$

leads to

$$\mathrm{Re}\,(\mathbf{V}e^{st}) = \mathrm{Re}\,(s L\mathbf{I}e^{st})$$

We now drop Re, thus considering the complex response to a complex forcing function, and suppress the superfluous factor e^{st}:

$$\mathbf{V} = s L\mathbf{I}$$

The ratio of the complex voltage to the complex current is once again the impedance. Since it depends in general upon the complex frequency $\mathbf{s}$, this functional dependence is sometimes indicated by writing

$$\mathbf{Z(s)} = \frac{\mathbf{V}}{\mathbf{I}} = sL$$

In a similar manner, the admittance of an inductor L is

$$\mathbf{Y(s)} = \frac{1}{sL}$$

We shall still call $\mathbf{V}$ and $\mathbf{I}$ *phasors*. Each of these complex quantities has an amplitude and phase angle which, along with a specific complex-frequency value, enable us to characterize the exponentially varying sinusoidal waveform completely. The phasor is still a frequency-domain description, but its application is not limited to the realm of radian frequencies. The frequency-domain equivalent of Fig. 13-2a is shown in Fig. 13-2b; phasor currents, phasor voltages, and impedances or admittances are now used.

The impedances of a resistor and a capacitor at a complex frequency $\mathbf{s}$ are obtained by a similar sequence of steps. Without going through the details, the results for the resistor, inductor, and capacitor are shown below:

	R	L	C
$\mathbf{Z(s)}$	R	sL	$\dfrac{1}{sC}$
$\mathbf{Y(s)}$	$\dfrac{1}{R}$	$\dfrac{1}{sL}$	sC

If we now reconsider the series RLC example of Fig. 13-1 in the frequency domain, the source voltage

$$v(t) = 60e^{-2t}\cos\,(4t + 10°)$$

Fig. 13-3: The frequency-domain equivalent of the series *RLC* circuit shown in Fig. 13-1.

$$60 \underline{/10°} \text{ V}$$
$$s = -2 + j4$$

$$2 \ \Omega \quad -6 + j12 \ \Omega$$

$$\frac{10}{-2 + j4} = -1 - j2 \ \Omega$$

is transformed to the phasor voltage

$$\mathbf{V} = 60\underline{/10°}$$

a phasor current **I** is assumed, and the impedance of each element at the complex frequency $\mathbf{s} = -2 + j4$ is determined and placed on a frequency-domain diagram, Fig. 13-3. The unknown current is now easily obtained by dividing the phasor voltage by the sum of the three impedances:

$$\mathbf{I} = \frac{60\underline{/10°}}{2 + (-6 + j12) + (-1 - j2)} = \frac{60\underline{/10°}}{-5 + j10} = 5.37\underline{/-106.6°} \quad \text{A}$$

Thus, the previous result is obtained, but much more easily and rapidly.

It is hardly necessary to say that all the techniques which we have used in the past to simplify frequency-domain analysis, such as mesh analysis, nodal analysis, superposition, source transformations, duality, Thévenin's theorem, and Norton's theorem, are still valid and useful. For example, the Thévenin equivalent of the network shown in Fig. 13-4*a* is obtained as follows:

The complex frequency is

$$\mathbf{s} = -5 + j10$$

The frequency-domain source voltage is

$$\mathbf{V}_s = 100\underline{/0°}$$

The inductor impedance is

$$\mathbf{Z}_L(\mathbf{s}) = 4(-5 + j10) = -20 + j40$$

The resistor impedance is

$$\mathbf{Z}_R(\mathbf{s}) = 20$$

The Thévenin impedance is the parallel equivalent of $\mathbf{Z}_L$ and $\mathbf{Z}_R$:

$$\mathbf{Z}_{th} = \frac{20(-20 + j40)}{20 - 20 + j40} = 20 + j10$$

Fig. 13-4: (*a*) A given two-terminal network. (*b*) The frequency-domain Thévenin equivalent.

$$4 \text{ H}$$
$$100e^{-5t} \cos 10t \text{ V}$$
$$20 \ \Omega$$
$$20 + j10 \ \Omega$$
$$-j50 \text{ V}$$
$$s = -5 + j10$$

(*a*)

(*b*)

The open-circuit voltage is

$$\mathbf{V}_{oc} = 100\underline{/0°}\,\frac{20}{20 - 20 + j40} = -j50$$

The frequency-domain Thévenin equivalent network is thus as shown in Fig. 13-4b. The return to the time domain is taken after some desired forced response has been determined in the frequency domain. If another 4-H inductor is placed across the open circuit, for example, the frequency-domain current is

$$\mathbf{I} = \frac{-j50}{20 + j10 - 20 + j40} = -1$$

corresponding to the time-domain current

$$i(t) = -e^{-5t}\cos 10t$$

Drill Problems

13-5. For the circuit of Fig. 13-5, find: (a) $\mathbf{V}_1(\mathbf{s})$; (b) $\mathbf{V}_2(\mathbf{s})$.

$$Ans:\; \frac{4s(s + 25)\mathbf{I}_s}{s^2 + 20s + 500};\; \frac{100s\mathbf{I}_s}{s^2 + 20s + 500}\;\text{V}$$

13-6. Find the phasor voltage $\mathbf{V}_1$ in Fig. 13-5 if $\mathbf{I}_s = 5\underline{/70°}$ A and $\mathbf{s} = :$ (a) 0; (b) ∞; (c) −20 Np/s; (d) $j20$ s^{-1}; (e) $-10 + j10$ s^{-1}.

Ans: 0; $20\underline{/70°}$; $4\underline{/-110°}$; $31.1\underline{/122.7°}$; $17.00\underline{/-121.3°}$ V

13-7. Let the time-domain current of Fig. 13-5 be $i_s(t) = 4e^{-10t}\cos(5t + 40°)$ A and find v_2 at $t = :$ (a) 0; (b) −0.25 s. Ans: −11.60; −76.6 V

Fig. 13-5: See Drill Probs. 13-5 to 13-7.

13-5
Frequency
response as a
function of σ

In Sec. 10-5 we considered the forced response of a circuit as a function of the radian frequency ω. We represented such quantities as impedance, admittance, specific voltages or currents, voltage and current gain, and transfer impedances and admittances also as functions of ω. Then we located the poles and zeros of these quantities and sketched the responses as functions of ω. Before we discuss the more general problem of frequency response as a function of the complex frequency s in the next section, let us devote a little time to the simpler problem of frequency response as a function of σ. This is a subject which is very useful in synthesizing RC and RL circuits, a popular topic in more advanced courses.

As a simple example we may select the series RL circuit excited by the frequency-domain voltage source $V_m\underline{/0°}$. The current is obtained as a function of s by dividing the source voltage by the input impedance,

$$\mathbf{I} = \frac{V_m\underline{/0°}}{R + sL}$$

We now set $\omega = 0$, $\mathbf{s} = \sigma + j0$, thus restricting ourselves to time-domain sources of the form

$$v_s = V_m e^{\sigma t}$$

and thus

$$I = \frac{V_m}{R + \sigma L}$$

or

$$I = \frac{V_m}{L} \frac{1}{\sigma + R/L} \tag{12}$$

Transforming to the time domain,

$$i(t) = \frac{V_m}{R + \sigma L} e^{\sigma t} \tag{13}$$

The necessary information about this forced response, however, is all contained in the frequency-domain description (12). As the neper frequency σ varies, a qualitative description of the response is easily provided. When σ is a large negative number, corresponding to a rapidly decreasing exponential function, the current response (12) is negative and relatively small in amplitude; it is, of course, also a rapidly decreasing (in magnitude) exponential function. As σ increases, becoming a smaller negative number, the magnitude of the negative response increases. We can obtain an arbitrarily large response by letting σ be sufficiently close to $-R/L$, a pole of the response. Thus, if $V_m = 1$ V, $L = 1$ H, and $R = 1 \, \Omega$, the pole is located at $\sigma = -1$ Np/s, and the amplitude of the current response, $V_m/(R + \sigma L) = 1/(\sigma + 1)$, will be greater in magnitude than 1 000 000 A whenever σ is within 10^{-6} Np/s of -1 Np/s, or $0 < |\sigma + 1| < 10^{-6}$.

As σ continues to increase, the next noteworthy point occurs when $\sigma = 0$. Since $v_s = V_m$, we are now faced with the dc case, and the forced response is obviously V_m/R, which agrees with (12). Positive values of σ must all provide positive amplitude responses, the larger amplitudes arising when σ is smaller. Finally, an infinite value of σ provides a zero-amplitude response, and thus establishes a zero. The only critical frequencies are the pole at $\sigma = -R/L$ and the zero at $\sigma = \pm\infty$.

This information may be presented quite easily by plotting $|I|$, the current magnitude, as a function of σ, as shown in Fig. 13-6. In preparation for the

Fig. 13-6: A plot of the magnitude of I versus the neper frequency σ for a series RL circuit excited by an exponential voltage source, $V_m e^{\sigma t}$. The only pole is located at $\sigma = -R/L$; the zero is located at $\sigma = \pm\infty$.

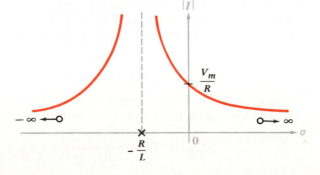

more general response information with changing **s** in the next section, we show the *magnitude* of I versus σ; the phase angle (not shown) is, of course, either $0°$ or $180°$.

Let us now turn our attention to the two critical frequencies of this forced response. The only finite critical frequency is the pole at $\sigma = -R/L$. Equation (12) implies that the application of a finite-amplitude forcing function at this frequency will cause a response of infinite amplitude. This is inconveniently large, of course, as well as being mathematically unsuitable. A much more informative approach to an understanding of the nature of the pole is based on a consideration of the natural response. For this series RL circuit, an appropriate form for the natural response is

$$i_n(t) = I_m e^{-Rt/L} \tag{14}$$

where I_m is the amplitude of the natural response at $t = 0$. We are not applying any forcing function at all, and the natural response is therefore the complete response. If we now interpret this response as the *forced response* produced by a zero-amplitude forcing function, we find that the zero-amplitude forcing function produces a nonzero-amplitude forced response. Instead of applying a forcing function of nonzero amplitude at the frequency of the pole and obtaining a response of infinite amplitude, we apply a forcing function of zero amplitude and see a response of finite amplitude. Moreover, the form of this forced response is identical with the form of the natural response.

As an example of this viewpoint, let us again consider our simple series circuit with $L = 1$ H, $R = 1$ Ω, and a pole at $\sigma = -1$ Np/s. The forcing function is $v_s = V_m e^{\sigma t}$, and we ask what amplitude V_m must have to cause a current of 1-A amplitude to be present, $i(t) = 1e^{\sigma t}$ A, at $\mathbf{s} = \sigma + j0$. Using (12) once more with $I = 1$, $R = 1$, and $L = 1$, we find that

$$1 = \frac{V_m}{\sigma + 1}$$

or

$$V_m = \sigma + 1$$

This is the amplitude of the forcing function that is required to maintain a current of 1-A amplitude in the circuit. We note that a zero-amplitude forcing function, or a short circuit, is sufficient at the frequency of the pole.

The relationship between the frequencies of the poles and the form of the natural response will arise in several examples discussed in this section, but we shall not discuss it thoroughly until the following section.

As a second example, let us consider the circuit drawn in Fig. 13-7a. The exponential current source

$$i_s(t) = 4e^{\sigma t} \quad \text{A}$$

is applied to the RC circuit, and the desired response is the output voltage $v_o(t)$. The input impedance of the network to the right of the source is

$$Z_{\text{in}}(\sigma) = 5 \left\| \left(5 + \frac{1}{0.01\,\sigma} \right) = \frac{5(5 + 100/\sigma)}{5 + 5 + 100/\sigma} \right.$$

(a)

(b)

Fig. 13-7: (a) An *RC* circuit excited by an exponential current source. (b) The magnitude of the voltage response $|V_o|$ possesses a pole at $\sigma = -10$ and a zero at $\sigma = -20$ Np/s.

or

$$Z_{\text{in}} = 2.5 \frac{\sigma + 20}{\sigma + 10}$$

Therefore, the frequency-domain output voltage is

$$V_o = I_s Z_{\text{in}} = \frac{10(\sigma + 20)}{\sigma + 10} \tag{15}$$

This last form of the forced response, written as a constant times the ratio of factors having the form $(\sigma + \sigma_1)$, is obviously well suited for the quick determination of the poles and zeros of the response. The voltage response indicates a pole at $\sigma = -10$ and a zero at $\sigma = -20$; infinite frequency is not a critical frequency. The response magnitude $|V_o|$ is plotted as a function of frequency in Fig. 13-7b. The pole at $\sigma = -10$ indicates that this forcing function, having an amplitude of 4 A, can cause an output voltage of arbitrarily large amplitude if the frequency is brought sufficiently close to -10 Np/s. Alternatively, we can provide an output of some given finite amplitude by using a vanishingly small input amplitude for which σ approaches -10 Np/s.

The zero at $\sigma = -20$ Np/s shows that no finite amplitude of the source at this frequency is capable of yielding any output voltage.

Finally, our desired time-domain output voltage is

$$v_o(t) = \frac{10(\sigma + 20)}{\sigma + 10} e^{\sigma t} \tag{16}$$

A more complicated response curve is obtained for the circuit shown in Fig. 13-8a. The current is easily found:

Fig. 13-8: (a) A series *RLC* circuit is driven by an exponential forcing function. (b) The resultant current-magnitude response curve shows zeros at $\sigma = 0$ and $\sigma = \pm\infty$, and poles at $\sigma = -5$ and -1 Np/s.

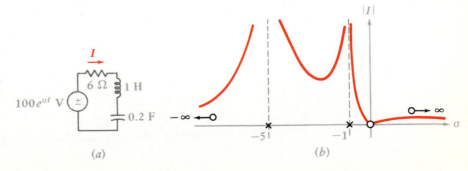

(a)

(b)

$$I = \frac{100}{6 + \sigma + 5/\sigma}$$

or
$$I = 100 \frac{\sigma}{(\sigma + 1)(\sigma + 5)} \qquad (17)$$

The response curve may be obtained by indicating the locations of all poles and zeros on the σ axis, and placing vertical asymptotes at the poles. When this is done, we find that a relative minimum of the forced response must exist between the two poles, and a relative maximum must be present at some frequency greater than zero. By differentiation, the locations of this minimum and maximum may be determined as $\sigma = -\sqrt{5}$ and $\sigma = \sqrt{5}$, respectively. The value of the response at the minimum turns out to be 65.5 and that of the maximum, 9.55. To emphasize the relationship between the time domain and frequency domain once again, the forced response at $\sigma = -3$ is found from (17) to be 75 A; hence, excitation of the network by the forcing function $v(t) = 100e^{-3t}$ produces the forced current response $i(t) = 75e^{-3t}$.

The two poles may again be used to construct the natural response,

$$i_n = Ae^{-t} + Be^{-5t}$$

The values of these two natural frequencies may be compared with those obtained using the methods of Sec. 7-6 for a series *RLC* circuit. That is, we let

$$i_n = A_1 e^{s_1 t} + A_2 e^{s_2 t}$$

where

$$s_{1,2} = -\frac{R}{2L} \pm \sqrt{\left(\frac{R}{2L}\right)^2 - \frac{1}{LC}}$$
$$= -1 \text{ and } -5$$

Thus, as the frequency of a nonzero forcing function approaches either of these two frequencies, $\sigma = -1$ or $\sigma = -5$, an arbitrarily large forced response results; a zero-amplitude forcing function may be associated with a finite-amplitude forced response which can be interpreted as the natural response of the circuit.

Drill Problems

13-8. Obtain an expression for $V_{\text{out}}/V_{\text{in}}$ as a function of σ for the circuit of Fig. 13-9a, and then: (a) determine all the critical frequencies of the ratio. Sketch $|V_{\text{out}}/V_{\text{in}}|$ versus σ and then: (b) evaluate $V_{\text{out}}/V_{\text{in}}$ at $\sigma = -12, -6$, and 2 Np/s.   *Ans:* $-2, -10, \infty$ Np/s; 1, -1.25, 0.417

13-9. For the circuit of Fig. 13-9a, find an expression for the input impedance seen by the source as a function of σ, and then: (a) identify all the critical frequencies. Sketch $|Z_{\text{in}}(\sigma)|$ versus σ and: (b) find the value of Z_{in} for $\sigma = -12, -9, -6, -1$, and 2 Np/s.
Ans: 0, $-2, -8, -10$ Np/s; 0.625, -1.167, 2, -1.929, and 3.6 MΩ

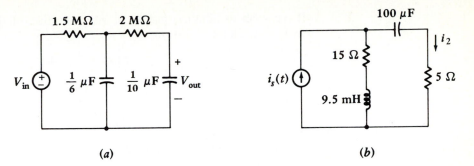

Fig. 13-9: See Drill
Probs. 13-8 to 13-10.

(a)   (b)

13-10. Let $i_s(t) = 4e^{-800t}$ mA in Fig. 13-9b, and find $i_2(t)$ at $t = $: (a) -1 ms; (b) 0; (c) 1 ms. *Ans: -659; -296; -133.0 mA*

13-11. Construct a function $Y_1(\sigma)$ that has poles at $\sigma = -200$ and -500 Np/s and zeros at $\sigma = -100$ and -300 Np/s; let $Y_1(\sigma)$ approach 10 m℧ as σ becomes very large and negative. A conductance of 10 m℧ is now placed in parallel with Y_1. Find all the critical frequencies of this parallel combination. *Ans: -171.9, -200, -378, -500 Np/s*

**13-6
The
complex-frequency
plane**

Now that we have considered the forced response of a circuit as ω varies (with $\sigma = 0$) and as σ varies (with $\omega = 0$), we are prepared to develop a more general graphical presentation by graphing quantities as functions of **s**; that is, we wish to show the response simultaneously as functions of both σ and ω.

Such a graphical portrayal of the forced response as a function of the complex frequency **s** is a useful, enlightening technique in the analysis of circuits, as well as in the design or synthesis of circuits. After we have developed the concept of the complex-frequency plane, or **s** plane, we shall see how quickly the behavior of a circuit can be approximated from a graphical representation of its critical frequencies in this **s** plane. The converse procedure is also very useful; if we are given a desired response curve (the frequency response of a filter, for example), it will be possible to decide upon the necessary location of its poles and zeros in the **s** plane and then to synthesize the filter. This synthesis problem is a subject for detailed study in subsequent courses and is not one which we shall do more than briefly examine here. The **s** plane is also the basic tool with which the possible presence of undesired oscillations is investigated in feedback amplifiers and automatic control systems.

Let us develop a method of obtaining circuit response as a function of **s** by extending the methods we have been using to find the response as a function of either σ or ω. To review these methods, we shall obtain the input or driving-point impedance of a network composed of a 3-Ω resistor in series with a 4-H inductor. As a function of **s**, we have

$$\mathbf{Z(s)} = 3 + 4\mathbf{s}$$

If we wish to obtain a graphical interpretation of the impedance variation with σ, we let $\mathbf{s} = \sigma + j0$,

$$\mathbf{Z}(\sigma) = 3 + 4\sigma$$

and recognize a zero at $\sigma = -\frac{3}{4}$ and a pole at infinity. These critical frequencies are marked on a σ axis, and after identifying the value of $\mathbf{Z}(\sigma)$ at some convenient noncritical frequency [perhaps $\mathbf{Z}(0) = 3$], it is easy to sketch $|\mathbf{Z}(\sigma)|$ versus σ.

In order to plot the response as a function of the radian frequency ω, we let $\mathbf{s} = 0 + j\omega$,

$$\mathbf{Z}(j\omega) = 3 + j4\omega$$

and then obtain both the magnitude and phase angle of $\mathbf{Z}(j\omega)$ as functions of ω:

$$|\mathbf{Z}(j\omega)| = \sqrt{9 + 16\omega^2}$$

$$\text{ang } \mathbf{Z}(j\omega) = \tan^{-1} \frac{4\omega}{3}$$

The magnitude function shows a single pole at infinity and a minimum at $\omega = 0$; it can be sketched readily as a curve of $|\mathbf{Z}(j\omega)|$ versus ω. The phase angle is an inverse tangent function, zero at $\omega = 0$ and $\pm 90°$ at $\omega = \pm\infty$; it is also easily presented as a plot of ang $\mathbf{Z}(j\omega)$ versus ω.

In graphing the response $\mathbf{Z}(j\omega)$ as a function of ω, two two-dimensional plots are required, magnitude and phase angle as functions of ω. When exponential excitation is assumed, we could present all the information on a single two-dimensional graph by allowing both positive and negative values of $\mathbf{Z}(\sigma)$ versus σ. However, we chose to plot the magnitude of $\mathbf{Z}(\sigma)$ in order that our sketches would compare more closely with those depicting the magnitude of $\mathbf{Z}(j\omega)$. The phase angle ($\pm 180°$ only) of $\mathbf{Z}(\sigma)$ was largely ignored. The important point to note is that there is only one independent variable, σ in the case of exponential excitation and ω in the sinusoidal case. Now let us consider what alternatives are available to us if we wish to plot a response as a function of $\mathbf{s}$.

The complex frequency $\mathbf{s}$ requires two parameters, σ and ω, for its complete specification. The response is also a complex function, and we must therefore consider sketching both the magnitude and phase angle as functions of $\mathbf{s}$. Either of these quantities, for example, the magnitude, is a function of the two parameters σ and ω, and we can only plot it in two dimensions as a family of curves, such as magnitude versus ω, with σ as parameter. Conversely, we could also show the magnitude versus σ, with ω as the parameter. Such a family of curves represents a tremendous amount of work, however, and this is what we are trying to avoid; it is also questionable whether we could ever draw any useful conclusions from the family of curves even after they were obtained.

A better method of representing the magnitude of some complex response graphically involves using a *three*-dimensional model. Although such a model is difficult to draw on a two-dimensional sheet of paper, we shall find

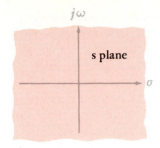

Fig. 13-10: The complex-frequency plane, or **s** plane.

that the model is not difficult to visualize; most of the drawing will be done mentally, where few supplies are needed and construction, correction, and erasures are quickly accomplished. Let us think of a σ axis and a $j\omega$ axis, perpendicular to each other, laid out on a horizontal surface such as the floor. The floor now represents a *complex-frequency plane*, or **s** *plane*, as sketched in Fig. 13-10. To each point in this plane there corresponds exactly one value of **s**, and with each value of **s** we may associate a single point in this complex plane.

Since we are already quite familiar with the type of time-domain function associated with a particular value of the complex frequency **s**, it is now possible to associate the functional form of a forcing function or forced response with a specific region in the **s** plane. The origin, for example, represents a dc quantity. Points lying on the σ axis represent exponential functions, decaying for $\sigma < 0$, increasing for $\sigma > 0$. Pure sinusoids are associated with points on the positive or negative $j\omega$ axis. The right half of the **s** plane, usually referred to simply as the RHP, contains points describing frequencies with positive real parts and thus corresponds to time-domain quantities which are exponentially increasing sinusoids, except on the σ axis. Correspondingly, points in the left half of the **s** plane (LHP) describe the frequencies of exponentially decreasing sinusoids, again with the exception of the negative σ axis. Figure 13-11 summarizes the relationship between the time domain and the various regions of the **s** plane.

Fig. 13-11: The nature of the time-domain function is sketched in the region of the complex-frequency plane to which it corresponds.

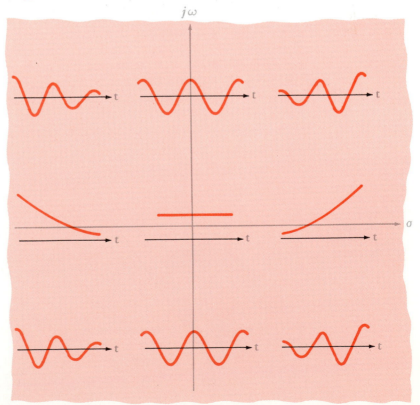

Let us now return to our search for an appropriate method of representing a response graphically as a function of the complex frequency **s**. The magnitude of the response may be represented by constructing, say, a plaster model whose height above the floor at every point corresponds to the magnitude of the response at that value of **s**. In other words, we have added a third axis, perpendicular to both the σ axis and the $j\omega$ axis and passing through the origin; this axis is labeled $|\mathbf{Z}|$, $|\mathbf{Y}|$, $|\mathbf{V_2}/\mathbf{V_1}|$, or with whatever symbol is appropriate. The response magnitude is determined for every value of **s**, and the resultant plot is a *surface* lying above (or just touching) the **s** plane.

We may try out these preliminary ideas by seeing what such a plaster model might look like. As an example, consider the admittance of the series combination of a 1-H inductor and a 3-Ω resistor,

$$\mathbf{Y(s)} = \frac{1}{\mathbf{s} + 3}$$

In terms of both σ and ω we have, as the magnitude,

$$|\mathbf{Y(s)}| = \frac{1}{\sqrt{(\sigma + 3)^2 + \omega^2}}$$

When $\mathbf{s} = -3 + j0$, the response magnitude is infinite; and when **s** is infinite, the magnitude of $\mathbf{Y(s)}$ is zero. Thus our model must be infinitely high over the point $(-3 + j0)$, and it must be zero height at all points infinitely far away from the origin. A cutaway view of such a model is shown in Fig. 13-12a.

Once the model is constructed, it is simple to visualize the variation of $|\mathbf{Y}|$ as a function of ω (with $\sigma = 0$) by cutting the model with a perpendicular plane containing the $j\omega$ axis. The model shown in Fig. 13-12a happens to be cut along this plane, and the desired plot of $|\mathbf{Y}|$ versus ω can be seen; the curve is also drawn in Fig. 13-12b. In a similar manner, a vertical plane containing the σ axis enables us to obtain $|\mathbf{Y}|$ versus σ (with $\omega = 0$), shown in Fig. 13-12c.

How might we obtain some qualitative response information without do-

Fig. 13-12: (*a*) A cutaway view of a plaster model whose top surface represents $|\mathbf{Y(s)}|$ for the series combination of a 1-H inductor and a 3-Ω resistor. (*b*) $|\mathbf{Y(s)}|$ as a function of ω. (*c*) $|\mathbf{Y(s)}|$ as a function of σ.

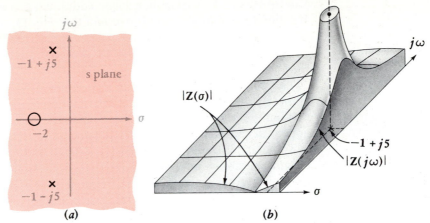

Fig. 13-13: (a) The pole-zero constellation of some impedance **Z(s)**. (b) A portion of the rubber-sheet model of the magnitude of **Z(s)**.

(a)   (b)

ing all this work? After all, most of us have neither the time nor the inclination to be good plasterers, and some more practical method is needed. Let us visualize the **s** plane once again as the floor and then imagine a larger rubber sheet laid on it. We now fix our attention on all the poles and zeros of the response. At each zero, the response is zero, the height of the sheet must be zero, and we therefore tack the sheet to the floor. At the value of **s** corresponding to each pole, we may prop up the sheet with a thin vertical rod. Zeros and poles at infinity must be treated by using a large-radius clamping ring or a high circular fence, respectively. If we have used an infinitely large, weightless, perfectly elastic sheet, tacked down with vanishingly small tacks, and propped up with infinitely long, zero-diameter rods, then the rubber sheet assumes a height which is exactly proportional to the magnitude of the response. Less accurate rubber-sheet models may actually be constructed in the laboratory, but their main advantage lies in the ease by which their construction may be visualized from a knowledge of the pole-zero locations of the response.

These comments may be illustrated by considering the configuration of the poles and zeros, sometimes called a *pole-zero constellation,* which locates all the critical frequencies of some frequency-domain quantity, say the impedance **Z(s)**. Such a pole-zero constellation is shown in Fig. 13-13a. If we visualize a rubber-sheet model, tacked down at **s** = −2 + *j*0 and propped up at **s** = −1 + *j*5 and −1 − *j*5, we should see a terrain whose distinguishing features are two mountains and one conical crater or depression. The portion of the model for the upper LHP is shown in Fig. 13-13b.

Let us now build up the expression for **Z(s)** which leads to this pole-zero configuration. The zero requires a factor of (**s** + 2) in the numerator, and the two poles require the factors (**s** + 1 − *j*5) and (**s** + 1 + *j*5) in the denominator. Except for a multiplying constant *k*, we now know the form of **Z(s)**:

$$\mathbf{Z(s)} = k \frac{\mathbf{s} + 2}{(\mathbf{s} + 1 - j5)(\mathbf{s} + 1 + j5)}$$

or

$$\mathbf{Z(s)} = k \frac{\mathbf{s} + 2}{\mathbf{s}^2 + 2\mathbf{s} + 26} \tag{18}$$

Let us select k by assuming a single additional fact about $\mathbf{Z}(s)$: let $\mathbf{Z}(0) = 1$. By direct substitution in (18), we find that k is 13, and therefore

$$\mathbf{Z}(s) = 13 \, \frac{s + 2}{s^2 + 2s + 26} \tag{19}$$

The plots $|\mathbf{Z}(\sigma)|$ versus σ and $|\mathbf{Z}(j\omega)|$ versus ω may be obtained exactly from (19), but the general form of the function is apparent from the pole-zero configuration and the rubber-sheet analogy. Portions of these two curves appear at the sides of the model shown in Fig. 13-13b.

Thus far, we have been using the s plane and the rubber-sheet model to obtain *qualitative* information about the variation of the *magnitude* of the frequency-domain function with frequency. It is possible, however, to get *quantitative* information concerning the variation of both the *magnitude* and *phase angle*. The method provides us with a powerful new tool.

Consider the representation of a complex frequency in polar form, as suggested by an arrow drawn from the origin of the s plane to the complex frequency under consideration. The length of the arrow is the magnitude of the frequency, and the angle that the arrow makes with the positive direction of the σ axis is the angle of the complex frequency. The frequency $\mathbf{s}_1 = -3 + j4 = 5\underline{/126.9°}$ is indicated in Fig. 13-14a.

It is also necessary to represent the difference between two values of s as an arrow or vector on the complex plane. Let us select a value of $\mathbf{s}$ that corresponds to a sinusoid $\mathbf{s} = j7$ and indicate it also as a vector, as shown in Fig. 13-14b. The *difference* $\mathbf{s} - \mathbf{s}_1$ is seen to be the vector drawn from the last-named point $\mathbf{s}_1$ to the first-named point $\mathbf{s}$; the vector $\mathbf{s} - \mathbf{s}_1$ is drawn in Fig. 13-14c. Note that $\mathbf{s}_1 + (\mathbf{s} - \mathbf{s}_1) = \mathbf{s}$. Numerically, $\mathbf{s} - \mathbf{s}_1 = j7 - (-3 + j4) = 3 + j3 = 4.24\underline{/45°}$, and this value agrees with the graphical difference.

Let us see how this graphical interpretation of the difference $(\mathbf{s} - \mathbf{s}_1)$ enables us to determine frequency response. Consider the admittance

$$\mathbf{Y}(s) = s + 2$$

Fig. 13-14: (*a*) The complex frequency $\mathbf{s}_1 = -3 + j4$ is indicated by drawing an arrow from the origin to $\mathbf{s}_1$. (*b*) The frequency $\mathbf{s} = j7$ is also represented vectorially. (*c*) The difference $\mathbf{s} - \mathbf{s}_1$ is represented by the vector drawn from $\mathbf{s}_1$ to $\mathbf{s}$.

(a)   (b)   (c)

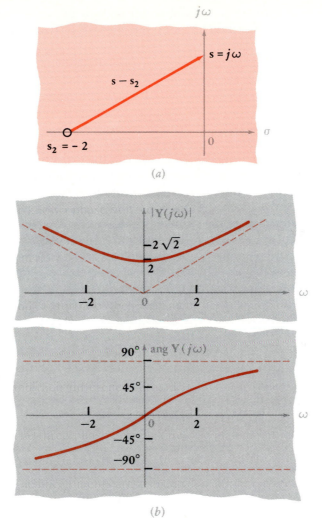

Fig. 13-15: (a) The vector representing the admittance $\mathbf{Y(s)} = \mathbf{s} + 2$ is shown for $\mathbf{s} = j\omega$. (b) Sketches of $|\mathbf{Y}(j\omega)|$ and ang $\mathbf{Y}(j\omega)$ as they might be obtained from the performance of the vector as $\mathbf{s}$ moves up or down the $j\omega$ axis from the origin.

This expression may be interpreted as the difference between some frequency of interest $\mathbf{s}$ and a zero location. Thus, the zero is present at $\mathbf{s}_2 = -2 + j0$, and the factor $\mathbf{s} + 2$, which may be written as $\mathbf{s} - \mathbf{s}_2$, is represented by the vector drawn from the zero location $\mathbf{s}_2$ to the frequency $\mathbf{s}$ at which the response is desired. If the sinusoidal response is desired, $\mathbf{s}$ must lie on the $j\omega$ axis, as illustrated in Fig. 13-15a. The magnitude of $\mathbf{s} + 2$ may now be visualized as ω varies from zero to infinity. When $\mathbf{s}$ is zero, the vector has a magnitude of 2 and an angle of 0°. Thus $\mathbf{Y}(0) = 2$. As ω increases, the magnitude increases, slowly at first, and then almost linearly with ω; the phase angle increases almost linearly at first, and then gradually approaches 90° as ω becomes infinite. At $\omega = 7$, $\mathbf{Y}(j7)$ has a magnitude of $\sqrt{2^2 + 7^2}$ and has a phase angle of $\tan^{-1}(3.5)$. The magnitude and phase of $\mathbf{Y(s)}$ are sketched as functions of ω in Fig. 13-15b.

Let us now construct a more realistic example by considering a frequency-domain function given by the quotient of two factors,

$$\mathbf{V(s)} = \frac{s + 2}{s + 3}$$

We again select a value of **s** which corresponds to sinusoidal excitation and draw the vectors **s** + 2 and **s** + 3, the first from the zero to the chosen point on the $j\omega$ axis and the second from the pole to the chosen point. The two vectors are sketched in Fig. 13-16a. The quotient of these two vectors has a magnitude equal to the quotient of the magnitudes and a phase angle equal to the difference of the numerator and denominator phase angles. An investigation of the variation of the magnitude of **V(s)** versus ω is made by allowing **s** to move from the origin up the $j\omega$ axis and considering the ratio of the distance from the zero to **s** $= j\omega$ and the distance from the pole to the same point on the $j\omega$ axis. The ratio evidently is $\frac{2}{3}$ at $\omega = 0$ and approaches unity as ω becomes infinite. A consideration of the difference of the two phase angles shows that ang $\mathbf{V}(j\omega)$ is 0° at $\omega = 0$, increases at first as ω increases since the angle of the vector **s** + 2 is greater than that of **s** + 3, and then decreases with a further increase in ω, finally approaching 0° at infinite frequency, where both vectors possess 90° angles. These results are sketched in Fig. 13-16b.

Fig. 13-16: (a) Vectors are drawn from the two critical frequencies of the voltage response **V(s)** = (**s** + 2)/(**s** + 3). (b) Sketches of the magnitude and phase angle of **V**($j\omega$) as obtained from the quotient of the two vectors shown in (a).

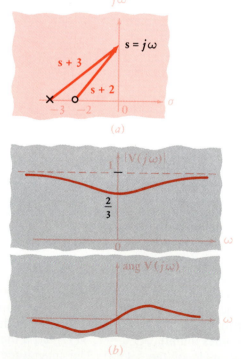

Although no quantitative markings are present on these sketches, it is important to note that they may be obtained easily. For example, the complex response at $\mathbf{s} = j4$ must be given by the ratio

$$\mathbf{V}(j4) = \frac{\sqrt{4 + 16}\underline{/\tan^{-1}\left(\frac{4}{2}\right)}}{\sqrt{9 + 16}\underline{/\tan^{-1}\left(\frac{4}{3}\right)}}$$

$$= \sqrt{\tfrac{20}{25}}\underline{/\tan^{-1} 2 - \tan^{-1}\left(\tfrac{4}{3}\right)}$$

$$= 0.894\underline{/10.3°}$$

In designing circuits to produce some desired response, the behavior of the vectors drawn from each critical frequency to a general point on the $j\omega$ axis is an important aid. For example, if it were necessary to increase the hump in the phase response of Fig. 13-16b, we can see that we must provide a greater difference in the angles of the two vectors. This may be achieved in Fig. 13-16a either by moving the zero closer to the origin or by locating the pole farther from the origin, or both.

The ideas we have been discussing to help in the graphical determination of the magnitude and angular variation of some frequency-domain function with frequency will be needed in the following chapter when we investigate the frequency performance of highly selective filters, or resonant circuits. These concepts are fundamental in obtaining a quick, clear understanding of the behavior of electrical networks and other engineering systems. The procedure is briefly summarized as follows:

1 Draw the pole-zero constellation of the frequency-domain function under consideration in the $\mathbf{s}$ plane, and locate a test point corresponding to the frequency at which the function is to be evaluated.
2 Draw an arrow from each pole and zero to the test point.
3 Determine the length of each pole arrow and zero arrow and the value of each pole-arrow angle and zero-arrow angle.
4 Divide the product of the zero-arrow lengths by the product of the pole-arrow lengths. This quotient is the magnitude of the frequency-domain function for the assumed frequency of the test point [within a multiplying constant, since $\mathbf{F}(\mathbf{s})$ and $k\mathbf{F}(\mathbf{s})$ have the same pole-zero constellations].
5 Subtract the sum of the pole-arrow angles from the sum of the zero-arrow angles. The resultant difference is the angle of the frequency-domain function, evaluated at the frequency of the test point. The angle does not depend upon the value of the real multiplying constant k.

Drill Problems

13-12. The series combination of 5 μF and 10 Ω is in parallel with the series combination of 2 mH and 10 Ω. Find $\mathbf{Z}_{in}(\mathbf{s})$, the input impedance of the parallel combination; identify all its critical frequencies; and draw the pole-zero configuration.
Ans: $10(\mathbf{s}^2 + 25\,000\mathbf{s} + 10^8)/(\mathbf{s}^2 + 10^4\mathbf{s} + 10^8)$;

$$-5, -20, -5 \pm j8.66 \text{ complex krad/s}$$

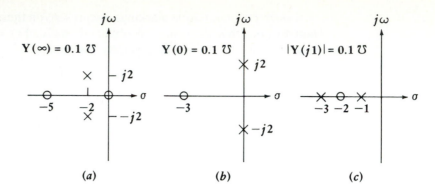

Fig. 13-17: See Drill Prob. 13-13.

(a)   (b)   (c)

13-13. Three pole-zero constellations are shown in Fig. 13-17. Each applies to an admittance. Obtain an expression for each admittance that is a ratio of polynomials in **s**.

$$Ans: \frac{s^2 + 5s}{10s^2 + 40s + 80}; \frac{2s + 6}{15s^2 + 60}; \frac{s + 2}{5s^2 + 20s + 15}$$

13-14. A plaster model of an impedance shows poles at $s = -1 \pm j5$ and zeros at $s = -3 \pm j4$. If the height of the model is 8 cm at the origin, find the height at $s = :$ (a) $j1$; (b) $3 + j1$; (c) ∞.   *Ans: 8.21; 10.69; 8.32 cm*

13-15. The pole-zero constellation of a voltage gain shows a zero on the negative σ axis at $s = z_1$ and a pole on the negative σ axis at $s = p_2$. The angle of the gain is 45° at $s = j20$ and also at $s = j40$ s^{-1}. Find: (a) z_1; (b) p_2.
      Ans: -11.23; -71.2 Np/s

**13-7
Natural response
and the s plane**

There is a tremendous amount of information contained in the pole-zero plot of some forced response in the **s** plane. We shall find out how a complete current response, natural plus forced, produced by an arbitrary forcing function can be quickly written from the pole-zero configuration of the forced current response and from the initial conditions; the method is similarly effective in finding the complete voltage response produced by an arbitrary source.

Let us introduce the method by considering the simplest example, a series *RL* circuit as shown in Fig. 13-18. A general voltage source $v_s(t)$ causes the current $i(t)$ to flow after closure of the switch at $t = 0$. The complete re-

Fig. 13-18: An example which illustrates the determination of the complete response through a knowledge of the critical frequencies of the impedance faced by the source.

sponse $i(t)$ for $t > 0$ is composed of a natural response and a forced response:

$$i(t) = i_n(t) + i_f(t) \tag{20}$$

We may find the forced response by working in the frequency domain, assuming, of course, that $v_s(t)$ has a functional form which we can transform to the frequency domain; if $v_s(t) = 1/(1 + t^2)$, we must proceed as best we can from the basic differential equation for the circuit. Here we have

$$\mathbf{I}_f(\mathbf{s}) = \frac{\mathbf{V}_s}{R + \mathbf{s}L}$$

or

$$\mathbf{I}_f(\mathbf{s}) = \frac{1}{L} \frac{\mathbf{V}_s}{\mathbf{s} + R/L} \tag{21}$$

and $i_f(t)$ is obtained by replacing $\mathbf{s}$, L, and R by their values, reinserting e^{st}, and taking the real part. The answer may even be obtained as a function of a general ω, σ, R, and L if desired.

Now let us consider the natural response. Of course, we know that the form will be a decaying exponential with the time constant L/R, but we may pretend that we are finding it for the first time. The form of the natural or *source-free* response is, by definition, independent of the forcing function; the forcing function contributes, along with the other initial conditions, only to the magnitude of the natural response. To find the proper form, replace all independent sources by their internal impedances; here, $v_s(t)$ is replaced by a short circuit. Now let us try to obtain this natural response as a limiting case of the forced response; we return to the frequency-domain expression (21) and obediently set $\mathbf{V}_s = 0$. On the surface, it appears that $\mathbf{I}(\mathbf{s})$ must also be zero, but this is not necessarily true if we are operating at a complex frequency that is a simple pole of $\mathbf{I}(\mathbf{s})$. That is, the denominator and the numerator may both be zero so that $\mathbf{I}(\mathbf{s})$ need not be zero.

Let us inspect this new idea from a slightly different vantage point. We fix our attention on the ratio of the desired forced response to the forcing function. Let us designate this ratio in general by $\mathbf{H}(\mathbf{s})$ and call it a *transfer function*. Here,

$$\frac{\mathbf{I}_f(\mathbf{s})}{\mathbf{V}_s} = \mathbf{H}(\mathbf{s}) = \frac{1}{L(\mathbf{s} + R/L)}$$

In this example, the transfer function is the input admittance faced by $\mathbf{V}_s$. We seek the natural or source-free response by setting $\mathbf{V}_s = 0$. However, $\mathbf{I}_f(\mathbf{s}) = \mathbf{V}_s \mathbf{H}(\mathbf{s})$, and, if $\mathbf{V}_s = 0$, a nonzero value for the current can be obtained only by operating at a pole of $\mathbf{H}(\mathbf{s})$. The poles of the transfer function therefore assume a special significance.

Returning to our series RL circuit, we see that the pole of the transfer function occurs when the operating frequency is $\mathbf{s} = -R/L + j0$. A finite current at this frequency thus represents the natural response

$$\mathbf{I}(\mathbf{s}) = A \qquad \text{at} \qquad \mathbf{s} = -\frac{R}{L} + j0$$

Fig. 13-19: The poles of the response, $I_1(s)$ or $V_2(s)$, produced by a voltage source V_s in (a) or a current source I_s in (b), determine the form of the natural response, $i_{1n}(t)$ or $v_{2n}(t)$, that occurs when V_s is replaced by a short circuit or I_s by an open circuit and some initial energy is available.

where A is an unknown constant. Transforming this natural response to the time domain,

$$i_n(t) = \text{Re } (Ae^{-Rt/L})$$

or

$$i_n(t) = Ae^{-Rt/L}$$

To complete this example, the total response is then

$$i(t) = Ae^{-Rt/L} + i_f(t)$$

and A may be determined once the initial conditions are specified for this circuit.

Now let us generalize these results. Figures 13-19a and b show single sources connected to networks containing no independent sources. The desired response, which might be some current $I_1(s)$ or some voltage $V_2(s)$, may be expressed by a transfer function that displays all the critical frequencies. To be specific, we select the response $V_2(s)$ in Fig. 13-19a:

$$\frac{V_2(s)}{V_s} = H(s) = k \frac{(s - s_1)(s - s_3) \cdots}{(s - s_2)(s - s_4) \cdots} \qquad (22)$$

The poles of $H(s)$ occur at $s = s_2, s_4, \ldots$, and thus a finite voltage $V_2(s)$ at each of these frequencies must be a possible functional form for the natural response. Therefore, we think of a zero-volt source (which is just a short circuit) applied to the input terminals; the natural response which occurs when the input terminals are short-circuited thus must have the form

$$v_{2n}(t) = A_2 e^{s_2 t} + A_4 e^{s_4 t} + \cdots$$

where each A must be evaluated in terms of the initial conditions (including the initial value of any voltage source applied at the input terminals).

To find the form of the natural response $i_{1n}(t)$ in Fig. 13-19a, we should determine the poles of the transfer function, $H(s) = I_1(s)/V_s$. The transfer functions applying to the situations depicted in Fig. 13-19b would obviously be $I_1(s)/I_s$ and $V_2(s)/I_s$, and their poles then determine the natural responses $i_{1n}(t)$ and $v_{2n}(t)$, respectively.

If the natural response is desired for a network that does not contain any independent sources, then a source V_s or I_s may be inserted at any convenient point, restricted only by the condition that the original network is obtained when the source is killed. The corresponding transfer function is then determined and its poles specify the natural frequencies. Note that the same frequencies must be obtained for any of the many source locations possible. If the network already contains a source, that source may be set equal to zero and another source inserted at a more convenient point.

Before we illustrate this method with several examples, completeness requires us to acknowledge two very special cases that might arise. One occurs when the network in Fig. 13-19a or b contains two or more parts that are isolated from each other. For example, we might have the parallel combination of three networks: R_1 in series with C, R_2 in series with L, and a short circuit. Obviously, a voltage source in series with R_1 and C cannot produce any current in R_2 and L; that transfer function would be zero. To find the form of the natural response of the inductor voltage, for example, the voltage source must be installed in the R_2L network. A case of this type can often be recognized by an inspection of the network before a source is installed, but, if it is not, then a transfer function equal to zero will be obtained. When $\mathbf{H(s)} = 0$, we obtain no information about the frequencies characterizing the natural response, and a more suitable location for the source must be used. One network of this type is present in the problems at the end of the chapter, but we forget which problem it is.

The other case involves multiple poles in the transfer function. That is, in (22) we might find that $\mathbf{s}_2$, $\mathbf{s}_4$, and $\mathbf{s}_{10}$ are identical. Transfer functions of this form are handled so much more easily by Laplace-transform techniques that we shall postpone consideration of all problems of this type until Chap. 19. However, it is interesting to note that the critically damped RLC circuits of Chap. 7 possess a double pole, and the form of their natural response provides a clue by which the curious might extrapolate and guess the results that we shall obtain later.

Let us now try out these techniques on two examples. The first circuit is source-free, and it appears as Fig. 13-20. We seek expressions for i_1 and i_2 for $t > 0$, given the initial conditions, $i_1(0) = i_2(0) = 11$ A. Let us install a voltage source $\mathbf{V}_s$ between points x and x' and find the transfer function, $\mathbf{H(s)} = \mathbf{I}_1(\mathbf{s})/\mathbf{V}_s$, which also happens to be the input admittance seen by the voltage source. We have

$$\mathbf{I}_1(\mathbf{s}) = \frac{\mathbf{V}_s}{2\mathbf{s} + 1 + 6\mathbf{s}/(3\mathbf{s} + 2)} = \frac{(3\mathbf{s} + 2)\mathbf{V}_s}{6\mathbf{s}^2 + 13\mathbf{s} + 2}$$

or

$$\mathbf{H(s)} = \frac{\mathbf{I}_1(\mathbf{s})}{\mathbf{V}_s} = \frac{\frac{1}{2}(\mathbf{s} + \frac{2}{3})}{(\mathbf{s} + 2)(\mathbf{s} + \frac{1}{6})}$$

Fig. 13-20: A circuit for which the natural responses i_1 and i_2 are desired.

Thus, i_1 must be of the form

$$i_1(t) = Ae^{-2t} + Be^{-t/6}$$

The solution is completed by using the given initial conditions to establish the values of A and B. Although we have been able to obtain the *form* of the solution rather rapidly, the procedure for evaluating the unknown constants in this second-order system is similar to that used for the *RLC* circuits back in Chap. 7; any saving in effort here must come later with a study of the Laplace transform. For practice, let us complete the solution for i_1. Since $i_1(0)$ is given as 11 A, then

$$11 = A + B$$

The necessary additional equation is obtained by differentiating i_1:

$$\left. \frac{di_1}{dt} \right|_{t=0} = -\frac{1}{2}[2i_2(0) + 1i_1(0)] = -\frac{22 + 11}{2} = -2A - \tfrac{1}{6}B$$

Thus, $A = 8$ and $B = 3$, and the desired solution is

$$i_1(t) = 8e^{-2t} + 3e^{-t/6}$$

The natural frequencies comprising i_2 are the same as those of i_1, and a similar procedure to evaluate the arbitrary constants leads to

$$i_2(t) = 12e^{-2t} - e^{-t/6}$$

As our last example we shall find the complete response $v(t)$ of the circuit shown in Fig. 13-21. The make-before-break switch ensures that all currents and voltages to the right of the switch are initially zero. At $t = 0$ the switch moves up, and the voltage across the 3-Ω resistor is to be found. This response is composed of both a forced and a natural response,

$$v(t) = v_f(t) + v_n(t)$$

Each may be found through a knowledge of the pole-zero configuration of the transfer function, $\mathbf{H(s)} = \mathbf{V(s)/I}_s$, which is also the input impedance of the portion of the network to the right of the switch. We have

$$\mathbf{V(s)} = \frac{\mathbf{I}_s}{\tfrac{1}{3} + 1/2\mathbf{s} + 1/(6\mathbf{s} + 12)}$$

or, after combining and factoring,

$$\mathbf{H(s)} = \frac{\mathbf{V(s)}}{\mathbf{I}_s} = \frac{3\mathbf{s(s} + 2)}{\mathbf{(s} + 1)(\mathbf{s} + 3)} \tag{23}$$

Fig. 13-21: A circuit whose complete response is to be found through an investigation of its critical frequencies.

The form of the natural response may now be written

$$v_n(t) = Ae^{-t} + Be^{-3t}$$

In order to find the forced response, the frequency-domain current source $\mathbf{I}_s(\mathbf{s}) = 1$ at $\mathbf{s} = -1 + j2$ may be multiplied by the input impedance, or the transfer function, evaluated at $\mathbf{s} = -1 + j2$,

$$\mathbf{V}(\mathbf{s}) = \mathbf{I}_s(\mathbf{s})\mathbf{H}(\mathbf{s}) = 3\,\frac{(-1 + j2)(1 + j2)}{j2(2 + j2)}$$

and thus

$$\mathbf{V}(\mathbf{s}) = 1.875\sqrt{2}\,\underline{/45°}$$

Transforming to the time domain, we have

$$v_f(t) = 1.875\sqrt{2}\,e^{-t}\cos(2t + 45°)$$

The complete response is therefore

$$v(t) = Ae^{-t} + Be^{-3t} + 1.875\sqrt{2}\,e^{-t}\cos(2t + 45°)$$

Since both inductor currents are initially zero, the initial source current of 1 A must flow through the 3-Ω resistor. Thus

$$v(0) = 3 = A + B + \frac{1.875\sqrt{2}}{\sqrt{2}} \tag{24}$$

Again it is necessary to differentiate and then to obtain an initial condition for dv/dt. We first find

$$\left.\frac{dv}{dt}\right|_{t=0} = 1.875\sqrt{2}\left(-\frac{2}{\sqrt{2}} - \frac{1}{\sqrt{2}}\right) - A - 3B$$

or

$$\left.\frac{dv}{dt}\right|_{t=0} = -5.625 - A - 3B \tag{25}$$

The initial value of this rate of change is obtained by analyzing the circuit. However, those rates of change which are most easily found are the derivatives of the inductor currents, for $v = L\,di/dt$, and the initial values of the inductor voltages should not be difficult to find. We therefore express the response $v(t)$ in terms of the resistor current,

$$v(t) = 3i_R$$

and then apply Kirchhoff's current law

$$v(t) = 3i_s - 3i_{L1} - 3i_{L2}$$

Now we may take the derivative,

$$\frac{dv}{dt} = 3\,\frac{di_s}{dt} - 3\,\frac{di_{L1}}{dt} - 3\,\frac{di_{L2}}{dt}$$

Differentiation of the source function and evaluation at $t = 0$ provide a value of -3 V/s for the first term; the second term is numerically $\frac{3}{2}$ of the initial voltage across the 2-H inductor, or -4.5 V/s; and the last term is -1.5 V/s. Thus,

$$\left.\frac{dv}{dt}\right|_{t=0} = -9$$

and we may now use (24) and (25) to determine the unknown amplitudes

$$A = 0 \qquad B = 1.125$$

The complete response is therefore

$$v(t) = 1.125e^{-3t} + 1.875\sqrt{2}e^{-t}\cos(2t + 45°)$$

The process which we must pursue to evaluate the amplitude coefficients of the natural response is a detailed one, except in those cases where the initial values of the desired response and its derivatives are obvious. However, we should not lose sight of the ease and rapidity with which the *form* of the natural response can be obtained.

Drill Problems

13-16. (*a*) If a current source, $i_1(t) = u(t)$, is present at *a-b* in Fig. 13-22 with the arrow entering *a*, find $\mathbf{H(s)} = \mathbf{V}_{cd}/\mathbf{I}_1$ and specify the natural frequencies present in $v_{cd}(t)$. (*b*) Repeat for the source $v_{ab} = u(t)$, $\mathbf{H(s)} = \mathbf{V}_{cd}/\mathbf{V}_{ab}$, and $v_{cd}(t)$. *Ans:* $800/(s + 500)$ Ω, -500 s^{-1}; $400/(s + 400)$, -400 s^{-1}

Fig. 13-22: See Drill Prob. 13-16.

13-17. Find the natural frequencies present in this natural response in the circuit of Fig. 13-23: (*a*) $v_1(t)$, if a source $i_s(t)$ is suddenly connected across *c* and *d*; (*b*) $i_2(t)$, if a source $v_s(t)$ is suddenly connected in place of the short circuit at *a-b*. *Ans:* $-0.750 \pm j0.661$; $-0.750 \pm j0.661$ s^{-1}

Fig. 13-23: See Drill Prob. 13-17.

13-18. In Fig. 13-24, the transfer function $\mathbf{I}_2/\mathbf{V}_1$ is given by $8s(s + 5)/(s + 20)$. (*a*) If $v_1 = 10e^{-8t}$ V, find the forced response $i_{2f}(t)$. (*b*) If $v_1 = 20$ V, find the forced response $i_{2f}(t)$. (*c*) If $v_1 = 10e^{-8t}u(t)$ V and $i_2(0^+) = 100$ A, find the complete response $i_2(t)$. *Ans:* $160e^{-8t}$; 0; $(160e^{-8t} - 60e^{-20t})u(t)$ A

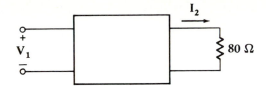

Fig. 13-24: See Drill Prob. 13-18.

13-8

A technique for synthesizing the voltage ratio,

H(s) = V_{out}/V_{in}

Much of the discussion in this chapter has been related to the poles and zeros of a transfer function. We have located them on the complex-frequency plane, we have used them to express transfer functions as ratios of factors or polynomials in **s**, we have calculated forced responses from them, and in the preceding section we have used their poles to establish the form of the natural response.

Now let us see how we might determine a network that can provide a desired transfer function. We consider only a small part of the general problem, working with a transfer function, $\mathbf{H(s)} = \mathbf{V_{out}(s)}/\mathbf{V_{in}(s)}$, as indicated in Fig. 13-25. For simplicity, we restrict **H(s)** to critical frequencies on the negative σ axis (including the origin). Thus, we shall consider transfer functions such as

$$\mathbf{H_1(s)} = \frac{10(s + 2)}{s + 5} \tag{26}$$

or

$$\mathbf{H_2(s)} = \frac{-5s}{(s + 8)^2} \tag{27}$$

or

$$\mathbf{H_3(s)} = 0.1s(s + 2) \tag{28}$$

Let us begin by finding the voltage gain of the network of Fig. 13-26, which contains an ideal op-amp. The voltage between the two input terminals of the op-amp is essentially zero, and the input impedance of the op-amp is essen-

Fig. 13-25: Given **H(s)** = V_{out}/V_{in}, we seek a network having a specified **H(s)**.

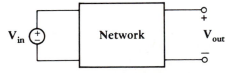

Fig. 13-26: For an ideal op-amp, **H(s)** = V_{out}/V_{in} = −Z_f/Z_1.

(a)

(b)

Fig. 13-27: (a) The transfer function, $\mathbf{H(s)} =$ $\mathbf{V}_{out}/\mathbf{V}_{in}$, has a pole at $\mathbf{s} =$ $-1/R_fC_f$. (b) Here, there is a zero at $\mathbf{s} = -1/R_1C_1$.

tially infinite. We therefore may set the sum of the currents entering the inverting input terminal equal to zero:

$$\frac{\mathbf{V}_{in}}{\mathbf{Z}_1} + \frac{\mathbf{V}_{out}}{\mathbf{Z}_f} = 0$$

or

$$\frac{\mathbf{V}_{out}}{\mathbf{V}_{in}} = -\frac{\mathbf{Z}_f}{\mathbf{Z}_1} \tag{29}$$

If $\mathbf{Z}_f$ and $\mathbf{Z}_1$ are both resistances, the circuit acts as an inverting amplifier, or possibly an attenuator. Our present interest, however, lies with those cases in which one of these impedances is a resistance while the other is an RC network.

In Fig. 13-27a, we let $\mathbf{Z}_1 = R_1$ while $\mathbf{Z}_f$ is the parallel combination of R_f and C_f. Therefore,

$$\mathbf{Z}_f = \frac{R_f/\mathbf{s}C_f}{R_f + (1/\mathbf{s}C_f)} = \frac{R_f}{1 + \mathbf{s}C_fR_f} = \frac{1/C_f}{\mathbf{s} + (1/R_fC_f)}$$

and

$$\mathbf{H(s)} = \frac{\mathbf{V}_{out}}{\mathbf{V}_{in}} = -\frac{\mathbf{Z}_f}{\mathbf{Z}_1} = -\frac{1/R_1C_f}{\mathbf{s} + (1/R_fC_f)} \tag{30}$$

We have a transfer function with a single (finite) critical frequency, a pole at $\mathbf{s} = -1/R_fC_f$.

Moving on to Fig. 13-27b, we now let $\mathbf{Z}_f$ be resistive while $\mathbf{Z}_1$ is an RC parallel combination,

$$\mathbf{Z}_1 = \frac{1/C_1}{\mathbf{s} + (1/R_1C_1)}$$

and

$$\mathbf{H(s)} = \frac{\mathbf{V}_{out}}{\mathbf{V}_{in}} = -\frac{\mathbf{Z}_f}{\mathbf{Z}_1} = -R_fC_1(\mathbf{s} + 1/R_1C_1) \tag{31}$$

The only finite critical frequency is a zero at $\mathbf{s} = -1/R_1C_1$.

For our ideal op-amps, the output or Thévenin impedance is zero and therefore $\mathbf{V}_{out}$ and $\mathbf{V}_{out}/\mathbf{V}_{in}$ are not functions of any load $\mathbf{Z}_L$ that may be placed across the output terminals. This includes the input to another op-amp. Therefore, we may connect circuits having poles and zeros at specified

Fig. 13-28: This network contains two ideal op-amps and gives the voltage transfer function, $H(s) = V_{out}/V_{in} = 10(s + 2)/(s + 5)$.

locations in cascade, where the output of one op-amp is connected directly to the input of the next, and thus generate any desired transfer function.

As an example, let us synthesize a circuit that will yield the transfer function of (26):

$$H(s) = \frac{V_{out}}{V_{in}} = \frac{10(s + 2)}{s + 5} \tag{26}$$

The pole at $s = -5$ may be obtained by a network of the form of Fig. 13-27a. Calling this network A, we have $1/R_{fA}C_{fA} = 5$. We arbitrarily select $R_{fA} = 100$ kΩ and therefore $C_{fA} = 2$ μF. For this portion of the complete circuit,

$$H_A(s) = -\frac{1/R_{1A}C_{fA}}{s + (1/R_{fA}C_{fA})} = -\frac{5 \times 10^5/R_{1A}}{s + 5}$$

Next we consider the zero at $s = -2$. From Fig. 13-27b, $1/R_{1B}C_{1B} = 2$, and, with $R_{1B} = 100$ kΩ, we have $C_{1B} = 5$ μF. Thus

$$H_B(s) = -R_{fB}C_{1B}(s + 1/R_{1B}C_{1B}) = -5 \times 10^{-6}R_{fB}(s + 2)$$

and $\quad H(s) = H_A(s)H_B(s) = 2.5\dfrac{R_{fB}}{R_{1A}}\dfrac{s + 2}{s + 5}$

We complete the design by letting $R_{fB} = 100$ kΩ and $R_{1A} = 25$ kΩ. The result is shown in Fig. 13-28. The capacitors in this circuit are fairly large, but this is a direct consequence of the low frequencies selected for the pole and zero of $H(s)$. If $H(s)$ were changed to $10(s + 2000)/(s + 5000)$, we could use 2- and 5-nF values.

Drill Problem

13-19. Specify suitable element values for Z_1 and Z_f in each of three cascaded stages to realize the transfer function given by Eq. (27), $H(s) = V_{out}/V_{in} = -5s/(s + 8)^2$.
Ans: 1 μF∥ ∞, 100 kΩ; 100 kΩ, 100 kΩ∥ 1.25 μF; 125 kΩ, 100 kΩ∥ 1.25 μF

Problems

1 Identify each of the complex frequencies present in the signal: (*a*) 2 (e^{-3t} + 12 sin 8t) cos 10t; (*b*) 2e^{-3t}(5 + 3e^{-8t}); (*c*) 4 + 3 sin (7t − 0.1). A phasor current is given as 2.1$\underline{/50°}$ A. Find the instantaneous value of the current at $t = 0.2$ s if $s = :$ (*d*) 0 + j8 s^{-1}; (*e*) 0 − j8 s^{-1}; (*f*) −8 s^{-1}.

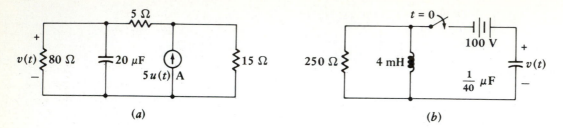

(a)   (b)

Fig. 13-29: See Prob. 2.

2 Determine all the complex frequencies associated with the total response $v(t)$ for $t > 0$ in the circuit shown in: (a) Fig. 13-29a; (b) Fig. 13-29b.

3 The complex phasor current **I** is specified by the equation, $(s + 250)$**I** $+ (6000$**I**$/s) = 20$. (a) Express **I** as the ratio of two polynomials in s. (b) Calculate **I** in polar form for $s = -200 + j100$ s^{-1}. (c) Compute $i(t)$ at $t = 2$ ms if $s = -200 + j100$ s^{-1}.

4 A current, $i = 8e^{-3t} \cos 4t$ A, flows through the series combination of 5 Ω and 2 H. Using the passive sign convention, the voltage across the series network is designated $v(t)$. (a) Find $v(t)$. (b) Calculate $v(t)$ at $t = 0.3$ s. (c) What value of inductance would cause the answer to part b to be zero?

5 (a) Find **Z**(s) for the network of Fig. 13-30. (b) Calculate **Z** at $s = -1 + j1$ s^{-1}. (c) If $s = \sigma + j0$, at what value of σ is $|$**Z**$|$ a minimum?

Fig. 13-30: See Prob. 5.

6 (a) Determine **Z**(s) as a ratio of two polynomials in s for the network shown in Fig. 13-31. (b) Evaluate **Z**$(-5 + j2)$ in rectangular form. (c) At what value or values of s is **Z**(s) $= 0$?

Fig. 13-31: See Prob. 6.

7 (a) The series combination of a 2-Ω resistor and an 8-H inductor is in parallel with the series combination of a 5-Ω resistor and a 50-mF capacitor. Find the impedance of the parallel combination, **Z**$_P$(s), as the ratio of two polynomials in s. (b) Find **Z**$_P(-0.5)$. (c) Find **Z**$_P(j2)$. (d) Find the admittance of the RL branch, **Y**$_{RL}$(s), as the ratio of two polynomials in s. (e) Find **Y**$_{RC}$(s) of the RC branch. (f) Show that **Y**$_{RL}$ + **Y**$_{RC}$ = 1/**Z**$_P$.

8 If $i_{s1}(t) = 2e^{-t} \cos 2t$ A and $i_{s2}(t) = e^{-t} \sin 2t$ A in Fig. 13-32, find $v_L(t)$.

Fig. 13-32: See Prob. 8.

9 For the network shown in Fig. 13-33: (a) find $\mathbf{Z(s)}$ as a ratio of polynomials in $\mathbf{s}$; (b) let $\mathbf{s} = j\omega$ and $\mathbf{Z}(j\omega) = R(j\omega) + jX(j\omega)$, and find $X(j\omega)$ as a ratio of polynomials in ω; (c) find $\mathbf{Y(s)}$; and (d) find $G(j\omega)$, where $\mathbf{Y}(j\omega) = G(j\omega) + jB(j\omega)$ and $\mathbf{s} = j\omega$.

Fig. 13-33: See Prob. 9.

10 (a) Let $v_s = 100e^{-2t} \cos 5t$ V in Fig. 13-34 and find $i_R(t)$. (b) Repeat if $v_s = 100e^{-2t} \cos \omega t$ V.

Fig. 13-34: See Prob. 10.

11 An input impedance has zeros at $\mathbf{s} = -2$ s^{-1} and $\mathbf{s} = \infty$, poles at $\mathbf{s} = 0$ and $\mathbf{s} = -3$ s^{-1}, and a value of 16 Ω at $\mathbf{s} = -2.5$ s^{-1}. (a) Write $\mathbf{Z(s)}$ as a ratio of polynomials in $\mathbf{s}$. (b) Plot $|\mathbf{Z}(\sigma)|$ vs. σ if $\mathbf{s} = \sigma + j0$.

12 (a) For the circuit of Fig. 13-35, determine $\mathbf{I(s)}$ as a function of $\mathbf{s}$. (b) State all the critical frequencies of $\mathbf{I(s)}$. (c) Plot $|\mathbf{I}(\sigma)|$ vs. σ. (d) What size capacitor should be added in series with the 2-Ω resistor to cause $\mathbf{I(s)}$ to have a pole at $\mathbf{s} = -5$ s^{-1}? (e) Find all critical frequencies of $\mathbf{I}_{\text{new}}(\mathbf{s})$.

Fig. 13-35: See Prob. 12.

13 (*a*) Find $\mathbf{V}_{out}(\mathbf{s})/\mathbf{I}_{in}(\mathbf{s})$ as the ratio of two polynomials in **s** for the circuit of Fig. 13-36. (*b*) Identify all the critical frequencies of $\mathbf{V}_{out}(\sigma)/\mathbf{I}_{in}(\sigma)$. (*c*) Plot $|\mathbf{V}_{out}(\sigma)/\mathbf{I}_{in}(\sigma)|$, $-5 \le \sigma \le 2$ Np/s.

Fig. 13-36: See Prob. 13.

14 The parallel combination of 2 Ω and 0.5 F is in series with a 0.4-H inductor and an ideal voltage source, $v_s(t) = 25e^{-2t}$ V. Find the average power absorbed by the capacitor during the time interval, $0.1 \le t \le 0.4$ s.

15 (*a*) Find $\mathbf{Z}(\mathbf{s})$ for the network shown in Fig. 13-37. (*b*) Draw the pole-zero constellation for $\mathbf{Z}(\mathbf{s})$. (*c*) Prepare plots of $|\mathbf{Z}(\sigma)|$ vs. σ with $\omega = 0$, and of $|\mathbf{Z}(j\omega)|$ vs. ω with $\sigma = 0$.

Fig. 13-37: See Prob. 15.

16 The voltage $\mathbf{V}(\mathbf{s}) = 6\mathbf{s}/[(\mathbf{s} + 2)(\mathbf{s} + 5)]$ has its magnitude represented as a plaster model. If the height of the model is 5 cm at $\mathbf{s} = 0 + j3$ s^{-1}: (*a*) what is the height at $\mathbf{s} = -4 + j5$ s^{-1}, and (*b*) at what point does the maximum height on the positive $j\omega$ axis occur?

17 A function of **s** has zeros at $\mathbf{s} = -1 \pm j3$ and $\mathbf{s} = -2 \pm j1$ s^{-1}, and poles at $\mathbf{s} = 0$ and $\mathbf{s} = -3 \pm j1$ s^{-1}. If $|\mathbf{F}(\mathbf{s})| = 12$ at $\mathbf{s} = j3$ s^{-1}: (*a*) show the pole-zero configuration in the **s** plane, and (*b*) find $\mathbf{F}(\mathbf{s})$ as the ratio of two polynomials in **s**.

18 An input impedance $\mathbf{Z}(\mathbf{s})$ has a zero at the origin and poles at $-1 \pm j10$ s^{-1}. As the sinusoidal operating frequency increases from 10 to 11 rad/s, by what factor does the length of the arrow increase that is drawn to the operating frequency from: (*a*) $-1 - j10$? (*b*) 0? (*c*) $-1 + j10$? (*d*) By what factor does $|\mathbf{Z}(\mathbf{s})|$ increase as the operating frequency increases from 10 to 11 rad/s?

19 For a certain network, $\mathbf{I}_{out}(\mathbf{s})/\mathbf{V}_{in}(\mathbf{s}) = \mathbf{H}(\mathbf{s}) = (2\mathbf{s}^2 + 18\mathbf{s} + 36)/(5\mathbf{s}^2 + 20\mathbf{s} + 100)$ A/V. (*a*) Show the critical frequencies of $\mathbf{H}(\mathbf{s})$ on the complex plane. (*b*) Find $\mathbf{H}(0)$ and $\mathbf{H}(\infty)$. (*c*) Sketch $|\mathbf{H}(\sigma)|$ vs. σ for $\omega = 0$ and $|\mathbf{H}(j\omega)|$ vs. ω for $\sigma = 0$. (*d*) If a plaster model of $|\mathbf{H}(\mathbf{s})|$ has a height of 2 in at the origin, what is its maximum height along the $j\omega$ axis, approximately? (Don't differentiate, it can ruin your day.)

20 The three-element network of Fig. 13-38 has an input impedance that includes a zero at $\mathbf{s} = -10 + j0$ s^{-1}. If a 20-Ω resistor is placed in series with the network, the zero shifts to $\mathbf{s} = -3.6 + j0$ s^{-1}. Find R and C.

Fig. 13-38: See Prob. 20.

21 (a) Determine the transfer function, $\mathbf{H(s)} = \mathbf{I}_1(\mathbf{s})/\mathbf{I}_s$, for the circuit shown in Fig. 13-39. (b) If $i_s(t)$ is a unit-step current, what will be the form of $i_1(t)$ for $t > 0$?

Fig. 13-39: See Prob. 21.

22 Let $\mathbf{V}_2/\mathbf{V}_1 = 7(\mathbf{s} + 8)/(\mathbf{s} + 3)$ for the network shown in Fig. 13-40. (a) If $v_1(t) = 4e^{-0.2t}$ V, find the forced response $v_{2f}(t)$. (b) If $v_1(t) = 6u(t)$ V and $v_2(0^-) = v_2(0^+) = 0$, find the complete response $v_2(t)$. (c) If $v_1(t) = 4e^{-0.2t}u(t)$ V and $v_2(0^-) = v_2(0^+) = 0$, find the complete response $v_2(t)$.

Fig. 13-40: See Prob. 22.

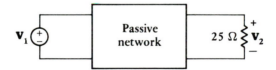

23 In the circuit of Fig. 13-41, the transfer function $\mathbf{H(s)} = \mathbf{I}_2/\mathbf{I}_1 = 5(\mathbf{s} + 20)(\mathbf{s} + 30)/[(\mathbf{s} + 10)(\mathbf{s} + 15)]$. Let $i_1(t) = u(t)$ A. (a) Find $i_{2f}(t)$. (b) If $di_2/dt = 60$ A/s at $t = 0^+$, find $i_2(t)$.

Fig. 13-41: See Prob. 23.

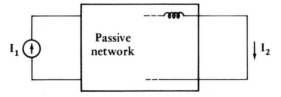

24 (a) Find $\mathbf{Z}_{in}(\mathbf{s})$ for the network given in Fig. 13-42. (b) If a voltage source were suddenly connected to the input terminals, what would be the form of the input current's natural response?

Fig. 13-42: See Prob. 24.

25 (a) Find $\mathbf{H(s)} = \mathbf{V}_C(s)/\mathbf{V}_s$ for the circuit shown in Fig. 13-43, and calculate its pole frequencies. (b) Find the exponential damping coefficient α, the resonant frequency ω_0, and the natural resonant frequency ω_d for the *RLC* circuit. Let $v_s(t) = u(t)$ V, and: (c) determine the forced response $v_{Cf}(t)$; (d) specify the form of the natural response $v_{Cn}(t)$; (e) find $v_C(0^+)$ and $v_C'(0^+)$; (f) write the complete response $v_C(t)$.

Fig. 13-43: See Prob. 25.

26 The passive network indicated in Fig. 13-44 has the pole-zero configuration shown and also $\mathbf{Z}(s) = 30\ \Omega$ at $s = -1$. Let $i_s(t) = 0.2u(t)$ A. (a) What value will $v_{in}(t)$ eventually assume? (b) Find $v_{in}(t)$ for $t > 0$ if $v_{in}(0^+) = 0$ and $v_{in}'(0^+) = 9$ V/s.

Fig. 13-44: See Prob. 26.

27 (a) Refer to Fig. 13-45 and find the transfer functions, $\mathbf{H}_1 = \mathbf{I}_o/\mathbf{V}_{s1}$ and $\mathbf{H}_2 = \mathbf{I}_o/\mathbf{V}_{s2}$. (b) If $v_{s1} = 10u(t)$ V and $v_{s2} = 5 + 25u(t)$ V, find $i_o(t)$.

Fig. 13-45: See Prob. 27.

28 An ideal voltage source $v_s(t)$, an open switch, and a 5-Ω resistor are in series with a passive two-terminal network having an input impedance, $\mathbf{Z}_A = (15s + 25)/(s + 5)\ \Omega$. The switch is closed at $t = 0$. If the current leaving the positive terminal of the source is $i_{in} = 2$ A at $t = 0^+$, determine $i_{in}(t)$ for $t > 0$ if $v_s(t) = :$ (a) 40 V; (b) $40e^{-0.5t}$ V.

29 Let R_f be 10 kΩ in Fig. 13-27a, and specify values for C_f and R_1 so that $\mathbf{H(s)} = \mathbf{V}_{out}/\mathbf{V}_{in} = :$ (a) $-10^5/(s + 10^4)$; (b) $-10^4/(s + 10^5)$; (c) -10; (d) -1; (e) $10^5/(s + 10^4)$ with two stages.

30 Let $R_f = 10$ kΩ in Fig. 13-27b, and specify values for R_1 and C_1 so that $\mathbf{H(s)} =$ $\mathbf{V}_{out}/\mathbf{V}_{in} = :$ (a) $-(1 + 10^{-4}\mathbf{s})$; (b) $-10(1 + 10^{-4}\mathbf{s})$; ($c$) -10; (d) -1; (e) $10(1 + 10^{-4}\mathbf{s})$ with two stages.

31 In the op-amp circuit of Fig. 13-26, find $\mathbf{H(s)} = \mathbf{V}_{out}/\mathbf{V}_{in}$ as a ratio of two polynomials in $\mathbf{s}$ if: (a) $\mathbf{Z}_1 = R_1$, $\mathbf{Z}_f = R_f + (1/\mathbf{s}C_f)$; ($b$) $\mathbf{Z}_1 = R_1 + (1/\mathbf{s}C_1)$, $\mathbf{Z}_f = R_f$; (c) $\mathbf{Z}_1 = R_1 + (1/\mathbf{s}C_1)$, $\mathbf{Z}_f = R_f + (1/\mathbf{s}C_f)$.

32 Use several (ideal) op-amps in cascade to realize the transfer function of Eq. (28) in Sec. 13-8. Specify values for all circuit elements external to the op-amps.

Chapter 14
Frequency Response

**14-1
Introduction**

Frequency response is a subject that has come up for consideration several times before. In Chap. 10 we discussed plots of admittance, impedance, current, and voltage as functions of ω, and the pole-zero concept was introduced as an aid in constructing and interpreting response curves. Response as a function of the neper frequency σ was discussed from the same standpoint in the last chapter. At that time, we also broadened our concept of frequency and introduced the complex frequency $\mathbf{s}$ and the $\mathbf{s}$ plane. We found that a plot of the critical frequencies of a response on the complex-frequency plane enabled us to tie together the forced response and the natural response; the critical frequencies themselves presented us almost directly with the form of the natural response, and the visualization of a three-dimensional rubber-sheet model or the performance of vectors drawn from each critical frequency to some test frequency gave us valuable information concerning the variation of the forced response with frequency.

In this chapter, we shall concentrate again on the forced response, and we shall particularly consider its variation with the radian frequency ω.

Why should we be so interested in the response to sinusoidal forcing functions when we so seldom encounter them in practice as such? The electric power industry is an exception, for the sinusoidal waveform appears throughout, although it is sometimes necessary to consider other frequencies introduced by the nonlinearity of some devices. But in most other electrical systems, the forcing functions and responses are not sinusoidal. In any system in which information is to be transmitted, the sinusoid by itself is almost valueless; it contains no information because its future values are exactly predictable from its past values. Moreover, once one period has been completed, any periodic nonsinusoidal waveform also contains no additional information.

Sinusoidal analysis, however, provides us with the response of a network as a function of ω, and the later work in Chaps. 18 and 19 will develop methods of determining network response to aperiodic signals (which *can* have a high information content) from the known sinusoidal frequency response.

The frequency response of a network provides useful information in its own right, however. Let us suppose that a certain forcing function is found to contain sinusoidal components having frequencies within the range of 10 to 100 Hz. Now let us imagine that this forcing function is applied to a network which has the property that all sinusoidal voltages with frequencies from zero to 200 Hz applied at the input terminals appear doubled in magnitude at the output terminals, with no change in phase angle. The output function is therefore an undistorted facsimile of the input function, but with twice the amplitude. If, however, the network has a frequency response such that the magnitudes of input sinusoids between 10 and 50 Hz are multiplied by a different factor than are those between 50 and 100 Hz, then the output would in general be distorted; it would no longer be a magnified version of the input. This distorted output might be desirable in some cases and undesirable in others. That is, the network frequency respone might be chosen deliberately to reject some frequency components of the forcing function, or to emphasize others.

The importance of frequency response is not limited to electrical systems, however; an understanding of electrical frequency response is certainly helpful in analyzing the frequency response of, say, a mechanical system. Suppose that we consider the launching of a space vehicle. The booster will subject the entire vehicle to extreme nonsinusoidal vibration, the mechanical forcing function, with frequency components from a few Hz up to perhaps 50 Hz. The structural members of every stage will be deflected or distorted, the mechanical response, by an amount which depends on the sinusoidal components of the vibrational forces and on the frequency response of the structure. From this information, a prediction as to whether or not the space vehicle can survive the rigors of the launch operation can be made. If the mechanical response at some vibration frequency is found to be excessive, then that component of the booster vibration must be reduced, a redesign of the supporting structure must be undertaken, or some vibration dampers (filters) must be provided.

This example should not make us think that it is undesirable to have the network response at some particular frequency be much larger than the response at all other frequencies. Such behavior is characteristic of tuned circuits or resonant circuits, as we shall see in this chapter. In discussing resonance we shall be able to apply all the methods we have discussed in presenting frequency response.

14-2
Parallel resonance

In this section we shall begin the study of a very important phenomenon which may occur in circuits containing both inductors and capacitors. The phenomenon is called resonance, and it may be loosely described as the condition existing in any physical system when a fixed-amplitude sinusoidal forcing function produces a response of maximum amplitude. However, we often speak of resonance as occurring even when the forcing function is not sinusoidal. The resonant system may be electrical, mechanical, hydraulic, acoustic, or some other kind, but we shall restrict out attention to electrical systems. We shall define resonance more exactly below.

Resonance is a familiar phenomenon. By jumping up and down on the bumper of an automobile, for example, the vehicle can be put into rather large oscillatory motion if the jumping is done *at the proper frequency* (about one jump per second), and if the shock absorbers are somewhat decrepit. However, if the jumping frequency is increased or decreased, the vibrational response of the automobile will be considerably less than it was before. A further illustration is furnished in the case of an opera singer who is able to shatter crystal goblets by means of a well-formed note *at the proper frequency*. In each of these examples, we are thinking of frequency as being adjusted until resonance occurs; it is also possible to adjust the size, shape, and material of the mechanical object being vibrated, but this may not be so easily accomplished physically.

The condition of resonance may or may not be desirable, depending upon the purpose which the physical system is to serve. In the automotive example above, a large amplitude of vibration may help to separate locked bumpers, but it would be somewhat disagreeable at 55 mi/h (88.5 km/h).

Let us now define resonance more carefully. In a two-terminal electrical network containing at least one inductor and one capacitor, we define *resonance* as the condition which exists when the input impedance of the network is purely resistive. Thus,

> a network is in *resonance* (or *resonant*) when the voltage and current at the network input terminals are in phase.

We shall also find that a maximum-amplitude response is produced in the network when it is in the resonant condition or *almost in the resonant condition*.

We first apply the definition of resonance to the parallel *RLC* network shown in Fig. 14-1. In many practical situations, this circuit is a very good approximation to the circuit we might build in the laboratory by connecting a physical inductor in parallel with a physical capacitor, where this parallel combination is driven by an energy source having a very high output impedance. The admittance offered to the ideal current source is

$$\mathbf{Y} = \frac{1}{R} + j\left(\omega C - \frac{1}{\omega L}\right) \tag{1}$$

and thus resonance occurs when

$$\omega C - \frac{1}{\omega L} = 0$$

The resonant condition may be achieved by adjusting L, C, or ω; we shall devote out attention to the case for which ω is the variable. Hence, the resonant frequency ω_0 is

$$\omega_0 = \frac{1}{\sqrt{LC}} \tag{2}$$

or

$$f_0 = \frac{1}{2\pi \sqrt{LC}} \tag{3}$$

This resonant frequency ω_0 is identical to the resonant frequency defined in Eq. (10), Chap. 7.

The pole-zero configuration of the admittance function can also be used to considerable advantage here. Given $\mathbf{Y(s)}$,

$$\mathbf{Y(s)} = \frac{1}{R} + \frac{1}{L\mathbf{s}} + C\mathbf{s}$$

or

$$\mathbf{Y(s)} = C \frac{\mathbf{s}^2 + \mathbf{s}/RC + 1/LC}{\mathbf{s}} \tag{4}$$

we may display the zeros of $\mathbf{Y(s)}$ by factoring the numerator,

$$\mathbf{Y(s)} = C \frac{(\mathbf{s} + \alpha - j\omega_d)(\mathbf{s} + \alpha + j\omega_d)}{\mathbf{s}}$$

where α and ω_d represent the same quantities that they did when we discussed the natural response of the parallel RLC circuit in Sec. 7-2. That is, α is the exponential damping coefficient,

$$\alpha = \frac{1}{2RC}$$

and ω_d is the natural resonant frequency (not the resonant frequency ω_0),

$$\omega_d = \sqrt{\omega_0^2 - \alpha^2}$$

The pole-zero constellation shown in Fig. 14-2a follows directly from the factored form.

Fig. 14-1: The parallel combination of a resistor, an inductor, and a capacitor, often referred to as a parallel resonant circuit.

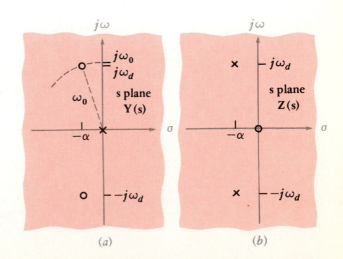

Fig. 14-2: (a) The pole-zero constellation of the input admittance of a parallel resonant circuit is shown on the **s** plane; $\omega_0^2 = \alpha^2 + \omega_d^2$. ($b$) The pole-zero constellation of the input impedance.

In view of the relationship among α, ω_d, and ω_0, it is apparent that the distance from the origin of the **s** plane to one of the admittance zeros is numerically equal to ω_0. Given the pole-zero configuration, the resonant frequency may therefore be obtained by purely graphical methods. We merely swing an arc, using the origin of the **s** plane as a center, through one of the zeros. The intersection of this arc and the positive $j\omega$ axis locates the point $\mathbf{s} = j\omega_0$. It is evident that ω_0 is slightly greater than the natural resonant frequency ω_d, but their ratio approaches unity as the ratio of ω_d to α increases.

Next let us examine the magnitude of the response, the voltage $\mathbf{V(s)}$ indicated in Fig. 14-1, as the frequency of the forcing function is varied. If we assume a constant-amplitude sinusoidal current source, the voltage response is proportional to the input impedance. This response can therefore be obtained from the pole-zero plot of the impedance $\mathbf{Z(s)}$, shown in Fig. 14-2b. The response obviously starts at zero, reaches a maximum value in the vicinity of the natural resonant frequency, and then drops again to zero as ω becomes infinite. The frequency response is sketched in Fig. 14-3. The maximum value of the response is indicated as R times the amplitude of the source current, implying that the maximum magnitude of the circuit impedance is R; moreover, the response maximum is shown to occur *exactly* at the resonant frequency ω_0. The two frequencies ω_1 and ω_2, which we shall later use as a measure of the width of the response curve, are also identified. Let us first show that the maximum impedance magnitude is R and that this maximum occurs at resonance.

The admittance, as specified by (1), possesses a constant conductance and a susceptance which has a minimum magnitude (zero) at resonance. The minimum admittance magnitude therefore occurs at resonance, and it is $1/R$. Hence, the maximum impedance magnitude is R, and it occurs at resonance.

At the resonant frequency, therefore, the voltage across the parallel resonant circuit of Fig. 14-1 is simply $\mathbf{I}R$, and we see that the *entire* source current $\mathbf{I}$ flows through the resistor. However, current is also present in L and C. For the former, $\mathbf{I}_{L,0} = \mathbf{I}R/j\omega_0 L$, while the capacitor current at resonance is $\mathbf{I}_{C,0} = j\omega_0 CR\mathbf{I}$. Since $1/\omega_0 C = \omega_0 L$ at resonance, we find that

$$\mathbf{I}_{C,0} = -\mathbf{I}_{L,0} = j\omega_0 CR\mathbf{I} \tag{5}$$

and $\mathbf{I}_{C,0} + \mathbf{I}_{L,0} = \mathbf{I}_{LC} = 0$.

The maximum value of the response magnitude and the frequency at which it occurs are not always found so easily. In less standard resonant circuits, we may find it necessary to express the magnitude of the response in analytical form, usually as the square root of the sum of the real part squared and the imaginary part squared; then we should differentiate this expression with respect to frequency, equate the derivative to zero, solve for the frequency of maximum response, and finally substitute this frequency into the magnitude expression to obtain the maximum-amplitude response. This procedure may be carried out for this simple case merely as a corroborative exercise, but as we have seen, it is not necessary.

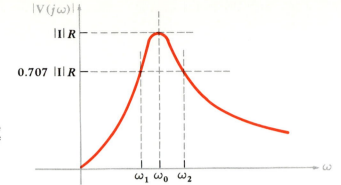

Fig. 14-3: The magnitude of the voltage response of a parallel resonant circuit is shown as a function of frequency.

It should be emphasized that, although the height of the response curve of Fig. 14-3 depends only upon the value of R for constant-amplitude excitation, the width of the curve or the steepness of the sides depends upon the other two element values also. We shall shortly relate the "width of the response curve" to a more carefully defined quantity, the bandwidth, but it will be helpful to express this relationship in terms of a very important parameter, the *quality factor Q*.[1]

We shall find that the sharpness of the response curve of any resonant circuit is determined by the maximum amount of energy that can be stored in the circuit, compared with the energy that is lost during one complete period of the response. We define Q as

$$Q = \text{quality factor} = 2\pi \frac{\text{maximum energy stored}}{\text{total energy lost per period}} \qquad (6)$$

The proportionality constant 2π is included in the definition in order to simplify the more useful expressions for Q which we shall now obtain. Since energy can be stored only in the inductor and the capacitor, and can be lost only in the resistor, we may express Q in terms of the instantaneous energy associated with each of the reactive elements and the average power dissipated in the resistor:

$$Q = 2\pi \frac{[w_L(t) + w_C(t)]_{\max}}{P_R T}$$

where T is the period of the sinusoidal frequency at which Q is being evaluated.

Now let us apply this definition to the parallel RLC circuit of Fig. 14-1 and determine the value of Q at the resonant frequency. This value of Q is denoted by Q_0. We select the current forcing function

$$i(t) = I_m \cos \omega_0 t$$

[1] This Q should not be confused with Q for charge or Q for reactive power.

and obtain the corresponding voltage response at resonance,

$$v(t) = Ri(t) = RI_m \cos \omega_0 t$$

Then the energy stored in the capacitor is

$$w_C(t) = \tfrac{1}{2}Cv^2 = \frac{I_m^2 R^2 C}{2} \cos^2 \omega_0 t$$

The instantaneous energy stored in the inductor becomes

$$w_L(t) = \tfrac{1}{2}Li_L^2 = \tfrac{1}{2}L \left(\frac{1}{L} \int_0^t v \, dt \right)^2$$

Thus

$$w_L(t) = \frac{I_m^2 R^2 C}{2} \sin^2 \omega_0 t$$

The total *instantaneous* stored energy is therefore constant,

$$w(t) = w_L(t) + w_C(t) = \frac{I_m^2 R^2 C}{2}$$

and this constant value must also be the maximum value. In order to find the energy lost in the resistor in one period, we take the average power absorbed by the resistor,

$$P_R = \tfrac{1}{2}I_m^2 R$$

and multiply by one period, obtaining

$$P_R T = \frac{1}{2f_0} I_m^2 R$$

We thus find the quality factor at resonance:

$$Q_0 = 2\pi \frac{I_m^2 R^2 C/2}{I_m^2 R/2f_0}$$

or $$Q_0 = 2\pi f_0 RC = \omega_0 RC \tag{7}$$

This equation [as well as the expressions in (8) below] holds only for the simple parallel *RLC* circuit of Fig. 14-1. Equivalent expressions for Q_0 which are often quite useful may be obtained by simple substitution:

$$Q_0 = R\sqrt{\frac{C}{L}} = \frac{R}{X_{C0}} = \frac{R}{X_{L0}} \tag{8}$$

It is apparent that Q_0 is a dimensionless constant which is a function of all three circuit elements in the parallel resonant circuit. The concept of Q, however, is not limited to electric circuits or even to electrical systems; it is useful in describing any resonant phenomenon. For example, let us consider a bouncing golf ball. If we assume a weight W and release the golf ball from a height h_1 above a very hard (lossless) horizontal surface, then the ball rebounds to some lesser height h_2. The energy stored initially is Wh_1, and the energy lost in one period is $W(h_1 - h_2)$. The Q_0 is therefore

$$Q_0 = 2\pi \frac{h_1 W}{(h_1 - h_2)W} = \frac{2\pi h_1}{h_1 - h_2}$$

A perfect golf ball would rebound to its original height and have an infinite Q_0; a more typical value is 35. It should be noted that the Q in this mechanical example has been calculated from the natural response and not from the forced response. The Q of an electric circuit may also be determined from a knowledge of the natural response, as illustrated by (10) and (11) below.

Another useful interpretation of Q is obtained when we inspect the inductor and capacitor currents at resonance, as given by (5),

$$\mathbf{I}_{C,0} = -\mathbf{I}_{L,0} = j\omega_0 CR\mathbf{I} = jQ_0\mathbf{I} \tag{9}$$

Note that each is Q_0 times the source current in amplitude and that they are 180° out of phase. Thus, if we apply 2 mA at the resonant frequency to a parallel resonant circuit with a Q_0 of 50, we find 2 mA in the resistor, and 100 mA in both the inductor and capacitor. A parallel resonant circuit can therefore act as a current amplifier, but not of course as a power amplifier, since it is a passive network.

Let us now relate the various parameters which we have associated with a parallel resonant circuit. The three parameters α, ω_d, and ω_0 were introduced much earlier in connection with the natural response. Resonance, by definition, is fundamentally associated with the forced response since it is defined in terms of a purely resistive input impedance, a sinusoidal steady-state concept. The two most important parameters of a resonant circuit are perhaps the resonant frequency ω_0 and the quality factor Q_0. Both the exponential damping coefficient and the natural resonant frequency may be expressed in terms of ω_0 and Q_0:

$$\alpha = \frac{1}{2RC} = \frac{1}{2(Q_0/\omega_0 C)C}$$

or

$$\alpha = \frac{\omega_0}{2Q_0} \tag{10}$$

and

$$\omega_d = \sqrt{\omega_0^2 - \alpha^2}$$

or

$$\omega_d = \omega_0 \sqrt{1 - \left(\frac{1}{2Q_0}\right)^2} \tag{11}$$

For future reference it may be helpful to note one additional relationship involving ω_0 and Q_0. The quadratic factor appearing in the numerator of (4),

$$\mathbf{s}^2 + \frac{1}{RC}\mathbf{s} + \frac{1}{LC}$$

may be written in terms of α and ω_0,

$$\mathbf{s}^2 + 2\alpha\mathbf{s} + \omega_0^2$$

In the field of system theory or automatic control theory, it is traditional to write this factor in a slightly different form that utilizes the dimensionless parameter ζ (zeta), called the *damping factor,*

$$s^2 + 2\zeta\omega_0 s + \omega_0{}^2$$

Comparison of these expressions enables us to relate ζ to our other parameters,

$$\zeta = \frac{\alpha}{\omega_0} = \frac{1}{2Q_0} \tag{12}$$

Let us now interpret Q_0 in terms of the pole-zero locations of the admittance $\mathbf{Y(s)}$ of the parallel *RLC* circuit. We shall keep ω_0 constant; this may be done, for example, by changing R while holding L and C constant. As Q_0 is increased, the relationship between α, Q_0, and ω_0 indicates that the two zeros must move closer to the $j\omega$ axis. The relationship between ω_d, ω_0, and Q_0 shows that the zeros must simultaneously move away from the σ axis. The exact nature of the movement becomes clearer when we remember that the point at which $\mathbf{s} = j\omega_0$ could be located on the $j\omega$ axis by swinging an arc, centered at the origin, through one of the zeros and over to the positive $j\omega$ axis; since ω_0 is to be held constant, the radius must be constant, and the zeros must therefore move along this arc toward the positive $j\omega$ axis as Q_0 increases.

The two zeros are indicated on Fig. 14-4, and the arrows show the path they take as R increases. When R is infinite, Q_0 is also infinite, and the two

Fig. 14-4: The two zeros of the admittance $\mathbf{Y(s)}$, located at $\mathbf{s} = -\alpha \pm j\omega_d$, provide a semicircular locus as R increases from $\frac{1}{2}\sqrt{L/C}$ to ∞.

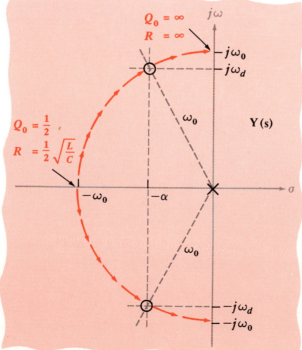

zeros are found at $s = \pm j\omega_0$ on the $j\omega$ axis. As R decreases, the zeros move toward the σ axis along the circular locus, joining to form a double zero on the σ axis at $s = -\omega_0$ when $R = \frac{1}{2}\sqrt{L/C}$ or $Q_0 = \frac{1}{2}$. This condition may be recalled as that for critical damping, so $\omega_d = 0$, $\alpha = \omega_0$. Lower values of R and lower values of Q_0 cause the zeros to separate and move in opposite directions on the negative σ axis, but these low values of Q_0 are not really typical of resonant circuits and we need not track them any further.

Later, we shall use the criterion, $Q_0 \geq 5$, to describe a high-Q circuit. When $Q_0 = 5$, the zeros are located at $s = -0.1\omega_0 \pm j0.9950\omega_0$, and thus ω_0 and ω_d differ only by one-half of 1 percent.

Drill Problems

14-1. Use the definition of resonance to find: (a) the resonant frequency of the network shown in Fig. 14-5; (b) $\mathbf{Z}_{in}$ at resonance.

Ans: 99.22 rad/s; 128.0 Ω

Fig. 14-5: See Drill Prob. 14-1.

14-2. The elements forming a parallel resonant circuit are $R = 6$ kΩ, $C = 0.1$ μF, and $L = 0.1$ H. Find: (a) ω_0; (b) Q_0; (c) ω_d; (d) α; (e) ζ.

Ans: 10 000 rad/s; 6; 9965 rad/s; 833 Np/s; 0.0833

14-3. Find the element values in a parallel resonant circuit having $Q_0 = 40$, $f_0 = 440$ Hz, and an admittance of 500 μ℧ at resonance.

Ans: 7.23 μF; 18.09 mH; 2 kΩ

14-3
More about parallel resonance

We continue our discussion of parallel resonance by defining half-power frequencies and bandwidth, and then we shall make good use of these new concepts in obtaining approximate response data for high-Q circuits.

The "width" of the resonant response curve, such as the one shown in Fig. 14-3, may now be defined more carefully and related to Q_0. Let us first define the two *half-power frequencies* ω_1 and ω_2 as those frequencies at which the magnitude of the input admittance of a parallel resonant circuit is greater than the magnitude at resonance by a factor of $\sqrt{2}$. Since the response curve of Fig. 14-3 displays the voltage produced across the parallel circuit by a sinusoidal current source as a function of frequency, the half-power frequencies also locate those points at which the voltage response is $1/\sqrt{2}$, or 0.707, times its maximum value. A similar relationship holds for the impedance magnitude. We shall select ω_1 as the *lower half-power frequency* and ω_2 as the *upper half-power frequency*. These names arise from the fact that a voltage which is 0.707 times the resonant voltage is equivalent to a squared voltage which is *one-half* the squared voltage at resonance.

Thus, at the half-power frequencies, the resistor absorbs one-half the power that it does at resonance.

The (half-power) bandwidth of a resonant circuit is defined as the difference of these two half-power frequencies.

$$\mathcal{B} = \omega_2 - \omega_1 \tag{13}$$

We think of this bandwidth as the "width" of the response curve, even though the curve actually extends from $\omega = 0$ to $\omega = \infty$. More exactly, the half-power bandwidth is measured by that portion of the response curve which is equal to or greater than 70.7 percent of the maximum value.

Now let us express the bandwidth $\mathcal{B}$ in terms of Q_0 and the resonant frequency. In order to do so, we first express the admittance of the parallel *RLC* circuit,

$$\mathbf{Y} = \frac{1}{R} + j\left(\omega C - \frac{1}{\omega L}\right)$$

in terms of Q_0,

$$\mathbf{Y} = \frac{1}{R} + j\,\frac{1}{R}\left(\frac{\omega\omega_0 CR}{\omega_0} - \frac{\omega_0 R}{\omega\omega_0 L}\right)$$

or

$$\mathbf{Y} = \frac{1}{R}\left[1 + jQ_0\left(\frac{\omega}{\omega_0} - \frac{\omega_0}{\omega}\right)\right] \tag{14}$$

We note again that the magnitude of the admittance at resonance is $1/R$, and then realize that an admittance magnitude of $\sqrt{2}/R$ can occur only when a frequency is selected such that the imaginary part of the bracketed quantity has a magnitude of unity. Thus

$$Q_0\left(\frac{\omega_2}{\omega_0} - \frac{\omega_0}{\omega_2}\right) = 1 \qquad \text{and} \qquad Q_0\left(\frac{\omega_1}{\omega_0} - \frac{\omega_0}{\omega_1}\right) = -1$$

Solving, we have

$$\omega_1 = \omega_0\left[\sqrt{1 + \left(\frac{1}{2Q_0}\right)^2} - \frac{1}{2Q_0}\right] \tag{15}$$

$$\omega_2 = \omega_0\left[\sqrt{1 + \left(\frac{1}{2Q_0}\right)^2} + \frac{1}{2Q_0}\right] \tag{16}$$

Although these expressions are somewhat unwieldy, their difference provides a very simple formula for the bandwidth:

$$\mathcal{B} = \omega_2 - \omega_1 = \frac{\omega_0}{Q_0} \tag{17}$$

Equations (15) and (16) may be multiplied by each other to show that ω_0 is exactly equal to the geometric mean of the half-power frequencies,

$$\omega_0^2 = \omega_1\omega_2$$

or

$$\omega_0 = \sqrt{\omega_1\omega_2}$$

Circuits possessing a higher Q_0 have a narrower bandwidth, or a sharper response curve; they have greater *frequency selectivity* or higher quality (factor).

Many resonant circuits are deliberately designed to have a large Q_0 in order to take advantage of the narrow bandwidth and high frequency selectivity associated with such circuits. When Q_0 is larger than about 5, it is possible to make some very useful approximations in the expressions for the upper and lower half-power frequencies and in the general expressions for the response in the neighborhood of resonance. Let us arbitrarily refer to a *high-Q* circuit as one in which Q_0 is equal to or greater than 5. The pole-zero configuration of $\mathbf{Y(s)}$ for a parallel *RLC* circuit having a Q_0 of about 5 is shown in Fig. 14-6. Since

$$\alpha = \frac{\omega_0}{2Q_0}$$

then

$$\alpha = \tfrac{1}{2}\mathcal{B}$$

and the locations of the two zeros may be approximated:

$$\mathbf{s}_{2,4} = -\alpha \pm j\omega_d \doteq -\tfrac{1}{2}\mathcal{B} \pm j\omega_0$$

Moreover, the locations of the two half-power frequencies (on the positive $j\omega$ axis) may also be determined in a concise approximate form:

$$\omega_{1,2} = \omega_0 \left[\sqrt{1 + \left(\frac{1}{2Q_0}\right)^2} \mp \frac{1}{2Q_0} \right] \doteq \omega_0 \left(1 \mp \frac{1}{2Q_0}\right)$$

or

$$\omega_{1,2} \doteq \omega_0 \mp \tfrac{1}{2}\mathcal{B} \tag{18}$$

Fig. 14-6: The pole-zero constellation of $\mathbf{Y(s)}$ for a parallel *RLC* circuit. The two zeros are exactly $\tfrac{1}{2}\mathcal{B}$ Np/s (or rad/s) to the left of the $j\omega$ axis and approximately $j\omega_0$ rad/s (or Np/s) from the σ axis. The upper and lower half-power frequencies are separated exactly $\mathcal{B}$ rad/s, and each is approximately $\tfrac{1}{2}\mathcal{B}$ rad/s away from the resonant frequency and the natural resonant frequency.

In a high-Q circuit, therefore, each half-power frequency is located approximately one-half bandwidth from the resonant frequency; this is indicated in Fig. 14-6.

The approximate relationships for ω_1 and ω_2 in (18) may be added to each other to show that ω_0 is approximately equal to the arithmetic mean of ω_1 and ω_2 in high-Q circuits,

$$\omega_0 \doteq \tfrac{1}{2}(\omega_1 + \omega_2)$$

Now let us visualize a test point slightly above $j\omega_0$ on the $j\omega$ axis. In order to determine the admittance offered by the parallel RLC network at this frequency, we construct the three vectors from the critical frequencies to the test point. If the test point is close to $j\omega_0$, then the vector from the pole is approximately $j\omega_0$ and that from the lower zero is nearly $j2\omega_0$. The admittance is therefore given approximately by

$$\mathbf{Y(s)} \doteq C\,\frac{(j2\omega_0)(\mathbf{s} - \mathbf{s}_2)}{j\omega_0} \doteq 2C(\mathbf{s} - \mathbf{s}_2) \qquad (19)$$

where C is the capacitance, as shown in (4). In order to determine a useful approximation for the vector $(\mathbf{s} - \mathbf{s}_2)$, let us consider an enlarged view of that portion of the $\mathbf{s}$ plane in the neighborhood of the zero $\mathbf{s}_2$ (Fig. 14-7).

In terms of its cartesian components, we see that

$$\mathbf{s} - \mathbf{s}_2 \doteq \tfrac{1}{2}\mathcal{B} + j(\omega - \omega_0)$$

where this expression would be exact if ω_0 were replaced by ω_d. We now substitute this equation into the approximation for $\mathbf{Y(s)}$ above and factor out $\tfrac{1}{2}\mathcal{B}$,

$$\mathbf{Y(s)} \doteq 2C(\tfrac{1}{2}\mathcal{B})\left(1 + j\,\frac{\omega - \omega_0}{\tfrac{1}{2}\mathcal{B}}\right)$$

or

$$\mathbf{Y(s)} \doteq \frac{1}{R}\left(1 + j\,\frac{\omega - \omega_0}{\tfrac{1}{2}\mathcal{B}}\right) \qquad (20)$$

Fig. 14-7: An enlarged portion of the pole-zero constellation for $\mathbf{Y(s)}$ of a high-Q_0 parallel RLC circuit.

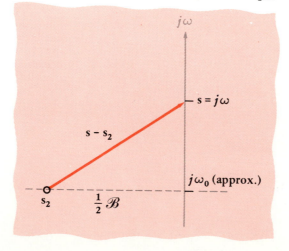

The fraction, $(\omega - \omega_0)/(\frac{1}{2}\mathcal{B})$, may be interpreted as the "number of half-bandwidths off resonance" and abbreviated by N. Thus,

$$\mathbf{Y(s)} \doteq \frac{1}{R}(1 + jN) \tag{21}$$

where

$$N = \frac{\omega - \omega_0}{\frac{1}{2}\mathcal{B}} \qquad \text{(number of half-bandwidths off resonance)} \tag{22}$$

At the upper half-power frequency, $\omega_2 \doteq \omega_0 + \frac{1}{2}\mathcal{B}$, $N = +1$, and we are one half-bandwidth above resonance. For the lower half-power frequency, $\omega_1 \doteq \omega_0 - \frac{1}{2}\mathcal{B}$, so that $N = -1$, locating us one half-bandwidth below resonance.

Equation (21) is much easier to use than the exact relationships we have had heretofore. It shows that the magnitude of the admittance is

$$|\mathbf{Y}(j\omega)| \doteq \frac{1}{R}\sqrt{1 + N^2} \tag{23}$$

while the angle of $\mathbf{Y}(j\omega)$ is given by the inverse tangent of N,

$$\text{ang } \mathbf{Y}(j\omega) \doteq \tan^{-1} N \tag{24}$$

As an example of the use of these approximations, let us determine the approximate value of the admittance of a parallel RLC network for which $R = 40$ kΩ, $L = 1$ H, and $C = 1/64$ μF. We find that $Q_0 = 5$, $\omega_0 = 8$ krad/s, $\mathcal{B} = 1.6$ krad/s, and $\frac{1}{2}\mathcal{B} = 0.8$ krad/s. Let us evaluate the admittance at $\omega = 8.2$ krad/s. Thus,

$$N = \frac{8.2 - 8}{0.8} = 0.25$$

and we see that we are operating 0.25 half-bandwidths above resonance. Next, we calculate

$$\text{ang } \mathbf{Y} \doteq \tan^{-1} 0.25 = 14.04°$$

and

$$|\mathbf{Y}| \doteq 25\sqrt{1 + (0.25)^2} \doteq 25.77 \quad \mu\mho$$

An exact calculation of the admittance shows that

$$\mathbf{Y}(j8200) = 25.75\underline{/13.87°} \quad \mu\mho$$

The approximate method therefore leads to values of admittance magnitude and angle that are quite accurate for this example.

Our intention is to use these approximations for high-Q circuits near resonance. We have already agreed that we shall let "high-Q" infer $Q_0 \geq 5$, but how near is "near"? It can be shown that the error in magnitude or phase is less than 5 percent if $Q_0 \geq 5$ and $0.9\omega_0 \leq \omega \leq 1.1\omega_0$. Although this narrow band of frequencies may seem to be prohibitively small, it is usually more than sufficient to contain the range of frequencies in which we are most interested. For example, an AM home radio usually contains a circuit tuned to a resonant frequency of 455 kHz with a half-power bandwidth of 10 kHz. This circuit must then have a value of 45.5 for Q_0, and the half-power frequencies are about 450 and 460 kHz. Our approximations, however, are valid from 409.5 to 500.5 kHz (with errors less than 5 percent), a range which covers essentially all the peaked portion of the response curve; only in the remote "tails" of the response curve do the approximations lead to unreasonably large errors.[2]

Let us conclude our coverage of the parallel resonant circuit by reviewing the various conclusions we have reached. The resonant frequency ω_0 is the frequency at which the imaginary part of the input admittance becomes zero or the angle of the admittance becomes zero. Then, $\omega_0 = 1/\sqrt{LC}$. The circuit's figure of merit Q_0 is defined as 2π times the ratio of the maximum energy stored in the circuit to the energy lost each period in the circuit. From this definition, we find that $Q_0 = \omega_0 RC$. The two half-power frequencies ω_1 and ω_2 are defined as the frequencies at which the admittance magnitude is $\sqrt{2}$ times the minimum admittance magnitude. These are also the frequencies at which the voltage response is 70.7 percent of the maximum response. The exact and approximate (for high Q_0) expressions for these two frequencies are

$$\omega_{1,2} = \omega_0 \left[\sqrt{1 + \left(\frac{1}{2Q_0}\right)^2} \mp \frac{1}{2Q_0} \right] \doteq \omega_0 \mp \tfrac{1}{2}\mathcal{B}$$

where $\mathcal{B}$ is the difference between the upper and the lower half-power frequencies. This half-power bandwidth is given by

$$\mathcal{B} = \omega_2 - \omega_1 = \frac{\omega_0}{Q_0}$$

The input admittance may also be expressed in approximate form for high-Q circuits,

$$\mathbf{Y} \doteq \frac{1}{R}(1 + jN) = \frac{1}{R}\sqrt{1 + N^2}\underline{/\tan^{-1} N}$$

where

$$N = \frac{\omega - \omega_0}{\tfrac{1}{2}\mathcal{B}}$$

The approximation is valid for frequencies which do not differ from the resonant frequency by more than one-tenth of the resonant frequency.

---

[2] At frequencies remote from resonance, we are often satisfied with very rough results; greater accuracy is not always necessary.

Drill Problems

14-4. A parallel resonant circuit has $f_0 = 600$ Hz and $Q_0 = 3$. Use (15) and (16) to calculate: (*a*) f_1; (*b*) f_2. Use (18) to calculate approximate values for: (*c*) f_1; (*d*) f_2. *Ans: 508.3; 708.3; 500; 700 Hz*

14-5. Element values of a parallel resonant circuit are 1000 Ω, 50 μF, and 80 mH. Use high-Q approximations to calculate $\mathbf{Z}_{in}$ at $\omega =$: (*a*) 485; (*b*) 497; (*c*) 501.23 rad/s. *Ans: $555\underline{/56.3°}$; $957.8\underline{/16.70°}$; $992.5\underline{/-7.01°}$ Ω*

14-6. Determine accurate answers for Drill Prob. 14-5. *Ans: $549\underline{/56.7°}$; $957.6\underline{/16.75°}$; $992.5\underline{/-7.00°}$ Ω*

14-4
Series resonance

Although we probably find less use for the series *RLC* circuit than we do for the parallel *RLC* circuit, it is still worthy of our attention. We shall consider the circuit shown in Fig. 14-8. It should be noted that the various circuit elements are given *s* (for series) subscripts for the time being in order to avoid confusing them with the parallel elements when the circuits are to be compared.

Our discussion of parallel resonance needed two sections of considerable length. We could now give the series *RLC* circuit the same kind of treatment, but it is much cleverer to avoid such needless repetition and use the duality principle. For simplicity, let us concentrate on the conclusions presented in the last paragraph of the preceding section on parallel resonance. The important results are contained there, and the use of dual language enables us to transcribe this paragraph to present the important results for the series *RLC* circuit.

Let us conclude our coverage of the series resonant circuit by summarizing the more important conclusions. The resonant frequency ω_{0s} is the frequency at which the imaginary part of the input impedance becomes zero or the angle of the impedance becomes zero. Then, $\omega_{0s} = 1/\sqrt{L_sC_s}$. The circuit's figure of merit Q_{0s} is defined as 2π times the ratio of the maximum energy stored in the circuit to the energy lost each period in the circuit. From this definition, we find that $Q_{0s} = \omega_{0s}L_s/R_s$. The two half-power frequencies ω_{1s} and ω_{2s} are defined as the frequencies at which the impedance magnitude is $\sqrt{2}$ times the minimum impedance magnitude. These are also the frequencies at which the current response is 70.7 percent of the maximum response. The exact and approximate (for high Q_{0s}) expressions for these two frequencies are

$$\omega_{1s,2s} = \omega_{0s}\left[\sqrt{1 + \left(\frac{1}{2Q_{0s}}\right)^2} \mp \frac{1}{2Q_{0s}}\right] \doteq \omega_{0s} \mp \tfrac{1}{2}\mathcal{B}_s$$

Fig. 14-8: A series resonant circuit.

where $\mathscr{B}_s$ is the difference between the upper and the lower half-power frequencies. This half-power bandwidth is given by

$$\mathscr{B}_s = \omega_{2s} - \omega_{1s} = \frac{\omega_{0s}}{Q_{0s}}$$

The input impedance may also be expressed in approximate form for high-Q_s circuits,

$$\mathbf{Z}_s \doteq R_s(1 + jN_s) = R_s\sqrt{1 + N_s^2}\,\underline{/\tan^{-1} N_s}$$

where

$$N_s = \frac{\omega - \omega_{0s}}{\frac{1}{2}\mathscr{B}_s}$$

The approximation is valid for frequencies which do not differ from the resonant frequency by more than one-tenth of the resonant frequency.

The series resonant circuit is characterized by a low impedance at resonance, while the parallel resonant circuit produces a high resonant impedance. The latter circuit provides inductor currents and capacitor currents at resonance which have amplitudes Q_0 times as great as the source current; the series resonant circuit provides inductor voltages and capacitor voltages which are greater than the source voltage by the factor Q_{0s}. The series circuit thus provides voltage amplification at resonance.

A comparison of our results for series and parallel resonance, as well as the exact and approximate expressions we have developed, appears in Table 14-1.

From this point on, we shall no longer identify series resonant circuits by use of an s subscript.

Drill Problems

14-7. A series resonant circuit includes a 1-μF capacitor and a 16-Ω resistor. If the bandwidth is 500 rad/s, determine: (a) ω_0; (b) Q_0; (c) L; (d) ω_1.
Ans: 5590 rad/s; 11.18; 32 mH; 5340 rad/s (approx.)

14-8. Use both accurate techniques and high-Q approximations to calculate the current amplitude in a series resonant circuit with $v_s = \cos \omega t$ V, $R = 20\ \Omega$, $\omega_0 = 200$ rad/s, and $Q_0 = 10$ if $\omega = :$ (a) 187; (b) 201.6 rad/s.
Ans: 29.83, 30.49; 49.38, 49.37 mA

14-5
Other resonant forms

The parallel and series *RLC* circuits of the previous two sections represent *idealized* resonant circuits; they are no more than useful approximate representations of a physical circuit which might be constructed by combining a coil of wire, a carbon resistor, and a tantalum capacitor in parallel or series. The degree of accuracy with which the idealized model fits the actual circuit depends on the operating frequency range, the Q of the circuit, the materials present in the physical elements, the element sizes, and many other factors. We are not studying the techniques of determining the best model for a given physical circuit, for this requires some knowledge of electromagnetic field

Table 14-1: A short summary of resonance

$$Q_0 = \omega_0 RC \qquad \alpha = \frac{1}{2RC}$$

$$|\mathbf{I}_L(j\omega_0)| = |\mathbf{I}_C(j\omega_0)| = Q_0|\mathbf{I}(j\omega_0)|$$

$$\mathbf{Y}_p = \frac{1}{R}\left[1 + jQ_0\left(\frac{\omega}{\omega_0} - \frac{\omega_0}{\omega}\right)\right]$$

$$Q_0 = \frac{\omega_0 L}{R} \qquad \alpha = \frac{R}{2L}$$

$$|\mathbf{V}_L(j\omega_0)| = |\mathbf{V}_C(j\omega_0)| = Q_0|\mathbf{V}(j\omega_0)|$$

$$\mathbf{Z}_s = R\left[1 + jQ_0\left(\frac{\omega}{\omega_0} - \frac{\omega_0}{\omega}\right)\right]$$

Exact expressions

$$\omega_0 = \frac{1}{\sqrt{LC}} = \sqrt{\omega_1\omega_2}$$

$$\omega_d = \sqrt{\omega_0^2 - \alpha^2} = \omega_0\sqrt{1 - \left(\frac{1}{2Q_0}\right)^2}$$

$$\omega_{1,2} = \omega_0\left[\sqrt{1 + \left(\frac{1}{2Q_0}\right)^2} \mp \frac{1}{2Q_0}\right]$$

$$N = \frac{\omega - \omega_0}{\frac{1}{2}\mathcal{B}}$$

$$\mathcal{B} = \omega_2 - \omega_1 = \frac{\omega_0}{Q_0} = 2\alpha$$

Approximate expressions

$$(Q_0 \geq 5 \qquad 0.9\omega_0 \leq |\omega| \leq 1.1\omega_0)$$

$$\omega_d \doteq \omega_0$$

$$\omega_{1,2} \doteq \omega_0 \mp \frac{1}{2}\mathcal{B}$$

$$\omega_0 \doteq \frac{1}{2}(\omega_1 + \omega_2)$$

$$\mathbf{Y}_p \doteq \frac{\sqrt{1 + N^2}}{R}\ \underline{/\tan^{-1}N} \qquad \mathbf{Z}_s \doteq R\sqrt{1 + N^2}\ \underline{/\tan^{-1}N}$$

Fig. 14-9: (*a*) A useful model of a physical network which consists of a physical inductor, capacitor, and resistor in parallel. (*b*) A network which can be equivalent to (*a*) over a narrow frequency band.

(*a*)   (*b*)

theory and the properties of materials; we are, however, concerned with the problem of reducing a more complicated model to one of the two simpler models with which we are more familiar.

The network shown in Fig. 14-9*a* is a reasonably accurate model for the parallel combination of a physical inductor, capacitor, and resistor. The resistor labeled R_1 is a hypothetical resistor that is included to account for the ohmic losses, core losses, and radiation losses of the physical coil. The losses in the dielectric within the physical capacitor, as well as the resistance of the physical resistor in the given *RLC* circuit, are accounted for by the resistor labeled R_2. In this model, there is no way to combine elements and produce a simpler model which is equivalent to the original model *for all frequencies*. We shall show, however, that a simpler equivalent may be constructed which is valid over a frequency band which is usually large enough to include all frequencies of interest. The equivalent will take the form of the network shown in Fig. 14-9*b*.

Before we learn how to develop such an equivalent circuit, let us first consider the given circuit, Fig. 14-9*a*. The resonant radian frequency for this network is *not* $1/\sqrt{LC}$, although if R_1 is sufficiently small it may be very close to this value. The definition of resonance is unchanged, and we may determine the resonant frequency by setting the imaginary part of the input admittance equal to zero:

$$\text{Im} \left[\mathbf{Y}(j\omega) \right] = \text{Im} \left(\frac{1}{R_2} + j\omega C + \frac{1}{R_1 + j\omega L} \right) = 0$$

Thus,

$$C = \frac{L}{R_1{}^2 + \omega^2 L^2}$$

and

$$\omega_0 = \sqrt{\frac{1}{LC} - \left(\frac{R_1}{L} \right)^2}$$

We note that ω_0 is less than $1/\sqrt{LC}$, but sufficiently small values of the ratio R_1/L may result in a negligible difference between ω_0 and $1/\sqrt{LC}$.

The maximum magnitude of the input impedance also deserves consideration. It is *not* R_2, and it does *not* occur at ω_0 (or at $\omega = 1/\sqrt{LC}$). The proof of these statements will not be shown because the expressions soon become algebraically cumbersome; the theory, however, is straightforward. Let us be content with a numerical example. We select the simple values $R_1 = 2\ \Omega$, $L = 1$ H, $C = \frac{1}{8}$ F, and $R_2 = 3\ \Omega$, and find the resonant frequency

$$\omega_0 = 2 \quad \text{rad/s}$$

and the input impedance at resonance

$$\mathbf{Z}(j2) = 1.714 \quad \Omega$$

At the frequency which would be the resonant frequency *if R_1 were zero,*

$$\frac{1}{\sqrt{LC}} = 2.83 \quad \text{rad/s}$$

the input impedance is

$$\mathbf{Z}(j2.83) = 1.947\underline{/-13.26°} \quad \Omega$$

However, the frequency at which the maximum impedance magnitude occurs, indicated by ω_m, is found to be

$$\omega_m = 3.26 \quad \text{rad/s}$$

and the impedance having the maximum magnitude is

$$\mathbf{Z}(j3.26) = 1.980\underline{/-21.4°} \quad \Omega$$

The impedance magnitude at resonance and the maximum magnitude differ by about 13 percent. Although it is true that such an error may be neglected occasionally in practice, it is too large to neglect on an exam in class. The later work in this section will show that the Q of the inductor-resistor combination at 2 rad/s is unity; this low value accounts for the 13 percent discrepancy.

In order to transform the given circuit of Fig. 14-9a into an equivalent of the form of that shown in Fig. 14-9b, we must discuss the Q of a simple series or parallel combination of a resistor and a reactor (inductor or capacitor). We first consider the series circuit shown in Fig. 14-10a. The Q of this network is again defined as 2π times the ratio of the maximum stored energy to the energy lost each period, but the Q may be evaluated at any frequency we choose. In other words, Q is a function of ω. It is true that we shall choose to evaluate it at a frequency which is, or apparently is, the resonant frequency of some network of which the series arm is a part. This frequency,

Fig. 14-10: (a) A series network which consists of a resistance R_s and an inductive or capacitive reactance X_s may be transformed into (b) a parallel network such that $\mathbf{Y}_s = \mathbf{Y}_p$ at one specific frequency. The reverse transformation is equally possible.

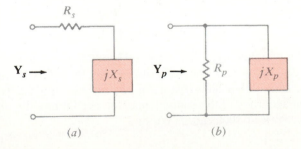

however, is not known until a more complete circuit is available. The eager reader is encouraged to show that the Q of this series arm is $|X_s|/R_s$, whereas the Q of the parallel network of Fig. 14-10b is $R_p/|X_p|$.

Let us now carry out the details necessary to find values for R_p and X_p so that the parallel network of Fig. 14-10b is equivalent to the series network of Fig. 14-10a at some single specific frequency. We equate $\mathbf{Y}_s$ and $\mathbf{Y}_p$,

$$\mathbf{Y}_s = \frac{1}{R_s + jX_s} = \frac{R_s - jX_s}{R_s^2 + X_s^2} = \mathbf{Y}_p = \frac{1}{R_p} - j\,\frac{1}{X_p}$$

and obtain

$$R_p = \frac{R_s^2 + X_s^2}{R_s} \qquad X_p = \frac{R_s^2 + X_s^2}{X_s}$$

Dividing these two expressions, we find

$$\frac{R_p}{X_p} = \frac{X_s}{R_s}$$

It follows that the Q's of the series and parallel networks must be equal:

$$Q_p = Q_s = Q$$

The transformation equations may therefore be simplified:

$$R_p = R_s(1 + Q^2) \tag{25}$$

$$X_p = X_s \left(1 + \frac{1}{Q^2}\right) \tag{26}$$

It is apparent that R_s and X_s may also be found if R_p and X_p are the given values; the transformation in either direction may be performed.

If $Q \geq 5$, little error is introduced by using the approximate relationships

$$R_p \doteq Q^2 R_s \tag{27}$$

$$X_p \doteq X_s \qquad (C_p \doteq C_s \text{ or } L_p \doteq L_s) \tag{28}$$

As an example, consider the series combination of a 100-mH inductor and a 5-Ω resistor. We shall perform the transformation at a frequency of 1000 rad/s, a value selected because it is approximately the resonant frequency of the network (not shown) of which this series arm is a part. We find that X_s is 100 Ω and Q is 20. Since the Q is sufficiently high, we use (27) and (28) to obtain

$$R_p \doteq Q^2 R_s = 2000 \quad \Omega \qquad L_p \doteq L_s = 100 \quad \text{mH}$$

The conclusion is that a 100-mH inductor in series with a 5-Ω resistor provides essentially the same input impedance as does a 100-mH inductor in parallel with a 2000-Ω resistor *at the frequency* 1000 rad/s. In order to check the accuracy of the equivalence, let us evaluate the input impedance for each network at 1000 rad/s. We find

$$\mathbf{Z}_s(j1000) = 5 + j100 = 100.1\underline{/87.1°}$$

$$\mathbf{Z}_p(j1000) = \frac{2000(j100)}{2000 + j100} = 99.9\underline{/87.1°}$$

and conclude that the approximation is exceedingly accurate at the transformation frequency. The accuracy at 900 rad/s is also reasonably good, because

$$\mathbf{Z}_s(j900) = 90.1\underline{/86.8°}$$

$$\mathbf{Z}_p(j900) = 89.9\underline{/87.4°}$$

If this inductor and series resistor had been used as part of a series *RLC* circuit for which the resonant frequency was 1000 rad/s, then the half-power bandwidth would have been

$$\mathcal{B} = \frac{\omega_0}{Q_0} = \frac{1000}{20} = 50$$

and the frequency of 900 rad/s would have represented a frequency that was 4 half-bandwidths off resonance. Thus the equivalent networks that we evaluated above would have been adequate for reproducing essentially all the peaked portion of the response curve.

As a further example of the replacement of a more complicated resonant circuit by an equivalent series or parallel *RLC* circuit, let us consider a problem in electronic instrumentation. The simple series *RLC* network in Fig. 14-11*a* is excited by a sinusoidal voltage source at the resonant frequency. The effective value of the source voltage is 0.5 V, and we wish to measure the effective value of the voltage across the capacitor with an

Fig. 14-11: (*a*) A given series resonant circuit in which the capacitor voltage is to be measured by a nonideal electronic voltmeter. (*b*) The effect of the voltmeter is included in the circuit; it reads V'_C volts. (*c*) A series resonant circuit is obtained when the parallel *RC* network in (*b*) is replaced by the series *RC* network which is equivalent at 10^5 rad/s.

(*a*)

(*b*)

(*c*)

electronic voltmeter (VM) having an internal resistance of 100 000 Ω. That is, an equivalent representation of the voltmeter is an ideal voltmeter in parallel with a 100 000-Ω resistor.

Before the voltmeter is connected, we find that the resonant frequency is 10^5 rad/s, $Q_0 = 50$, the current is 25 mA, and the rms capacitor voltage is 25 V. As indicated at the end of Sec. 14-4, this voltage is Q_0 times the applied voltage. Thus, if the voltmeter were ideal, it would read 25 V when connected across the capacitor.

However, when the actual voltmeter is connected, the circuit shown in Fig. 14-11b results. In order to obtain a series RLC circuit, it is now necessary to replace the parallel RC network by a series RC network. Let us assume that the Q of this RC network is sufficiently high so that the equivalent series capacitor will be the same as the given parallel capacitor. We do this in order to approximate the resonant frequency of the final series RLC circuit. Thus, if the series RLC circuit also contains a 0.01-μF capacitor, the resonant frequency remains 10^5 rad/s. We need to know this estimated resonant frequency in order to calculate the Q of the parallel RC network; it is

$$Q = \frac{R_p}{|X_p|} = \omega R_p C_p = 10^5 (10^5)(10^{-8}) = 100$$

Since this value is greater than 5, our vicious circle of assumptions is justified, and the equivalent series RC network consists of the capacitor

$$C_s = 0.01 \quad \mu\text{F}$$

and the resistor

$$R_s \doteq \frac{R_p}{Q^2} = 10 \quad \Omega$$

Hence, the equivalent circuit of Fig. 14-11c is obtained. The resonant Q of this circuit is now only 33.3, and thus the voltage across the capacitor in the circuit of Fig. 14-11c is $16\frac{2}{3}$ V. But we need to find $|\mathbf{V}'_c|$, the voltage across the series RC combination; we obtain

$$|\mathbf{V}'_c| = \frac{0.5}{30} |10 - j1000| = 16.7 \quad \text{V}$$

The capacitor voltage and $|\mathbf{V}'_c|$ are essentially equal since the voltage across the 10-Ω resistor is quite small.

The final conclusion must be that an apparently good voltmeter may produce a severe effect on the response of a high-Q resonant circuit. A similar effect may occur when a nonideal ammeter is inserted in the circuit.

We conclude this section with a technical fable.

Once upon a time there was a student whom we shall call Pat, and a professor identified simply as Dr. Noe.

In the laboratory one afternoon, Dr. Noe gave Pat three practical circuit devices: a resistor, an inductor, and a capacitor, having nominal element values of 20 Ω, 20 mH, and 1 μF. The student was asked to connect a variable-frequency voltage source to the series combination of these three elements, to measure the resultant voltage across the resistor as a function of frequency, and then to calculate numerical values for the resonant frequency, the Q at resonance, and the half-power bandwidth. The student was also asked to predict the results of the experiment before making the measurements.

Pat, who was not one of the more spectacular performers in the class, formed a mental picture of the equivalent circuit for this problem like that shown in Fig. 14-12a, and then calculated:

$$f_0 = \frac{1}{2\pi\sqrt{LC}} = \frac{1}{2\pi\sqrt{20 \times 10^{-3} \times 10^{-6}}} = 1125 \quad \text{Hz}$$

$$Q_0 = \frac{\omega_0 L}{R} = 7.07$$

$$\mathscr{B} = \frac{f_0}{Q_0} = 159 \quad \text{Hz}$$

Next, Pat made the measurements that Dr. Noe requested, calculated the numerical values, compared them to the predicted values, and then felt a strong urge to transfer to the business school. The results were:

$$f_0 = 1000 \quad \text{Hz}$$

$$Q_0 = 0.625$$

$$\mathscr{B} = 1600 \quad \text{Hz}$$

Even Pat knew that discrepancies of this magnitude could not be charac-

Fig. 14-12: (a) A first model for a 20-mH inductor, a 1-μF capacitor, and a 20-Ω resistor in series with a voltage generator. (b) An improved model in which more accurate values are used and the losses in the inductor and capacitor are acknowledged. (c) The final model also contains the output resistance of the voltage source.

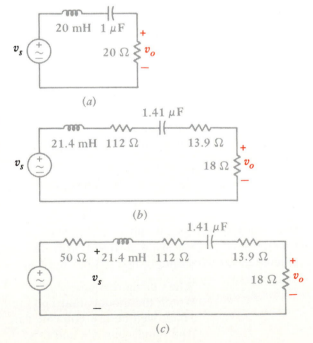

terized as being "within engineering accuracy" or "due to meter errors." Sadly, the results were handed to the professor.

Remembering many past errors in judgment, some of which were even self-made, Dr. Noe smiled benevolently and called Pat's attention to the Q-meter (or impedance bridge) which is present in most well-equipped laboratories, and suggested that it might be used to find out what these practical circuit elements really looked like at some convenient frequency near resonance, say, 1000 Hz.

Upon doing so, Pat discovered that the resistor had a measured value of 18 Ω, the inductor was 21.4 mH with a Q of 1.2, while the capacitor had a capacitance of 1.41 μF and a dissipation factor (the reciprocal of Q) equal to 0.123.

So, with the hope that springs eternal within the human breast, Pat reasoned that a better model for the practical inductor would be 21.4 mH in series with $\omega L/Q = 112 \; \Omega$, while a more appropriate model for the capacitor would be 1.41 μF in series with $1/\omega CQ = 13.9 \; \Omega$. Using these data, Pat prepared the modified circuit model shown as Fig. 14-12b and calculated a new set of predicted values:

$$f_0 = \frac{1}{2\pi \sqrt{21.4 \times 10^{-3} \times 1.41 \times 10^{-6}}} = 916 \quad \text{Hz}$$

$$Q_0 = \frac{2\pi \times 916 \times 21.4 \times 10^{-3}}{143.9} = 0.856$$

$$\mathcal{B} = \frac{916}{0.856} = 1070 \quad \text{Hz}$$

Since these results were much closer to the measured values, Pat was much happier. Dr. Noe, however, being a stickler for detail, looked askance at the differences in the predicted and measured values for both Q_0 and the bandwidth. "Have you," Dr. Noe asked, "given any consideration to the output impedance of the voltage source?" "Not yet," said Pat, trotting back to the laboratory bench.

It turned out that the output impedance in question was 50 Ω, so Pat added this value to the circuit diagram, as shown in Fig. 14-12c. Using the new equivalent resistance value of 193.9 Ω, improved values for Q_0 and $\mathcal{B}$ were then obtained:

$$Q_0 = 0.635$$

$$\mathcal{B} = 1442 \quad \text{Hz}$$

Since all the theoretical and experimental values now agreed within 10 percent, Pat was once again suffused with the joys of studying electrical engineering. Dr. Noe simply nodded her head agreeably as she moralized:

> When using real devices,
> Watch the models that you choose;
> Think well before you calculate,
> And mind your Z's and Q's!

Drill Problems

14-9. At a frequency of 800 rad/s, find the parallel equivalent of each network in Figs. 14-13*a* and *b* and the series equivalents for *c* and *d*.
   Ans: 1 H, 32 kΩ; 10 μF, 3.91 kΩ; 10 mH, 0.16 Ω; 0.2 μF, 39.1 Ω

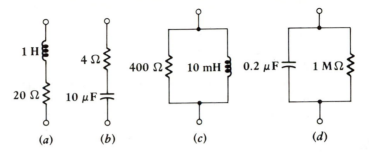

Fig. 14-13: See Drill
Prob. 14-9.

(*a*)   (*b*)   (*c*)   (*d*)

14-10. Find the Q at $\omega = 10^6$ rad/s of: (*a*) a 400-pF capacitor in parallel with 2 MΩ; (*b*) a 2-Ω resistor in series with the network of *a*; (*c*) a 1-MΩ resistor in parallel with the network of *b*.   *Ans:* 800; 488; 220

14-11. A series resonant circuit consists of a 0.4-Ω resistor in series with an inductor, modeled by 20 mH in parallel with 2 kΩ, in series with a capacitor, modeled by 50 μF in parallel with 10 kΩ. (*a*) What is the (approximate) resonant frequency? At this frequency, find the Q of the: (*b*) inductor; (*c*) capacitor; (*d*) series resonant circuit.   *Ans:* 1 krad/s; 100; 500; 31.25

**14-6
Scaling**

Some of the examples and problems which we have been solving have involved circuits containing passive element values ranging around a few ohms, a few henrys, and a few farads. The applied frequencies were a few radians per second. These particular numerical values were used not because they are those commonly met in practice, but because arithmetic manipulations are so much easier than they would be if it were necessary to carry along various powers of 10 throughout the calculations. The scaling procedures that will be discussed in this section enable us to analyze networks composed of practical-sized elements by scaling the element values to permit more convenient numerical calculations. We shall consider both magnitude scaling and frequency scaling.

   Let us select the parallel resonant circuit shown in Fig. 14-14*a* as our

Fig. 14-14: (*a*) A parallel resonant circuit used as an example to illustrate magnitude and frequency scaling. (*b*) The magnitude of the input impedance is shown as a function of frequency.

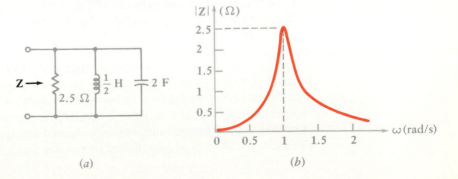

(*a*)   (*b*)

example. The impractical element values lead to the unlikely response curve drawn as Fig. 14-14b; the maximum impedance is 2.5 Ω, the resonant frequency is 1 rad/s, Q_0 is 5, and the bandwidth is 0.2 rad/s. These numerical values are much more characteristic of the electrical analog of some mechanical system than they are of any basically electrical device. We have convenient numbers with which to calculate but an impractical circuit to construct.

Let us assume that our goal is to scale this network in such a way as to provide an impedance maximum of 5000 Ω at a resonant frequency of 5×10^6 rad/s, or 796 kHz. In other words, we may use the same response curve shown in Fig. 14-14b if every number on the ordinate scale is increased by a factor of 2000 and every number on the abscissa scale is increased by a factor of 5×10^6. We shall treat this as two problems: (1) scaling in magnitude by a factor of 2000 and (2) scaling in frequency by a factor of 5×10^6.

Magnitude scaling is defined as the process by which the *impedance* of a two-terminal network is increased by a factor of K_m, the frequency remaining constant. The factor K_m is real and positive; it may be greater or smaller than unity. We shall understand that the shorter statement, "the network is scaled in magnitude by a factor of 2," infers that the impedance of the new network is to be *twice* that of the old network at any frequency. Let us now determine how we must scale each type of passive element. To increase the input impedance of a network by a factor of K_m, it is sufficient to increase the impedance of each element in the network by this same factor. Thus, a resistor R must be replaced by a resistor $K_m R$. Each inductor must also exhibit an impedance which is K_m times as great at any frequency. In order to increase an impedance sL by a factor of K_m when s remains constant, the inductor L must be replaced by an inductor $K_m L$. In a similar manner, each capacitor C must be replaced by a capacitor C/K_m. In summary, these changes will produce a network which is scaled in magnitude by a factor of K_m:

$$\left. \begin{array}{l} R \rightarrow K_m R \\[2mm] L \rightarrow K_m L \\[2mm] C \rightarrow \dfrac{C}{K_m} \end{array} \right\} \quad \text{magnitude scaling}$$

When each element in the network of Fig. 14-14a is scaled in magnitude by a factor of 2000, the network shown in Fig. 14-15a results. The response curve shown in Fig. 14-15b indicates that no change in the previously drawn response curve need be made other than a change in the scale of the ordinate.

Let us now take this new network and scale it in frequency. We define *frequency scaling* as the process by which the frequency at which any impedance occurs is increased by a factor of K_f. Again, we shall make use of the shorter expression, "the network is scaled in frequency by a factor of 2," to infer that the same impedance is now obtained at a frequency twice as great. Frequency scaling is accomplished by scaling each passive element in fre-

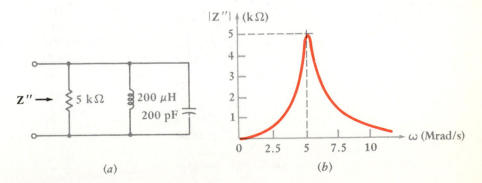

Fig. 14-15: (a) The network of Fig. 14-14a after being scaled in magnitude by a factor $K_m = 2000$. (b) The corresponding response curve.

(a) (b)

quency. It is apparent that no resistor is affected. The impedance of any inductor is sL, and if this same impedance is to be obtained at a frequency K_f times as great, then the inductor L must be replaced by one having an inductance of L/K_f. Similarly, a capacitor C is to be replaced by one having a capacitance C/K_f. Thus, if a network is to be scaled in frequency by a factor of K_f, then the changes necessary in each passive element are

$$\left. \begin{array}{l} R \to R \\ L \to \dfrac{L}{K_f} \\ C \to \dfrac{C}{K_f} \end{array} \right\} \text{ frequency scaling}$$

When each element of the magnitude-scaled network of Fig. 14-15a is scaled in frequency by a factor of 5×10^6, the network of Fig. 14-16a is obtained. The corresponding response curve is shown in Fig. 14-16b.

The circuit elements in this last network have values which are easily achieved in physical circuits; the network can actually be built and tested. It follows that, if the original network of Fig. 14-14a were actually an analog of some mechanical resonant system, we could have scaled this analog in both magnitude and frequency in order to achieve a network which we might construct in the laboratory; tests that are expensive or inconvenient to run

Fig. 14-16: (a) The network of Fig. 14-15a after being scaled in frequency by a factor $K_f = 5 \times 10^6$. (b) The corresponding response curve.

(a) (b)

on the mechanical system could then be made on the scaled electrical system, and the results should then be "unscaled" and converted into mechanical units to complete the analysis.

The effect of magnitude scaling or frequency scaling on the pole-zero constellation of an impedance is not very difficult to ascertain, but its determination offers such an excellent opportunity to review the meaning and significance of a pole-zero plot in the s plane that it forms the basis for one of the drill problems below.

An impedance which is given as a function of s may also be scaled in magnitude or frequency, and this may be done without any knowledge of the specific elements out of which the two-terminal network is composed. In order to scale $\mathbf{Z}(s)$ in magnitude, the definition of magnitude scaling shows that it is only necessary to multiply $\mathbf{Z}(s)$ by K_m in order to obtain the magnitude-scaled impedance. Thus, the impedance of the parallel resonant circuit shown in Fig. 14-14a is

$$\mathbf{Z}(s) = \frac{s}{2s^2 + 0.4s + 2}$$

or

$$\mathbf{Z}(s) = 0.5 \frac{s}{(s + 0.1 + j0.995)(s + 0.1 - j0.995)}$$

The impedance $\mathbf{Z}'(s)$ of the magnitude-scaled network is

$$\mathbf{Z}'(s) = K_m \mathbf{Z}(s)$$

and, therefore,

$$\mathbf{Z}'(s) = 1000 \frac{s}{(s + 0.1 + j0.995)(s + 0.1 - j0.995)}$$

If $\mathbf{Z}'(s)$ is now to be scaled in frequency by a factor of 5×10^6, then $\mathbf{Z}''(s)$ and $\mathbf{Z}'(s)$ are to provide identical values of impedance if $\mathbf{Z}''(s)$ is evaluated at a frequency K_f times that at which $\mathbf{Z}'(s)$ is evaluated. After some careful cerebral activity, this conclusion may be stated concisely in functional notation:

$$\mathbf{Z}''(s) = \mathbf{Z}'\left(\frac{s}{K_f}\right)$$

Note that we obtain $\mathbf{Z}''(s)$ by replacing every s in $\mathbf{Z}'(s)$ by s/K_f. The analytic expression for the impedance of the network shown in Fig. 14-16a must therefore be

$$\mathbf{Z}''(s) = 1000 \frac{s/(5 \times 10^6)}{[s/(5 \times 10^6) + 0.1 + j0.995][s/(5 \times 10^6) + 0.1 - j0.995]}$$

or

$$\mathbf{Z}''(s) = \frac{5 \times 10^9 \times s}{(s + 0.5 \times 10^6 + j4.975 \times 10^6)(s + 0.5 \times 10^6 - j4.975 \times 10^6)}$$

Although scaling is a process normally applied to passive elements, dependent sources may also be scaled in magnitude and frequency. We assume that the output of any source is given as $k_x v_x$ or $k_y i_y$, where k_x has the dimensions of an admittance for a dependent current source and is dimensionless for a dependent voltage source, while k_y has the dimensions of ohms for a dependent voltage source and is dimensionless for a dependent current source. If the network containing the dependent source is scaled in magnitude by K_m, then it is only necessary to treat k_x or k_y as if it were the type of element consistent with its dimensions. That is, if k_x (or k_y) is dimensionless, it is left unchanged; if it is an admittance, it is divided by K_m; and if it is an impedance, it is multiplied by K_m. Frequency scaling does not affect the dependent sources.

Drill Problems

14-12. A series resonant circuit is defined by $C = 0.01$ F, $\mathcal{B} = 2.5$ rad/s, and $\omega_0 = 20$ rad/s. Find the values of R and L if the network is scaled in: (a) magnitude by a factor of 8; (b) frequency by a factor of 10; (c) magnitude by a factor of 8 and frequency by a factor of 10.

Ans: 5, 2; 0.625, 0.025; 5, 0.2 (Ω, H)

14-13. Two characteristics of the pole-zero plot of a certain transfer function $\mathbf{H(s)}$ are a pole at $\mathbf{s} = -3 + j6$ s^{-1} and a magnitude at the origin of 24. The network to which the transfer function applies is then scaled by $K_m = 12$, $K_f = 15$. Specify the pole location and the magnitude at $\mathbf{s} = 0$ if $\mathbf{H(s)} = $: (a) $\mathbf{V_2/I_1}$; (b) $\mathbf{I_2/V_1}$; (c) $\mathbf{V_2/V_1}$; (d) $\mathbf{I_2/I_1}$.

Ans: $-45 + j90$ s^{-1}, 288 Ω; $-45 + j90$ s^{-1}, 2 ℧; $-45 + j90$ s^{-1}, 24; $-45 + j90$ s^{-1}, 24

14-14. Scale the network shown in Fig. 14-17 by $K_m = 4$ and $K_f = 5$. (a) Give parameter values for all three elements, and (b) find $\mathbf{Z}_{in}$ for the scaled network. *Ans:* 0.05 V_1, 0.32 H, 0.005 F; $0.32(\mathbf{s}^2 + 625)/(\mathbf{s} + 10)$ Ω

Fig. 14-17: See Drill Prob. 14-14.

14-7
Bode diagrams

This chapter is concerned with frequency response, as the title promises, and we have already investigated several topics that we illustrated with plots showing some responses as functions of frequency. Curves of this type first began to appear back in Chap. 10 when the sinusoidal forcing function was introduced, and then reappeared in Chap. 13 as we sought to understand complex frequency.

In this section we shall discover a quick method of obtaining an approximate picture of the amplitude and phase variation of a given transfer function as functions of ω. Accurate curves may of course be plotted after calculating values with a programmable calculator or computer; curves may also be produced directly on the computer. Our object here, however, is to obtain a better picture of the response than we could visualize from a pole-zero plot, but yet not mount an all-out computational offensive.

The approximate response curve we construct is called an *asymptotic plot,* or a *Bode plot* or *Bode diagram* after its developer, Hendrik W. Bode,[3] who was an electrical engineer and mathematician with the Bell Telephone Laboratories. Both the magnitude and phase curves are shown using a logarithmic frequency scale for the abscissa, and the magnitude itself is also shown in logarithmic units called *decibels* (dB). We define the value of $|\mathbf{H}(j\omega)|$ in dB as follows:

$$H_{dB} = 20 \log |\mathbf{H}(j\omega)| \tag{29}$$

where the common logarithm (base 10) is used. A different definition is used for power ratios, but we shall not need it here. The inverse operation is

$$|\mathbf{H}(j\omega)| = 10^{(H_{dB}/20)} \tag{30}$$

Before we actually begin a detailed discussion of the technique for drawing Bode diagrams, it will help to gain some feeling for the size of the decibel unit, to learn a few of its important values, and to recall some of the properties of the logarithm. Since $\log 1 = 0$, $\log 2 = 0.301\ 03$, and $\log 10 = 1$, we note the correspondences:

$$|\mathbf{H}(j\omega)| = 1 \quad \Leftrightarrow \quad H_{dB} = 0$$

$$|\mathbf{H}(j\omega)| = 2 \quad \Leftrightarrow \quad H_{dB} \doteq 6 \text{ dB}$$

$$|\mathbf{H}(j\omega)| = 10 \quad \Leftrightarrow \quad H_{dB} = 20 \text{ dB}$$

Moreover, $\log 10^n = n$, and thus $10^n \Leftrightarrow 20n$ dB, so that 1000 corresponds to 60 dB, while 0.01 is represented as -40 dB. Using only the values given above, we may also find that $20 \log 5 = 20 \log \frac{10}{2} = 20 \log 10 - 20 \log 2 = 20 - 6 = 14$ dB, and thus $5 \Leftrightarrow 14$ dB. Also, $\log \sqrt{x} = \frac{1}{2} \log x$, and therefore $\sqrt{2} \Leftrightarrow 3$ dB, and $1/\sqrt{2} \Leftrightarrow -3$ dB.[4]

We shall write our transfer functions in terms of **s**, substituting $\mathbf{s} = j\omega$ when we are ready to find the magnitude or phase angle. Our next step is to factor $\mathbf{H}(\mathbf{s})$ to display its poles and zeros. We first consider a zero at $\mathbf{s} = -a$, written in a standardized form as

$$\mathbf{H}(\mathbf{s}) = 1 + \frac{\mathbf{s}}{a}$$

The Bode diagram for this function consists of the two asymptotic curves approached by H_{dB} for very large and very small values of ω. Thus,

$$|\mathbf{H}(j\omega)| = \left| 1 + \frac{j\omega}{a} \right| = \sqrt{1 + \frac{\omega^2}{a^2}}$$

---

[3] Sounds like "OK."

[4] Note that we are being slightly dishonest by using $20 \log 2 = 6$ dB rather than 6.02 dB.

and

$$H_{dB} = 20 \log \left| 1 + \frac{j\omega}{a} \right| = 20 \log \sqrt{1 + \frac{\omega^2}{a^2}}$$

When $\omega \ll a$,

$$H_{dB} \doteq 20 \log 1 = 0 \qquad (\omega \ll a)$$

This simple asymptote is shown in Fig. 14-18. It is drawn as a solid line for $\omega < a$, and as a broken line for $\omega > a$.

When $\omega \gg a$,

$$H_{dB} \doteq 20 \log \frac{\omega}{a} \qquad (\omega \gg a)$$

At $\omega = a$, $H_{dB} = 0$; at $\omega = 10a$, $H_{dB} = 20$ dB; and at $\omega = 100a$, $H_{dB} = 40$ dB. Thus, the value of H_{dB} increases 20 dB for every 10-fold increase in frequency. The asymptote thus has a slope of 20 dB/decade. Since H_{dB} increases by 6 dB when ω doubles, an alternate value for the slope is 6 dB/octave. The high-frequency asymptote is also shown on Fig. 14-18, a solid line for $\omega > a$, and a broken line for $\omega < a$. Note that the two asymptotes intersect at $\omega = a$, the frequency of the zero. This frequency is also described as the *corner, break,* 3-dB, or *half-power frequency.*

The Bode plot represents the response in terms of two asymptotes, both straight lines, and easily drawn.

Now let us see how much error is embodied in our asymptotic response curve. At the corner frequency,

$$H_{dB} = 20 \log \sqrt{1 + \frac{a^2}{a^2}} = 3 \text{ dB}$$

as compared to an asymptotic value of 0 dB. At $\omega = 0.5a$, we have

$$H_{dB} = 20 \log \sqrt{1.25} = 1 \text{ dB}$$

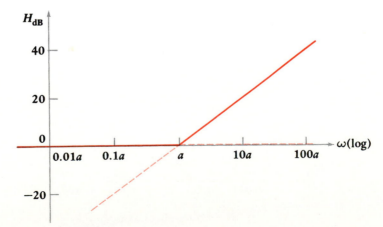

Fig. 14-18: The Bode amplitude plot for **H(s)** = 1 + **s**/a consists of the low- and high-frequency asymptotes, shown in solid color. They intersect on the abscissa at the corner frequency.

Thus, the exact response is represented by a smooth curve that lies 3 dB above the asymptotic response at $\omega = a$, and 1 dB above it at $\omega = 0.5a$ (and also at $\omega = 2a$). This information can always be used to smooth out the corner if a more exact result is desired.

We now need the phase response for the simple zero,

$$\text{ang } \mathbf{H}(j\omega) = \text{ang } \left(1 + \frac{j\omega}{a}\right) = \tan^{-1}\frac{\omega}{a}$$

This expression is also represented by its asymptotes, although three straight-line segments are required. For $\omega \ll a$, ang $\mathbf{H}(j\omega) \doteq 0°$, and we use this as our asymptote when $\omega < 0.1a$,

$$\text{ang } \mathbf{H}(j\omega) = 0° \qquad (\omega < 0.1a)$$

At the high end, $\omega \gg a$, we have ang $\mathbf{H}(j\omega) \doteq 90°$, and we use this above $\omega = 10a$:

$$\text{ang } \mathbf{H}(j\omega) = 90° \qquad (\omega > 10a)$$

Since the angle is 45° at $\omega = a$, we now construct the straight-line asymptote extending from 0° at $\omega = 0.1a$, through 45° at $\omega = a$, to 90° at $\omega = 10a$. This straight line has a slope of 45°/decade. It is shown as a solid curve in Fig. 14-19, while the exact angle response is shown as a broken line. The maximum differences between the asymptotic and true responses are $\pm5.71°$ at $\omega = 0.1a$ and $10a$. Errors of $\pm5.29°$ also occur at $\omega = 0.394a$ and $2.54a$; the error is zero at $\omega = 0.159a$, a, and $6.31a$.

We next consider a simple pole,

$$\mathbf{H}(s) = \frac{1}{1 + s/a}$$

Since this is the reciprocal of a zero, the logarithmic operation leads to a Bode plot which is the negative of that obtained previously. The amplitude is

Fig. 14-19: The asymptotic angle response for $\mathbf{H}(s) = 1 + s/a$ is shown as the three straight-line segments in solid color. The endpoints of the ramp are 0° at 0.1a and 90° at 10a.

(a)

(b)

Fig. 14-20: The asymptotic diagrams are shown for: (a) **H(s)** = **s**, and (b) **H(s)** = 1/**s**. Both are infinitely long straight lines passing through 0 dB at $\omega = 1$, and having slopes of ±20 dB/decade.

0 dB up to $\omega = a$, and then the slope is -20 dB/decade for $\omega > a$. The angle plot is 0° for $\omega < 0.1a$, $-90°$ for $\omega > 10a$, $-45°$ at $\omega = a$, and it has a slope of $-45°$/decade when $0.1a < \omega < 10a$.

Another term that can appear in **H(s)** is a factor of **s** in the numerator or denominator. If **H(s)** = **s**, then

$$H_{dB} = 20 \log |\omega|$$

Thus, we have an infinite straight line passing through 0 dB at $\omega = 1$ and having a slope everywhere of 20 dB/decade. This is shown in Fig. 14-20a. If the **s** factor occurs in the denominator, a straight line is obtained having a slope of -20 dB/decade and passing through 0 dB at $\omega = 1$, as shown in Fig. 14-20b.

Another simple term found in **H(s)** is the multiplying constant K. This yields a Bode plot which is a horizontal straight line lying $20 \log |K|$ dB above the abscissa.

Let us now consider an example in which several of these different factors are combined in **H(s)**. We choose the input impedance of the network shown in Fig. 14-21:

$$\mathbf{Z}_{in}(\mathbf{s}) = \mathbf{H}(\mathbf{s}) = 20 + 0.2\mathbf{s}$$

Putting this in standard form, we have

$$\mathbf{H}(\mathbf{s}) = 20 \left(1 + \frac{\mathbf{s}}{100} \right)$$

Fig. 14-21: If **H(s)** is selected as **Z**$_{in}$(**s**) for this network, then the Bode plot for H_{dB} is shown in Fig. 14-22b.

20 Ω

H(s) = **Z**$_{in}$ (s) →

0.2 H

Fig. 14-22: (a) The Bode plots for the factors of **H(s)** = 20(1 + **s**/100) are sketched individually. (b) The composite Bode plot is shown as the sum of the plots of (a).

The two factors comprising **H(s)** are a zero at $\omega = 100$, and a constant, equivalent to 20 log 20 = 26 dB. Each of these is sketched lightly in Fig. 14-22a. Since we are working with the logarithm of **H(s)**, we next *add* the Bode plots corresponding to the individual factors. The resultant magnitude plot appears as Fig. 14-22b. No attempt has been made to smooth out the corner with a +3 dB correction at $\omega = 100$ rad/s.

The Bode phase plot is merely the angle corresponding to the zero at $\omega = 100$. We therefore have 0° below $\omega = 10$, 90° above $\omega = 1000$, 45° at $\omega = 100$, and a 45°/decade slope for $10 < \omega < 1000$. The result is shown as Fig. 14-23.

Fig. 14-23: The Bode phase plot for **H(s)** = 20(1 + **s**/100).

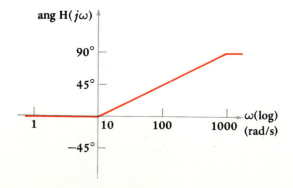

Fig. 14-24: If $\mathbf{H(s)} = \mathbf{V}_{out}/\mathbf{V}_{in}$, this amplifier is found to have the Bode amplitude plot shown in Fig. 14-25b, and the phase plot shown in Fig. 14-26.

As a second example, we choose a circuit similar to an amplifier with both low- and high-frequency limitations. Working from the circuit diagram of Fig. 14-24, we find

$$\mathbf{H(s)} = \frac{\mathbf{V}_{out}}{\mathbf{V}_{in}} = \frac{4000}{5000 + 10^6/20s} \left(-\frac{1}{200}\right) \frac{(10^8/s)5000}{5000 + 10^8/s}$$

$$= \frac{8 \times 10^4 s}{10^6 + 10^5 s} \left(-\frac{1}{200}\right) \frac{5 \times 10^{11}}{10^8 + 5000s}$$

which simplifies to

$$\mathbf{H(s)} = \frac{-2s}{(1 + s/10)(1 + s/20\ 000)}$$

We see a constant, $20 \log |-2| = 6$ dB, poles at $\omega = 10$ and $\omega = 20\ 000$, and a linear factor $\mathbf{s}$. Each of these is sketched in Fig. 14-25a, and the four sketches are added to give the Bode magnitude plot in Fig. 14-25b.

Fig. 14-25: (a) Individual Bode magnitude sketches are made for the factors (-2), $(\mathbf{s})$, $(1 + \mathbf{s}/10)^{-1}$, and $(1 + \mathbf{s}/20\ 000)^{-1}$. (b) The four separate plots of (a) are added to give the Bode magnitude plot for the amplifier of Fig. 14-24.

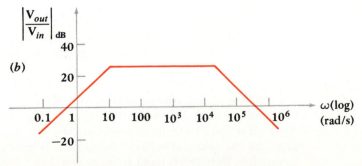

Before we construct the phase plot for this amplifier, let us take a few moments to investigate several of the details of the magnitude plot.

First, it is wise not to rely too heavily on graphical addition of the individual magnitude plots. Instead, the exact value of the combined magnitude plot may be found easily at selected points by considering the asymptotic value of each factor of $\mathbf{H(s)}$ at the point in question. For example, in the flat region of Fig. 14-25b between $\omega = 10$ and $\omega = 20\ 000$, we are below the corner at $\omega = 20\ 000$ and we represent $(1 + \mathbf{s}/20\ 000)$ by 1, but we are above $\omega = 10$, so $(1 + \mathbf{s}/10)$ is represented as $\omega/10$. Hence.

$$H_{\mathrm{dB}} = 20 \log \left| \frac{-2\omega}{(\omega/10)1} \right| = 20 \log 20 = 26 \text{ dB} \qquad (10 < \omega < 20\ 000)$$

We might also wish to know the frequency at which the asymptotic response crosses the abscissa at the high end. The two corners are expressed here as $\omega/10$ and $\omega/20\ 000$; thus

$$H_{\mathrm{dB}} = 20 \log \left| \frac{-2\omega}{(\omega/10)(\omega/20\ 000)} \right| = 20 \log |400\ 000/\omega|$$

Since $H_{\mathrm{dB}} = 0$ at the abscissa crossing, $400\ 000/\omega = 1$, and $\omega = 400\ 000$ rad/s.

Many times we may not need an accurate Bode plot drawn on printed semilog paper. Instead we construct a rough logarithmic frequency axis on simple lined paper. After selecting the interval for a decade, say a distance L extending from $\omega = \omega_1$ to $\omega = 10\omega_1$ (where ω_1 is usually an integral power of 10), we let x locate the distance that ω lies to the right of ω_1, so that $x/L = \log(\omega/\omega_1)$. Of particular help is the knowledge that $x = 0.3L$ when $\omega = 2\omega_1$, $x = 0.6L$ at $\omega = 4\omega_1$, and $x = 0.7L$ at $\omega = 5\omega_1$.

Now let us finish up the example of Fig. 14-24 by drawing the phase plot for

$$\mathbf{H(s)} = \frac{-2\mathbf{s}}{(1 + \mathbf{s}/10)(1 + \mathbf{s}/20\ 000)}$$

or

$$\mathbf{H}(j\omega) = \frac{-j2\omega}{(1 + j\omega/10)(1 + j\omega/20\ 000)} \qquad (31)$$

The angle of the numerator is a constant, $-90°$, and the remaining factors are represented as the sum of the angles contributed by simple poles at $\omega = 10$ and $\omega = 20\ 000$. These three terms appear as broken-line asymptotic curves in Fig. 14-26, and their sum is shown as the solid curve. An equivalent representation is obtained if the curve is shifted upward by $360°$.

Exact values can also be obtained for the asymptotic phase response. For example, at $\omega = 10^4$ rad/s, the angle in Fig. 14-26 is obtained from the numerator and denominator terms in (31). The numerator angle is $-90°$. The angle for the pole at $\omega = 10$ is $-90°$ since ω is greater than 10 times the corner frequency. Between 0.1 and 10 times the corner frequency, we recall that the

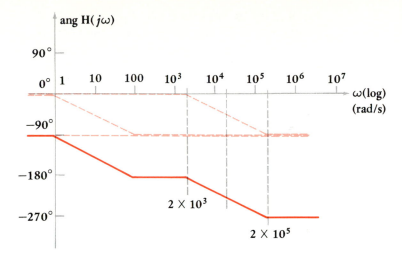

Fig. 14-26: The solid curve displays the asymptotic phase response of the amplifier shown in Fig. 14-24.

slope is $-45°$/decade for a simple pole. For the pole at 20 000 rad/s, we therefore have an angle, $-45° \log (\omega/0.1a) = -45° \log [10\,000/(0.1 \times 20\,000)] = -31.5°$. The algebraic sum of these three contributions is $-90° - 90° - 31.5° = -211.5°$, a value which appears to be moderately near the asymptotic phase curve of Fig. 14-26.

The zeros and poles that we have been considering are all first-order terms, such as $s^{\pm1}$, $(1 + 0.2s)^{\pm1}$, and so forth. We may extend our analysis to higher-order poles and zeros very easily, however. A term $s^{\pm n}$ yields a magnitude response that passes through $\omega = 1$ with a slope of $\pm20n$ dB/decade; the phase response is a constant angle of $\pm90°n$. Also, a multiple zero, $(1 + s/a)^n$, must represent the sum of n of the magnitude-response curves or n of the phase-response curves of the simple zero. We therefore obtain an asymptotic magnitude plot that is 0 dB for $\omega < a$ and has a slope of $20n$ dB/decade when $\omega > a$; the error is $-3n$ dB at $\omega = a$, and $-n$ dB at $\omega = 0.5a$ and $2a$. The phase plot is $0°$ for $\omega < 0.1a$, $90°n$ for $\omega > 10a$, $45°n$ at $\omega = a$, a straight line with a slope of $45°n$/decade for $0.1a < \omega < 10a$, and has errors as large as $\pm5.71°n$ at two frequencies.

The asymptotic magnitude and phase curves for a factor such as $(1 + s/20)^{-3}$ may be drawn quickly, but the relatively large errors associated with the higher powers should be kept in mind.

The last type of factor that we need to consider represents a conjugate complex pair of poles or zeros. We adopt the following as the standard form for a pair of zeros,

$$\mathbf{H(s)} = 1 + 2\zeta \left(\frac{\mathbf{s}}{\omega_0}\right) + \left(\frac{\mathbf{s}}{\omega_0}\right)^2$$

The quantity ζ is the damping factor introduced in Sec. 14-2, and we shall see shortly that ω_0' is the corner frequency of the asymptotic response.[5]

[5] It is also the resonant frequency if $\zeta = 0$.

If $\zeta = 1$, we see that $\mathbf{H(s)} = 1 + 2(\mathbf{s}/\omega_0) + (\mathbf{s}/\omega_0)^2 = (1 + \mathbf{s}/\omega_0)^2$, a second-order zero such as we have just considered. If $\zeta > 1$, then $\mathbf{H(s)}$ may be factored to show two simple zeros. For example, if $\zeta = 1.25$, then $\mathbf{H(s)} = 1 + 2.5(\mathbf{s}/\omega_0) + (\mathbf{s}/\omega_0)^2 = (1 + \mathbf{s}/2\omega_0)(1 + \mathbf{s}/0.5\omega_0)$, and we again have a familiar situation.

A new case arises when $0 \le \zeta < 1$. There is no need to find values for the conjugate complex pair of roots. Instead, we determine the low- and high-frequency asymptotic values for both the magnitude and phase response, and then apply a correction that depends on the value of ζ.

For the magnitude response, we have

$$H_{\text{dB}} = 20 \log |\mathbf{H}(j\omega)| = 20 \log \left| 1 + j2\zeta \left(\frac{\omega}{\omega_0}\right) - \left(\frac{\omega}{\omega_0}\right)^2 \right| \qquad (32)$$

When $\omega \ll \omega_0$, $H_{\text{dB}} \doteq 20 \log|1| = 0$ dB. This is the low-frequency asymptote. Next, if $\omega \gg \omega_0$, only the squared term is important, and $H_{\text{dB}} \doteq 20 \log|-(\omega/\omega_0)^2| = 40 \log(\omega/\omega_0)$. We have a slope of $+40$ dB/decade. This is the high-frequency asymptote and the two asymptotes intersect at 0 dB, $\omega = \omega_0$. The solid curve in Fig. 14-27 shows this asymptotic representation of the magnitude response. However, a correction must be applied in the neighborhood of the corner frequency. We let $\omega = \omega_0$ in (32) and have

$$H_{\text{dB}} = 20 \log \left| j2\zeta \left(\frac{\omega_0}{\omega_0}\right) \right| = 20 \log(2\zeta)$$

Fig. 14-27: Bode amplitude plots are shown for $\mathbf{H(s)} = 1 + 2\zeta(\mathbf{s}/\omega_0) + (\mathbf{s}/\omega_0)^2$ for several values of the damping factor ζ.

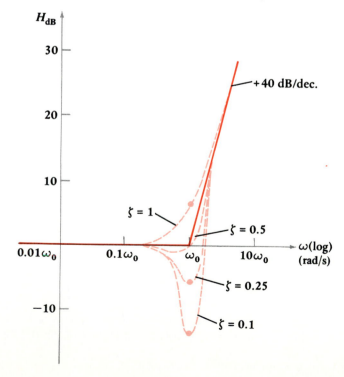

If $\zeta = 1$, a limiting case, the correction is $+6$ dB; for $\zeta = 0.5$, no correction is required; and if $\zeta = 0.1$, the correction is -14 dB. Knowing this one correction value is often sufficient to draw a satisfactory asymptotic magnitude response. Figure 14-27 shows more accurate curves for $\zeta = 1, 0.5, 0.25$, and 0.1, as calculated from (31). For example, if $\zeta = 0.25$, then the exact value of H_{dB} at $\omega = 0.5\omega_0$ is

$$H_{\text{dB}} = 20 \log|1 + j0.25 - 0.25| = 20 \log \sqrt{0.75^2 + 0.25^2} = -2.0 \text{ dB}$$

The negative peaks do not show a minimum value exactly at $\omega = \omega_0$, as we can see by the curve for $\zeta = 0.5$. The valley is always found at a slightly lower frequency.

If $\zeta = 0$, then $\mathbf{H}(j\omega_0) = 0$ and $H_{\text{dB}} = -\infty$. Bode plots are not usually drawn for this situation.

Our last task is to draw the asymptotic phase response for $\mathbf{H}(j\omega) = 1 + j2\zeta(\omega/\omega_0) - (\omega/\omega_0)^2$. Below $\omega = 0.1\omega_0$, we let ang $\mathbf{H}(j\omega) = 0°$; above $\omega = 10\omega_0$, we have ang $\mathbf{H}(j\omega) \doteq \text{ang}[-(\omega/\omega_0)^2] = 180°$. At the corner frequency, ang $\mathbf{H}(j\omega_0) = \text{ang}(j2\zeta) = 90°$. In the interval, $0.1\omega_0 < \omega < 10\omega_0$, we begin with the straight line shown as a solid curve in Fig. 14-28. It extends from $(0.1\omega_0, 0°)$, through $(\omega_0, 90°)$, and terminates at $(10\omega_0, 180°)$; it has a slope of $90°/\text{decade}$.

We must now provide some correction to this basic curve for various values of ζ. From (32), we have

$$\text{ang } \mathbf{H}(j\omega) = \tan^{-1} \frac{2\zeta(\omega/\omega_0)}{1 - (\omega/\omega_0)^2}$$

One accurate value above and one below $\omega = \omega_0$ may be sufficient to give an approximate shape to the curve. If we take $\omega = 0.5\omega_0$, we find ang $\mathbf{H}(j0.5\omega_0) = \tan^{-1}(4\zeta/3)$, while the angle is $180° - \tan^{-1}(4\zeta/3)$ at $\omega = 2\omega_0$. Phase curves are shown as broken lines in Fig. 14-28 for $\zeta = 1, 0.5, 0.25$, and 0.1; heavy dots identify accurate values at $\omega = 0.5\omega_0$ and $\omega = 2\omega_0$.

Fig. 14-28: The straight-line approximation to the phase characteristic for $\mathbf{H}(j\omega) = 1 + j2\zeta(\omega/\omega_0) - (\omega/\omega_0)^2$ is shown as a solid curve, and the true phase response is shown for $\zeta = 1, 0.5, 0.25$, and 0.1 as broken lines.

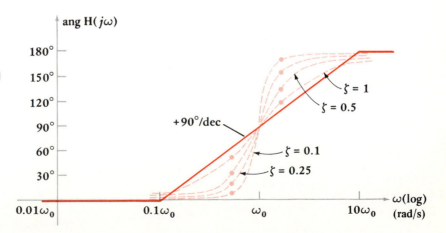

If the quadratic factor appears in the denominator, both the magnitude and phase curves are the negatives of those just discussed.

We conclude with an example that contains both linear and quadratic factors,

$$H(s) = \frac{100\ 000s}{(s + 1)(10\ 000 + 20s + s^2)}$$

Let's consider the quadratic factor first and arrange it in a form so that we can see the value of ζ. We begin by dividing the second-order factor by its constant term, 10 000:

$$H(s) = \frac{10s}{(1 + s)(1 + 0.002s + 0.0001s^2)}$$

An inspection of the s^2 term next shows that ω_0 is 100. Then the linear term of the quadratic is written to display the factor 2, the factor (s/ω_0), and finally the factor ζ,

$$H(s) = \frac{10s}{(1 + s)[1 + 2(0.1)(s/100) + (s/100)^2]}$$

We see that $\zeta = 0.1$.

The asymptotes of the magnitude response curve are sketched in lightly in Fig. 14-29: 20 dB for the factor of 10, an infinite straight line through $\omega = 1$ with a $+20$ dB/decade slope for the s factor, a corner at $\omega = 1$ for the simple pole, and a corner at $\omega_0 = 100$ with a slope of -40 dB/decade for the second-order term in the denominator. Adding these four curves and supplying a correction of $+14$ dB for the quadratic factor leads to the heavy curve of Fig. 14-29.

The phase response contains three components: $+90°$ for the factor s; $0°$ for $\omega < 0.1$, $-90°$ for $\omega > 10$, and $-45°/$decade for the simple pole at $\omega = 1$; and $0°$ for $\omega < 10$, $-180°$ for $\omega > 1000$, and $-90°/$decade for the quadratic

Fig. 14-29: The Bode magnitude plot of the transfer function, **H(s)** =

$$\frac{100\ 000s}{(s + 1)(10\ 000 + 20s + s^2)}$$

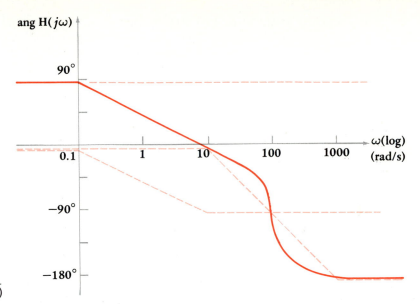

Fig. 14-30: The Bode phase plot of the transfer function, **H(s)** =

$$\frac{100\ 000\mathbf{s}}{(\mathbf{s} + 1)(10\ 000 + 20\mathbf{s} + \mathbf{s}^2)}$$

factor. The addition of these three asymptotes plus some improvement for $\zeta = 0.1$ is shown as the solid curve in Fig. 14-30.

Drill Problems

14-15. Find H_{dB} at $\omega = 10$ rad/s if $\mathbf{H(s)} = :$ (a) $150/(\mathbf{s} + 7)$; (b) $520\mathbf{s}$; (c) $1/(\mathbf{s} + 16)$. Find $|\mathbf{H}(j\omega)|$ if $H_{\mathrm{dB}} = :$ (d) 37 dB; (e) -65.4 dB.
Ans: 21.8 dB; 74.3 dB; -25.5 dB; 70.8; 5.37×10^{-4}

14-16. Draw a Bode magnitude plot for $\mathbf{H(s)} = :$ (a) $\mathbf{s}/(\mathbf{s} + 10)$; (b) $(10\mathbf{s} + 1000)/\mathbf{s}$; (c) $(2\mathbf{s} + 40)/(20\mathbf{s} + 40)$.
Ans: (a) 0 dB, $\omega > 10$; $+20$ dB/dec, $\omega < 10$ (b) 20 dB, $\omega > 100$; -20 dB/dec, $\omega < 100$ (c) 0 dB, $\omega < 2$; -20 dB, $\omega > 20$; -20 dB/dec, $2 < \omega < 20$

14-17. Draw the Bode phase plot for each $\mathbf{H(s)}$ given in Prob. D14-16.
Ans: (a) 90°, $\omega < 1$; 0°, $\omega > 100$; $-45°$/dec, $1 < \omega < 100$ (b) $-90°$, $\omega < 10$; 0°, $\omega > 1000$; $+45°$/dec, $10 < \omega < 1000$ (c) 0°, $\omega < 0.2$ and $\omega > 200$; $-45°$, $2 < \omega < 20$; $-45°$/dec, $0.2 < \omega < 2$; $+45°$/dec, $20 < \omega < 200$

14-18. Let $\mathbf{H(s)} = (\mathbf{s}^2 + 10\mathbf{s} + 400)/(1000\mathbf{s}^2)$. After sketching the Bode amplitude plot, calculate a value for: (a) ω when $H_{\mathrm{dB}} = 0$; (b) H_{dB} as $\omega \rightarrow \infty$; (c) H_{dB} at $\omega = 2$.
Ans: 0.632 rad/s; -60 dB; -20 dB

Problems

1 For the resonant circuit shown in Fig. 14-31, find: (a) ω_0; (b) f_0; (c) Q_0; (d) the critical frequencies of $\mathbf{Z}_{\mathrm{in}}(\mathbf{s})$; (e) $|\mathbf{I}_L|$ if $i = 5 \sin \omega_0 t$ mA is connected to the input; (f) $|\mathbf{I}_R|$ if $i = 5 \sin \omega_0 t$ mA is connected to the input.

Fig. 14-31: See Prob. 1.

2 Design a parallel resonant circuit that will have admittance zeros at $s = -20 \pm j160$ s^{-1} and an admittance magnitude of 2 m℧ at $s = j160$ s^{-1}.

3 Find values for R, L, and C in the circuit of Fig. 14-1 if: (a) $\mathbf{I} = 2\underline{/0°}$ μA, $|\mathbf{V}|_{max} = 125$ mV, $Q_0 = 40$, and $f_0 = 1$ MHz; (b) $\mathbf{Z}_{in}$ seen by the current source is $8 + j0$ kΩ at $\omega = 2000$ rad/s and $4\underline{/\theta}$ kΩ at $\omega = 1800$ rad/s.

4 The pole-zero plot for the input admittance of a parallel resonant circuit shows zeros at $-100 \pm j1600$ s^{-1}, and a pole at the origin. If the value of C is 20 μF, find: (a) Q_0; (b) ω_0; (c) α; (d) ζ; (e) R; (f) L. (g) At what frequencies is $|\mathbf{Y}_{in}| = 2/R$?

5 (a) Find the exact resonant frequency of the network shown in Fig. 14-32, and (b) the value of $\mathbf{Z}_{in}$ at that frequency.

Fig. 14-32: See Prob. 5.

6 Design a parallel resonant circuit so that a variable capacitor can adjust the resonant frequency over the AM broadcast band, 535 to 1605 kHz, with $Q_0 \leq 45$ at any frequency in the band. Let $R = 25$ kΩ and give values for L, C_{min}, and C_{max}.

7 First derive an expression for the input admittance of the network appearing in Fig. 14-33, and then use it to determine ω_0 and Q_0.

Fig. 14-33: See Prob. 7.

8 A parallel resonant circuit has $\omega_0 = 10$ krad/s, $Q_0 = 8$, and an impedance at resonance of 500 Ω. (a) Find $\mathcal{B}$. (b) Determine N at 9.2 krad/s. (c) Use approximate methods to determine $\mathbf{Z}_{in}$ at 9.2 krad/s. (d) Determine the true value of $\mathbf{Z}_{in}$ at that frequency. (e) Calculate the percentage error for $|\mathbf{Z}_{in}|$ and ang $\mathbf{Z}_{in}$.

9 Determine the element values, Q_0 and $\mathcal{B}$, for a parallel RLC circuit for which $C = 50$ μF and: (a) $\omega_0 = 100$ and $\omega_2 = 120$ rad/s; (b) $\omega_0 = 100$ and $\omega_1 = 80$ rad/s; (c) $\omega_1 = 80$ and $\omega_2 = 120$ rad/s.

10 (a) Use approximate methods to prepare a sketch of $|\mathbf{V}|$ vs. ω, $0.9\omega_0 \leq \omega \leq 1.1\omega_0$, for the parallel resonant circuit of Fig. 14-1 if $R = 8$ kΩ, $L = 0.25$ H, $C = 0.25$ μF, and $\mathbf{I} = 2\underline{/0°}$ mA. Find $\mathbf{V}$ at 3250 rad/s by: (b) approximate methods; (c) exact methods.

11 For the circuit shown in Fig. 14-34: (a) use approximate methods to plot a curve of $|\mathbf{V}_L|$ vs. ω, $920 \leq \omega \leq 1080$ rad/s; (b) find the exact value of $\mathbf{V}_L$ at $\omega = 1050$ rad/s.

12 A parallel resonant circuit has $Q_0 = 20$ and is resonant at $\omega_0 = 10$ krad/s. If $\mathbf{Z}_{in} = 5$ kΩ at $\omega = \omega_0$, what is the width of the frequency band about resonance for which $|\mathbf{Z}_{in}| \geq 3$ kΩ?

Fig. 14-34: See Prob. 11.

13 A parallel resonant circuit is driven by a current source such that the voltage across it has an amplitude of 100 V at resonance. If $f_0 = 500$ Hz and $Q_0 = 25$, use approximate relationships to find: (*a*) the voltage amplitude at 495 Hz; (*b*) the frequency at which $|\mathbf{V}_{out}| = 50$ V; (*c*) the frequency at which the phase angle of the voltage is 50° greater than it is at resonance.

14 Refer to the network shown in Fig. 14-35 and: (*a*) sketch $|\mathbf{V}_{AG}|$ vs. ω if a current source, $\mathbf{I}_s = 1\underline{/0°}$ mA, is applied at *B-G*; (*b*) sketch $|\mathbf{V}_{BG}|$ vs. ω if the current source is applied at *A-G*.

Fig. 14-35: See Prob. 14.

15 Use accurate methods on the network of Fig. 14-36 to find: (*a*) ω_0; (*b*) $\mathbf{Y}_{in,0}$; (*c*) ω_1; (*d*) ω_2; (*e*) $\mathcal{B}$; (*f*) $\omega_0/\mathcal{B}$.

Fig. 14-36: See Prob. 15.

16 (*a*) Calculate values for ω_0, Q_0, and $\mathcal{B}$ for the circuit of Fig. 14-37. (*b*) Determine I_m at $\omega = \omega_0 \pm 10n$ rad/s, where $n = 0, 1, 2$, and 3, and plot the results. Use both the approximate and exact formulas.

Fig. 14-37: See Prob. 16.

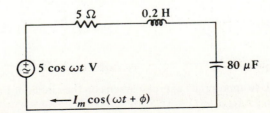

17 A 5-mV sinusoidal signal is applied to a series RLC circuit containing 2 Ω, 20 μH, and a capacitor C. (a) Find the value of C that will cause $|V_C|$, the magnitude of the capacitor voltage, to be a maximum if the source is operating at 50 Mrad/s. What are the values of Q_0 and $|V_C|_{max}$? (b) Find the frequency at which the source should operate to cause $|V_C|$ to be a maximum if $C = 10$ pF. Find Q_0 and $|V_C|_{max}$. (c) Find the value of $|V_C|$ if $\omega = 33$ Mrad/s and $C = 45$ pF.

18 (a) Find ω_0 and Q_0 for the series resonant circuit of Fig. 14-38. (b) Sketch $|V_C|$ vs. ω.

Fig. 14-38: See Prob. 18.

19 If the circuit shown in Fig. 14-39 were constructed in the lab, would you be willing to put your hands across the capacitor? Plot $|V_C|$ vs. ω to help you decide.

Fig. 14-39: See Prob. 19.

20 (a) Calculate element values for a series resonant circuit having $f_0 = 5$ kHz, $\mathscr{B} = 180$ Hz, and a capacitive reactance at resonance of 600 Ω. (b) Apply a source voltage of 1-V amplitude and calculate the amplitude of the inductor voltage at 4.8, 5, and 5.2 kHz using exact methods.

21 The input impedance $\mathbf{Z}_{in}(\mathbf{s})$ of a certain network has a pole at the origin, zeros at $-300 \pm j1000$ s^{-1}, and a pole at infinity. Specify the simplest configuration of the network completely if it is also true that: (a) $\mathbf{Z}_{in}(-200) = -5 + j0$ Ω; (b) $\mathbf{Z}_{in}(\mathbf{s}) \rightarrow$ s/50 as $\mathbf{s} \rightarrow \infty$; (c) Re[$\mathbf{Z}_{in}(-100 + j500)$] = 2 Ω.

22 In the circuit of Fig. 14-40, use good approximations to determine: (a) the resonant frequency; (b) the quality factor at resonance. (c) Sketch $|\mathbf{I}|$ vs. ω, showing the approximate behavior near resonance, and the exact behavior as $\omega \rightarrow 0$ and $\omega \rightarrow \infty$.

Fig. 14-40: See Prob. 22.

23 Use appropriate approximations on the circuit of Fig. 14-41 to find: (a) Q_0; (b) $\mathscr{B}$; (c) $v_2(t)$ if $v_s(t) = 12.5 \cos 13\,000t$ V.

Fig. 14-41: See Prob. 23.

24 For the circuit of Fig. 14-42, make a sketch of: (*a*) $|\mathbf{V}_{out}|$ vs. ω; (*b*) ang $\mathbf{V}_{out}$ vs. ω.

Fig. 14-42: See Prob. 24.

25 (*a*) Show whether the circuit of Fig. 14-43 is similar to a parallel or to a series resonant circuit. (*b*) Find ω_0 and Q_0.

Fig. 14-43: See Prob. 25.

26 A 2-nF capacitor, modeled by the network of Fig. 14-44*a*, a 100-μH inductor, modeled by *b*, and a 10-kΩ resistor, modeled by *c*, are all in parallel with a 1-mA current source. Plot the magnitude of the voltage across the 10-kΩ resistor vs. ω.

Fig. 14-44: See Prob. 26.

(*a*)

(*b*)

(*c*)

Fig. 14-45: See Prob. 27.

(a)

(b)

27 Use approximate methods to find ω_0 and Q_0 for each of the networks shown in Fig. 14-45.

28 The circuit of Fig. 14-46 shows a Butterworth filter operating between a 75-Ω source and a 75-Ω load. The filter essentially passes all frequencies between $f = 0$ and its cutoff frequency, $f_0 = 1$ MHz, while rejecting all frequencies above the cutoff frequency. Scale the filter, the source, and the load so that the circuit operates between a 300-Ω source and load, and has a cutoff frequency of 60 MHz. Draw the scaled circuit.

Fig. 14-46: See Prob. 28.

29 (a) Find $\mathbf{Z}_{in}(\mathbf{s})$ for the network of Fig. 14-47. (b) Write an expression for $\mathbf{Z}_{in}(\mathbf{s})$ after it has been scaled by $K_m = 2$, $K_f = 5$. (c) Scale the network by $K_m = 2$, $K_f = 5$, and draw the new network.

Fig. 14-47: See Prob. 29.

Fig. 14-48: See Prob. 30.

30 (a) Sketch $|\mathbf{I}|$ vs. ω for the circuit shown in Fig. 14-48. Suitable approximations are encouraged. (b) Scale the circuit so that all critical frequencies are increased by a factor of 1000, while the current amplitude is increased by a factor of 20. (c) Sketch $|\mathbf{I}|$ vs. ω for the scaled circuit.

31 A network composed entirely of ideal R's, L's, and C's has a pair of input terminals to which a sinusoidal current source $\mathbf{I}_s$ is connected, and a pair of open-circuited output terminals at which a voltage $\mathbf{V}_{\text{out}}$ is defined. If $\mathbf{I}_s = 1\underline{/0°}$ A at $\omega = 100$ rad/s, then $\mathbf{V}_{\text{out}} = 60\underline{/40°}$ V. Specify $\mathbf{V}_{\text{out}}$ for each condition described below. If it is impossible to determine the value of $\mathbf{V}_{\text{out}}$, write OTSK.[3] (a) $\mathbf{I}_s = 1\underline{/0°}$ A, $\omega = 1000$ rad/s; (b) $\mathbf{I}_s = 1\underline{/60°}$ A, $\omega = 100$ rad/s; (c) $\mathbf{I}_s = 10\underline{/0°}$ A, $\omega = 100$ rad/s; (d) the network is scaled by $K_m = 20$, $\mathbf{I}_s = 10\underline{/0°}$ A, $\omega = 100$ rad/s; (e) $K_m = 20$, $K_f = 10$, $\mathbf{I}_s = 10\underline{/0°}$ A, $\omega = 1000$ rad/s.

32 Scale the network of Fig. 14-49 and draw the result if: (a) the circuit is scaled in magnitude so that the new L is 5 mH; (b) the circuit is scaled in frequency so that the new L is 5 mH; (c) the circuit is scaled in magnitude and frequency so that the new L is 5 mH and the new C is 0.2 μF.

Fig. 14-49: See Prob. 32.

33 (a) Show that the transfer function for the ideal-op-amp circuit of Fig. 14-50 is $\mathbf{H}(\mathbf{s}) = \mathbf{V}_o(\mathbf{s})/\mathbf{V}_g(\mathbf{s}) = -600s/(s^2 + 3s + 900)$. (b) Sketch $|\mathbf{H}(j\omega)|$ vs. ω, $27 \leq \omega \leq 33$ rad/s. (c) Find ω_{max} where $|\mathbf{H}(j\omega_{\text{max}})|_{\text{max}}$ occurs. (d) Determine the half-power bandwidth of the response $|\mathbf{H}(j\omega)|$. (e) Scale the network so that the source has an output resistance of 75 Ω and $|\mathbf{H}(j\omega)|_{\text{max}}$ occurs at 10 krad/s.

Fig. 14-50: See Prob. 33.

[3] Only The Shadow Knows.

34 The ideal-op-amp circuit of Fig. 14-51 contains no inductors, and it is interesting to compare Probs. 33 and 34. (*a*) Show that the transfer function for the ideal-op-amp circuit of Fig. 14-51 is $\mathbf{H}(s) = \mathbf{V}_o(s)/\mathbf{V}_g(s) = -600s/(s^2 + 3s + 900)$. (*b*) Sketch $|\mathbf{H}(j\omega)|$ vs. ω, $27 \le \omega \le 33$ rad/s. (*c*) Find ω_{max} where $|\mathbf{H}(j\omega_{max})|_{max}$ occurs. (*d*) Determine the half-power bandwidth of the response $|\mathbf{H}(j\omega)|$. (*e*) Scale the network so that the source has an output resistance of 75 Ω and $|\mathbf{H}(j\omega)|_{max}$ occurs at 10 krad/s.

Fig. 14-51: See Prob. 34.

35 If $\mathbf{G}(s) = 10^4s/[(s + 1)(s + 10\ 000)]$: (*a*) prepare a Bode amplitude plot and use the asymptotes to determine the values of ω at which $|\mathbf{G}(j\omega)| = |\mathbf{G}(j100)|/10$; (*b*) calculate the exact values of ω at which $|\mathbf{G}(j\omega)| = |\mathbf{G}(j100)|/10$.

36 Construct asymptotic magnitude and phase plots for $\mathbf{H}(s) =$: (*a*) $5000(s + 10)/(s^2 + 100s)$; (*b*) $0.2(s + 100)^2/s$.

37 Draw asymptotic amplitude and phase plots for $\mathbf{H}(s) = (s + 1)(s + 1000)^2/[(s + 10)^2(s + 10\ 000)]$.

38 Let $\mathbf{H}(s) = 40s^2/(s + 5)$, and construct magnitude and phase diagrams, giving numerical coordinates for all corners and intercepts on the magnitude plot if the origin is at $H_{dB} = 0$, $\omega = 1$ rad/s.

39 Draw the Bode diagram for the input admittance of the network shown in Fig. 14-52. Give both magnitude and phase.

Fig. 14-52: See Prob. 39.

40 Let $\mathbf{H}(s) = 1000s/(s^2 + 10s + 400)$ and construct asymptotic plots for both magnitude and phase. Also sketch in better approximations to the true curves.

41 Express $\mathbf{H}(s)$ as the ratio of two polynomials in s if the Bode magnitude plot is that shown in Fig. 14-53.

Fig. 14-53: See Prob. 41.

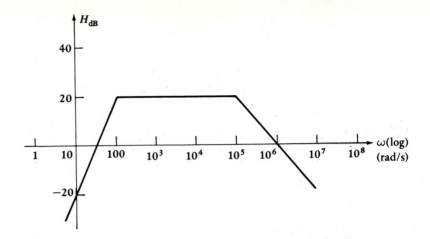

42 Draw the pole-zero plot and construct the asymptotic magnitude plot for $|\mathbf{I}_R/\mathbf{I}_s|$ in the circuit of Fig. 14-54 if $R = :$ (a) 500 Ω; (b) 400 Ω; (c) 200 Ω.

Fig. 14-54: See Prob. 42.

PART FIVE

TWO-PORT NETWORKS

Chapter 15
Magnetically Coupled Circuits

15-1
Introduction

Several hundred pages ago the inductor was introduced as a circuit element and defined in terms of the voltage across it and the time rate of change of the current through it. Strictly speaking, our definition was of "self-inductance," but, loosely speaking, "inductance" is the commonly used term. Now we need to consider mutual inductance, a property which is associated mutually with two or more coils which are physically close together. A circuit element called the "mutual inductor" does not exist; furthermore, mutual inductance is not a property which is associated with a single pair of terminals, but instead is defined with reference to two pairs of terminals.

Mutual inductance results through the presence of a common magnetic flux which links two coils. It may be defined in terms of this common magnetic flux, just as we might have defined self-inductance in terms of the magnetic flux about the single coil. However, we have agreed to confine our attention to circuit concepts, and such quantities as magnetic flux and flux linkages are mentioned only in passing; we cannot define these quantities easily or accurately at this time, and we must accept them as nebulous concepts which are useful in establishing some background only.

The physical device whose operation is based inherently on mutual inductance is the transformer. The 60-Hz power systems use many transformers, ranging in size from the dimensions of a living room to those of the living-room wastebasket. They are used to change the amplitude of the voltage, increasing it for more economical transmission, and then decreasing it for safer operation of home or industrial electrical equipment. Most radios contain one or more transformers, as do television receivers, hi-fi systems, some telephones, automobiles, and electrified railroads (BART and Lionel both).

We first define mutual inductance and study the methods whereby its effects are included in the circuit equations. We conclude with a study of the important characteristics of a linear transformer and an important approximation to a good iron-core transformer which is known as an ideal transformer.

15-2
Mutual inductance

When we defined inductance, we did so by specifying the relationship between the terminal voltage and current,

$$v(t) = L \frac{di(t)}{dt}$$

where the passive sign convention is assumed. We learned, however, that the physical basis for such a current-voltage characteristic rests upon (1) the production of a magnetic flux by a current, the flux being proportional to the current in linear inductors, and (2) the production of a voltage by the time-varying magnetic field, the voltage being proportional to the time rate of change of the magnetic field or the magnetic flux. The proportionality between voltage and time rate of change of current thus becomes evident.

Mutual inductance results from a slight extension of this same argument. A current flowing in one coil establishes a magnetic flux about that coil and also about a second coil which is in its vicinity; the time-varying flux surrounding the second coil produces a voltage across the terminals of this second coil; this voltage is proportional to the time rate of change of the current flowing through the first coil. Figure 15-1a shows a simple model of two coils L_1 and L_2 sufficiently close together that the flux produced by a current $i_1(t)$ flowing through L_1 establishes an open-circuit voltage $v_2(t)$ across the terminals of L_2. Without considering the proper algebraic sign for the relationship, we define the *coefficient of mutual inductance,* or simply *mutual inductance* M_{21},

$$v_2(t) = M_{21} \frac{di_1(t)}{dt} \tag{1}$$

The order of the subscripts on M_{21} indicates that a voltage response is produced at L_2 by a current source at L_1. If the system is reversed, as indicated in Fig. 15-1b, and a voltage response is produced at L_1 by a current source at L_2, then we have

$$v_1(t) = M_{12} \frac{di_2(t)}{dt} \tag{2}$$

Two coefficients of mutual inductance are not necessary, however; we shall use energy relationships a little later to prove that M_{12} and M_{21} are equal. Thus, $M_{12} = M_{21} = M$. The existence of mutual coupling between two coils is indicated by a double-headed arrow, as shown in Figs. 15-1a and b.

Mutual inductance is measured in henrys and, like resistance, inductance, and capacitance, is always positive.[1] The voltage $M \, di/dt$, however, may appear as either a positive or a negative quantity in the same way that $v = -Ri$ is useful.

The inductor is a two-terminal element, and we are able to use the passive

[1] Mutual inductance is not universally assumed to be positive. It is particularly convenient to allow it to "carry its own sign" when three or more coils are involved and each coil interacts with each other coil. We shall restrict our attention to the more important simple case of two coils.

Fig. 15-1: (a) A current i_1 at L_1 produces an open-circuit voltage v_2 at L_2. (b) A current i_2 at L_2 produces an open-circuit voltage v_1 at L_1.

sign convention in order to select the correct sign for the voltage $L\,di/dt$, $j\omega L\mathbf{I}$, or $s L\mathbf{I}$. If the current enters the terminal at which the positive voltage reference is located, then the positive sign is used. Mutual inductance, however, cannot be treated in exactly the same way because four terminals are involved. The choice of a correct sign is established by use of one of several possibilities which include the "dot convention," or an extension of the dot convention which involves the use of a larger variety of special symbols, or by an examination of the particular way in which each coil is wound. We shall use the dot convention and merely look briefly at the physical construction of the coils; the use of other special symbols is not necessary when only two coils are coupled.

The dot convention makes use of a large dot placed at one end of each of the two coils which are mutually coupled. We determine the sign of the mutual voltage as follows:

> A current entering the dotted terminal of one coil produces an open-circuit voltage between the terminals of the second coil which is sensed in the direction indicated by a positive voltage reference at the dotted terminal of this second coil.

Thus, in Fig. 15-2a, i_1 enters the dotted terminal of L_1, v_2 is sensed posi-

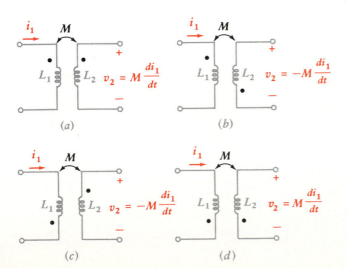

(a)   (b)

(c)   (d)

Fig. 15-2: Current entering the dotted terminal of one coil produces a voltage which is sensed positively at the dotted terminal of the second coil. Current entering the undotted terminal of one coil produces a voltage which is sensed positively at the undotted terminal of the second coil.

tively at the dotted terminal of L_2, and $v_2 = M\, di_1/dt$. We have found previously that it is often not possible to select voltages or currents throughout a circuit so that the passive sign convention is everywhere satisfied; the same situation arises with mutual coupling. For example, it may be more convenient to represent v_2 by a positive voltage reference at the undotted terminal, as shown in Fig. 15-2b: then $v_2 = -M\, di_1/dt$. Currents which enter the dotted terminal are also not always available, as indicated by Figs. 15-2c and d. We note then that:

> A current entering the *undotted* terminal of one coil provides a voltage which is positively sensed at the *undotted* terminal of the second coil.

We have as yet considered only a mutual voltage present across an *open-circuited* coil. In general, a nonzero current will be flowing in each of the two coils, and a mutual voltage will be produced in each coil because of the current flowing in the other coil. *This mutual voltage is present independently of and in addition to any voltage of self-induction.* In other words, the voltage across the terminals of L_1 will be composed of two terms, $L_1\, di_1/dt$ and $M\, di_2/dt$, each carrying a sign depending on the current directions, the assumed voltage sense, and the placement of the two dots. In the portion of a circuit drawn in Fig. 15-3a, currents i_1 and i_2 are shown, each arbitrarily assumed entering the dotted terminal. The voltage across L_1 is thus composed of two parts,

$$v_1 = L_1 \frac{di_1}{dt} + M \frac{di_2}{dt}$$

as is the voltage across L_2,

$$v_2 = L_2 \frac{di_2}{dt} + M \frac{di_1}{dt}$$

In Fig. 15-3b the currents and voltages are not selected with the object of obtaining all positive terms for v_1 and v_2. By inspecting only the reference symbols for i_1 and v_1, it is apparent that the passive sign convention is not satisfied and the sign of $L_1\, di_1/dt$ must therefore be negative. An identical conclusion is reached for the term $L_2\, di_2/dt$. The mutual term of v_2 is signed

Fig. 15-3: (a) Since the pairs v_1, i_1 and v_2, i_2 each satisfy the passive sign convention, the voltages of self-induction are both positive; since i_1 and i_2 each enter dotted terminals, and since v_1 and v_2 are both positively sensed at the dotted terminals, the voltages of mutual induction are also both positive. (b) Since the pairs v_1, i_1 and v_2, i_2 are not sensed according to the passive sign convention, the voltages of self-induction are both negative; since i_1 enters the dotted terminal and v_2 is positively sensed at the dotted terminal, the mutual term of v_2 is positive; and since i_2 enters the undotted terminal and v_1 is positively sensed at the undotted terminal, the mutual term of v_1 is also positive.

(a)   (b)

Fig. 15-4: The physical construction of two mutually coupled coils. From a consideration of the direction of magnetic flux produced by each coil, it is shown that dots may be placed either on the upper terminal of each coil or on the lower terminal of each coil.

by inspecting the direction of i_1 and v_2; since i_1 enters the dotted terminal and v_2 is sensed positive at the dotted terminal, the sign of $M\,di_1/dt$ must be positive. Finally, i_2 enters the undotted terminal of L_2 and v_1 is sensed positive at the undotted terminal of L_1; hence, the mutual portion of v_1, $M\,di_2/dt$, must also be positive. Thus, we have

$$v_1 = -L_1\frac{di_1}{dt} + M\frac{di_2}{dt} \qquad v_2 = -L_2\frac{di_2}{dt} + M\frac{di_1}{dt}$$

The same considerations lead to identical choices of signs for excitation at a complex frequency **s**,

$$\mathbf{V}_1 = -\mathbf{s}L_1\mathbf{I}_1 + \mathbf{s}M\mathbf{I}_2 \qquad \mathbf{V}_2 = -\mathbf{s}L_2\mathbf{I}_2 + \mathbf{s}M\mathbf{I}_1$$

or in the sinusoidal steady state where $\mathbf{s} = j\omega$,

$$\mathbf{V}_1 = -j\omega L_1\mathbf{I}_1 + j\omega M\mathbf{I}_2 \qquad \mathbf{V}_2 = -j\omega L_2\mathbf{I}_2 + j\omega M\mathbf{I}_1$$

Before we apply the dot convention to the analysis of a numerical example, we can gain a more complete understanding of the dot symbolism by looking at the physical basis for the convention. The meaning of the dots is now interpreted in terms of magnetic flux. Two coils are shown wound on a cylindrical form in Fig. 15-4, and the direction of each winding is evident. Let us assume that the current i_1 is positive and increasing with time. The magnetic flux that i_1 produces within the form has a direction which may be found by the right-hand rule: when the right hand is wrapped around the coil with the fingers pointing in the direction of current flow, the thumb indicates the direction of the flux within the coil. Thus i_1 produces a flux which is directed downward; since i_1 is increasing with time, the flux, which is proportional to i_1, is also increasing with time. Turning now to the second coil, let us also think of i_2 as positive and increasing; the application of the right-hand rule shows that i_2 also produces a magnetic flux which is directed downward and is increasing. In other words, the assumed currents i_1 and i_2 produce *additive* fluxes.

The voltage across the terminals of any coil results from the time rate of change of the flux linking that coil. The voltage across the terminals of the first coil is therefore greater with i_2 flowing than it would be if i_2 were zero. Thus i_2 induces a voltage in the first coil which has the same sense as the self-induced voltage in that coil. The sign of the self-induced voltage is known from the passive sign convention, and the sign of the mutual voltage is thus obtained.

The dot convention merely enables us to suppress the physical construction of the coils by placing a dot at one terminal of each coil such that currents entering dot-marked terminals produce additive fluxes. It is apparent that there are always two possible locations for the dots, because both dots may always be moved to the other ends of the coils and additive fluxes will still result.

Fig. 15-5: A circuit containing mutual inductance in which the voltage ratio V_2/V_1 is desired.

$V_1 = 10 \angle 0°$ V

$\omega = 10$ rad/s

Let us now consider the circuit outlined in Fig. 15-5. We desire the ratio of the output voltage across the 400-Ω resistor to the source voltage. Two conventional mesh currents are established, and KVL is applied to each mesh. In the left mesh, the sign of the mutual term is determined by applying the dot convention. Since I_2 enters the undotted terminal of L_2, the mutual voltage across L_1 must have the positive reference at the undotted terminal. Thus,

$$I_1(1 + j10) - j90I_2 = 10$$

The sign of the mutual term in the second mesh is determined in a similar manner. Since I_1 enters the dot-marked terminal, the mutual term in the right mesh has its + reference at the dotted terminal of the 100-H inductor. Thus, we may write

$$I_2(400 + j1000) - j90I_1 = 0$$

The two equations may be solved by determinants (or a simple elimination of I_1),

$$I_2 = \frac{\begin{vmatrix} 1 + j10 & 10 \\ -j90 & 0 \end{vmatrix}}{\begin{vmatrix} 1 + j10 & -j90 \\ -j90 & 400 + j1000 \end{vmatrix}}$$

and

$$I_2 = 0.1724 \angle -16.7°$$

and thus

$$\frac{V_2}{V_1} = \frac{400(0.1724 \angle -16.7°)}{10}$$

or

$$\frac{V_2}{V_1} = 6.90 \angle -16.7°$$

The output voltage is greater in magnitude than the input voltage, and thus a voltage gain is possible with mutual coupling just as it is in a resonant circuit. The voltage gain available in this circuit, however, is present over a relatively wide frequency range; in a moderately high Q resonant circuit, the voltage step-up is proportional to the Q and occurs only over a frequency range which is inversely proportional to Q. Let us see if this point can be made clear by reference to the complex plane.

We find $I_2(s)$ for this particular circuit,

$$I_2(s) = \frac{\begin{vmatrix} 1+s & V_1 \\ -9s & 0 \end{vmatrix}}{\begin{vmatrix} 1+s & -9s \\ -9s & 400+100s \end{vmatrix}} = \frac{9sV_1}{19s^2 + 500s + 400}$$

and thus obtain the ratio of output to input voltage as a function of s,

$$\frac{V_2}{V_1} = \frac{3600s}{19s^2 + 500s + 400} = 189.5\frac{s}{(s+0.826)(s+25.5)}$$

The pole-zero plot of this transfer function is shown in Fig. 15-6. The location of the pole at $s = -25.5$ is distorted in order to show both poles clearly; the ratio of the distances of the two poles from the origin is actually about $30:1$. An inspection of this plot shows that the transfer function is zero at zero frequency but that as soon as ω is greater than, say, 1 or 2 the ratio of the distances from the zero and the pole nearer the origin is essentially unity. Thus the voltage gain is affected only by the distant pole on the negative σ axis, and this distance does not increase appreciably until ω approaches 15 or 20.

These tentative conclusions are verified by the response curve of Fig. 15-7, which shows that the magnitude of the voltage gain is greater than 0.707 times its maximum value from $\omega = 0.78$ to $\omega = 27$.

The circuit is still passive, except for the voltage source, and the voltage

Fig. 15-6: A pole-zero plot of the transfer function V_2/V_1 for the circuit shown in Fig. 15-5. The plot is useful in showing that the magnitude of the transfer function is relatively large from $\omega = 1$ or 2 to $\omega = 15$ or 20 rad/s.

Fig. 15-7: The voltage gain $|V_2/V_1|$ of the circuit shown in Fig. 15-5 is plotted as a function of ω. The voltage gain is greater than 5 from about $\omega = 0.75$ to approximately $\omega = 28$ rad/s.

Fig. 15-8: A three-mesh circuit with mutual coupling may be analyzed most easily by loop or mesh currents.

gain must not be mistakenly interpreted as a power gain. At $\omega = 10$, the voltage gain is 6.90, but the ideal voltage source, possessing a voltage of 10 V, delivers a total power of 8.07 W of which only 5.94 W reaches the 400-Ω resistor. The ratio of the output power to the source power, which we may define as the power gain, is thus 0.736.

Let us consider briefly one additional example, illustrated in Fig. 15-8. The circuit contains three meshes, and three mesh currents are assigned. Applying Kirchhoff's voltage law to the first mesh, a positive sign for the mutual term is assured by selecting $(\mathbf{I}_3 - \mathbf{I}_2)$ as the current through the second coil. Thus,

$$5\mathbf{I}_1 + 7s(\mathbf{I}_1 - \mathbf{I}_2) + 2s(\mathbf{I}_3 - \mathbf{I}_2) = \mathbf{V}_1$$

or
$$(5 + 7s)\mathbf{I}_1 - 9s\mathbf{I}_2 + 2s\mathbf{I}_3 = \mathbf{V}_1 \qquad (3)$$

The second mesh requires two self-inductance terms and two mutual-inductance terms; the equation cannot be written carelessly. We obtain

$$7s(\mathbf{I}_2 - \mathbf{I}_1) + 2s(\mathbf{I}_2 - \mathbf{I}_3) + \frac{1}{s}\mathbf{I}_2 + 6s(\mathbf{I}_2 - \mathbf{I}_3) + 2s(\mathbf{I}_2 - \mathbf{I}_1) = 0$$

or
$$-9s\mathbf{I}_1 + \left(17s + \frac{1}{s}\right)\mathbf{I}_2 - 8s\mathbf{I}_3 = 0 \qquad (4)$$

Finally, for the third mesh

$$6s(\mathbf{I}_3 - \mathbf{I}_2) + 2s(\mathbf{I}_1 - \mathbf{I}_2) + 3\mathbf{I}_3 = 0$$

or
$$2s\mathbf{I}_1 - 8s\mathbf{I}_2 + (3 + 6s)\mathbf{I}_3 = 0 \qquad (5)$$

Equations (3) to (5) may be solved by any of the conventional methods.

Drill Problems

15-1. In the circuit of Fig. 15-9, let $i_1 = -\beta e^{-20t}$ A and $i_2 = 1.8e^{-20t}$ A. Find: (a) v_2; (b) v_1; (c) v_s. *Ans: $-180e^{-20t}$; $12e^{-20t}$; $-138e^{-20t}$ V*

Fig. 15-9: See Drill Probs. 15-1 to 15-3.

15-2. Let $v_s = 100 \cos 40t$ V in the circuit of Fig. 15-9, and find: (a) $i_1(t)$; (b) $i_2(t)$. *Ans:* $1.343 \cos(40t - 38.3°)$; $0.391 \cos(40t + 155.8°)$ A

15-3. A 0.005-F capacitor is inserted in series with the 100-Ω resistor of Fig. 15-9. Determine all the critical frequencies of $\mathbf{H(s)} = \mathbf{I_2(s)/I_1(s)}$. *Ans:* 0 (double); -2.76; -7.24 s^{-1}

**15-3
Energy
considerations**

Let us now consider the energy stored in a pair of mutually coupled inductors. The results will be useful in several different ways. We shall first justify our assumption that $M_{12} = M_{21}$, and we may then determine the maximum possible value of the mutual inductance between two given inductors.

A pair of coupled coils is shown in Fig. 15-3a with currents, voltages, and polarity dots indicated. In order to show that $M_{12} = M_{21}$ we begin by letting all currents and voltages be zero, thus establishing zero initial energy storage in the network. We first open-circuit the right terminal pair and increase i_1 from zero to some constant value I_1 at time $t = t_1$. The power entering the network from the left at any instant is

$$v_1 i_1 = L_1 \frac{di_1}{dt} i_1$$

and that entering from the right is

$$v_2 i_2 = 0$$

since $i_2 = 0$.

The energy stored within the network when $i_1 = I_1$ is thus

$$\int_0^{t_1} v_1 i_1 \, dt = \int_0^{I_1} L_1 i_1 \, di_1 = \tfrac{1}{2} L_1 I_1^2$$

We now hold i_1 constant, $i_1 = I_1$, and let i_2 change from zero at $t = t_1$ to some constant value I_2 at $t = t_2$. The energy delivered from the right source is thus

$$\int_{t_1}^{t_2} v_2 i_2 \, dt = \int_0^{I_2} L_2 i_2 \, di_2 = \tfrac{1}{2} L_2 I_2^2$$

However, even though the value of i_1 remains constant, the left source also delivers energy to the network during this time interval,

$$\int_{t_1}^{t_2} v_1 i_1 \, dt = \int_{t_1}^{t_2} M_{12} \frac{di_2}{dt} i_1 \, dt$$

$$= M_{12} I_1 \int_0^{I_2} di_2$$

$$= M_{12} I_1 I_2$$

The total energy stored in the network when both i_1 and i_2 have reached constant values is

$$W_{total} = \tfrac{1}{2} L_1 I_1^2 + \tfrac{1}{2} L_2 I_2^2 + M_{12} I_1 I_2$$

Now we may establish the same final currents in this network by allowing the currents to reach their final values in the reverse order, that is, first increasing i_2 from zero to I_2 and then holding i_2 constant while i_1 increases from zero to I_1. If the total energy stored is calculated for this experiment, the result is found to be

$$W_{total} = \tfrac{1}{2}L_1I_1^2 + \tfrac{1}{2}L_2I_2^2 + M_{21}I_1I_2$$

The only difference is the interchange of the mutual inductances M_{21} and M_{12}. The initial and final conditions in the network are the same, however, and the two values of the stored energy must be identical. Thus,

$$M_{12} = M_{21} = M$$

and
$$W = \tfrac{1}{2}L_1I_1^2 + \tfrac{1}{2}L_2I_2^2 + MI_1I_2 \tag{6}$$

If one current enters a dot-marked terminal while the other leaves a dot-marked terminal, the sign of the mutual energy term is reversed:

$$W = \tfrac{1}{2}L_1I_1^2 + \tfrac{1}{2}L_2I_2^2 - MI_1I_2 \tag{7}$$

Although (6) and (7) were derived by treating the final values of the two currents as constants, it is apparent that these "constants" may have any value, and the energy expressions correctly represent the energy stored when the *instantaneous* values of i_1 and i_2 are I_1 and I_2, respectively. In other words, lowercase symbols might just as well be used,

$$w(t) = \tfrac{1}{2}L_1[i_1(t)]^2 + \tfrac{1}{2}L_2[i_2(t)]^2 \pm M[i_1(t)][i_2(t)] \tag{8}$$

The only assumption upon which (8) is based is the logical establishment of a zero-energy reference level when both currents are zero.

Equation (8) may now be used to establish an upper limit for the value of M. Since $w(t)$ represents the energy stored within a *passive* network, it cannot be negative for any values of i_1, i_2, L_1, L_2, or M. Let us assume first that i_1 and i_2 are either both positive or both negative; their product is therefore positive. From (8), the only case in which the energy could possibly be negative is

$$w = \tfrac{1}{2}L_1i_1^2 + \tfrac{1}{2}L_2i_2^2 - Mi_1i_2$$

which we may write, by completing the square,

$$w = \tfrac{1}{2}(\sqrt{L_1}\,i_1 - \sqrt{L_2}\,i_2)^2 + \sqrt{L_1L_2}\,i_1i_2 - Mi_1i_2$$

Now the energy cannot be negative; the right side of this equation therefore cannot be negative. The first term, however, may be as small as zero, and thus the sum of the last two terms cannot be negative. Hence,

$$\sqrt{L_1L_2} \geq M$$

or
$$M \leq \sqrt{L_1L_2} \tag{9}$$

There is, therefore, an upper limit to the possible magnitude of the mutual inductance; it can be no larger than the geometric mean of the inductances of the two coils between which the mutual inductance exists. Although we have derived this inequality on the assumption that i_1 and i_2 carried the same algebraic sign, a similar development is possible if the signs are opposite; it is only necessary to select the positive sign in (8).

We might also have demonstrated the truth of (9) from a physical consideration of the magnetic coupling; if we think of i_2 as being zero and the current i_1 as establishing the magnetic flux linking both L_1 and L_2, it is apparent that the flux linking L_2 cannot be greater than the flux linking L_1, which represents the total flux. Qualitatively, then, there is an upper limit to the magnitude of the mutual inductance possible between two given inductors. For example, if $L_1 = 1$ H and $L_2 = 10$ H, then $M \leq 3.16$ H.

The degree to which M approaches its maximum value is exactly described by the coefficient of coupling. We define the *coefficient of coupling*, symbolized as k,

$$k = \frac{M}{\sqrt{L_1 L_2}} \tag{10}$$

It is evident that

$$0 \leq k \leq 1$$

The larger values of the coefficient of coupling are obtained with coils which are physically closer, which are wound or oriented to provide a larger common magnetic flux, or which are provided with a common path through a material which serves to concentrate and localize the magnetic flux (a high-permeability material). Coils having a coefficient of coupling close to unity are said to be tightly coupled.

Drill Problems

15-4. In Fig. 15-3b, let $L_1 = 3.2$ H, $L_2 = 5$ H, $k = 0.5$, $i_1 = 2i_2 = 4 \cos (100t - 60°)$ A, and find values at $t = 0$ for: (a) v_1; (b) v_2; (c) the total energy stored in the system. *Ans:* -762 V; -173.2 V; 4.9 J

15-5. Find the total energy stored in the network of Fig. 15-10 at $t = 0$ if the 1.2-H inductor is: (a) open-circuited; (b) short-circuited; (c) connected across a 7-Ω resistor. *Ans:* 15; 11.25; 12.50 J

Fig. 15-10: See Drill Prob. 15-5.

15-4
The linear
transformer

We are now ready to apply our knowledge of magnetic coupling to an analytical description of the performance of two specific practical devices, each of which may be represented by a model containing mutual inductance. Both of the devices are *transformers,* a term which we may define as a network containing two or more coils which are deliberately coupled magnetically. In

Fig. 15-11: A linear transformer containing a source in the primary circuit and a load in the secondary circuit. Resistance is also included in both the primary and secondary.

Fig. 15-11: A linear transformer containing a source in the primary circuit and a load in the secondary circuit. Resistance is also included in both the primary and secondary.

this section we shall consider the linear transformer, which happens to be an excellent model for the practical linear transformer used at radio frequencies, or higher frequencies. In the following section we shall consider the ideal transformer, which is an idealized unity-coupled model of a physical transformer that has a core made of some magnetic material, usually iron alloy.

In Fig. 15-11 a transformer is shown with two mesh currents identified. The first mesh, usually containing the source, is called the *primary*, while the second mesh, usually containing the load, is known as the *secondary*. The inductors labeled L_1 and L_2 are also referred to as the primary and secondary, respectively, of the transformer. We shall assume that the transformer is linear. This infers that magnetic material causing a nonlinear flux-versus-current relationship is not employed. Without such material, however, it is difficult to achieve a coefficient of coupling greater than a few tenths. The two resistors serve to account for the resistance of the wire out of which the primary and secondary coils are wound, and any other losses.

In many applications, the linear transformer is used with a tuned, or resonant, secondary; the primary winding is also operated often in a resonant condition by replacing the ideal voltage source by a current source in parallel with a large resistance and a capacitance. The analysis of such a single-tuned or double-tuned circuit is a fairly lengthy process, and we shall not undertake it at this time. It may be pointed out, however, that the secondary response is characterized by the familiar resonance curve for relatively small coefficients of coupling but that a greater control of the shape of the response curve as a function of frequency becomes possible for larger coefficients of coupling. Response curves which possess flatter tops and sharper drops on each side may be achieved in the double-tuned circuit.

Consider the input impedance offered at the terminals of the primary circuit. The two mesh equations are

$$\mathbf{V}_s = \mathbf{I}_1(R_1 + sL_1) - \mathbf{I}_2 sM \tag{11}$$

$$0 = -\mathbf{I}_1 sM + \mathbf{I}_2(R_2 + sL_2 + \mathbf{Z}_L) \tag{12}$$

We may simplify by defining

$$\mathbf{Z}_{11} = R_1 + sL_1 \qquad \mathbf{Z}_{22} = R_2 + sL_2 + \mathbf{Z}_L$$

and thus

$$\mathbf{V}_s = \mathbf{I}_1\mathbf{Z}_{11} - \mathbf{I}_2 sM \tag{13}$$

$$0 = -\mathbf{I}_1 sM + \mathbf{I}_2\mathbf{Z}_{22} \tag{14}$$

Solving the second equation for $\mathbf{I}_2$ and inserting this in the first equation enables us to find the input impedance,

$$\mathbf{Z}_{in} = \frac{\mathbf{V}_s}{\mathbf{I}_1} = \mathbf{Z}_{11} - \frac{\mathbf{s}^2 M^2}{\mathbf{Z}_{22}} \tag{15}$$

Before manipulating this expression any further, we can draw several exciting conclusions. In the first place, this result is independent of the location of the dots on either winding, for if either dot is moved to the other end of the coil, the result is a change in sign of each term involving M in (11) to (14). This same effect could be obtained by replacing M by $(-M)$, and such a change cannot affect the input impedance, as (15) demonstrates. We also may note in (15) that the input impedance is simply $\mathbf{Z}_{11}$ if the coupling is reduced to zero. As the coupling is increased from zero, the input impedance differs from $\mathbf{Z}_{11}$ by an amount $-\mathbf{s}^2 M^2/\mathbf{Z}_{22}$, termed the *reflected impedance*. The nature of this change is more evident if examined under steady-state sinusoidal operation. Letting $\mathbf{s} = j\omega$,

$$\mathbf{Z}_{in}(j\omega) = \mathbf{Z}_{11}(j\omega) + \frac{\omega^2 M^2}{R_{22} + jX_{22}}$$

and rationalizing the reflected impedance,

$$\mathbf{Z}_{in} = \mathbf{Z}_{11} + \frac{\omega^2 M^2 R_{22}}{R_{22}^2 + X_{22}^2} + \frac{-j\omega^2 M^2 X_{22}}{R_{22}^2 + X_{22}^2}$$

Since $\omega^2 M^2 R_{22}/(R_{22}^2 + X_{22}^2)$ must be positive, it is evident that the presence of the secondary increases the losses in the primary circuit. In other words, the presence of the secondary might be accounted for in the primary circuit by increasing the value of R_1. Moreover, the reactance which the secondary *reflects* into the primary circuit has a sign which is opposite to that of X_{22}, the net reactance around the secondary loop. This reactance X_{22} is the sum of ωL_2 and X_L; it is necessarily positive for inductive loads and either positive or negative for capacitive loads, depending on the magnitude of the load reactance.

Let us consider the effects of this reflected reactance and resistance by considering the special case in which both the primary and secondary are identical series-resonant circuits. Thus, $R_1 = R_2 = R$, $L_1 = L_2 = L$, and the load impedance $\mathbf{Z}_L$ is produced by a capacitance C, identical to a capacitance inserted in series in the primary circuit. The series-resonant frequency of either the primary or secondary alone is thus $\omega_0 = 1/\sqrt{LC}$. At this resonant frequency, the net secondary reactance is zero, the net primary reactance is zero, and no reactance is reflected into the primary by the secondary. The input impedance is therefore a pure resistance, and a resonant condition is present. At a slightly higher frequency, the net primary and secondary reactances are both inductive, and the reflected reactance is therefore capacitive. If the magnetic coupling is sufficiently large, the input impedance may once again be a pure resistance; another resonant condition is achieved, but at a frequency slightly higher than ω_0. A similar condition

Fig. 15-12: A given transformer which may be replaced by the equivalent networks shown in either Fig. 15-13 or Fig. 15-14.

will also occur at a frequency slightly below ω_0. Each circuit alone is then capacitive, the reflected reactance is inductive, and cancellation may again occur.

Although we are leaping at conclusions, it seems possible that these three adjacent resonances, each similar to a series resonance, may permit a relatively large primary current to flow from the voltage source. The large primary current in turn provides a large induced voltage in the secondary circuit, and a large secondary current. The large secondary current is present over a band of frequencies extending from slightly below to slightly above ω_0, and thus a maximum response is achieved over a wider range of frequencies than is possible in a simple resonant circuit. Such a response curve is obviously desirable if the primary source is some intelligence signal containing energy distributed throughout a band of frequencies, rather than at a single frequency. Such signals are present in AM and FM radio, television, telemetry, radar, and all other communication systems.

It is often convenient to replace a transformer by an equivalent network in the form of a T or π. If we separate the primary and secondary resistances from the transformer, only the pair of mutually coupled inductors remains, as shown in Fig. 15-12. Note that the two lower terminals of the transformer are connected together to form a three-terminal network. We do this because both of our equivalent networks are also three-terminal networks. The differential equations describing this circuit are, once again,

$$v_1 = L_1 \frac{di_1}{dt} + M \frac{di_2}{dt} \tag{16}$$

and

$$v_2 = M \frac{di_1}{dt} + L_2 \frac{di_2}{dt} \tag{17}$$

The form of these two equations is familiar and may be easily interpreted in terms of mesh analysis. Let us select a clockwise i_1 and a counterclockwise i_2 so that i_1 and i_2 are exactly identifiable with the currents in Fig. 15-12. The terms $M\,di_2/dt$ in (16) and $M\,di_1/dt$ in (17) indicate that the two meshes must then have a common *self*-inductance M. Since the total inductance around the left mesh is L_1, a self-inductance of $L_1 - M$ must be inserted in the first mesh, but not in the second mesh. Similarly, a self-inductance of $L_2 - M$ is required in the second mesh, but not in the first mesh. The resultant equivalent network is shown in Fig. 15-13. The equivalence is guaranteed by the identical pairs of equations relating v_1, i_1, v_2, and i_2 for the two networks.

If either of the dots on the windings of the given transformer is placed on the opposite end of its coil, the sign of the mutual terms in (16) and (17) will be negative. This is analogous to replacing M by $-M$, and such a replacement in the network of Fig. 15-13 leads to the correct equivalent for this case. The three self-inductance values are now $L_1 + M$, $-M$, and $L_2 + M$.

The inductances in the T equivalent are all self-inductances; no mutual inductance is present. It is possible that negative values of inductance may be obtained for the equivalent circuit, but this is immaterial if our only desire is a mathematical analysis; the actual construction of the equivalent network is of course impossible in any form involving a negative inductance. How-

Fig. 15-13: The **T** equivalent of the transformer shown in Fig. 15-12.

ever, there are times when procedures for synthesizing networks to provide a desired transfer function lead to circuits containing a T network having a negative inductance; this network may then be realized by use of an appropriate linear transformer.

The equivalent π network is not obtained as easily; it is more complicated, and it is not used as much. We develop it by solving (17) for di_2/dt and substituting the result into (16),

$$v_1 = L_1 \frac{di_1}{dt} + \frac{M}{L_2} v_2 - \frac{M^2}{L_2} \frac{di_1}{dt}$$

or

$$\frac{di_1}{dt} = \frac{L_2}{L_1 L_2 - M^2} v_1 - \frac{M}{L_1 L_2 - M^2} v_2$$

If we now integrate from 0 to t, we obtain

$$i_1 - i_1(0)u(t) = \frac{L_2}{L_1 L_2 - M^2} \int_0^t v_1 \, dt - \frac{M}{L_1 L_2 - M^2} \int_0^t v_2 \, dt \tag{18}$$

In a similar fashion, we also have

$$i_2 - i_2(0)u(t) = \frac{-M}{L_1 L_2 - M^2} \int_0^t v_1 \, dt + \frac{L_1}{L_1 L_2 - M^2} \int_0^t v_2 \, dt \tag{19}$$

Equations (18) and (19) may be interpreted as a pair of nodal equations. A step-current source must be installed at each node in order to provide the proper initial conditions. The factors multiplying each integral are evidently the inverses of certain equivalent inductances. Thus, the second coefficient in (18), $M/(L_1 L_2 - M^2)$, is $1/L_B$, or the reciprocal of the inductance extending between nodes 1 and 2, as shown on the equivalent π network, Fig. 15-14. The first coefficient in (18), $L_2/(L_1 L_2 - M^2)$, is $1/L_A + 1/L_B$. Thus,

$$\frac{1}{L_A} = \frac{L_2}{L_1 L_2 - M^2} - \frac{M}{L_1 L_2 - M^2}$$

Fig. 15-14: The π network which is equivalent to the transformer shown in Fig. 15-12.

or

$$L_A = \frac{L_1 L_2 - M^2}{L_2 - M}$$

Finally,

$$L_C = \frac{L_1 L_2 - M^2}{L_1 - M}$$

No magnetic coupling is present among the inductors in the equivalent circuit, and the initial currents *in the three self-inductances* are zero.

We may compensate for a reversal of either dot in the given transformer by merely changing the sign of M in the equivalent network. Also, just as we found in the equivalent T, negative self-inductances may appear in the equivalent network.

Drill Problems

15-6. Element values in a linear transformer are $R_1 = 2\,\Omega$, $L_1 = 50$ mH, $R_2 = 10\,\Omega$, $L_2 = 150$ mH, and $M = 60$ mH. Determine $\mathbf{Z}_{in}$ at $\omega = 2000$ rad/s if the load $\mathbf{Z}_L$ is: (*a*) another 150-mH inductor; (*b*) 300 Ω; (*c*) 5/3 μF.

Ans: 2.40 + j76.0; 26.0 + j76.8; 1442 + j100 Ω

15-7. Find the three inductances present in the equivalent T network if $L_1 = 4$, $L_2 = 7$, and $M = 5$ H in the transformer of: (*a*) Fig. 15-12; (*b*) Fig. 15-15.

Ans: −1, 5, 2; 9, −5, 12 H

Fig. 15-15: See Drill Prob. 15-7.

15-8. If $L_1 = 20$, $L_2 = 90$, and $M = 40$ mH in a lossless linear transformer, draw equivalent T's for both dot arrangements and then find the six equivalent inductance values seen at the primary terminals with the secondary short-circuited, open-circuited, and also connected in parallel with the primary.

Ans: 2.22, 20, 6.67; 2.22, 20, 1.053 mH

15-5
The ideal transformer

An ideal transformer is a useful approximation of a very tightly coupled transformer in which the coefficient of coupling is almost unity and both the primary and secondary inductive reactances are extremely large in comparison with the terminating impedances. These characteristics are closely approached by most well-designed iron-core transformers over a reasonable range of frequencies for a reasonable range of terminal impedances. The approximate analysis of a circuit containing an iron-core transformer may be achieved very simply by replacing that transformer by an ideal transformer; the ideal transformer may be thought of as a first-order model of an iron-core transformer.

One new concept arises with the ideal transformer, the *turns ratio a*. The self-inductance of either the primary or secondary coil is proportional to the square of the number of turns of wire forming the coil. This relationship is valid only if all the flux established by the current flowing in the coil links all the turns. In order to develop this result logically, it is necessary to utilize magnetic-field concepts, a subject which is not included in our discussion of circuit analysis. However, a qualitative argument may suffice. If a current I flows through a coil of N turns, then N times the magnetic flux of a single-turn coil will be produced. If we think of the N turns as being coincident, then all the flux certainly links all the turns. As the current and flux change with time, a voltage is then induced *in each turn* which is N times larger than that caused by a single-turn coil. Finally, the voltage induced *in the N-turn coil* must be N^2 times the single-turn voltage. Thus, the proportionality between inductance and the square of the numbers of turns arises. It follows that

$$\frac{L_2}{L_1} = \frac{N_2^2}{N_1^2} = a^2$$

where

$$a = \frac{N_2}{N_1}$$

Figure 15-16 shows an ideal transformer to which a secondary load is connected. The ideal nature of the transformer is established by several conventions: the use of the vertical lines between the two coils to indicate the iron laminations present in many iron-core transformers, the unity value of the coupling coefficient, and the presence of the symbol $1 : a$, suggesting a turns ratio of N_1 to N_2.

Let us analyze this transformer in the sinusoidal steady state in order that we may interpret our assumptions in the simplest context. The two mesh equations are

$$\mathbf{V}_1 = \mathbf{I}_1 j\omega L_1 - \mathbf{I}_2 j\omega M \tag{20}$$

$$0 = -\mathbf{I}_1 j\omega M + \mathbf{I}_2(\mathbf{Z}_L + j\omega L_2) \tag{21}$$

We first determine the input impedance of an ideal transformer. Although we shall let the self-inductance of each winding become infinite, the input impe-

Fig. 15-16: An ideal transformer is connected to a general load impedance.

dance will remain finite. By solving (21) for $\mathbf{I}_2$ and substituting into (20), we obtain

$$\mathbf{V}_1 = \mathbf{I}_1 j\omega L_1 + \mathbf{I}_1 \frac{\omega^2 M^2}{\mathbf{Z}_L + j\omega L_2}$$

and

$$\mathbf{Z}_{in} = \frac{\mathbf{V}_1}{\mathbf{I}_1} = j\omega L_1 + \frac{\omega^2 M^2}{\mathbf{Z}_L + j\omega L_2}$$

Since $k = 1$, $M^2 = L_1 L_2$, and

$$\mathbf{Z}_{in} = j\omega L_1 + \frac{\omega^2 L_1 L_2}{\mathbf{Z}_L + j\omega L_2}$$

We now must let both L_1 and L_2 tend to infinity. Their ratio, however, remains finite, as specified by the turns ratio. Thus,

$$L_2 = a^2 L_1$$

and

$$\mathbf{Z}_{in} = j\omega L_1 + \frac{\omega^2 a^2 L_1^2}{\mathbf{Z}_L + j\omega a^2 L_1}$$

Now if we let L_1 become infinite, both of the terms on the right side of the above expression become infinite, and the result is indeterminate. It is necessary to first combine these two terms,

$$\mathbf{Z}_{in} = \frac{j\omega L_1 \mathbf{Z}_L - \omega^2 a^2 L_1^2 + \omega^2 a^2 L_1^2}{\mathbf{Z}_L + j\omega a^2 L_1} \tag{22}$$

or

$$\mathbf{Z}_{in} = \frac{j\omega L_1 \mathbf{Z}_L}{\mathbf{Z}_L + j\omega a^2 L_1} \tag{23}$$

Now as L_1 becomes infinite, it is apparent that $\mathbf{Z}_{in}$ becomes

$$\mathbf{Z}_{in} = \frac{\mathbf{Z}_L}{a^2} \tag{24}$$

for finite $\mathbf{Z}_L$.

This result has some interesting implications, and at least one of them appears to contradict one of the characteristics of the linear transformer. It should not, of course, since the linear transformer represents the more general case. The input impedance of the ideal transformer is an impedance which is proportional to the load impedance, the proportionality constant being the reciprocal of the square of the turns ratio. In other words, if the load impedance is a capacitive impedance, then the input impedance is a capacitive impedance. In the linear transformer, however, the reflected impedance suffered a sign change in its reactive part; a capacitive load led to an inductive contribution to the input impedance. The explanation of this occurrence is achieved by first realizing that $\mathbf{Z}_L/a^2$ is *not* the reflected impedance, although it is often loosely called by that name. The true reflected impedance is infinite in the ideal transformer; otherwise it could not "cancel'' the infinite impedance of the primary inductance. This cancellation occurs in the numerator of (22). The impedance $\mathbf{Z}_L/a^2$ represents a small term which is the amount by which an exact cancellation does not occur. The true reflected impedance in the ideal transformer *does* change sign in its reactive part; as the primary and secondary inductances become infinite, however, the effect of the infinite primary-coil reactance and the infinite, but negative, reflected reactance of the secondary coil is one of cancellation.

The first important characteristic of the ideal transformer is therefore its capability to change the magnitude of an impedance, or to change impedance level. An ideal transformer having 100 primary turns and 10 000 secondary turns has a turns ratio of 10 000/100, or 100. Any impedance placed across the secondary then appears at the primary terminals reduced in magnitude by a factor of 100^2, or 10 000. A 20 000-Ω resistor looks like 2 Ω, a 200-mH inductor looks like 20 μH, and a 100-pF capacitor looks like 1 μF. If the primary and secondary windings are interchanged, then $a = 0.01$ and the load impedance is apparently increased in magnitude. In practice, this exact change in magnitude does not always occur, for we must remember that as we took the last step in our derivation and allowed L_1 to become infinite in (23), it was necessary to neglect $\mathbf{Z}_L$ in comparison with $j\omega L_2$. Since L_2 can never be infinite, it is evident that the ideal transformer model will become invalid if the load impedances are very large.

A practical example of the use of an iron-core transformer as a device for changing impedance level is in the coupling of an audio power amplifier to a speaker system. In order to achieve maximum power transfer, we know that the resistance of the load should be equal to the internal resistance of the source; the speaker usually has an impedance magnitude (often assumed to be a resistance) of only a few ohms, while the power amplifier possesses an internal resistance of several thousand ohms. An ideal transformer in which $N_2 < N_1$ is called for. For example, if the amplifier internal impedance is 4000 Ω and the speaker impedance is 8 Ω, then we desire that

$$\mathbf{Z}_g = 4000 = \frac{\mathbf{Z}_L}{a^2} = \frac{8}{a^2}$$

or

$$a = \frac{1}{22.4}$$

and thus

$$\frac{N_1}{N_2} = 22.4$$

There is also a simple relationship between the primary and secondary currents $\mathbf{I}_1$ and $\mathbf{I}_2$ in an ideal transformer. From (21)

$$\frac{\mathbf{I}_2}{\mathbf{I}_1} = \frac{j\omega M}{\mathbf{Z}_L + j\omega L_2}$$

We allow L_2 to become infinite,

$$\frac{\mathbf{I}_2}{\mathbf{I}_1} = \frac{j\omega M}{j\omega L_2} = \sqrt{\frac{L_1}{L_2}}$$

or

$$\frac{\mathbf{I}_2}{\mathbf{I}_1} = \frac{1}{a} \qquad (25)$$

The ratio of the primary and secondary currents is the turns ratio. If we have $N_2 > N_1$, then $a > 1$, and it is apparent that the larger current flows in the winding with the fewer number of turns. In other words,

$$N_1\mathbf{I}_1 = N_2\mathbf{I}_2$$

It should also be noted that the current ratio is the negative of the turns ratio if either current is reversed or if either dot location is changed. In the example above in which an ideal transformer was used to change the impedance level to match a speaker efficiently to a power amplifier, an rms current of 50 mA at 1000 Hz in the primary causes an rms current of 1.12 A at 1000 Hz in the secondary. The power delivered to the speaker is $1.12^2(8)$, or 10 W, and the power delivered to the transformer by the power amplifier is $(0.05)^2 4000$, or 10 W. The result is comforting, since the ideal transformer contains neither an active device which can generate power nor any resistor which can absorb power.

Since the power delivered to the ideal transformer is identical with that delivered to the load, whereas the primary and secondary currents are related by the turns ratio, it is obvious that the primary and secondary voltages must also be related to the turns ratio. If we define the secondary voltage, or load voltage,

$$\mathbf{V}_2 = \mathbf{I}_2\mathbf{Z}_L$$

and the primary voltage as the voltage across L_1, then

$$\mathbf{V}_1 = \mathbf{I}_1\mathbf{Z}_{in} = \mathbf{I}_1 \frac{\mathbf{Z}_L}{a^2}$$

The ratio of the two voltages is

$$\frac{\mathbf{V}_2}{\mathbf{V}_1} = a^2 \frac{\mathbf{I}_2}{\mathbf{I}_1}$$

or

$$\frac{\mathbf{V}_2}{\mathbf{V}_1} = a = \frac{N_2}{N_1} \qquad (26)$$

The ratio of the secondary to primary voltage is equal to the turns ratio. This ratio may also be negative if either voltage is reversed or either dot location is changed.

Combining the voltage and current ratios, (25) and (26),

$$\mathbf{V}_2\mathbf{I}_2 = \mathbf{V}_1\mathbf{I}_1$$

and we see that the primary and secondary complex voltamperes are equal.

The magnitude of this product is usually specified as a maximum allowable value on power transformers. If the load has a phase angle θ,

$$\mathbf{Z}_L = |\mathbf{Z}_L|\underline{/\theta}$$

then $\mathbf{V}_2$ leads $\mathbf{I}_2$ by an angle θ. Moreover, the input impedance is $\mathbf{Z}_L/a^2$, and thus $\mathbf{V}_1$ also leads $\mathbf{I}_1$ by the same angle θ. If we let the voltage and current represent rms values, then $|\mathbf{V}_2|\,|\mathbf{I}_2|\cos\theta$ must equal $|\mathbf{V}_1|\,|\mathbf{I}_1|\cos\theta$, and all the power delivered to the primary terminals reaches the load; none is absorbed by or delivered to the ideal transformer.

The characteristics of the ideal transformer which we have obtained have all been determined by frequency-domain analysis. They are certainly true in the sinusoidal steady state, but we have no reason to believe that they are correct for the complete response. Actually, they are applicable in general, and the demonstration that this statement is true is much simpler than the frequency-domain analysis we have just completed. Our analysis, however, has served to point out the specific approximations which must be made on a more exact model of an actual transformer in order to obtain an ideal transformer. For example, we have seen that the reactance of the secondary winding must be much greater in magnitude than the impedance of any load which is connected to the secondary. Some feeling for those operating conditions under which a transformer ceases to behave as an ideal transformer is thus achieved.

Let us now determine how the time-domain quantities v_1 and v_2 are related in the ideal transformer. Returning to the circuit shown in Fig. 15-12 and the two equations, (16) and (17), describing it, we may solve the second equation for di_2/dt and substitute into the first equation,

$$v_1 = L_1 \frac{di_1}{dt} + \frac{M}{L_2} v_2 - \frac{M^2}{L_2} \frac{di_1}{dt}$$

However, for unity coupling, $M^2 = L_1 L_2$, and thus

$$v_1 = \frac{M}{L_2} v_2 = \sqrt{\frac{L_1}{L_2}} v_2 = \frac{1}{a} v_2$$

The relationship between primary and secondary voltage is thus found to apply to the complete time-domain response.

An expression relating primary and secondary current in the time domain is most quickly obtained by dividing (16) throughout by L_1,

$$\frac{v_1}{L_1} = \frac{di_1}{dt} + \frac{M}{L_1} \frac{di_2}{dt} = \frac{di_1}{dt} + a \frac{di_2}{dt}$$

and then invoking one of the hypotheses underlying the ideal transformer: L_1 must be infinite. If we assume that v_1 is not infinite, then

$$\frac{di_1}{dt} = -a \frac{di_2}{dt}$$

Integrating,

$$i_1 = -ai_2 + A$$

where A is a constant of integration which does not vary with time. Thus, if we neglect any direct currents in the two windings and fix our attention only on the time-varying portion of the response,

$$i_1 = -ai_2$$

The minus sign arises, of course, from the placement of the dots and selection of the current directions in Fig. 15-12.

The same current and voltage relationships are therefore obtained in the time domain as were obtained previously in the frequency domain, provided that dc components are ignored. The time-domain results are more general, but they have been obtained by a less informative process.

The characteristics of the ideal transformer which we have established may be utilized to simplify circuits in which ideal transformers appear. Let us assume, for purposes of illustration, that everything to the left of the primary terminals has been replaced by its Thévenin equivalent, as has the network to the right of the secondary terminals. We thus consider the circuit shown in Fig. 15-17. Excitation at any complex frequency s is assumed.

Thévenin's or Norton's theorems may now be used to achieve an equivalent circuit which does not contain a transformer. For example, let us determine the Thévenin equivalent of the network to the left of the secondary terminals. Open-circuiting the secondary, $I_2 = 0$ and therefore $I_1 = 0$ (remember L_1 is infinite). No voltage appears across Z_{g1}, and thus $V_1 = V_{s1}$ and $V_{2oc} = aV_{s1}$. The Thévenin impedance is obtained by killing V_{s1} and utilizing the square of the turns ratio, being careful to use the reciprocal turns ratio since we are looking in at the secondary terminals. Thus, $Z_{th2} = Z_{g1}a^2$. As a check on our equivalent, let us also determine the short-circuit secondary current I_{2sc}. With the secondary short-circuited, the primary generator faces an impedance of Z_{g1} and, thus, $I_1 = V_{s1}/Z_{g1}$. Therefore, $I_{2sc} = V_{s1}/aZ_{g1}$. The ratio of the open-circuit voltage to the short-circuit current is a^2Z_{g1}, as it should be. The Thévenin equivalent of the transformer and the primary circuit is shown in the circuit of Fig. 15-18.

Each primary voltage may therefore be multiplied by the turns ratio, each primary current divided by the turns ratio, and each primary impedance multiplied by the square of the turns ratio, these modified voltages, currents, and impedances replacing the given voltages, currents, and impedances plus

Fig. 15-17: The networks connected to the primary and secondary terminals of an ideal transformer are represented by their Thévenin equivalents.

Fig. 15-18: The Thévenin equivalent of the network to the left of the secondary terminals in Fig. 15-17 is used to simplify that circuit.

the transformer. If either dot is interchanged, the equivalent may be obtained by using the negative of the turns ratio.

Note that this equivalence, as illustrated by Fig. 15-18, is possible only if the network connected to the two primary terminals, as well as that connected to the two secondary terminals, can be replaced by their Thévenin equivalents. That is, each must be a two-terminal network. For example, if we cut the two primary leads at the transformer, the circuit must be divided into two separate networks; there can be no element or network bridging across the transformer between primary and secondary, such as those shown in Figs. 15-46, 15-49, or 15-50.

A similar analysis of the transformer and the secondary network shows that everything to the right of the primary terminals may be replaced by an identical network without the transformer, each voltage being divided by a, each current being multiplied by a, and each impedance being divided by a^2. A reversal of either winding corresponds to the use of a turns ratio of $-a$.

As a simple example of this application of equivalent circuits, consider the circuit given in Fig. 15-19. Let $a = 10$. The input impedance is 10 000/100, or 100 Ω. Thus $I_1 = 0.25$ A, $V_1 = 25$ V, and the source delivers 12.5 W, of which 6.25 W is dissipated in the internal resistance of the source and 6.25 W is delivered to the load. This is the condition for maximum power transfer to the load. If the secondary circuit and the ideal transformer are removed by the use of the Thévenin equivalent, the simplified circuit of Fig. 15-20a is

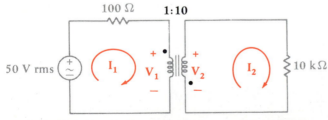

Fig. 15-19: A simple circuit in which a resistive load is matched to the source impedance by means of an ideal transformer.

Fig. 15-20: The circuit of Fig. 15-19 is simplified by replacing (a) the transformer and secondary circuit by its Thévenin equivalent or (b) the transformer and primary circuit by its Thévenin equivalent.

obtained. The primary current and voltage are now immediately evident. If, instead, the network to the left of the secondary terminals is replaced by its Thévenin equivalent, the simpler circuit of Fig. 15-20b is obtained. The presence of the minus sign on the equivalent source should be verified. The corresponding Norton equivalents may also be obtained easily.

Drill Problems

15-9. In the circuit of Fig. 15-21, let $\mathbf{Z}_L = 80 - j60$ Ω and find the average power delivered to $\mathbf{Z}_L$ if: (a) $\mathbf{I}_2 = 2\underline{/60°}$ A rms; (b) $\mathbf{V}_2 = 100\underline{/30°}$ V rms; (c) $\mathbf{I}_1 = 0.5\underline{/20°}$ A rms; (d) $\mathbf{V}_1 = 625\underline{/120°}$ V rms; (e) $\mathbf{V}_s = 1200\underline{/0°}$ V rms.
Ans: 320; 80; 500; 125; 256 W

Fig. 15-21: See Drill Prob. 15-9.

15-10. Find the average power delivered to each resistor in the circuit of Fig. 15-22. *Ans: 11.44; 18.01; 8.29; 37.89 W*

Fig. 15-22: See Drill Prob. 15-10.

Problems

1 The physical construction of three pairs of coupled coils is shown in Fig. 15-23. Show two different possible locations for the two dots on each pair of coils.

Fig. 15-23: See Prob. 1.

(a)

(b)

(c)

2　In the circuit shown in Fig. 15-24: (*a*) find $v_1(t)$ and $v_2(t)$ if $i_1 = 4 \cos 3t$ A and $i_2 = 0$; (*b*) find $v_1(t)$ and $v_2(t)$ if $i_1 = 4 \cos 3t$ A and $i_2 = 3 \cos 3t$ A; (*c*) find $\mathbf{V}_1$ and $\mathbf{V}_2$ in polar form if $\mathbf{I}_1 = 4\underline{/0°}$ A, $\mathbf{I}_2 = 5\underline{/-90°}$ A, and $\omega = 3$ rad/s.

Fig. 15-24: See Probs. 2
and 3.

3　For the circuit shown in Fig. 15-24, let $\mathbf{V}_1 = 40\underline{/0°}$ V, $\mathbf{V}_2 = 60\underline{/90°}$ V, $\mathbf{s} = -2 + j3$ s^{-1}, and find $\mathbf{I}_1$ and $\mathbf{I}_2$ in polar form.

4　(*a*) Refer to Fig. 15-25 and find $\mathbf{H}(\mathbf{s}) = \mathbf{I}_2(\mathbf{s})/\mathbf{V}_s(\mathbf{s})$. (*b*) At what value of ω is $|\mathbf{H}(j\omega)|$ a maximum?

Fig. 15-25: See Prob. 4.

5　Let $\mathbf{s} = j10$ s^{-1} for the arrangement of coupled coils shown in Fig. 15-26 and find $\mathbf{Z}_{in}$ if the missing dot is at: (*a*) *A*; (*b*) *B*.

Fig. 15-26: See Prob. 5.

6　In Fig. 15-27, the source current $i_1(t) = 10\sqrt{2} \cos 2t$ A. (*a*) Find $i_2(t)$. (*b*) Find the phasor-voltage ratio, $\mathbf{V}_2/\mathbf{V}_1$.

Fig. 15-27: See Prob. 6.

Fig. 15-28: See Prob. 7.

1000 $\underline{/30°}$ V

1 kΩ

$j2.25$ kΩ

$j0.75$ kΩ

$j1$ kΩ

$-j0.5$ kΩ

7 For the circuit shown in Fig. 15-28, find: (*a*) the power factor at which the source is operating, specifying whether it is leading or lagging; (*b*) the average power delivered to the 1-kΩ resistor.

8 Using the mesh currents assigned in Fig. 15-29, let $\omega = 10$ rad/s, and: (*a*) write the three mesh equations in standard form; (*b*) find the average power being supplied by $\mathbf{V}_s$ if $\mathbf{V}_s = 100\underline{/0°}$ V.

Fig. 15-29: See Prob. 8.

10 Ω

$\mathbf{I}_3$

$\frac{1}{500}$ F

20 Ω

$\mathbf{V}_s$ $\mathbf{I}_1$ 3 H $\mathbf{I}_2$

3 H

4 H

9 It is possible to arrange three coils physically in such a way that there is mutual coupling between coils *A* and *B* and between *B* and *C*, but not between *A* and *C*. Such an arrangement is shown in Fig. 15-30. Find $v(t)$.

Fig. 15-30: See Prob. 9.

8 H

120 cos ωt
V

4 H

5 H

10 H $v(t)$

+

−

6 H

10 With reference to Fig. 15-31, what value of L_2 will cause exactly 5 W average power to be delivered to the 8-Ω speaker at $\omega = 1$ krad/s?

Fig. 15-31: See Prob. 10.

$k = 0.8$

20 $\underline{/0°}$
V rms

50 mH L_2

8 Ω (speaker)

11 Let $v_s = 100\sqrt{2} \cos 10^3 t$ V in Fig. 15-32. Find the average power dissipated in each resistor if $k = :$ (a) 0.1; (b) 1. (c) Sketch the average power delivered to the 40-Ω resistor as a function of k, $0 \le k \le 1$.

Fig. 15-32: See Prob. 11.

12 Determine the total energy stored in the system of Fig. 15-33 if the missing dot is at: (a) A, and $I_2 = 2$ A; (b) B, and $I_2 = 2$ A; (c) A, and $I_2 = -2$ A; (d) B, and $I_2 = -2$ A.

Fig. 15-33: See Prob. 12.

13 (a) Find $i_x(t)$ and $v(t)$ in Fig. 15-34 if $i_1 = 5 \cos 40t$ A and $i_2 = 2 \sin 40t$ A. (b) Is the direction of average positive power flow to the left or to the right in the center of the circuit?

Fig. 15-34: See Prob. 13.

14 If $i_1 = 2 \cos 10t$ A in Fig. 15-35, find the total energy stored in the network at $t = 0$ if: (a) a-b is open-circuited; (b) a-b is short-circuited.

Fig. 15-35: See Prob. 14.

15 After having been open for a long time, the switch appearing in the circuit of Fig. 15-36 is closed at $t = 0$. (a) Find the energy stored in the system of inductors at $t = 0^-$. (b) At what time has half this energy been dissipated in the resistor?

Fig. 15-36: See Prob. 15.

16 Find two frequencies other than 0 or 10 rad/s at which $\mathbf{Z}_{in} = R_{in} + j0$ for the network of Fig. 15-37.

Fig. 15-37: See Prob. 16.

17 Calculate the inductance presented at the specified terminals in Fig. 15-38 if these terminals are connected together: (a) L_{AD}, B to C; (b) L_{AC}, B to D; (c) L_{AB}, C to D; (d) L_{CD}, A to B; (e) L_{AB}, A to C, and B to D; (f) L_{AB}, no other connections; (g) L_{CD}, no other connections. (h) Find another connection that will provide a different value of L.

Fig. 15-38: See Prob. 17.

18 Find the equivalent inductance seen at terminals 1–2 in Fig. 15-39 if the following terminals are connected together: (a) none; (b) A to B; (c) B to C; (d) A to C.

Fig. 15-39: See Prob. 18.

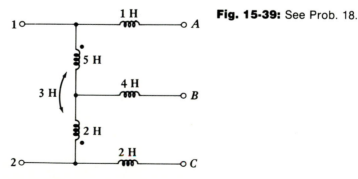

19 For the circuit of Fig. 15-40: (a) find the transfer function, $\mathbf{H(s)} = \mathbf{I}_L(\mathbf{s})/\mathbf{V}_s(\mathbf{s})$; ($b$) specify the form of the natural response, $i_{Ln}(t)$; (c) let $v_s = 100u(t)$ V and write two differential equations in terms of the variables i_s and i_L; (d) specify values for $i_L(0^+)$ and $i'_L(0^+)$; (e) find $i_L(t)$.

Fig. 15-40: See Prob. 19.

20 In Fig. 15-11, let $R_1 = 2\ \Omega$, $R_2 = 0$, and $\mathbf{Z}_L = 40 + j0\ \Omega$. If $\omega = 1$ krad/s, determine $\mathbf{V}_L/\mathbf{V}_s$ if: (a) $L_1 = 1$ mH, $L_2 = 25$ mH, and $k = 1$; (b) $L_1 = 1$ H, $L_2 = 25$ H, and $k = 0.99$; (c) $L_1 = 1$ H, $L_2 = 25$ H, and $k = 1$.

21 In Fig. 15-11, let $R_1 = R_2 = 0$, $L_1 = 1$ H, $L_2 = 4$ H, $k = 1$, and $\omega = 1000$ rad/s. Find the equivalent one- or two-element series network (R, L, and C values) seen at the input terminals if $\mathbf{Z}_L$ is: (a) $100\ \Omega$; (b) 0.1 H; (c) $10\ \mu$F.

22 At $\omega = 10$ krad/s, find $\mathbf{Z}_{in}$ for the network shown in Fig. 15-41.

Fig. 15-41: See Prob. 22.

23 A transformer has a primary inductance L_1, a secondary inductance L_2, and unity coupling. A 10-Ω resistive load is connected across L_2. Find the input impedance viewed at the terminals of L_1 at $\omega = 50$ rad/s if: (a) $L_1 = 10$ H and $L_2 = 40$ H; (b) L_1 and L_2 are very large, but $L_2/L_1 = 4$.

24 (a) What value should be used for the turns ratio N_2/N_1 of the ideal transformer in the circuit of Fig. 15-42 so that maximum average power is delivered to the 8-Ω resistor? (b) Find the value of $P_{8,max}$. (c) What value of N_2/N_1 will result in maximum dissipation in the 300-Ω resistor?

Fig. 15-42: See Prob. 24.

25 Find the average power absorbed by each resistor in the circuit of Fig. 15-43.

Fig. 15-43: See Prob. 25.

Fig. 15-44: See Prob. 26.

26 For the circuit of Fig. 15-44, find the average power absorbed by each resistor and generated by each source.

27 (*a*) Find the average power delivered to each 10-Ω resistor in Fig. 15-45. (*b*) Repeat after connecting *A* to *C* and *B* to *D*.

Fig. 15-45: See Prob. 27.

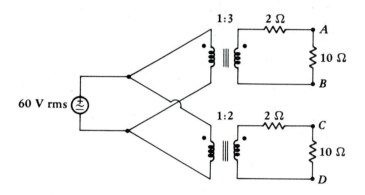

28 Find the voltages V_8 and V_{20} in the circuit shown in Fig. 15-46.

Fig. 15-46: See Prob. 28.

29 (*a*) Find the Thévenin equivalent of the network in Fig. 15-47. (*b*) Repeat if one dot is moved to the other end of its winding.

Fig. 15-47: See Prob. 29.

Fig. 15-48: See Prob. 30.

30 Select values for the turns ratios a and b in Fig. 15-48 so that the ideal source supplies 1 kW, half of which is delivered to the 100-Ω load.

31 Find the power being generated by the ideal source in the circuit of Fig. 15-49.

Fig. 15-49: See Prob. 31.

32 In the circuit of Fig. 15-50, find the power being dissipated in each resistor and being generated by the source.

Fig. 15-50: See Prob. 32.

Chapter 16
Two-Port Networks

16-1
Introduction

A general network having two pairs of terminals, one perhaps labeled the "input terminals" and the other the "output terminals," is a very important building block in electronic systems, communication systems, automatic control systems, transmission and distribution systems, or other systems in which an electrical signal or electrical energy enters the input terminals, is acted upon by the network, and leaves via the output terminals. The output terminal pair may very well connect with the input terminal pair of another network. A pair of terminals at which a signal may enter or leave a network is called a *port,* and a network having only one such pair of terminals is called a *one-port network,* or simply a *one port.* No connections may be made to any other nodes internal to the one port, and it is therefore evident that i_a must equal i_b in the one port shown in Fig. 16-1a. When more than one pair of terminals is present, the network is known as a *multiport network.* The two-port network to which this chapter is principally devoted is shown in Fig. 16-1b. The currents in the two leads comprising each port must be equal, and it follows that $i_a = i_b$ and $i_c = i_d$ in the two port shown in Fig. 16-1b. Sources and loads must be connected directly across the two terminals of a port if the methods of this chapter are to be used. In other words, each port can only be connected to a one-port network or to a port of another multiport network. For example, no device may be connected between terminals a and c of the two-port network in Fig. 16-1b. If such a circuit must be analyzed, general loop or nodal equations should usually be written.

Fig. 16-1: (a) A one-port network. (b) A two-port network.

(a)   (b)

The special methods of analysis which are developed for two-port networks, or simply two ports, emphasize the current and voltage relationships at the terminals of the networks and suppress the specific nature of the currents and voltages within the networks. Our introductory study should serve to acquaint us with a number of important parameters and their use in simplifying and systematizing linear two-port network analysis.

16-2
One-port networks

Some of the introductory study of one- and two-port networks is accomplished best by using a generalized network notation and the abbreviated nomenclature for determinants introduced in Appendix 1. Thus, if we write a set of loop equations for a passive network,

$$\mathbf{Z}_{11}\mathbf{I}_1 + \mathbf{Z}_{12}\mathbf{I}_2 + \mathbf{Z}_{13}\mathbf{I}_3 + \cdots + \mathbf{Z}_{1N}\mathbf{I}_N = \mathbf{V}_1$$
$$\mathbf{Z}_{21}\mathbf{I}_1 + \mathbf{Z}_{22}\mathbf{I}_2 + \mathbf{Z}_{23}\mathbf{I}_3 + \cdots + \mathbf{Z}_{2N}\mathbf{I}_N = \mathbf{V}_2$$
$$\mathbf{Z}_{31}\mathbf{I}_1 + \mathbf{Z}_{32}\mathbf{I}_2 + \mathbf{Z}_{33}\mathbf{I}_3 + \cdots + \mathbf{Z}_{3N}\mathbf{I}_N = \mathbf{V}_3 \tag{1}$$
$$\cdots\cdots\cdots\cdots\cdots\cdots\cdots\cdots\cdots\cdots\cdots$$
$$\mathbf{Z}_{N1}\mathbf{I}_1 + \mathbf{Z}_{N2}\mathbf{I}_2 + \mathbf{Z}_{N3}\mathbf{I}_3 + \cdots + \mathbf{Z}_{NN}\mathbf{I}_N = \mathbf{V}_N$$

then the coefficient of each current will be an impedance $\mathbf{Z}_{ij}(\mathbf{s})$, and the circuit determinant, or determinant of the coefficients, is

$$\Delta_{\mathbf{Z}} = \begin{vmatrix} \mathbf{Z}_{11} & \mathbf{Z}_{12} & \mathbf{Z}_{13} & \cdots & \mathbf{Z}_{1N} \\ \mathbf{Z}_{21} & \mathbf{Z}_{22} & \mathbf{Z}_{23} & \cdots & \mathbf{Z}_{2N} \\ \mathbf{Z}_{31} & \mathbf{Z}_{32} & \mathbf{Z}_{33} & \cdots & \mathbf{Z}_{3N} \\ \cdots & \cdots & \cdots & \cdots & \cdots \\ \mathbf{Z}_{N1} & \mathbf{Z}_{N2} & \mathbf{Z}_{N3} & \cdots & \mathbf{Z}_{NN} \end{vmatrix} \tag{2}$$

where N loops have been assumed, the currents appear in subscript order in each equation, and the order of the equations is the same as that of the currents. We also assume that KVL is applied so that the sign of each $\mathbf{Z}_{ii}$ term ($\mathbf{Z}_{11}$, $\mathbf{Z}_{22}$, . . . , $\mathbf{Z}_{NN}$) is positive; the sign of any $\mathbf{Z}_{ij}(i \neq j)$ or mutual term may be either positive or negative, depending on the reference directions assigned to $\mathbf{I}_i$ and $\mathbf{I}_j$.

If there are dependent sources within the network, then all the coefficients in the loop equations may not be resistances or impedances, as we saw in Chapter 3. Even so, we shall continue to refer to the circuit determinant as $\Delta_{\mathbf{Z}}$.

The use of minor notation (Appendix 1) enables the input or driving-point impedance at the terminals of a *one-port* network to be expressed very concisely. The result is also applicable to a *two-port* network if one of the two ports is terminated in a passive impedance, including an open or short circuit.

Fig. 16-2: (*a*) An ideal voltage source $\mathbf{V}_1$ is connected to the single port of a linear one-port network containing no independent sources; $\mathbf{Z}_{in} = \Delta_{\mathbf{z}}/\Delta_{11}$. (*b*) A resistive one port used as an example. (*c*) A one port containing a dependent source used as an example.

Let us suppose that the one-port network shown in Fig. 16-2*a* is composed entirely of passive elements and dependent sources; linearity is also assumed. An ideal voltage source $\mathbf{V}_1$ is connected to the port, and the source current is identified as the current in loop 1. By the familiar procedure, then,

$$\mathbf{I}_1 = \frac{\begin{vmatrix} \mathbf{V}_1 & \mathbf{Z}_{12} & \mathbf{Z}_{13} & \cdots & \mathbf{Z}_{1N} \\ 0 & \mathbf{Z}_{22} & \mathbf{Z}_{23} & \cdots & \mathbf{Z}_{2N} \\ 0 & \mathbf{Z}_{32} & \mathbf{Z}_{33} & \cdots & \mathbf{Z}_{3N} \\ \cdots\cdots\cdots\cdots\cdots\cdots \\ 0 & \mathbf{Z}_{N2} & \mathbf{Z}_{N3} & \cdots & \mathbf{Z}_{NN} \end{vmatrix}}{\begin{vmatrix} \mathbf{Z}_{11} & \mathbf{Z}_{12} & \mathbf{Z}_{13} & \cdots & \mathbf{Z}_{1N} \\ \mathbf{Z}_{21} & \mathbf{Z}_{22} & \mathbf{Z}_{23} & \cdots & \mathbf{Z}_{2N} \\ \mathbf{Z}_{31} & \mathbf{Z}_{32} & \mathbf{Z}_{33} & \cdots & \mathbf{Z}_{3N} \\ \cdots\cdots\cdots\cdots\cdots\cdots \\ \mathbf{Z}_{N1} & \mathbf{Z}_{N2} & \mathbf{Z}_{N3} & \cdots & \mathbf{Z}_{NN} \end{vmatrix}}$$

or, more concisely,

$$\mathbf{I}_1 = \frac{\mathbf{V}_1 \Delta_{11}}{\Delta_{\mathbf{z}}}$$

Thus,

$$\mathbf{Z}_{in} = \frac{\mathbf{V}_1}{\mathbf{I}_1} = \frac{\Delta_{\mathbf{z}}}{\Delta_{11}} \qquad\qquad (3)$$

For the one-port resistive network shown in Fig. 16-2b,

$$\Delta_{\mathbf{z}} = \begin{vmatrix} 10 & -10 & 0 & 0 \\ -10 & 17 & -2 & -5 \\ 0 & -2 & 7 & -1 \\ 0 & -5 & -1 & 26 \end{vmatrix} = 9680 \quad \Omega^4$$

and

$$\Delta_{11} = \begin{vmatrix} 17 & -2 & -5 \\ -2 & 7 & -1 \\ -5 & -1 & 26 \end{vmatrix} = 2778 \quad \Omega^3$$

Thus,

$$\mathbf{Z}_{in} = 9680/2778 = 3.48 \quad \Omega$$

Upon changing the circuit by including a dependent source, Fig. 16-2c, the four mesh equations are written

$$10\mathbf{I}_1 - 10\mathbf{I}_2 = \mathbf{V}_1$$

$$-10\mathbf{I}_1 + 17\mathbf{I}_2 - 2\mathbf{I}_3 - 5\mathbf{I}_4 = 0$$

$$-2\mathbf{I}_2 + 7\mathbf{I}_3 - \mathbf{I}_4 = 0$$

$$\mathbf{I}_4 = -0.5\mathbf{I}_a = -0.5(\mathbf{I}_4 - \mathbf{I}_3) \qquad \text{or} \qquad -0.5\mathbf{I}_3 + 1.5\mathbf{I}_4 = 0$$

and

$$\Delta_{\mathbf{z}} = \begin{vmatrix} 10 & -10 & 0 & 0 \\ -10 & 17 & -2 & -5 \\ 0 & -2 & 7 & -1 \\ 0 & 0 & -0.5 & 1.5 \end{vmatrix} = 590 \quad \Omega^3$$

while

$$\Delta_{11} = \begin{vmatrix} 17 & -2 & -5 \\ -2 & 7 & -1 \\ 0 & -0.5 & 1.5 \end{vmatrix} = 159 \quad \Omega^2$$

giving

$$\mathbf{Z}_{in} = 590/159 = 3.71 \quad \Omega$$

We may also select a similar procedure using nodal equations, yielding the input admittance:

$$\mathbf{Y}_{in} = \frac{1}{\mathbf{Z}_{in}} = \frac{\Delta_\mathbf{Y}}{\Delta_{11}} \tag{4}$$

where Δ_{11} now refers to the minor of $\Delta_\mathbf{Y}$. For the example of Fig. 16-2*b*, we order the node voltages $\mathbf{V}_1$, $\mathbf{V}_2$, and $\mathbf{V}_3$ from left to right, select the reference at the bottom node, and have

$$\Delta_\mathbf{Y} = \begin{vmatrix} 0.35 & -0.2 & -0.05 \\ -0.2 & 1.7 & -1 \\ -0.05 & -1 & 1.3 \end{vmatrix} = 0.347 \ \mho^3$$

$$\Delta_{11} = \begin{vmatrix} 1.7 & -1 \\ -1 & 1.3 \end{vmatrix} = 1.21 \ \mho^2$$

so that

$$\mathbf{Y}_{in} = 0.347/1.21 = 0.287 \ \mho$$

which corresponds to

$$\mathbf{Z}_{in} = 1/0.287 = 3.48 \ \Omega$$

once again.

Problems 4 and 5 at the end of the chapter give one ports that can be built using operational amplifiers so that negative resistances may be obtained from networks whose only passive circuit elements are resistors, and inductors may be simulated with only resistors and capacitors.

Drill Problems

16-1. Find the input impedance of the network shown in Fig. 16-3 if it is formed into a one-port network by breaking it at terminals: (*a*) *a-a'*; (*b*) *b-b'*; (*c*) *c-c'*.   *Ans:* 4.97; 8.21; 7.39 Ω

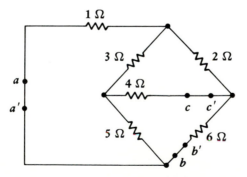

Fig. 16-3: See Drill Prob. 16-1.

16-2. Write a set of nodal equations, calculate $\Delta_\mathbf{Y}$, and then find the input admittance of the network shown in Fig. 16-4 as seen between nodes: (*a*) 1 and the reference; (*b*) 2 and the reference.   *Ans:* 0.131 25; 0.700 $\mho$

Fig. 16-4: See Drill Prob. 16-2.

16-3 Admittance parameters

Let us now turn our attention to two-port networks. We shall assume in all that follows that the network is composed of linear elements and contains no independent sources; dependent sources are permissible. Further conditions will also be placed on the network in some special cases.

We shall consider the two port as it is shown in Fig. 16-5; the voltage and current at the input terminals are V_1 and I_1, and V_2 and I_2 are specified at the output port. The directions of I_1 and I_2 are both customarily selected as *into* the network at the upper conductors (and out at the lower conductors). Since the network is linear and contains no independent sources within it, I_1 may be considered to be the superposition of two components, one caused by V_1 and the other by V_2. When the same argument is applied to I_2, we may begin with the set of equations

$$I_1 = y_{11}V_1 + y_{12}V_2 \tag{5}$$

$$I_2 = y_{21}V_1 + y_{22}V_2 \tag{6}$$

where the **y**'s are no more than proportionality constants, or unknown coefficients, for the present. They are called the **y** parameters and they are defined by (5) and (6).

The **y** parameters, as well as other sets of parameters we shall define later in the chapter, may be represented concisely as matrices.[1] Here, we define the (2 by 1) column matrix [**I**],

$$[\mathbf{I}] = \begin{bmatrix} \mathbf{I}_1 \\ \mathbf{I}_2 \end{bmatrix} \tag{7}$$

the (2 by 2) square matrix of the **y** parameters,

$$[\mathbf{Y}] = \begin{bmatrix} \mathbf{y}_{11} & \mathbf{y}_{12} \\ \mathbf{y}_{21} & \mathbf{y}_{22} \end{bmatrix} \tag{8}$$

and the (2 by 1) column matrix [**V**],

$$[\mathbf{V}] = \begin{bmatrix} \mathbf{V}_1 \\ \mathbf{V}_2 \end{bmatrix} \tag{9}$$

Fig. 16-5: A general two port with terminal voltages and currents specified. The two port is composed of linear elements, including dependent sources, but not containing any independent sources.

---

[1] Our use of matrices is quite elementary. A brief discussion of the necessary techniques appears in Appendix 2.

Thus, we may write the matrix equation, $[\mathbf{I}] = [\mathbf{y}][\mathbf{V}]$, or

$$\begin{bmatrix} \mathbf{I}_1 \\ \mathbf{I}_2 \end{bmatrix} = \begin{bmatrix} \mathbf{y}_{11} & \mathbf{y}_{12} \\ \mathbf{y}_{21} & \mathbf{y}_{22} \end{bmatrix} \begin{bmatrix} \mathbf{V}_1 \\ \mathbf{V}_2 \end{bmatrix}$$

and matrix multiplication of the right side gives us the equality

$$\begin{bmatrix} \mathbf{I}_1 \\ \mathbf{I}_2 \end{bmatrix} = \begin{bmatrix} \mathbf{y}_{11}\mathbf{V}_1 + \mathbf{y}_{12}\mathbf{V}_2 \\ \mathbf{y}_{21}\mathbf{V}_1 + \mathbf{y}_{22}\mathbf{V}_2 \end{bmatrix}$$

These (2 by 1) matrices must be equal, element by element, and thus we are led to the defining equations, (5) and (6).

The most useful and informative way to attach a physical meaning to the **y** parameters is through a direct inspection of (5) and (6). Consider (5), for example; if we let $\mathbf{V}_2$ be zero, then we see that $\mathbf{y}_{11}$ must be given by the ratio of $\mathbf{I}_1$ to $\mathbf{V}_1$. We therefore describe $\mathbf{y}_{11}$ as the admittance measured at the input terminals with the output terminals *short-circuited* ($\mathbf{V}_2 = 0$). Since there can be no question as to which terminals are short-circuited, $\mathbf{y}_{11}$ is best described as the *short-circuit input admittance*. Alternatively, we might describe $\mathbf{y}_{11}$ as the reciprocal of the input impedance measured with the output terminals short-circuited, but a description as an admittance is obviously more direct. It is not the *name* of the parameter that is important; rather it is the conditions which must be applied to (5) or (6), and hence to the network, that are most meaningful; when the conditions are determined, the parameter can be found directly from an analysis of the circuit (or by experiment on the physical circuit). Each of the **y** parameters may be described as a current-voltage ratio with either $\mathbf{V}_1 = 0$ (the input terminals short-circuited) or $\mathbf{V}_2 = 0$ (the output terminals short-circuited):

$$\mathbf{y}_{11} = \frac{\mathbf{I}_1}{\mathbf{V}_1}\bigg|_{\mathbf{V}_2=0} \tag{10}$$

$$\mathbf{y}_{12} = \frac{\mathbf{I}_1}{\mathbf{V}_2}\bigg|_{\mathbf{V}_1=0} \tag{11}$$

$$\mathbf{y}_{21} = \frac{\mathbf{I}_2}{\mathbf{V}_1}\bigg|_{\mathbf{V}_2=0} \tag{12}$$

$$\mathbf{y}_{22} = \frac{\mathbf{I}_2}{\mathbf{V}_2}\bigg|_{\mathbf{V}_1=0} \tag{13}$$

Because each parameter is an admittance which is obtained by short-circuiting either the output or input port, the **y** parameters are known as the *short-circuit admittance parameters*. The specific name of $\mathbf{y}_{11}$ is the *short-circuit input admittance*, $\mathbf{y}_{22}$ is the *short-circuit output admittance*, and $\mathbf{y}_{12}$ and $\mathbf{y}_{21}$ are the *short-circuit transfer admittances*.

Consider the resistive two port shown in Fig. 16-6*a*. The values of the parameters may be easily established by applying (10) to (13), which we obtained directly from the defining equations (5) and (6). To determine $\mathbf{y}_{11}$,

Fig. 16-6: (a) A resistive two port. (b) The resistive two port is terminated with specific one ports.

we short-circuit the output and find the ratio of I_1 to V_1. This may be done by letting $V_1 = 1$ V, for then $y_{11} = I_1$. By inspection of Fig. 16-6a, it is apparent that 1 V applied at the input with the output short-circuited will cause an input current of $(\frac{1}{5} + \frac{1}{10})$, or 0.3 A. Hence,

$$y_{11} = 0.3 \quad \mho$$

In order to find y_{12}, we short-circuit the input terminals and apply 1 V at the output terminals. The input current flows through the short circuit and is $-\frac{1}{10}$ A. Thus

$$y_{12} = -0.1 \quad \mho$$

By similar methods,

$$y_{21} = -0.1 \quad \mho \qquad y_{22} = 0.15 \quad \mho$$

The describing equations for this two port in terms of the admittance parameters are, therefore,

$$I_1 = 0.3V_1 - 0.1V_2 \tag{14}$$

$$I_2 = -0.1V_1 + 0.15V_2 \tag{15}$$

and
$$[\mathbf{y}] = \begin{bmatrix} 0.3 & -0.1 \\ -0.1 & 0.15 \end{bmatrix} \quad \text{(all } \mho)$$

It is not necessary to find these parameters one at a time by using (10) to (13), however. Let us assign node voltages V_1 and V_2 and write the expressions for I_1 and I_2 in terms of them. We have

$$I_1 = \frac{V_1}{5} + \frac{V_1 - V_2}{10} = 0.3V_1 - 0.1V_2$$

and

$$I_2 = \frac{V_2 - V_1}{10} + \frac{V_2}{20} = -0.1V_1 + 0.15V_2$$

These equations are identical with (14) and (15), and the four **y** parameters may be read from them directly. In general, it is easier to use (10), (11), (12), or (13) when only one parameter is desired. If we need all of them, however, it is usually easier to assign $\mathbf{V}_1$ and $\mathbf{V}_2$ to the input and output nodes, to assign other node-to-reference voltages at any interior nodes, and then to carry through with the general solution.

In order to see what use might be made of such a system of equations, let us now terminate each port with some specific one-port network. The simple example of Fig. 16-6b shows a practical current source connected to the input port and a resistive load to the output port. A relationship must now exist between $\mathbf{V}_1$ and $\mathbf{I}_1$ that is independent of the two-port network. This relationship may be determined solely from this external circuit. If we apply KCL (or write a single nodal equation) at the input,

$$\mathbf{I}_1 = 15 - 0.1\mathbf{V}_1$$

For the output, Ohm's law yields

$$\mathbf{I}_2 = -0.25\mathbf{V}_2$$

Substituting these expressions for $\mathbf{I}_1$ and $\mathbf{I}_2$ into (14) and (15), we have

$$15 = 0.4\mathbf{V}_1 - 0.1\mathbf{V}_2$$

$$0 = -0.1\mathbf{V}_1 + 0.4\mathbf{V}_2$$

from which are obtained

$$\mathbf{V}_1 = 40 \quad \text{V} \qquad \mathbf{V}_2 = 10 \quad \text{V}$$

The input and output currents are also easily found,

$$\mathbf{I}_1 = 11 \quad \text{A} \qquad \mathbf{I}_2 = -2.5 \quad \text{A}$$

and the complete terminal characteristics of this resistive two port are then known.

The advantages of two-port analysis do not show up very strongly for such a simple example, but it should be apparent that once the **y** parameters are determined for a more complicated two port the performance of the two port for different terminal conditions is easily determined; it is only necessary to relate $\mathbf{V}_1$ to $\mathbf{I}_1$ at the input and $\mathbf{V}_2$ to $\mathbf{I}_2$ at the output.

In the example just concluded, $\mathbf{y}_{12}$ and $\mathbf{y}_{21}$ were both found to be -0.1 ℧. It is not difficult to show that this equality is also obtained if three general impedances $\mathbf{Z}_A$, $\mathbf{Z}_B$, and $\mathbf{Z}_C$ are contained in this π network. It is somewhat more difficult to determine the specific conditions which are necessary in order that $\mathbf{y}_{12} = \mathbf{y}_{21}$, but the use of determinant notation is of some help. Let us see if the relationships (10) to (13) can be expressed in terms of the impedance determinant and its minors.

Since our concern is with the two port and not with the specific networks with which it is terminated, we shall let $\mathbf{V}_1$ and $\mathbf{V}_2$ be represented by two ideal voltage sources. Equation (10) is applied by letting $\mathbf{V}_2 = 0$ (thus short-circuiting the output) and finding the input admittance. The network now, however, is simply a one port, and the input impedance of a one port was found in the previous section. We select loop 1 to include the input terminals, and let $\mathbf{I}_1$ be that loop current; we identify $(-\mathbf{I}_2)$ as the loop current in loop 2 and assign the remaining loop currents in any convenient manner. Thus,

$$\mathbf{Z}_{in}\big|_{\mathbf{V}_2=0} = \frac{\Delta_{\mathbf{z}}}{\Delta_{11}}$$

and, therefore,

$$\mathbf{y}_{11} = \frac{\Delta_{11}}{\Delta_{\mathbf{z}}} \tag{16}$$

Similarly,

$$\mathbf{y}_{22} = \frac{\Delta_{22}}{\Delta_{\mathbf{z}}} \tag{17}$$

In order to find $\mathbf{y}_{12}$, we let $\mathbf{V}_1 = 0$, and find $\mathbf{I}_1$ as a function of $\mathbf{V}_2$. We find that $\mathbf{I}_1$ is given by the ratio

$$\mathbf{I}_1 = \frac{\begin{vmatrix} 0 & \mathbf{Z}_{12} & \cdots & \mathbf{Z}_{1N} \\ -\mathbf{V}_2 & \mathbf{Z}_{22} & \cdots & \mathbf{Z}_{2N} \\ 0 & \mathbf{Z}_{32} & \cdots & \mathbf{Z}_{3N} \\ \cdots\cdots\cdots\cdots\cdots\cdots \\ 0 & \mathbf{Z}_{N2} & \cdots & \mathbf{Z}_{NN} \end{vmatrix}}{\begin{vmatrix} \mathbf{Z}_{11} & \mathbf{Z}_{12} & \cdots & \mathbf{Z}_{1N} \\ \mathbf{Z}_{21} & \mathbf{Z}_{22} & \cdots & \mathbf{Z}_{2N} \\ \cdots\cdots\cdots\cdots\cdots\cdots \\ \mathbf{Z}_{N1} & \mathbf{Z}_{N2} & \cdots & \mathbf{Z}_{NN} \end{vmatrix}}$$

Thus,

$$\mathbf{I}_1 = -\frac{(-\mathbf{V}_2)\Delta_{21}}{\Delta_{\mathbf{z}}}$$

and

$$\mathbf{y}_{12} = \frac{\Delta_{21}}{\Delta_{\mathbf{z}}} \tag{18}$$

In a similar manner, we may show that

$$\mathbf{y}_{21} = \frac{\Delta_{12}}{\Delta_{\mathbf{z}}} \tag{19}$$

The equality of $\mathbf{y}_{12}$ and $\mathbf{y}_{21}$ is thus contingent on the equality of the two minors of $\Delta_{\mathbf{Z}}$, Δ_{12} and Δ_{21}. These two minors are

$$\Delta_{21} = \begin{vmatrix} \mathbf{Z}_{12} & \mathbf{Z}_{13} & \mathbf{Z}_{14} & \cdots & \mathbf{Z}_{1N} \\ \mathbf{Z}_{32} & \mathbf{Z}_{33} & \mathbf{Z}_{34} & \cdots & \mathbf{Z}_{3N} \\ \mathbf{Z}_{42} & \mathbf{Z}_{43} & \mathbf{Z}_{44} & \cdots & \mathbf{Z}_{4N} \\ \cdots & \cdots & \cdots & \cdots & \cdots \\ \mathbf{Z}_{N2} & \mathbf{Z}_{N3} & \mathbf{Z}_{N4} & \cdots & \mathbf{Z}_{NN} \end{vmatrix} \qquad \Delta_{12} = \begin{vmatrix} \mathbf{Z}_{21} & \mathbf{Z}_{23} & \mathbf{Z}_{24} & \cdots & \mathbf{Z}_{2N} \\ \mathbf{Z}_{31} & \mathbf{Z}_{33} & \mathbf{Z}_{34} & \cdots & \mathbf{Z}_{3N} \\ \mathbf{Z}_{41} & \mathbf{Z}_{43} & \mathbf{Z}_{44} & \cdots & \mathbf{Z}_{4N} \\ \cdots & \cdots & \cdots & \cdots & \cdots \\ \mathbf{Z}_{N1} & \mathbf{Z}_{N3} & \mathbf{Z}_{N4} & \cdots & \mathbf{Z}_{NN} \end{vmatrix}$$

Their equality is shown by first interchanging the rows and columns of one minor, say Δ_{21}, an operation which any college algebra book proves is valid, and then letting every mutual impedance $\mathbf{Z}_{ij}$ be replaced by $\mathbf{Z}_{ji}$. Thus, we set

$$\mathbf{Z}_{12} = \mathbf{Z}_{21} \qquad \mathbf{Z}_{23} = \mathbf{Z}_{32} \qquad \text{etc.}$$

This equality of $\mathbf{Z}_{ij}$ and $\mathbf{Z}_{ji}$ is certainly obvious for the three passive elements, the resistor, capacitor, and inductor, and it is also true for mutual inductance, as we proved in the preceding chapter. However, it is not true for every type of device which we may wish to include inside a two-port network. Specifically, it is not true in general for a dependent source, and it is not true for the gyrator, a useful model for Hall-effect devices and for waveguide sections containing ferrites. Over a narrow range of radian frequencies, the gyrator provides an additional phase shift of 180° for a signal passing from the output to the input over that for a signal in the forward direction, and thus $\mathbf{y}_{12} = -\mathbf{y}_{21}$. The most common type of passive element, however, which leads to the inequality of $\mathbf{Z}_{ij}$ and $\mathbf{Z}_{ji}$ is the nonlinear element.

Any device for which $\mathbf{Z}_{ij} = \mathbf{Z}_{ji}$ is called a *bilateral element,* and a circuit which contains only bilateral elements is called a *bilateral circuit.* We have therefore shown that an important property of a bilateral two port is

$$\mathbf{y}_{12} = \mathbf{y}_{21}$$

and this property is glorified by stating it as the *reciprocity theorem:*

> In any passive linear bilateral network, if the single voltage source $\mathbf{V}_x$ in branch x produces the current response $\mathbf{I}_y$ in branch y, then the removal of the voltage source from branch x and its insertion in branch y will produce the current response $\mathbf{I}_y$ in branch x.

A simple way of stating the theorem is to say that the interchange of an ideal voltage source and an ideal ammeter in any passive linear bilateral circuit will not change the ammeter reading.

If we had been working with the admittance determinant of the circuit and had proved that the minors Δ_{21} and Δ_{12} of the admittance determinant $\Delta_{\mathbf{Y}}$ were equal, then we should have obtained the reciprocity theorem in its dual form:

> In any passive linear bilateral network, if the single current source $\mathbf{I}_x$ between nodes x and x' produces the voltage response $\mathbf{V}_y$ between nodes y and y', then the removal of the current source from nodes x and x' and its insertion between nodes y and y' will produce the voltage response $\mathbf{V}_y$ between nodes x and x'.

In other words, the interchange of an ideal current source and an ideal voltmeter in any passive linear bilateral circuit will not change the voltmeter reading.

Two ports containing dependent sources receive emphasis in the following section.

Drill Problems

16-3. By applying the appropriate 1-V sources and short circuits in the resistive circuit of Fig. 16-7, determine: (*a*) $\mathbf{y}_{11}$; (*b*) $\mathbf{y}_{21}$; (*c*) $\mathbf{y}_{22}$; (*d*) $\mathbf{y}_{12}$.

Ans: 27.3; -18.18; 50.5; -18.18 m℧

16-4. In the circuit of Fig. 16-7, let $\mathbf{I}_1$ and $\mathbf{I}_2$ represent ideal current sources. Assign the nodal voltage $\mathbf{V}_1$ at the input, $\mathbf{V}_2$ at the output, and $\mathbf{V}_x$ from the central node to the common reference node. Write three nodal equations, eliminate $\mathbf{V}_x$ to obtain two equations, and then rearrange these equations into the form of (5) and (6) so that all four admittance parameters may be read directly from the equations.

$$Ans: \begin{bmatrix} 27.3 & -18.18 \\ -18.18 & 50.5 \end{bmatrix} \quad \text{(all m℧)}$$

Fig. 16-7: See Drill Probs. 16-3 and 16-4.

16-5. Find [**y**] for the two port shown in Fig. 16-8.

$$Ans: \begin{bmatrix} 0.2 & -0.24 \\ -0.12 & 0.194 \end{bmatrix} \quad \text{(all ℧)}$$

Fig. 16-8: See Drill Prob. 16-5.

16-4
Some equivalent networks

In analyzing electronic circuits, it is usually necessary to replace the active device (and perhaps some of its associated passive circuitry) with an equivalent two port containing only three or four impedances. The validity of the equivalent may be restricted to small signal amplitudes and a single frequency, or perhaps a limited range of frequencies. The equivalent is also a linear approximation of a nonlinear circuit. However, if we are faced with a network containing a number of resistors, capacitors, inductors, plus a transistor labeled 2N3823, then we cannot analyze the circuit by any of the techniques we have studied previously; the transistor must first be replaced by a linear model, just as we replaced the op-amp by a linear model. The **y** parameters provide one such model in the form of a two-port network that is often used at high frequencies. Another common linear model for a transistor appears in Sec. 16-6.

The two basic equations which determine the short-circuit admittance parameters,

$$\mathbf{I}_1 = \mathbf{y}_{11}\mathbf{V}_1 + \mathbf{y}_{12}\mathbf{V}_2 \tag{20}$$

$$\mathbf{I}_2 = \mathbf{y}_{21}\mathbf{V}_1 + \mathbf{y}_{22}\mathbf{V}_2 \tag{21}$$

have the form of a pair of nodal equations written for a circuit containing two nonreference nodes. The determination of an equivalent circuit that leads to (20) and (21) is made more difficult by the inequality, in general, of $\mathbf{y}_{12}$ and $\mathbf{y}_{21}$; it helps to resort to a little trickery in order to obtain a pair of equations that possesses equal mutual coefficients. Let us add and subtract $\mathbf{y}_{12}\mathbf{V}_1$ (the term we would like to see present in the second equation above) on the right side of (21),

$$\mathbf{I}_2 = \mathbf{y}_{12}\mathbf{V}_1 + \mathbf{y}_{22}\mathbf{V}_2 + (\mathbf{y}_{21} - \mathbf{y}_{12})\mathbf{V}_1 \tag{22}$$

or $\quad\quad \mathbf{I}_2 - (\mathbf{y}_{21} - \mathbf{y}_{12})\mathbf{V}_1 = \mathbf{y}_{12}\mathbf{V}_1 + \mathbf{y}_{22}\mathbf{V}_2 \tag{23}$

The right sides of (20) and (23) now show the proper symmetry for a bilateral circuit; the left side of (23) may be interpreted as the algebraic sum of two current sources, one an independent source $\mathbf{I}_2$ entering node 2, and the other a dependent source $(\mathbf{y}_{21} - \mathbf{y}_{12})\mathbf{V}_1$ leaving node 2.

Let us now "read" the equivalent network from (20) and (23). We first provide a reference node, and then a node labeled $\mathbf{V}_1$ and one labeled $\mathbf{V}_2$. From (20), we establish the current $\mathbf{I}_1$ flowing into node 1, we supply a mutual admittance $(-\mathbf{y}_{12})$ between nodes 1 and 2, and we supply an admittance of $(\mathbf{y}_{11} + \mathbf{y}_{12})$ between node 1 and the reference node. With $\mathbf{V}_2 = 0$, the ratio of $\mathbf{I}_1$ to $\mathbf{V}_1$ is then $\mathbf{y}_{11}$, as it should be. Now consider (23); we cause the current $\mathbf{I}_2$ to flow into the second node, we cause the current $(\mathbf{y}_{21} - \mathbf{y}_{12})\mathbf{V}_1$ to leave the node, we note that the proper admittance $(-\mathbf{y}_{12})$ exists between the nodes, and we complete the circuit by installing the admittance $(\mathbf{y}_{22} + \mathbf{y}_{12})$ from node 2 to the reference node. The completed circuit is shown in Fig. 16-9a.

Another form of equivalent network is obtained by subtracting and adding $\mathbf{y}_{21}\mathbf{V}_2$ in (20); this equivalent circuit is shown in Fig. 16-9b.

If the two port is bilateral, then $\mathbf{y}_{12} = \mathbf{y}_{21}$, and either of the equivalents

Fig. 16-9: (a) and (b) Two ports which are equivalent to any general linear two port. The dependent source in (a) depends on V_1, and that in (b) depends on V_2. (c) An equivalent for a bilateral network.

reduces to a simple passive π network. The dependent source disappears. This equivalent of the bilateral two port is shown in Fig. 16-9c.

There are several uses to which these equivalent circuits may be put. In the first place, we have succeeded in showing that an equivalent of any complicated linear two port *exists*. It does not matter how many nodes or loops are contained within the network; the equivalent is no more complex than the circuits of Fig. 16-9. One of these may be much simpler to use than the given circuit if we are interested only in the terminal characteristics of the given network.

The three-terminal network shown in Fig. 16-10a is often referred to as a Δ

Fig. 16-10: The three-terminal Δ network of (a) and the three-terminal Y network in (b) are equivalent if the six impedances satisfy the conditions of the Y-Δ (or π-T) transformation, Eqs. (24) to (29).

$$Z_A = \frac{Z_1 Z_2 + Z_2 Z_3 + Z_3 Z_1}{Z_2} \quad (24)$$

$$Z_B = \frac{Z_1 Z_2 + Z_2 Z_3 + Z_3 Z_1}{Z_3} \quad (25)$$

$$Z_C = \frac{Z_1 Z_2 + Z_2 Z_3 + Z_3 Z_1}{Z_1} \quad (26)$$

$$Z_1 = \frac{Z_A Z_B}{Z_A + Z_B + Z_C} \quad (27)$$

$$Z_2 = \frac{Z_B Z_C}{Z_A + Z_B + Z_C} \quad (28)$$

$$Z_3 = \frac{Z_C Z_A}{Z_A + Z_B + Z_C} \quad (29)$$

of impedances, while that in Fig. 16-10b is called a Y. One network may be replaced by the other if certain specific relationships between the impedances are satisfied, and these interrelationshps may be established by use of the **y** parameters. We find that

$$\mathbf{y}_{11} = \frac{1}{\mathbf{Z}_A} + \frac{1}{\mathbf{Z}_B} = \frac{1}{\mathbf{Z}_1 + \mathbf{Z}_2\mathbf{Z}_3/(\mathbf{Z}_2 + \mathbf{Z}_3)}$$

$$\mathbf{y}_{12} = \mathbf{y}_{21} = -\frac{1}{\mathbf{Z}_B} = \frac{-\mathbf{Z}_3}{\mathbf{Z}_1\mathbf{Z}_2 + \mathbf{Z}_2\mathbf{Z}_3 + \mathbf{Z}_3\mathbf{Z}_1}$$

$$\mathbf{y}_{22} = \frac{1}{\mathbf{Z}_C} + \frac{1}{\mathbf{Z}_B} = \frac{1}{\mathbf{Z}_2 + \mathbf{Z}_1\mathbf{Z}_3/(\mathbf{Z}_1 + \mathbf{Z}_3)}$$

These equations may be solved for $\mathbf{Z}_A$, $\mathbf{Z}_B$, and $\mathbf{Z}_C$ in terms of $\mathbf{Z}_1$, $\mathbf{Z}_2$, and $\mathbf{Z}_3$,

$$\mathbf{Z}_A = \frac{\mathbf{Z}_1\mathbf{Z}_2 + \mathbf{Z}_2\mathbf{Z}_3 + \mathbf{Z}_3\mathbf{Z}_1}{\mathbf{Z}_2} \tag{24}$$

$$\mathbf{Z}_B = \frac{\mathbf{Z}_1\mathbf{Z}_2 + \mathbf{Z}_2\mathbf{Z}_3 + \mathbf{Z}_3\mathbf{Z}_1}{\mathbf{Z}_3} \tag{25}$$

$$\mathbf{Z}_C = \frac{\mathbf{Z}_1\mathbf{Z}_2 + \mathbf{Z}_2\mathbf{Z}_3 + \mathbf{Z}_3\mathbf{Z}_1}{\mathbf{Z}_1} \tag{26}$$

or for the inverse relationships,

$$\mathbf{Z}_1 = \frac{\mathbf{Z}_A\mathbf{Z}_B}{\mathbf{Z}_A + \mathbf{Z}_B + \mathbf{Z}_C} \tag{27}$$

$$\mathbf{Z}_2 = \frac{\mathbf{Z}_B\mathbf{Z}_C}{\mathbf{Z}_A + \mathbf{Z}_B + \mathbf{Z}_C} \tag{28}$$

$$\mathbf{Z}_3 = \frac{\mathbf{Z}_C\mathbf{Z}_A}{\mathbf{Z}_A + \mathbf{Z}_B + \mathbf{Z}_C} \tag{29}$$

These equations enable us to transform easily between the equivalent Y and Δ networks, a process known as the Y-Δ transformation (or π-T transformation if the networks are drawn in the forms of those letters). In going from Y to Δ, (24) to (26), first find the value of the common numerator as the sum of the products of the impedances in the Y taken two at a time. Each impedance in the Δ is then found by dividing the numerator by the impedance of that element in the Y which has no common node with the desired Δ element. Conversely, given the Δ, first sum the three impedances around the Δ; then divide the product of the two Δ impedances having a common node with the desired Y element by that sum.

These transformations are often useful in simplifying passive networks, particularly resistive ones, thus avoiding the need for any mesh or nodal analysis. Thus, the input resistance of the circuit in Fig. 16-11a is found to be 159/71 Ω after making the sequence of transformations indicated.

As an example of the use of an equivalent circuit with a dependent generator, let us consider the two port shown in Fig. 16-12. This circuit may be

Fig. 16-11: (a) A given resistive network whose input resistance is desired. (b) The upper Δ is replaced by an equivalent Y. (c) and (d) Series and parallel combinations give the equivalent input resistance 159/71 Ω.

considered as an approximate linear equivalent of a transistor circuit in which the emitter terminal is the bottom node, the base terminal is the upper input node, and the collector terminal is the upper output node; a 2000-Ω resistor is connected between collector and base for some special application and makes the analysis of the circuit more difficult.

There are several ways we might think about this circuit. If we recognize it as being in the form of the equivalent circuit shown in Fig. 16-9a, then we may immediately determine the values of the **y** parameters. If recognition is not immediate, then the **y** parameters may be determined for the two port by applying the relationships (10) to (13). We also might avoid any use of two-port analysis methods and write equations directly for the circuit as it stands. Let us compare the network with the equivalent circuit. Thus, we first obtain

$$\mathbf{y}_{12} = -\frac{1}{2000} = -0.5 \quad \text{m}\mho$$

and hence,

$$\mathbf{y}_{11} = \frac{1}{500} - \left(-\frac{1}{2000}\right) = 2.5 \quad \text{m}\mho$$

Then

$$\mathbf{y}_{22} = \frac{1}{10\,000} - \left(-\frac{1}{2000}\right) = 0.6 \quad \text{m}\mho$$

and

$$\mathbf{y}_{21} = 0.0395 + \left(-\frac{1}{2000}\right) = 39 \quad \text{m}\mho$$

The following equations must then apply:

$$\mathbf{I}_1 = 2.5\mathbf{V}_1 - 0.5\mathbf{V}_2 \quad \text{mA} \tag{30}$$

$$\mathbf{I}_2 = 39\mathbf{V}_1 + 0.6\mathbf{V}_2 \quad \text{mA} \tag{31}$$

Fig. 16-12: The linear equivalent circuit of a transistor in common-emitter configuration with resistive feedback between collector and base. It is used as a two-port example.

where we are now using units of mA, m℧, kΩ, and V. We make use of these two equations by analyzing the performance of this two port when a current source of $1\underline{/0°}$ mA is provided at the input and a 0.5-kΩ (2-m℧) load is connected to the output. The terminating networks thus give us the following specific information relating $\mathbf{I}_1$ to $\mathbf{V}_1$ and $\mathbf{I}_2$ to $\mathbf{V}_2$:

$$\mathbf{I}_1 = 1 \qquad \mathbf{I}_2 = -2\mathbf{V}_2$$

Substituting into (30) and (31), we may obtain two equations relating $\mathbf{V}_1$ and $\mathbf{V}_2$:

$$1 = 2.5\mathbf{V}_1 - 0.5\mathbf{V}_2 \qquad 0 = 39\mathbf{V}_1 + 2.6\mathbf{V}_2$$

Thus,

$$\mathbf{V}_1 = 0.1 \quad \text{V} \qquad \mathbf{V}_2 = -1.5 \quad \text{V}$$

$$\mathbf{I}_1 = 1 \quad \text{mA} \qquad \mathbf{I}_2 = 3 \quad \text{mA}$$

From these data, it is easy to calculate the voltage gain

$$\mathbf{G}_V = \frac{\mathbf{V}_2}{\mathbf{V}_1} = -15$$

the current gain

$$\mathbf{G}_I = \frac{\mathbf{I}_2}{\mathbf{I}_1} = 3$$

and, if sinusoidal excitation is assumed, the power gain

$$G_P = \frac{P_{out}}{P_{in}} = \frac{\text{Re}[-\frac{1}{2}\mathbf{V}_2\mathbf{I}_2^*]}{\text{Re}[\frac{1}{2}\mathbf{V}_1\mathbf{I}_1^*]} = 45$$

The device might be termed either a voltage, a current, or a power amplifier, since all the gains are greater than unity. If the 2-kΩ resistor were removed, the power gain would rise to 354.

The input and output impedances of the amplifier are often desired in order that maximum power transfer may be achieved to or from an adjacent two port. The input impedance is the ratio of input voltage to current:

$$\mathbf{Z}_{in} = \frac{\mathbf{V}_1}{\mathbf{I}_1} = 0.1 \quad \text{kΩ}$$

This is the impedance offered to the current source when the 500-Ω load is connected to the output. (With the output short-circuited, the input impedance is necessarily $1/\mathbf{y}_{11}$, or 400 Ω.) It should be noted that the input impedance *cannot* be determined by replacing every source by its internal impedance and then combining resistances or conductances. In the circuit given, this procedure would yield a value of 416 Ω. The error, of course, comes from treating the dependent source as an independent source. If we think of the input impedance as being numerically equal to the input voltage produced by an input current of 1 A, the application of the 1-A source produces some input voltage $\mathbf{V}_1$, and the strength of the dependent source

$(0.0395\mathbf{V}_1)$ cannot be zero. We should recall that, when we obtain the Thévenin equivalent impedance of a circuit containing a dependent source along with one or more independent sources, we must replace the independent sources by short circuits or open circuits, but a dependent source must not be killed. Of course, if the voltage or current on which the dependent source depends is zero, the dependent source will itself be inactive; occasionally a circuit may be simplified by recognizing such an occurrence.

The output impedance is just another term for the Thévenin impedance appearing in the Thévenin equivalent circuit of that portion of the network faced by the load. In our circuit, which we have assumed driven by a $1\underline{/0°}$-mA current source, we therefore replace this source by an open circuit, leave the dependent source alone, and seek the *input* impedance seen looking to the left from the output terminals. This impedance may be found by several methods (none of which consists of combining all the elements in series or in parallel); let us apply $1\underline{/0°}$ mA at the output terminals and determine $\mathbf{V}_2$. We place these requirements on (30) and (31), and obtain

$$0 = 2.5\mathbf{V}_1 - 0.5\mathbf{V}_2 \qquad 1 = 39\mathbf{V}_1 + 0.6\mathbf{V}_2$$

Solving,

$$\mathbf{V}_2 = 0.119 \quad \text{V}$$

and thus

$$\mathbf{Z}_{out} = 0.119 \quad \text{k}\Omega$$

An alternative procedure might be to find the open-circuit output voltage and the short-circuit output current; the ratio is the Thévenin impedance or $\mathbf{Z}_{out}$. With an input current $\mathbf{I}_1 = 1$ mA and with the load short-circuited ($\mathbf{V}_2 = 0$), we find that

$$\mathbf{I}_1 = 1 = 2.5\mathbf{V}_1 - 0 \qquad \mathbf{I}_2 = 39\mathbf{V}_1 + 0$$

and thus

$$\mathbf{I}_{2sc} = 15.6 \quad \text{mA}$$

Again, with $\mathbf{I}_1 = 1$ mA and with the output open-circuited ($\mathbf{I}_2 = 0$), we have

$$1 = 2.5\mathbf{V}_1 - 0.5\mathbf{V}_2 \qquad 0 = 39\mathbf{V}_1 + 0.6\mathbf{V}_2$$

and

$$\mathbf{V}_{2oc} = -1.857 \quad \text{V}$$

The assumed directions of $\mathbf{V}_2$ and $\mathbf{I}_2$ therefore result in a Thévenin or output impedance

$$\mathbf{Z}_{out} = -\frac{\mathbf{V}_{2oc}}{\mathbf{I}_{2sc}} = -\frac{-1.857}{15.6} = 0.119 \quad \text{k}\Omega$$

as before.

We now have enough information to enable us to draw the Thévenin or Norton equivalent of the two port of Fig. 16-12 (driven by a $1\underline{/0°}$-mA current source and terminated in a 500-Ω load, although this termination is of no consequence in the output equivalent circuits) as offered to the 500-Ω load.

Fig. 16-13: (*a*) The Norton equivalent of the network (Fig. 16-12) to the left of the output terminals, with $I_1 = 1\underline{/0°}$ mA. (*b*) The Thévenin equivalent of that portion of the network to the right of the input terminals, if $I_2 = -2V_2$ mA.

(*a*)   (*b*)

The Thévenin and Norton equivalents offered to the current source are also readily drawn. Thus, the Norton equivalent presented to the load must contain a current source equal to the short-circuit current I_{2sc} in parallel with the output impedance; this equivalent is shown in Fig. 16-13*a*. The Thévenin equivalent offered to the $1\underline{/0°}$-mA input source must consist solely of the input impedance, as drawn in Fig. 16-13*b*.

Before leaving the **y** parameters, we should recognize their usefulness in describing the parallel connection of two ports, as indicated in Fig. 16-14. When we first defined a port in Sec. 16-1, we noted that the currents entering and leaving the two terminals of a port had to be equal, and there could be no external connections made that bridged between ports. Apparently the parallel connection shown in Fig. 16-14 violates this condition. However, if each two port has a reference node that is common to its input and output port, and if the two ports are connected in parallel so that they have a common reference node, then all ports remain ports after the connection. Thus, for the A network,

$$[I_A] = [y_A][V_A]$$

where

$$[I_A] = \begin{bmatrix} I_{A1} \\ I_{A2} \end{bmatrix} \quad \text{and} \quad [V_A] = \begin{bmatrix} V_{A1} \\ V_{A2} \end{bmatrix}$$

and for the B network,

$$[I_B] = [y_B][V_B]$$

But, $[V_A] = [V_B] = [V]$, and $[I] = [I_A] + [I_B]$. Thus

$$[I] = ([y_A] + [y_B])[V]$$

Fig. 16-14: The parallel connection of two two-port networks. If both inputs and outputs have the same reference node, then $[y] = [y_A] + [y_B]$.

and we see that each **y** parameter of the parallel network is given as the sum of the corresponding parameters of the individual networks,

$$[\mathbf{y}] = [\mathbf{y}_A] + [\mathbf{y}_B] \tag{32}$$

This may obviously be extended to any number of two ports connected in parallel.

Drill Problems

16-6. An amplifier is represented as a two-port network with $y_{11} = 0.5$, $y_{12} = -0.3$, $y_{21} = 25$, and $y_{22} = 0.2$, all in m℧. A source having an internal resistance of 500 Ω is connected to the input, and an admittance of 0.5 m℧ is at the output. Calculate: (a) $\mathbf{G}_V$; (b) $\mathbf{G}_I$; (c) G_P. *Ans:* -35.7; 1.592; 56.9

16-7. Determine the input and output impedance of the amplifier described in Drill Prob. 16-6. The source and load are in place. *Ans:* 89.2; 312.5 Ω

16-8. Find the four **y** parameters and also the output impedance of the terminated two port shown in Fig. 16-15.

$$Ans: \begin{bmatrix} 0.1875 & -0.006\ 25 \\ 5 & 0.5 \end{bmatrix} \quad \text{(all ℧); } 1.9 \ \Omega$$

Fig. 16-15: See Drill Prob. 16-8.

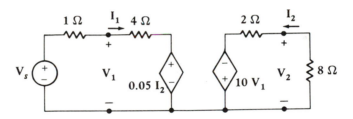

16-9. Use Y-Δ and Δ-Y transformations to determine R_{in} for each network shown in Fig. 16-16. *Ans:* 8.06; 8.75 Ω

Fig. 16-16: See Drill Prob. 16-9.

(a)

Each $R = 10 \ \Omega$

(b)

16-5
Impedance parameters

The concept of two-port parameters has been introduced in terms of the short-circuit admittance parameters. There are other sets of parameters, however, and each set is associated with a particular class of networks for which its use provides the simplest analysis. We shall consider three other types of parameters, the open-circuit impedance parameters which are the subject of this section, and the hybrid and the transmission parameters which are discussed in following sections.

We begin again with the general linear two port which does not contain any independent sources; the currents and voltages are assigned as before (Fig. 16-5). Now let us consider the voltage V_1 as the response produced by two current sources I_1 and I_2. We thus write for V_1

$$V_1 = z_{11}I_1 + z_{12}I_2 \tag{33}$$

and for V_2

$$V_2 = z_{21}I_1 + z_{22}I_2 \tag{34}$$

or

$$[V] = \begin{bmatrix} V_1 \\ V_2 \end{bmatrix} = [z][I] = \begin{bmatrix} z_{11} & z_{12} \\ z_{21} & z_{22} \end{bmatrix} \begin{bmatrix} I_1 \\ I_2 \end{bmatrix} \tag{35}$$

Of course, in using these equations it is not necessary that I_1 and I_2 be current sources; nor is it necessary that V_1 and V_2 be voltage sources. In general, we may have any networks terminating the two port at either end. As the equations are written, we probably think of V_1 and V_2 as given quantities, or independent variables, and I_1 and I_2 as unknowns, or dependent variables. The six ways in which two equations may be written to relate these four quantities define the different systems of parameters. We study the four most important of these six systems of parameters.

The most informative description of the **z** parameters, defined in (33) and (34), is obtained by setting each of the currents equal to zero. Thus

$$z_{11} = \left. \frac{V_1}{I_1} \right|_{I_2=0} \tag{36}$$

$$z_{12} = \left. \frac{V_1}{I_2} \right|_{I_1=0} \tag{37}$$

$$z_{21} = \left. \frac{V_2}{I_1} \right|_{I_2=0} \tag{38}$$

$$z_{22} = \left. \frac{V_2}{I_2} \right|_{I_1=0} \tag{39}$$

Since zero current results from an open-circuit termination, the **z** parameters are known as the *open-circuit impedance parameters*. They are easily related to the short-circuit admittance parameters by solving (33) and (34) for I_1 and I_2:

$$I_1 = \frac{\begin{vmatrix} V_1 & z_{12} \\ V_2 & z_{22} \end{vmatrix}}{\begin{vmatrix} z_{11} & z_{12} \\ z_{21} & z_{22} \end{vmatrix}}$$

or

$$I_1 = \left(\frac{z_{22}}{z_{11}z_{22} - z_{12}z_{21}} \right) V_1 + \left(-\frac{z_{12}}{z_{11}z_{22} - z_{12}z_{21}} \right) V_2$$

Using determinant notation, and being careful that the subscript is a lower-case $\mathbf{z}$, we assume that $\Delta_\mathbf{z} \neq 0$ and obtain

$$\mathbf{y}_{11} = \frac{\Delta_{11}}{\Delta_\mathbf{z}} = \frac{\mathbf{z}_{22}}{\Delta_\mathbf{z}} \qquad \mathbf{y}_{12} = -\frac{\Delta_{21}}{\Delta_\mathbf{z}} = -\frac{\mathbf{z}_{12}}{\Delta_\mathbf{z}}$$

and from solving for $\mathbf{I}_2$,

$$\mathbf{y}_{21} = -\frac{\Delta_{12}}{\Delta_\mathbf{z}} = -\frac{\mathbf{z}_{21}}{\Delta_\mathbf{z}} \qquad \mathbf{y}_{22} = \frac{\Delta_{22}}{\Delta_\mathbf{z}} = \frac{\mathbf{z}_{11}}{\Delta_\mathbf{z}}$$

In a similar manner, the $\mathbf{z}$ parameters may be expressed in terms of the admittance parameters. Transformations of this nature are possible between any of the various parameter systems, and quite a collection of occasionally useful formulas may be obtained. Transformations between the $\mathbf{y}$, $\mathbf{z}$, $\mathbf{h}$, and $\mathbf{t}$ parameters are given in Table 16-1 as a helpful reference.

Table 16-1: Transformations between $\mathbf{y}$, $\mathbf{z}$, $\mathbf{h}$, and $\mathbf{t}$ parameters

		$\mathbf{y}$		$\mathbf{z}$		$\mathbf{h}$		$\mathbf{t}$	
$\mathbf{y}$	$\mathbf{y}_{11}$	$\mathbf{y}_{12}$	$\dfrac{\mathbf{z}_{22}}{\Delta_\mathbf{z}}$	$\dfrac{-\mathbf{z}_{12}}{\Delta_\mathbf{z}}$	$\dfrac{1}{\mathbf{h}_{11}}$	$\dfrac{-\mathbf{h}_{12}}{\mathbf{h}_{11}}$	$\dfrac{\mathbf{t}_{22}}{\mathbf{t}_{12}}$	$\dfrac{-\Delta_\mathbf{t}}{\mathbf{t}_{12}}$
	$\mathbf{y}_{21}$	$\mathbf{y}_{22}$	$\dfrac{-\mathbf{z}_{21}}{\Delta_\mathbf{z}}$	$\dfrac{\mathbf{z}_{11}}{\Delta_\mathbf{z}}$	$\dfrac{\mathbf{h}_{21}}{\mathbf{h}_{11}}$	$\dfrac{\Delta_\mathbf{h}}{\mathbf{h}_{11}}$	$\dfrac{-1}{\mathbf{t}_{12}}$	$\dfrac{\mathbf{t}_{11}}{\mathbf{t}_{12}}$
$\mathbf{z}$	$\dfrac{\mathbf{y}_{22}}{\Delta_\mathbf{y}}$	$\dfrac{-\mathbf{y}_{12}}{\Delta_\mathbf{y}}$	$\mathbf{z}_{11}$	$\mathbf{z}_{12}$	$\dfrac{\Delta_\mathbf{h}}{\mathbf{h}_{22}}$	$\dfrac{\mathbf{h}_{12}}{\mathbf{h}_{22}}$	$\dfrac{\mathbf{t}_{11}}{\mathbf{t}_{21}}$	$\dfrac{\Delta_\mathbf{t}}{\mathbf{t}_{21}}$
	$\dfrac{-\mathbf{y}_{21}}{\Delta_\mathbf{y}}$	$\dfrac{\mathbf{y}_{11}}{\Delta_\mathbf{y}}$	$\mathbf{z}_{21}$	$\mathbf{z}_{22}$	$\dfrac{-\mathbf{h}_{21}}{\mathbf{h}_{22}}$	$\dfrac{1}{\mathbf{h}_{22}}$	$\dfrac{1}{\mathbf{t}_{21}}$	$\dfrac{\mathbf{t}_{22}}{\mathbf{t}_{21}}$
$\mathbf{h}$	$\dfrac{1}{\mathbf{y}_{11}}$	$\dfrac{-\mathbf{y}_{12}}{\mathbf{y}_{11}}$	$\dfrac{\Delta_\mathbf{z}}{\mathbf{z}_{22}}$	$\dfrac{\mathbf{z}_{12}}{\mathbf{z}_{22}}$	$\mathbf{h}_{11}$	$\mathbf{h}_{12}$	$\dfrac{\mathbf{t}_{12}}{\mathbf{t}_{22}}$	$\dfrac{\Delta_\mathbf{t}}{\mathbf{t}_{22}}$
	$\dfrac{\mathbf{y}_{21}}{\mathbf{y}_{11}}$	$\dfrac{\Delta_\mathbf{y}}{\mathbf{y}_{11}}$	$\dfrac{-\mathbf{z}_{21}}{\mathbf{z}_{22}}$	$\dfrac{1}{\mathbf{z}_{22}}$	$\mathbf{h}_{21}$	$\mathbf{h}_{22}$	$\dfrac{-1}{\mathbf{t}_{22}}$	$\dfrac{\mathbf{t}_{21}}{\mathbf{t}_{22}}$
$\mathbf{t}$	$\dfrac{-\mathbf{y}_{22}}{\mathbf{y}_{21}}$	$\dfrac{-1}{\mathbf{y}_{21}}$	$\dfrac{\mathbf{z}_{11}}{\mathbf{z}_{21}}$	$\dfrac{\Delta_\mathbf{z}}{\mathbf{z}_{21}}$	$\dfrac{-\Delta_\mathbf{h}}{\mathbf{h}_{21}}$	$\dfrac{-\mathbf{h}_{11}}{\mathbf{h}_{21}}$	$\mathbf{t}_{11}$	$\mathbf{t}_{12}$
	$\dfrac{-\Delta_\mathbf{y}}{\mathbf{y}_{21}}$	$\dfrac{-\mathbf{y}_{11}}{\mathbf{y}_{21}}$	$\dfrac{1}{\mathbf{z}_{21}}$	$\dfrac{\mathbf{z}_{22}}{\mathbf{z}_{21}}$	$\dfrac{-\mathbf{h}_{22}}{\mathbf{h}_{21}}$	$\dfrac{-1}{\mathbf{h}_{21}}$	$\mathbf{t}_{21}$	$\mathbf{t}_{22}$

For all parameter sets: $\Delta_\mathbf{p} = \mathbf{p}_{11}\mathbf{p}_{22} - \mathbf{p}_{12}\mathbf{p}_{21}$.

If the two port is a bilateral network, reciprocity is present; it is easy to show that this results in the equality of $\mathbf{z}_{12}$ and $\mathbf{z}_{21}$.

Equivalent circuits may again be obtained from an inspection of (33) and (34); their construction is facilitated by adding and subtracting either $\mathbf{z}_{12}\mathbf{I}_1$ to (34) or $\mathbf{z}_{21}\mathbf{I}_2$ to (33). Each of these equivalent circuits contains a dependent voltage source.

Let us leave the derivation of such an equivalent to some leisure moment, such as during the next hourly exam, and consider next an example of a rather general nature. Can we construct a general Thévenin equivalent of the two port, as viewed from the output terminals? It is necessary first to assume a specific input circuit configuration, and we shall select an independent voltage source $\mathbf{V}_s$ (positive sign at top) in series with a generator impedance $\mathbf{Z}_g$. Thus

$$\mathbf{V}_s = \mathbf{V}_1 + \mathbf{I}_1 \mathbf{Z}_g$$

Combining this result with (33) and (34), we may eliminate $\mathbf{V}_1$ and $\mathbf{I}_1$ and obtain

$$\mathbf{V}_2 = \frac{\mathbf{z}_{21}}{\mathbf{z}_{11} + \mathbf{Z}_g} \mathbf{V}_s + \left(\mathbf{z}_{22} - \frac{\mathbf{z}_{12}\mathbf{z}_{21}}{\mathbf{z}_{11} + \mathbf{Z}_g} \right) \mathbf{I}_2$$

The Thévenin equivalent circuit may be drawn directly from this equation; it is shown in Fig. 16-17. The output impedance, expressed in terms of the $\mathbf{z}$ parameters, is

$$\mathbf{Z}_{out} = \mathbf{z}_{22} - \frac{\mathbf{z}_{12}\mathbf{z}_{21}}{\mathbf{z}_{11} + \mathbf{Z}_g}$$

If the generator impedance is zero, the simpler expression

$$\mathbf{Z}_{out} = \frac{\mathbf{z}_{11}\mathbf{z}_{22} - \mathbf{z}_{12}\mathbf{z}_{21}}{\mathbf{z}_{11}} = \frac{\Delta_{\mathbf{z}}}{\Delta_{22}} = \frac{1}{\mathbf{y}_{22}} \qquad (\mathbf{Z}_g = 0)$$

is obtained. For this special case, the output *admittance* is identical to $\mathbf{y}_{22}$, as indicated by the basic relationship of (13).

As a numerical example, let us use a set of impedance parameters representative of a transistor operating in the common-emitter configuration.

$$[\mathbf{z}] = \begin{bmatrix} 10^3 & 10 \\ -10^6 & 10^4 \end{bmatrix} \qquad \text{(all } \Omega\text{)}$$

We shall consider the two port as driven by an ideal sinusoidal voltage source $\mathbf{V}_s$ in series with a 500-Ω resistor and terminated in a 10-kΩ load resistor. The two describing equations for the two port are

$$\mathbf{V}_1 = 10^3 \mathbf{I}_1 + 10 \mathbf{I}_2 \qquad (40)$$

$$\mathbf{V}_2 = -10^6 \mathbf{I}_1 + 10^4 \mathbf{I}_2 \qquad (41)$$

and the characterizing equations of the input and output networks are

$$\mathbf{V}_s = 500 \mathbf{I}_1 + \mathbf{V}_1 \qquad (42)$$

$$\mathbf{V}_2 = -10^4 \mathbf{I}_2 \qquad (43)$$

From these last four equations, we may easily obtain expressions for $\mathbf{V}_1$, $\mathbf{I}_1$, $\mathbf{V}_2$, and $\mathbf{I}_2$ in terms of $\mathbf{V}_s$,

$$\mathbf{V}_1 = 0.75\mathbf{V}_s \qquad \mathbf{I}_1 = \frac{\mathbf{V}_s}{2000}$$

$$\mathbf{V}_2 = -250\mathbf{V}_s \qquad \mathbf{I}_2 = \frac{\mathbf{V}_s}{40}$$

From this information, it is simple to determine the voltage gain,

$$\mathbf{G}_V = \frac{\mathbf{V}_2}{\mathbf{V}_1} = -333$$

the current gain,

$$\mathbf{G}_I = \frac{\mathbf{I}_2}{\mathbf{I}_1} = 50$$

the power gain,

$$G_P = \frac{\mathrm{Re}[-\frac{1}{2}\mathbf{V}_2\mathbf{I}_2^*]}{\mathrm{Re}[\frac{1}{2}\mathbf{V}_1\mathbf{I}_1^*]} = 16\ 700$$

and the input impedance,

$$\mathbf{Z}_{in} = \frac{\mathbf{V}_1}{\mathbf{I}_1} = 1500 \quad \Omega$$

The output impedance may be obtained by referring to Fig. 16-17,

$$\mathbf{Z}_{out} = \mathbf{z}_{22} - \frac{\mathbf{z}_{12}\mathbf{z}_{21}}{\mathbf{z}_{11} + \mathbf{Z}_g} = 16.7 \quad \mathrm{k}\Omega$$

In accordance with the predictions of the maximum power transfer theorem, the power gain reaches a maximum value when $\mathbf{Z}_L = \mathbf{Z}_{out} = 16.7\ \mathrm{k}\Omega$; that maximum value is 17 045.

The **y** parameters are useful when two ports are interconnected in parallel, and, in a dual manner, the **z** parameters simplify the problem of a series connection of networks, shown in Fig. 16-18. Note that the series connection is *not* the same as the cascade connection that we shall discuss later in connection with the transmission parameters. If each two port has a com-

Fig. 16-18: The series connection of two two-port networks is made by connecting the four common reference nodes together; then, [**z**] = [**z**$_A$] + [**z**$_B$].

mon reference node for its input and output, and if these references are connected together as indicated in Fig. 16-18, then I_1 flows through the input ports of the two networks in series. A similar statement holds for I_2. Thus, ports remain ports after the interconnection. It follows that $[I] = [I_A] = [I_B]$ and

$$[V] = [V_A] + [V_B] = [z_A][I_A] + [z_B][I_B] = ([z_A] + [z_B])[I] = [z][I]$$

where

$$[z] = [z_A] + [z_B] \tag{44}$$

so that $z_{11} = z_{11A} + z_{11B}$, and so forth.

Drill Problems

16-10. Find the four z parameters for the two port shown in: (a) Fig. 16-19a; (b) Fig. 16-19b.

$$Ans: \begin{bmatrix} 84 & 72 \\ 72 & 96 \end{bmatrix} \text{ (all } \Omega); \qquad \begin{bmatrix} 84 & 72 \\ 72 & 96 \end{bmatrix} \text{ (all } \Omega)$$

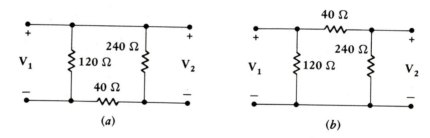

Fig. 16-19: (a) and (b) See Drill Prob. 16-10. (c) See Drill Prob. 16-11. (d) See Drill Prob. 16-12.

16-11. Determine all four open-circuit impedance parameters for the network of Fig. 16-19c.

$$Ans: \begin{bmatrix} 5.71 & -4.29 \\ 2.14 & 2.14 \end{bmatrix} \text{ (all } \Omega)$$

16-12. For the amplifier shown in Fig. 16-19d, find: (a) G_I; (b) G_V; (c) G_P; (d) Z_{in}; (e) Z_{out}. $Ans: -2; 2.22; 4.44; 45 \Omega; 94.4 \Omega$

16-13. Use Table 16-1 to find: (a) $[z]$, if $[y] = \begin{bmatrix} 10 & -2 \\ 100 & 20 \end{bmatrix}$ (all m℧); (b) $[y]$, if $[z] = \begin{bmatrix} 8 & 6 \\ 2 & 4 \end{bmatrix}$ (all Ω).

$$Ans: \begin{bmatrix} 50 & 5 \\ -250 & 25 \end{bmatrix} \text{ (all } \Omega); \qquad \begin{bmatrix} 0.2 & -0.3 \\ -0.1 & 0.4 \end{bmatrix} \text{ (all } ℧)$$

**16-6
Hybrid parameters**

The use of the hybrid parameters is well suited to transistor circuits because these parameters are among the most convenient to measure experimentally for a transistor. The difficulty in measuring, say, the open-circuit impedance parameters arises when a parameter such as z_{21} must be measured. A known sinusoidal current is easily supplied at the input terminals, but because of the exceedingly high output impedance of the transistor circuit, it is difficult to open-circuit the output terminals and yet supply the necessary dc biasing voltages and measure the sinusoidal output voltage. A short-circuit current measurement at the output terminals is much simpler to instrument.

The hybrid parameters are defined by writing the pair of equations relating V_1, I_1, V_2, and I_2 as if V_1 and I_2 were the independent variables:

$$V_1 = h_{11}I_1 + h_{12}V_2 \tag{45}$$

$$I_2 = h_{21}I_1 + h_{22}V_2 \tag{46}$$

or

$$\begin{bmatrix} V_1 \\ I_2 \end{bmatrix} = [h] \begin{bmatrix} I_1 \\ V_2 \end{bmatrix} \tag{47}$$

The nature of the parameters is made clear by first setting $V_2 = 0$,

$$h_{11} = \frac{V_1}{I_1}\bigg|_{V_2=0} = \text{short-circuit input impedance}$$

and

$$h_{21} = \frac{I_2}{I_1}\bigg|_{V_2=0} = \text{short-circuit forward current gain}$$

and then letting $I_1 = 0$,

$$h_{12} = \frac{V_1}{V_2}\bigg|_{I_1=0} = \text{open-circuit reverse voltage gain}$$

and

$$h_{22} = \frac{I_2}{V_2}\bigg|_{I_1=0} = \text{open-circuit output admittance}$$

Since the parameters represent an impedance, an admittance, a voltage gain, and a current gain, it is understandable that they are called the "hybrid" parameters.

The subscript designations for these parameters are often simplified when they are applied to transistors. Thus, h_{11}, h_{12}, h_{21}, and h_{22} become h_i, h_r, h_f, and h_o, respectively, where the subscripts denote input, reverse, forward, and output.

In order to illustrate the ease with which these parameters may be evaluated, consider the bilateral resistive circuit drawn in Fig. 16-20. With the

Fig. 16-20: A bilateral network for which the **h** parameters are found; $h_{12} = -h_{21}$.

Fig. 16-21: The four **h** parameters are referred to a two port. The pertinent equations are: $V_1 = h_{11}I_1 + h_{12}V_2$, $I_2 = h_{21}I_1 + h_{22}V_2$.

output short-circuited ($V_2 = 0$), the application of a 1-A source at the input ($I_1 = 1$ A) produces an input voltage of 3.4 V ($V_1 = 3.4$ V); hence, $h_{11} = 3.4$ Ω. Under these same conditions, the output current is easily obtained by current division, $I_2 = -0.4$ A; thus, $h_{21} = -0.4$. The remaining two parameters are obtained with the input open-circuited ($I_1 = 0$). Let us apply a voltage of 1 V at the output terminals ($V_2 = 1$ V). The response at the input terminals is 0.4 V ($V_1 = 0.4$ V), and thus $h_{12} = 0.4$. The current delivered by this source at the output terminals is 0.1 A ($I_2 = 0.1$ A), and therefore $h_{22} = 0.1$ $\mho$. It is a consequence of the reciprocity theorem that $h_{12} = -h_{21}$ for a bilateral network.

The circuit shown in Fig. 16-21 is a direct translation of the two defining equations (45) and (46). The first represents KVL about the input loop, while the second is obtained from KCL at the upper output node. This circuit is also a popular transistor equivalent circuit. Let us assume some reasonable values for the common-emitter configuration: $h_{11} = 1200$ Ω, $h_{12} = 2 \times 10^{-4}$, $h_{21} = 50$, $h_{22} = 50 \times 10^{-6}$ $\mho$, a voltage generator of $1\underline{/0°}$ mV in series with 800 Ω, and a 5-kΩ load. For the input,

$$10^{-3} = (1200 + 800)I_1 + 2 \times 10^{-4}V_2$$

and at the output,

$$I_2 = -2 \times 10^{-4}V_2 = 50I_1 + 50 \times 10^{-6}V_2$$

Solving,

$$I_1 = 0.510 \quad \mu A \qquad V_1 = 0.592 \quad mV$$

$$I_2 = 20.4 \quad \mu A \qquad V_2 = -102 \quad mV$$

Through the transistor we have a current gain of 40, a voltage gain of -172, and a power gain of 6880. The input impedance to the transistor is 1160 Ω, and a few more calculations show that the output impedance is 22.2 kΩ.

Hybrid parameters may be added directly when two ports are connected in series at the input and in parallel at the output. This is called a series-parallel interconnection, and it is not used very often.

Drill Problems

16-14. Find [**h**] for the two-port network of: (a) Fig. 16-22a; (b) Fig. 16-22b.

Ans: $\begin{bmatrix} R_B & 1 \\ -1 & 1/R_C \end{bmatrix}$; $\begin{bmatrix} R_A R_B/(R_A + R_B) & R_A/(R_A + R_B) \\ -R_A/(R_A + R_B) & 1/(R_A + R_B) \end{bmatrix}$

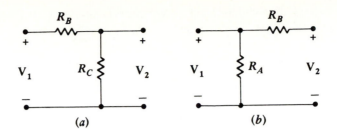

Fig. 16-22: See Drill Prob. 16-14.

(a)   (b)

16-15. Find [**h**] if: (a) [**y**] = $\begin{bmatrix} 10 & -2 \\ 100 & 20 \end{bmatrix}$ (all m℧); (b) [**z**] = $\begin{bmatrix} 8 & 6 \\ 2 & 4 \end{bmatrix}$ (all Ω)

Ans: $\begin{bmatrix} 100\ \Omega & 0.2 \\ 10 & 0.04\ ℧ \end{bmatrix}$; $\begin{bmatrix} 5\ \Omega & 1.5 \\ -0.5 & 0.25\ ℧ \end{bmatrix}$

16-16. Find [**h**] at $\omega = 10^8$ rad/s for the high-frequency transistor equivalent circuit of Fig. 16-23.

Ans: $\begin{bmatrix} 0.857\underline{/-31.0°}\ \text{k}\Omega & 0.1715\underline{/59.0°} \\ 8.58\underline{/-32.1°} & 1.888\underline{/61.1°}\ \text{m}℧ \end{bmatrix}$

2 pF

Fig. 16-23: See Drill Prob. 16-16.

16-17. Find the input and output impedances of an amplifier having $\mathbf{h}_{11} = 2\ \Omega$, $\mathbf{h}_{12} = 1$, $\mathbf{h}_{21} = 5$, and $\mathbf{h}_{22} = 2\ ℧$, if it is driven by a source having an internal resistance of 4 Ω and is terminated in that load which will draw maximum power from the amplifier. *Ans:* 0.421; 0.857 Ω

**16-7
Transmission
parameters**

The last set of two-port parameters that we shall consider is called the **t** *parameters*, the **ABCD** *parameters*, or simply the *transmission parameters*. They are defined by

$$\mathbf{V}_1 = \mathbf{t}_{11}\mathbf{V}_2 - \mathbf{t}_{12}\mathbf{I}_2 \tag{48}$$

$$\mathbf{I}_1 = \mathbf{t}_{21}\mathbf{V}_2 - \mathbf{t}_{22}\mathbf{I}_2 \tag{49}$$

or $\qquad \begin{bmatrix} \mathbf{V}_1 \\ \mathbf{I}_1 \end{bmatrix} = [\mathbf{t}] \begin{bmatrix} \mathbf{V}_2 \\ -\mathbf{I}_2 \end{bmatrix} \tag{50}$

where $\mathbf{V}_1$, $\mathbf{V}_2$, $\mathbf{I}_1$, and $\mathbf{I}_2$ are defined as usual (Fig. 16-5). The minus signs that appear in (48) and (49) should be associated with the output current, as $(-\mathbf{I}_2)$. Thus, both $\mathbf{I}_1$ and $-\mathbf{I}_2$ are directed to the right, the direction of energy or signal transmission.

Other widely used nomenclature for this set of parameters is

$$
\begin{bmatrix} t_{11} & t_{12} \\ t_{21} & t_{22} \end{bmatrix} = \begin{bmatrix} A & B \\ C & D \end{bmatrix}
\tag{51}
$$

Note that there are no minus signs in the **t** or **ABCD** matrices.

Looking again at (48) to (50), we see that the quantities on the left, often thought of as the given or independent variables, are the input voltage and current, V_1 and I_1; the dependent variables, V_2 and I_2, are the output quantities. Thus, the transmission parameters provide a direct relationship between input and output. Their major use arises in transmission-line analysis and in cascaded networks.

Let us find the **t** parameters for the bilateral resistive two port of Fig. 16-24a. To illustrate one possible procedure for finding a single parameter, consider

$$
t_{12} = \left. \frac{V_1}{-I_2} \right|_{V_2=0}
$$

Therefore, we short-circuit the output ($V_2 = 0$) and set $V_1 = 1$ V, as shown in Fig. 16-24b. Note that we cannot set the denominator equal to unity by placing a 1-A current source at the output; we already have a short circuit there. The equivalent resistance offered to the 1-V source is $R_{eq} = 2 + (4 \| 10)$ Ω, and we then use current division to get

$$
-I_2 = \frac{1}{2 + (4 \| 10)} \times \frac{10}{10 + 4} = \frac{5}{34} \quad A
$$

Hence,

$$
t_{12} = \frac{1}{-I_2} = \frac{34}{5} = 6.8 \quad \Omega
$$

If it is necessary to find all four parameters, we write any convenient pair of equations using all four terminal quantities, V_1, V_2, I_1, and I_2. From Fig. 16-24a, we have two mesh equations:

$$
V_1 = 12I_1 + 10I_2 \tag{52}
$$

$$
V_2 = 10I_1 + 14I_2 \tag{53}
$$

Solving (53) for I_1, we get

$$
I_1 = 0.1V_2 - 1.4I_2
$$

so that $t_{21} = 0.1$ ℧ and $t_{22} = 1.4$. Substituting the expression for I_1 into (52), we find

$$
V_1 = 12(0.1V_2 - 1.4I_2) + 10I_2 = 1.2V_2 - 6.8I_2
$$

and $t_{11} = 1.2$ and $t_{12} = 6.8$ Ω, once again.

Fig. 16-24: (a) A two-port resistive network for which the **t** parameters are to be found. (b) To find t_{12}, set $V_2 = 0$ and $V_1 = 1$ V; then $t_{12} = 1/(-I_2) = 6.8$ Ω.

For reciprocal networks, the determinant of the **t** matrix is equal to unity,

$$\Delta_t = t_{11}t_{22} - t_{12}t_{21} = 1 \tag{54}$$

In the resistive example above, $\Delta_t = 1.2 \times 1.4 - 6.8 \times 0.1 = 1$. Good!

We conclude our two-port discussion by connecting two ports in cascade, as illustrated for two networks in Fig. 16-25. Terminal voltages and currents are indicated for each two port, and the corresponding **t**-parameter relationships are, for network A,

$$\begin{bmatrix} V_1 \\ I_1 \end{bmatrix} = [t_A]\begin{bmatrix} V_2 \\ -I_2 \end{bmatrix} = [t_A]\begin{bmatrix} V_3 \\ I_3 \end{bmatrix}$$

and for network B,

$$\begin{bmatrix} V_3 \\ I_3 \end{bmatrix} = [t_B]\begin{bmatrix} V_4 \\ -I_4 \end{bmatrix}$$

Combining these results, we have

$$\begin{bmatrix} V_1 \\ I_1 \end{bmatrix} = [t_A][t_B]\begin{bmatrix} V_4 \\ -I_4 \end{bmatrix}$$

Therefore, the **t** parameters for the cascaded networks are found by the matrix product,

$$[t] = [t_A][t_B]$$

This product is *not* obtained by multiplying corresponding elements in the two matrices. If necessary, review the correct procedure for matrix multiplication in Appendix 2.

Fig. 16-25: When two-port networks A and B are cascaded, the **t**-parameter matrix for the combined network is given by the matrix product, $[t] = [t_A][t_B]$.

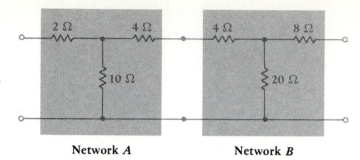

Fig. 16-26: A cascaded connection for which [**t**] =

$$\begin{bmatrix} 1.2 & 6.8 \\ 0.1 & 1.4 \end{bmatrix} \begin{bmatrix} 1.2 & 13.6 \\ 0.05 & 1.4 \end{bmatrix} = \begin{bmatrix} 1.78 & 25.84 \\ 0.19 & 3.32 \end{bmatrix}$$

Network *A*   Network *B*

As an example of such a calculation, consider the cascaded connection shown in Fig. 16-26. Network *A* is the two port of Fig. 16-24,

$$[t_A] = \begin{bmatrix} 1.2 & 6.8 \ \Omega \\ 0.1 \ \text{℧} & 1.4 \end{bmatrix}$$

while network *B* has resistance values twice as large, so that

$$[t_B] = \begin{bmatrix} 1.2 & 13.6 \ \Omega \\ 0.05 \ \text{℧} & 1.4 \end{bmatrix}$$

For the combined network,

$$[t] = [t_A][t_B] = \begin{bmatrix} 1.2 & 6.8 \\ 0.1 & 1.4 \end{bmatrix} \begin{bmatrix} 1.2 & 13.6 \\ 0.05 & 1.4 \end{bmatrix}$$

$$= \begin{bmatrix} 1.2 \times 1.2 + 6.8 \times 0.05 & 1.2 \times 13.6 + 6.8 \times 1.4 \\ 0.1 \times 1.2 + 1.4 \times 0.05 & 0.1 \times 13.6 + 1.4 \times 1.4 \end{bmatrix}$$

and $$[t] = \begin{bmatrix} 1.78 & 25.84 \ \Omega \\ 0.19 \ \text{℧} & 3.32 \end{bmatrix}$$

Drill Problems

16-18. Find [**t**] for the two port shown in: (*a*) Fig. 16-27*a*; (*b*) Fig. 16-27*b*.

$$Ans: \begin{bmatrix} 2.43 & 6.2 \ \Omega \\ 0.467 \ \text{℧} & 1.6 \end{bmatrix}; \begin{bmatrix} 0.556 & 3.33 \ \Omega \\ -0.1111 \ \text{℧} & 0.333 \end{bmatrix}$$

Fig. 16-27: See Drill Prob. 16-18.

(*a*)   (*b*)

16-19. Given $t_{11} = 3$, $t_{12} = 2\,\Omega$, $t_{21} = 1\,\mho$, and $t_{22} = 2$, find: (*a*) [**y**] for this two port; (*b*) [**y**] if three of the given networks are in cascade.

$$Ans: \begin{bmatrix} 1 & -2 \\ -0.5 & 1.5 \end{bmatrix} \text{ (all } \mho); \quad \begin{bmatrix} 0.524 & -1.524 \\ -0.0238 & 1.024 \end{bmatrix} \text{ (all } \mho)$$

Problems

1 Using nodal analysis and Eq. (4), find the input admittance of the network shown in Fig. 16-2*c* if the arrow in the dependent current source is reversed in direction.

2 Find $\mathbf{Z}_{in}$ for the one port of Fig. 16-28 by: (*a*) determining $\Delta_\mathbf{Z}$; (*b*) finding $\Delta_\mathbf{Y}$ and $\mathbf{Y}_{in}$ first, and then $\mathbf{Z}_{in}$.

Fig. 16-28: See Prob. 2.

3 Find the output impedance of the network shown in Fig. 16-29 as a function of **s**.

Fig. 16-29: See Prob. 3.

4 Assume that the op-amp shown in Fig. 16-30 is ideal and find R_{in}.

Fig. 16-30: See Prob. 4.

Fig. 16-31: See Prob. 5.

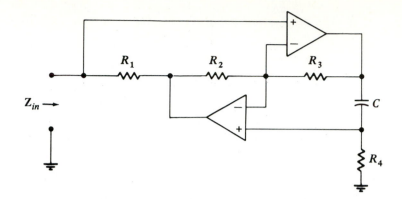

5 (a) Assume that both op-amps appearing in the circuit of Fig. 16-31 are ideal and find $\mathbf{Z}_{in}$. (b) If $R_1 = 1$ kΩ, $R_2 = 2$ kΩ, $R_3 = 3$ kΩ, $R_4 = 4$ kΩ, and $C = 100$ pF, show that $\mathbf{Z}_{in} = j\omega L_{in}$, where $L_{in} = 0.6$ mH.

6 The impedance determinant of the one-port network shown in Fig. 16-2a is

$$\Delta_{\mathbf{Z}} = \begin{vmatrix} 8 & -1 & -3 \\ -1 & 10 & -2 \\ -3 & -2 & 6 \end{vmatrix}, \text{ where all units are ohms. } (a) \text{ Determine } \mathbf{Z}_{in}. \ (b) \text{ What}$$

will $\mathbf{Z}_{in}$ become if a 3-Ω resistor is inserted in series with the branch common to meshes 2 and 3?

7 Find $\mathbf{Z}_{in}$ for the one-port network shown in Fig. 16-32 by: (a) writing mesh equations and finding $\Delta_{\mathbf{Z}}$; (b) writing nodal equations and finding $\Delta_{\mathbf{Y}}$.

Fig. 16-32: See Prob. 7.

8 Find all the critical frequencies of the input impedance $\mathbf{Z}_{in}(\mathbf{s})$ for the one port shown in Fig. 16-33.

Fig. 16-33: See Prob. 8.

Fig. 16-34: See Prob. 9.

9 Find the four **y** parameters for the two port of Fig. 16-34: (*a*) one (or perhaps two) at a time by the judicious use of 1-V sources and short circuits; (*b*) all at once by finding the currents supplied by sources V_1 and V_2.

10 (*a*) Find y_{12} and y_{21} for the network of Fig. 16-35. (*b*) To what value should the factor 10 in $10I_1$ be changed to provide a reciprocal network?

Fig. 16-35: See Prob. 10.

11 (*a*) What value of *k* in the two port of Fig. 16-36 will produce a reciprocal network? (*b*) Let *k* = 0.8, connect a 5-Ω load at the right and a 50-V source at the left, and determine the power delivered to the load.

Fig. 16-36: See Prob. 11.

12 Find [**y**(s)] for the two port of Fig. 16-37.

Fig. 16-37: See Prob. 12.

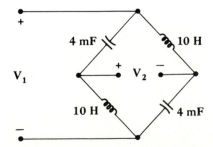

13 A two-port network has the **y** parameters: $y_{11} = 10$, $y_{12} = -5$, $y_{21} = -20$, and $y_{22} = 2$, all in m℧. Find the new [**y**] if a 100-Ω resistor is connected: (*a*) across the

input; (b) across the output; (c) between the upper input and output terminals inside the two port.

14 Complete the table given as part of Fig. 16-38 and also give values for the **y** parameters.

Fig. 16-38: See Prob. 14.

	V_{s1} (V)	V_{s2} (V)	I_1 (A)	I_2 (A)
Exp't # 1	50	100	−1	27
Exp't # 2	100	50	7	24
Exp't # 3	200	0		
Exp't # 4			20	0
Exp't # 5			10	30

15 A two port is characterized by the equations, $12v_1 - 3i_1 - 6i_2 + 4v_2 = 0$ and $5i_1 = 6v_1 - 2i_2 + 0.1v_2$. (a) Find [**y**]. (b) Find [**y**]$_{new}$ after a 5-Ω resistor is added in series with the input.

16 (a) Find R_{in} in Fig. 16-39a by making Y-Δ or Δ-Y transformations and using resistance combinations. (b) Find $\mathbf{Z}_{in}$ for Fig. 16-39b by using T-π and π-T transformations and impedance combinations.

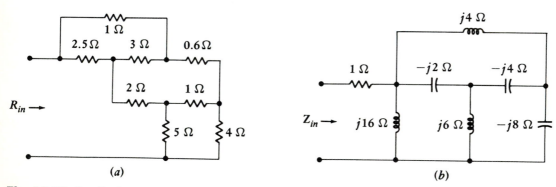

(a)   (b)

Fig. 16-39: See Prob. 16.

17 A four-sided pyramid has a 60-Ω resistor along each of its eight edges. What resistance would be measured between: (a) adjacent corners of the base? (b) diagonally opposite corners of the base? (c) the apex and one corner of the base?

18 (a) Find the equivalent π network for the T network shown in Fig. 16-40a. (b) Find R_{in} for the network of Fig. 16-40b by using Y-Δ and Δ-Y transformations.

Fig. 16-40: See Prob. 18.

(a)   (b)

Fig. 16-41: See Probs. 19 and 20.

19 Let $R_s = 2\ k\Omega$, $\mathbf{y}_{11} = 1\ m\mho$, $\mathbf{y}_{12} = -1\ m\mho$, $\mathbf{y}_{21} = 20\ m\mho$, $\mathbf{y}_{22} = 2\ m\mho$, and $R_L = 0.5$ $k\Omega$ for the circuit shown in Fig. 16-41. Determine numerical values for: (a) voltage gain, $\mathbf{G}_V = \mathbf{V}_2/\mathbf{V}_1$; (b) current gain, $\mathbf{I}_2/\mathbf{I}_1$; (c) current gain, $\mathbf{I}_2/\mathbf{I}_s$; (d) power gain, $G_P = P_L/P_s$; (e) input impedance, $\mathbf{Z}_{in} = \mathbf{V}_1/\mathbf{I}_1$; (f) output impedance, $\mathbf{Z}_{out} = \mathbf{V}_2/\mathbf{I}_2$ with $\mathbf{I}_s = 0$ and $R_L = \infty$; (g) reverse voltage gain, $\mathbf{G}_{VR} = \mathbf{V}_1/\mathbf{V}_2$ with $\mathbf{I}_s = 0$ and $R_L = \infty$.

20 The input network of Fig. 16-41 is replaced by $\mathbf{V}_s$ in series with $R_s = 2\ k\Omega$. If $R_L = 20\ k\Omega$ and $[\mathbf{y}] = \begin{bmatrix} 0.4 & -0.002 \\ -5 & 0.04 \end{bmatrix}$ (m$\mho$) for the two port, find: (a) $\mathbf{V}_2/\mathbf{V}_1$; (b) $\mathbf{V}_2/\mathbf{V}_s$; (c) $\mathbf{I}_2/\mathbf{I}_1$; (d) $\mathbf{V}_1/\mathbf{I}_1$; (e) R_{out}.

21 (a) Find $[\mathbf{y}]_a$ for the two port of Fig. 16-42a. (b) Find $[\mathbf{y}]_b$ for Fig. 16-42b. (c) Draw the network that is obtained when these two ports are connected in parallel, and show that $[\mathbf{y}]$ for this network is equal to $[\mathbf{y}]_a + [\mathbf{y}]_b$.

Fig. 16-42: See Prob. 21.

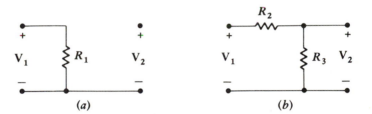

(a)   (b)

22 (a) Draw an equivalent circuit in the form of Fig. 16-9a for which $[\mathbf{y}]_A = \begin{bmatrix} 1.5 & -1 \\ 4 & 3 \end{bmatrix}$ (m$\mho$). (b) If two of these two ports are connected in parallel, draw the new equivalent circuit and show that $[\mathbf{y}]_{new} = 2[\mathbf{y}]_A$.

23 (a) Find the four **z**-parameters at $\omega = 10^8$ rad/s for the transistor, high-frequency equivalent circuit shown in Fig. 16-43. (b) Which of these four parameter values are functions of frequency?

Fig. 16-43: See Prob. 23.

24 A sinusoidal source, 100 V rms in series with 50 Ω, is connected to the input of a two-port network for which $[\mathbf{z}] = \begin{bmatrix} 200 & 20 \\ 400 & 100 \end{bmatrix}$ (Ω), and a load, $R_L = 250\ \Omega$,

is placed at the output. Find the Thévenin equivalent circuit seen at the: (*a*) input terminals of the two port; (*b*) output terminals of the two port.

25 For the two port shown in Fig. 16-44: (*a*) find [**z**]; (*b*) place V_s in series with $R_s = 200\ \Omega$ at the input and $R_L = 4\ k\Omega$ at the output, and determine $G_V = V_2/V_1$ and $R_{in} = V_1/I_1$.

Fig. 16-44: See Prob. 25.

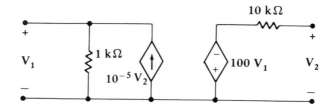

26 Let $[\mathbf{z}] = \begin{bmatrix} 10 & -2 \\ 40 & 20 \end{bmatrix}$ (Ω) for a certain two port. (*a*) A voltage source, $V_s = 1\underline{/0°}$ V, in series with 2 Ω is connected to the input. Find the Thévenin equivalent circuit seen at the output terminals. (*b*) Calculate the insertion power gain, which is the power delivered to the load, $R_L = 25\ \Omega$, divided by the power that would be delivered to the load if the two port were replaced by short circuits connecting each input terminal to the corresponding output terminal. (*c*) Calculate the reverse voltage gain, $G_{VR} = V_1/V_2$, if an independent voltage source V_2 is applied to the output and V_s is set equal to zero.

27 A two-port network for which $[\mathbf{z}] = \begin{bmatrix} 10 & 2 \\ 40 & 5 \end{bmatrix}$ (Ω) is terminated in $\mathbf{Z}_L = \mathbf{z}_{22}$ and driven by a 20-V rms source in series with $\mathbf{Z}_s = \mathbf{z}_{11}$. (*a*) How much power is delivered to $\mathbf{Z}_L$? (*b*) What value should $\mathbf{Z}_L$ have to receive a maximum power? (*c*) What is the value of this maximum power?

28 For the network given in Fig. 16-45, find [**h**(**s**)].

Fig. 16-45: See Prob. 28.

29 A certain two port is characterized by the following pair of equations: $2V_1 + 4I_2 = I_1$ and $8I_2 = V_2 + 6V_1$. Find the value of: (*a*) y_{11}; (*b*) z_{21}; (*c*) h_{21}.

30 The two identical stages of the amplifier shown in Fig. 16-46 each have $[\mathbf{H}] = \begin{bmatrix} 1\ k\Omega & 0.001 \\ 100 & 0.1\ m\mho \end{bmatrix}$. What amplitude of the sinusoidal input voltage V_s is required to obtain a 25-V amplitude for V_o?

Fig. 16-46: See Prob. 30.

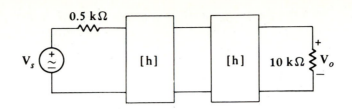

31 A two port for which $[\mathbf{h}] = \begin{bmatrix} 10\ \Omega & -1 \\ 4 & 0.05\ \mho \end{bmatrix}$ is terminated by $R_L = 10\ \Omega$ and driven by an ideal sinusoidal voltage source, $V_s = 100$ V rms, in series with $R_s = 2\ \Omega$. Find the average power delivered to: (a) $R_s = 2\ \Omega$; (b) $R_L = 10\ \Omega$; (c) the entire two-port network.

32 Find $[\mathbf{h}]$ for the two port illustrated in Fig. 16-47.

Fig. 16-47: See Prob. 32.

33 In Fig. 16-48, a 2-Ω and a 5-Ω resistor are connected as shown at the input of the two port whose **h** parameters are given. Find $[\mathbf{h}]$ for the composite network.

Fig. 16-48: See Prob. 33.

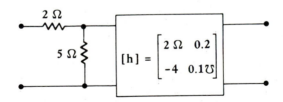

34 Find the **t** parameters for: (a) the two port shown in Fig. 16-49; (b) a two port for which $[\mathbf{h}] = \begin{bmatrix} 20\ \Omega & 0.3 \\ -4 & 0.1\ \mho \end{bmatrix}$

Fig. 16-49: See Prob. 34.

35 (*a*) Find $[\mathbf{t}]_a$ for the two port of Fig. 16-50*a*. (*b*) Find $[\mathbf{t}]_b$ for Fig. 16-50*b*. (*c*) Find $[\mathbf{t}]_a[\mathbf{t}]_b$ and show that $[\mathbf{t}]$ for the cascaded combination is the same. (*d*) Find $[\mathbf{t}]_b[\mathbf{t}]_a$.

Fig. 16-50: See Prob. 35.

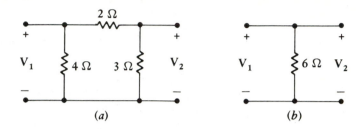

(*a*)   (*b*)

36 A two port contains a single 1-Ω resistor connected from the upper input terminal to the upper output terminal; the lower terminals are connected directly to each other. (*a*) Find $[\mathbf{t}]_a$ for this network. (*b*) Show that the **t** parameters for a similarly connected 5-Ω resistor can be obtained by $([\mathbf{t}]_a)^5$.

37 Find $[\mathbf{t}]$ for the two port shown in Fig. 16-51.

Fig. 16-51: See Prob. 37.

38 (*a*) Determine $[\mathbf{t}]$ for the networks shown in Figs. 16-52*a*, *b*, and *c*. (*b*) By using the rules for connecting two ports in cascade, find $[\mathbf{t}]$ for the network of Fig. 16-52*d*.

Fig. 16-52: See Prob. 38.

(*a*)   (*b*)   (*c*)

(*d*)

Fig. 16-53: See Prob. 39.

39 (a) Find [t] for the active two port shown in Fig. 16-53. (b) Use the method suggested by Prob. 38 to find the new **t** parameters if a 20-Ω resistor is connected across the output.

40 A two-port network, having $[\mathbf{t}] = \begin{bmatrix} 10^{-2} & 10^{-1}\ \Omega \\ 10^{-5}\ \mho & 10^{-4} \end{bmatrix}$, is terminated by $R_L = 100\ \Omega$ and driven by a sinusoidal source $\mathbf{V}_s$ in series with $R_s = 100\ \Omega$. Find $\mathbf{G}_V$, $\mathbf{G}_I$, $\mathbf{G}_P$, $\mathbf{Z}_{in}$, and $\mathbf{Z}_{out}$. Is the two port reciprocal?

PART SIX
SIGNAL ANALYSIS

Chapter 17
Fourier Analysis

17-1 Introduction

In this chapter we continue our introduction to circuit analysis by studying periodic functions in both the time and frequency domains.

We know that the complete response of a linear circuit to an arbitrary forcing function is composed of the sum of a forced response and a natural response. The natural response was initially considered in Chaps. 5 to 7 but, with few exceptions, only simple series or parallel RL, RC, and RLC circuits were examined. Later, the complex-frequency concept in Chap. 13 provided us with a general method of obtaining the natural response; we discovered that we could write the form of the natural response after locating the poles of an appropriate transfer function of the network. Thus, we found a powerful general method for determining the natural response.

A second general method for finding the natural response (as well as the forced response) is based on state-variable theory, a subject which will probably occupy some of our time next year.

Now let us consider our status with respect to the forced response. We are able to find the forced response in any purely resistive linear circuit, regardless of the nature of the forcing function, but this can hardly be classed as a scientific breakthrough. If the circuit contains energy-storage elements, then we can certainly find the forced response for those circuits and forcing functions to which we can apply the impedance concept; that is, whenever the forcing function is direct current, exponential, sinusoidal, or exponentially varying sinusoidal in form.

In this chapter we consider forcing functions which are *periodic* and have functional natures which satisfy certain mathematical restrictions that are characteristic of any function which we can generate in the laboratory. Any such function may be represented as the sum of an infinite number of sine and cosine functions which are harmonically related. Therefore, since the forced response to each sinusoidal component may be determined easily by sinusoidal steady-state analysis, the response of the linear network to the general periodic forcing function may be obtained by superposing the partial responses.

Some feeling for the validity of representing a general periodic function by an infinite sum of sine and cosine functions may be gained by considering a simple example. Let us first assume a cosine function of radian frequency ω_0,

$$v_1(t) = 2 \cos \omega_0 t$$

where

$$\omega_0 = 2\pi f_0$$

and the period T is

$$T = \frac{1}{f_0} = \frac{2\pi}{\omega_0}$$

Although T does not usually carry a zero subscript, it is the period of the fundamental frequency. The *harmonics* of this sinusoid have frequencies $n\omega_0$, where ω_0 is the fundamental frequency and $n = 1, 2, 3, \ldots$. The frequency of the first harmonic is the fundamental frequency. Next let us select a third harmonic voltage

$$v_{3a}(t) = \cos 3\omega_0 t$$

The fundamental $v_1(t)$, the third harmonic $v_{3a}(t)$, and the sum of these two waves are shown as functions of time in Fig. 17-1a. It should be noted that the sum is periodic with period $T = 2\pi/\omega_0$.

The form of the resultant periodic function changes as the phase and amplitude of the third harmonic component change. Thus, Fig. 17-1b shows the effect of combining $v_1(t)$ and a third harmonic of slightly larger amplitude,

$$v_{3b}(t) = 1.5 \cos 3\omega_0 t$$

By shifting the phase of the third harmonic,

$$v_{3c}(t) = \sin 3\omega_0 t$$

the sum, shown in Fig. 17-1c, takes on a still different character. In all cases, the period of the resultant waveform is the same as the period of the fundamental waveform. The nature of the waveform depends on the amplitude and phase of every possible harmonic component, and we shall find it possible to generate waveforms which have extremely nonsinusoidal characteristics by an appropriate combination of sinusoidal functions.

After we have become familiar with the use of the sum of an infinite number of sine and cosine functions to represent a periodic waveform, we shall consider the frequency-domain representation of a general nonperiodic waveform in the next chapter.

Drill Problem

17-1. Let the third harmonic voltage used in the text example be expressed as $v_3 = V_{m3} \cos 3\omega_0 t$. (a) Find the largest possible value of V_{m3} so that $v(t) = 2 \cos \omega_0 t + v_3(t)$ will cross the time axis only once in the interval $0 < \omega_0 t < \pi$. (b) Evaluate $v(t)$ at $\omega_0 t = 1$ rad if $V_m = 1$. *Ans:* 0.667; 0.0906

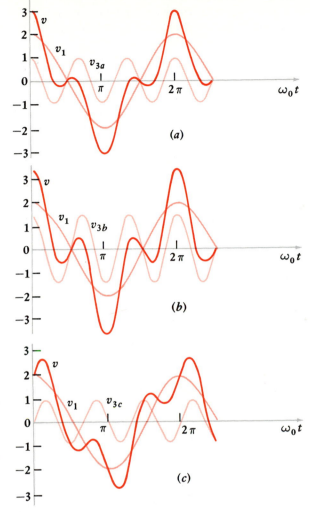

Fig. 17-1: Several of the infinite number of different waveforms which may be obtained by combining a fundamental and a third harmonic. The fundamental is $v_1 = 2 \cos \omega_0 t$ and the third harmonic is: (a) $v_{3a} = \cos 3\omega_0 t$; (b) $v_{3b} = 1.5 \cos 3\omega_0 t$; (c) $v_{3c} = \sin 3\omega_0 t$.

17-2
Trigonometric form
of the Fourier
series

We first consider a *periodic* function $f(t)$, defined in Sec. 11-3 by the functional relationship,

$$f(t) = f(t + T)$$

where T is the period. We further assume that the function $f(t)$ satisfies the following properties:

1 $f(t)$ is single-valued everywhere; that is, $f(t)$ satisfies the mathematical definition of a function.

2 The integral $\int_{t_0}^{t_0+T} |f(t)|\, dt$ exists (i.e., is not infinite) for any choice of t_0.

3 $f(t)$ has a finite number of discontinuities in any one period.

4 $f(t)$ has a finite number of maxima and minima in any one period.

We shall consider $f(t)$ to represent a voltage or current waveform, and any voltage or current waveform which we can actually produce must satisfy these conditions. Certain mathematical functions which we might hypothesize may not satisfy these conditions, but we shall assume that the four conditions listed above are always satisfied.

Given such a periodic function $f(t)$, the Fourier theorem[1] states that $f(t)$ may be represented by the infinite series

$$f(t) = a_0 + a_1 \cos \omega_0 t + a_2 \cos 2\omega_0 t + \cdots + b_1 \sin \omega_0 t + b_2 \sin 2\omega_0 t + \cdots$$

$$= a_0 + \sum_{n=1}^{\infty} (a_n \cos n\omega_0 t + b_n \sin n\omega_0 t) \tag{1}$$

where the fundamental frequency ω_0 is related to the period T by

$$\omega_0 = \frac{2\pi}{T}$$

and where a_0, a_n, and b_n are constants which depend upon n and $f(t)$. Equation (1) is the trigonometric form of the *Fourier series for $f(t)$*, and the process of determining the values of the constants a_0, a_n, and b_n is called *Fourier analysis*. Our object is not the proof of this theorem, but only a simple development of the procedures of Fourier analysis and a feeling that the theorem is plausible.

Before we discuss the evaluation of the constants appearing in the Fourier series, let us collect a set of useful trigonometric integrals. We shall let both n and k represent any element of the set of integers 1, 2, 3, ... In the following integrals, 0 and T are used as the integration limits, but it is understood that any interval of one period is equally correct. Since the average value of a sinusoid over one period is zero,

$$\int_0^T \sin n\omega_0 t \, dt = 0 \tag{2}$$

and

$$\int_0^T \cos n\omega_0 t \, dt = 0 \tag{3}$$

It is also a simple matter to show that the following three definite integrals are zero:

$$\int_0^T \sin k\omega_0 t \cos n\omega_0 t \, dt = 0 \tag{4}$$

$$\int_0^T \sin k\omega_0 t \sin n\omega_0 t \, dt = 0 \qquad k \neq n \tag{5}$$

$$\int_0^T \cos k\omega_0 t \cos n\omega_0 t \, dt = 0 \qquad k \neq n \tag{6}$$

Those cases which are excepted in (5) and (6) are also easily evaluated; we obtain

---

[1] Jean-Baptiste Joseph Fourier published this theorem in 1822. Some rather unbelievable pronunciations of this French name come from American students; it should rhyme with "poor pay."

$$\int_0^T \sin^2 n\omega_0 t \, dt = \frac{T}{2} \tag{7}$$

$$\int_0^T \cos^2 n\omega_0 t \, dt = \frac{T}{2} \tag{8}$$

The evaluation of the unknown constants may now be accomplished readily. We first attack a_0. If we integrate each side of the Fourier series (1) over a full period, we obtain

$$\int_0^T f(t) \, dt = \int_0^T a_0 \, dt + \int_0^T \sum_{n=1}^{\infty} (a_n \cos n\omega_0 t + b_n \sin n\omega_0 t) \, dt$$

But every term in the summation is of the form of (2) or (3), and thus

$$\int_0^T f(t) \, dt = a_0 T$$

or
$$a_0 = \frac{1}{T} \int_0^T f(t) \, dt \tag{9}$$

This constant a_0 is simply the average value of $f(t)$ over a period, and we therefore describe it as the dc component of $f(t)$.

To evaluate one of the cosine coefficients, say a_k, the coefficient of $\cos k\omega_0 t$, we first multiply each side of (1) by $\cos k\omega_0 t$ and then integrate both sides of the equation over a full period:

$$\int_0^T f(t) \cos k\omega_0 t \, dt = \int_0^T a_0 \cos k\omega_0 t \, dt + \int_0^T \sum_{n=1}^{\infty} a_n \cos k\omega_0 t \cos n\omega_0 t \, dt$$
$$+ \int_0^T \sum_{n=1}^{\infty} b_n \cos k\omega_0 t \sin n\omega_0 t \, dt$$

From (3), (4), and (6) we note that every term on the right side of the equation is zero except for the single term where $k = n$. That term is evaluated by (8) and we have found a_k, or a_n:

$$a_n = \frac{2}{T} \int_0^T f(t) \cos n\omega_0 t \, dt \tag{10}$$

This result is *twice* the average value of the product $f(t) \cos n\omega_0 t$ over a period.

In a similar way, we obtain b_k by multiplying by $\sin k\omega_0 t$, integrating over a period, noting that all but one of the terms on the right side are zero, and performing that single integration by (7). The result is

$$b_n = \frac{2}{T} \int_0^T f(t) \sin n\omega_0 t \, dt \tag{11}$$

which is *twice* the average value of $f(t) \sin n\omega_0 t$ over a period.

Equations (9) to (11) now enable us to determine values for a_0 and all the a_n and b_n in the Fourier series (1):

$$f(t) = a_0 + \sum_{n=1}^{\infty} (a_n \cos n\omega_0 t + b_n \sin n\omega_0 t) \qquad (1)$$

$$\omega_0 = \frac{2\pi}{T} = 2\pi f_0$$

$$a_0 = \frac{1}{T} \int_0^T f(t)\, dt \qquad (9)$$

$$a_n = \frac{2}{T} \int_0^T f(t) \cos n\omega_0 t\, dt \qquad (10)$$

$$b_n = \frac{2}{T} \int_0^T f(t) \sin n\omega_0 t\, dt \qquad (11)$$

Let us consider a numerical example. The "half-sinusoidal" waveform shown in Fig. 17-2a represents the voltage response obtained at the output of a half-wave rectifier circuit, a nonlinear circuit whose purpose is to convert a sinusoidal input voltage to a (pulsating) dc output voltage. In order to represent this voltage as a Fourier series, we must first determine the period and then express the graphical voltage as an analytical function of time. From the graph, the period is seen to be

$$T = 0.4 \text{ s}$$

and thus

$$f_0 = 2.5 \text{ Hz}$$

and

$$\omega_0 = 5\pi \text{ rad/s}$$

Fig. 17-2: (a) The output of a half-wave rectifier to which a sinusoidal input is applied. (b) The discrete line spectrum of the waveform in (a).

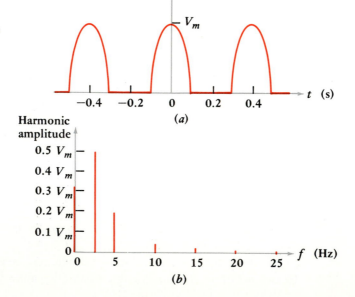

With these three quantities determined, we now seek an appropriate expression for $f(t)$ or $v(t)$ which is valid throughout the period. Obtaining this equation or set of equations proves to be the most difficult part of Fourier analysis for many students. The source of the difficulty is apparently either the inability to recognize the given curve, carelessness in determining multiplying constants within the functional expression, or negligence in not writing the complete expression. In this example, the statement of the problem infers that the functional form is a sinusoid whose amplitude is V_m. The radian frequency has already been determined as 5π, and only the positive portion of the cosine wave is present. The functional expression over the period $t = 0$ to $t = 0.4$ is therefore

$$v(t) = \begin{cases} V_m \cos 5\pi t & 0 \leq t \leq 0.1 \\ 0 & 0.1 \leq t \leq 0.3 \\ V_m \cos 5\pi t & 0.3 \leq t \leq 0.4 \end{cases}$$

It is evident that the choice of the period extending from $t = -0.1$ to $t = 0.3$ will result in fewer equations and, hence, fewer integrals:

$$v(t) = \begin{cases} V_m \cos 5\pi t & -0.1 \leq t \leq 0.1 \\ 0 & 0.1 \leq t \leq 0.3 \end{cases} \tag{12}$$

This form is preferable, although either description will yield the correct results.

The zero-frequency component is easily obtained:

$$a_0 = \frac{1}{0.4} \int_{-0.1}^{0.3} v(t)\, dt$$

$$= \frac{1}{0.4} \left[\int_{-0.1}^{0.1} V_m \cos 5\pi t\, dt + \int_{0.1}^{0.3} (0)\, dt \right]$$

and
$$a_0 = \frac{V_m}{\pi} \tag{13}$$

Notice that integration over an entire period must be broken up into subintervals of the period, in each of which the functional form of $v(t)$ is known.

The amplitude of a general cosine term is

$$a_n = \frac{2}{0.4} \int_{-0.1}^{0.1} V_m \cos 5\pi t \cos 5\pi n t\, dt$$

The form of the function we obtain upon integrating is different when n is unity than it is for any other choice of n. If $n = 1$, we have

$$a_1 = 5V_m \int_{-0.1}^{0.1} \cos^2 5\pi t\, dt = \frac{V_m}{2} \tag{14}$$

whereas if n is not equal to unity, we find

$$a_n = 5V_m \int_{-0.1}^{0.1} \cos 5\pi t \cos 5\pi nt \; dt$$

$$= 5V_m \int_{-0.1}^{0.1} \tfrac{1}{2}[\cos 5\pi(1+n)t + \cos 5\pi(1-n)t] \; dt$$

or $\qquad a_n = \dfrac{2V_m}{\pi} \dfrac{\cos(\pi n/2)}{1-n^2} \qquad n \neq 1 \qquad\qquad$ (15)

Some of the details of the integration have been left out for those who prefer to work out the small tedious steps for themselves. It should be pointed out, incidentally, that the expression for a_n when $n \neq 1$ will yield the correct result for $n = 1$ in the limit.

A similar integration shows that $b_n = 0$ for any value of n, and the Fourier series thus contains no sine terms. The Fourier series is therefore obtained from (13), (14), and (15):

$$v(t) = \frac{V_m}{\pi} + \frac{V_m}{2} \cos 5\pi t + \frac{2V_m}{3\pi} \cos 10\pi t - \frac{2V_m}{15\pi} \cos 20\pi t$$

$$+ \frac{2V_m}{35\pi} \cos 30\pi t - \cdots \quad (16)$$

In Fig. 17-2a, $v(t)$ is shown as a function of time; in (12), $v(t)$ is expressed as an analytical function of time. Either of these representations is a time-domain representation. Equation (16), the Fourier series representation of $v(t)$, is also a time-domain expression, but it may be transformed easily into a frequency-domain representation. For example, we could locate the points in the **s** plane that represent the frequencies present in (16). The result would be a mark at the origin and symmetrical marks on the positive and negative $j\omega$ axis. A more customary method of presenting this information, and one which shows the amplitude of each frequency component, is by a *line spectrum*. A line spectrum for (16) is shown in Fig. 17-2b; the amplitude of each frequency component is indicated by the length of the vertical line located at the corresponding frequency. We also speak of this spectrum as a *discrete* spectrum because any finite frequency interval contains only a finite number of frequency components.

One note of caution must be injected. The example we have considered contains no sine terms, and the amplitude of the nth harmonic is therefore $|a_n|$. If b_n is not zero, then the amplitude of the component at a frequency $n\omega_0$ must be $\sqrt{a_n^2 + b_n^2}$. This is the general quantity which we must show in a line spectrum. When we discuss the complex form of the Fourier series, we shall see that this quantity is obtained more directly.

In addition to the amplitude spectrum, we also may construct a discrete *phase spectrum*. At the frequency $n\omega_0$, we combine the cosine and sine terms to determine the phase angle ϕ_n:

$$a_n \cos n\omega_0 t + b_n \sin n\omega_0 t = \sqrt{a_n^2 + b_n^2} \cos\left(n\omega_0 t + \tan^{-1}\frac{-b_n}{a_n}\right)$$

$$= \sqrt{a_n^2 + b_n^2} \cos(n\omega_0 t + \phi_n)$$

or
$$\phi_n = \tan^{-1} \frac{-b_n}{a_n}$$

In (16), $\phi_n = 0°$ or $180°$ for every n.

The Fourier series obtained for this example includes no sine terms and no odd harmonics (except the fundamental) among the cosine terms. It is possible to anticipate the absence of certain terms in a Fourier series, before any integrations are performed, by an inspection of the symmetry of the given time function. We shall investigate the use of symmetry in the following section.

Drill Problems

17-2. A periodic waveform $f(t)$ is described as follows: $f(t) = 5, 0 < t < 0.1$; $f(t) = -3, 0.1 < t < 0.3$; $f(t) = 0, 0.3 < t < 0.5$; $T = 0.5$. Evaluate: (a) a_0; (b) a_4; (c) b_1.   *Ans:* -0.200; -0.746; 0.0321

17-3. Write the Fourier series for the three voltage waveforms shown in Fig. 17-3.

$$Ans: \frac{4}{\pi} \left(\sin \pi t + \frac{1}{3} \sin 3\pi t + \frac{1}{5} \sin 5\pi t + \cdots \right) V;$$

$$\frac{4}{\pi} \left(\cos \pi t - \frac{1}{3} \cos 3\pi t + \frac{1}{5} \cos 5\pi t - \cdots \right) V;$$

$$\frac{8}{\pi^2} \left(\sin \pi t - \frac{1}{9} \sin 3\pi t + \frac{1}{25} \sin 5\pi t - \cdots \right) V$$

(a)

(b)

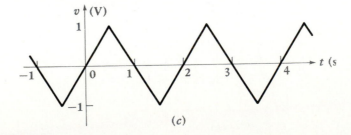

(c)

Fig. 17-3: See Drill Probs. 17-3 and 17-7.

17-3
The use of
symmetry

The two types of symmetry which are most readily recognized are *even-function symmetry* and *odd-function symmetry*, or simply *even symmetry* and *odd symmetry*. We say that $f(t)$ possesses the property of even symmetry if

$$f(t) = f(-t) \tag{17}$$

Such functions as t^2, cos $3t$, ln (cos t), $\sin^2 7t$, and a constant C all possess even symmetry; the replacement of t by $(-t)$ does not change the value of any of these functions. This type of symmetry may also be recognized graphically, for if $f(t) = f(-t)$ then mirror symmetry exists about the $f(t)$ axis. The function shown in Fig. 17-4a possesses even symmetry; if the figure were to be folded along the $f(t)$ axis, then the portions of the graph of the function for positive and negative time would fit exactly, one on top of the other.

We define odd symmetry by stating that if odd symmetry is a property of $f(t)$, then

$$f(t) = -f(-t) \tag{18}$$

In other words, if t is replaced by $(-t)$, then the negative of the given function is obtained; for example, t, sin t, t cos $70t$, $t\sqrt{1 + t^2}$, and the function sketched in Fig. 17-4b are all odd functions and possess odd symmetry. The graphical characteristics of odd symmetry are apparent if the portion of $f(t)$ for $t > 0$ is rotated about the positive t axis and the resultant figure is then rotated about the $f(t)$ axis; the two curves will fit exactly, one on top of the other. That is, we now have symmetry about the origin, rather than about the $f(t)$ axis as we did for even functions.

Having definitions for even and odd symmetry, we should note that the product of two functions with even symmetry, or of two functions with odd

Fig. 17-4: (a) A waveform showing even symmetry. (b) A waveform showing odd symmetry.

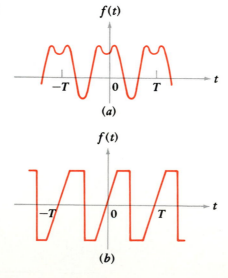

(a)

(b)

symmetry, yields a function with even symmetry. Furthermore, the product of an even and an odd function gives a function with odd symmetry.

Now let us investigate the effects that even symmetry produces in a Fourier series. If we think of the expression which equates an even function $f(t)$ and the sum of an infinite number of sine and cosine functions, then it is apparent that the sum must also be an even function. A sine wave, however, is an odd function, and no sum of sine waves can produce any even function other than zero (which is both even and odd). It is thus plausible that the Fourier series of any even function is composed of only a constant and cosine functions. Let us now show carefully that $b_n = 0$. We have

$$b_n = \frac{2}{T} \int_{-T/2}^{T/2} f(t) \sin n\omega_0 t \, dt$$

$$= \frac{2}{T} \left[\int_{-T/2}^{0} f(t) \sin n\omega_0 t \, dt + \int_{0}^{T/2} f(t) \sin n\omega_0 t \, dt \right]$$

Now let us replace the variable t in the first integral by $-\tau$, or $\tau = -t$, and make use of the fact that $f(t) = f(-t) = f(\tau)$:

$$b_n = \frac{2}{T} \left[\int_{T/2}^{0} f(-\tau) \sin(-n\omega_0 \tau)(-d\tau) + \int_{0}^{T/2} f(t) \sin n\omega_0 t \, dt \right]$$

$$= \frac{2}{T} \left[-\int_{0}^{T/2} f(\tau) \sin n\omega_0 \tau \, d\tau + \int_{0}^{T/2} f(t) \sin n\omega_0 t \, dt \right]$$

But the symbol we use to identify the variable of integration cannot affect the value of the integral. Thus,

$$\int_{0}^{T/2} f(\tau) \sin n\omega_0 \tau \, d\tau = \int_{0}^{T/2} f(t) \sin n\omega_0 t \, dt$$

and
$$b_n = 0 \quad \text{(even sym.)} \tag{19}$$

No sine terms are present. Therefore, if $f(t)$ shows even symmetry, then $b_n = 0$; conversely, if $b_n = 0$, then $f(t)$ must have even symmetry.

A similar examination of the expression for a_n leads to an integral over the *half period* extending from $t = 0$ to $t = \frac{1}{2}T$:

$$a_n = \frac{4}{T} \int_{0}^{T/2} f(t) \cos n\omega_0 t \, dt \quad \text{(even sym.)} \tag{20}$$

The fact that a_n may be obtained for an even function by taking "twice the integral over half the range" should seem logical.

A function having odd symmetry can contain no constant term or cosine terms in its Fourier expansion. Let us prove the second part of this statement. We have

$$a_n = \frac{2}{T} \int_{-T/2}^{T/2} f(t) \cos n\omega_0 t \, dt$$

$$= \frac{2}{T} \left[\int_{-T/2}^{0} f(t) \cos n\omega_0 t \, dt + \int_{0}^{T/2} f(t) \cos n\omega_0 t \, dt \right]$$

and we now let $t = -\tau$ in the first integral,

$$a_n = \frac{2}{T}\left[\int_{T/2}^{0} f(-\tau) \cos(-n\omega_0\tau)(-d\tau) + \int_{0}^{T/2} f(t) \cos n\omega_0 t\, dt\right]$$

$$= \frac{2}{T}\left[\int_{0}^{T/2} f(-\tau) \cos n\omega_0\tau\, d\tau + \int_{0}^{T/2} f(t) \cos n\omega_0 t\, dt\right]$$

But $f(-\tau) = -f(\tau)$, and therefore

$$a_n = 0 \qquad \text{(odd sym.)} \qquad\qquad (21)$$

A similar, but simpler, proof shows that

$$a_0 = 0 \qquad \text{(odd sym.)}$$

With odd symmetry, therefore, $a_n = 0$ and $a_0 = 0$, and, conversely, if $a_n = 0$ and $a_0 = 0$, odd symmetry is present.

The values of b_n may again be obtained by integrating over half the range:

$$b_n = \frac{4}{T}\int_{0}^{T/2} f(t) \sin n\omega_0 t\, dt \qquad \text{(odd sym.)} \qquad\qquad (22)$$

Examples of even and odd symmetry are afforded by Drill Prob. 17-3, preceding this section. In both parts (a) and (b), a square wave of the same amplitude and period is the given function. The time origin, however, is selected to provide odd symmetry in (a) and even symmetry in (b), and the resultant series contain, respectively, only sine terms and only cosine terms. It is also noteworthy that the point at which $t = 0$ could be selected to provide neither even nor odd symmetry; the determination of the coefficients of the terms in the Fourier series then takes at least twice as long.

The Fourier series for both of these square waves have one other interesting characteristic; neither contains any even *harmonics*.[2] That is, the only frequency components present in the series have frequencies which are odd multiples of the fundamental frequency; a_n and b_n are zero for n even. This result is caused by another type of symmetry, called half-wave symmetry. We shall say that $f(t)$ possesses *half-wave symmetry* if

$$f(t) = -f(t - \tfrac{1}{2}T)$$

or the equivalent expression,

$$f(t) = -f(t + \tfrac{1}{2}T)$$

Except for a change of sign, each half cycle is like the adjacent half cycles. Half-wave symmetry, unlike even and odd symmetry, is not a function of the choice of the point $t = 0$. Thus, we can state that the square wave (Fig. 17-3a or b) shows half-wave symmetry. Neither waveform shown in Fig. 17-4 has

---

[2] Constant vigilance is required to avoid confusion between an even function and an even harmonic, or an odd function and an odd harmonic. For example, b_{10} is the coefficient of an even harmonic, and it is zero if $f(t)$ is an even function.

(a)

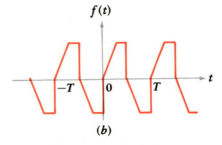

(b)

Fig. 17-5: (a) A waveform somewhat similar to the one shown in Fig. 17-4a, but possessing half-wave symmetry. (b) A waveform somewhat similar to the one shown in Fig. 17-4b, but possessing half-wave symmetry.

half-wave symmetry, but the two somewhat similar functions plotted in Fig. 17-5 do possess half-wave symmetry.

It may be shown that the Fourier series of any function which has half-wave symmetry contains only odd harmonics. Let us consider the coefficients a_n. We have again

$$a_n = \frac{2}{T} \int_{-T/2}^{T/2} f(t) \cos n\omega_0 t \; dt$$

$$= \frac{2}{T} \left[\int_{-T/2}^{0} f(t) \cos n\omega_0 t \; dt + \int_{0}^{T/2} f(t) \cos n\omega_0 t \; dt \right]$$

which we may represent as

$$a_n = \frac{2}{T} (I_1 + I_2)$$

Now we substitute the new variable $\tau = t + \frac{1}{2}T$ into integral I_1:

$$I_1 = \int_{0}^{T/2} f(\tau - \tfrac{1}{2}T) \cos n\omega_0(\tau - \tfrac{1}{2}T) \; d\tau$$

$$= \int_{0}^{T/2} -f(\tau) \left(\cos n\omega_0\tau \cos \frac{n\omega_0 T}{2} + \sin n\omega_0\tau \sin \frac{n\omega_0 T}{2} \right) d\tau$$

But $\omega_0 T$ is 2π, and thus

$$\sin \frac{n\omega_0 T}{2} = \sin n\pi = 0$$

Hence

$$I_1 = -\cos n\pi \int_{0}^{T/2} f(\tau) \cos n\omega_0\tau \; d\tau$$

After noting the form of I_2, we therefore may write

$$a_n = \frac{2}{T}(1 - \cos n\pi) \int_0^{T/2} f(t) \cos n\omega_0 t \, dt$$

The factor $(1 - \cos n\pi)$ indicates that a_n is zero if n is even. Thus,

$$
\left.
\begin{aligned}
a_n &= \frac{4}{T} \int_0^{T/2} f(t) \cos n\omega_0 t \, dt && n \text{ odd} \\
&= 0 && n \text{ even}
\end{aligned}
\right\} \quad \text{(half-wave sym.)} \quad (23)
$$

A similar investigation shows that b_n is also zero for all even n and, therefore,

$$
\left.
\begin{aligned}
b_n &= \frac{4}{T} \int_0^{T/2} f(t) \sin n\omega_0 t \, dt && n \text{ odd} \\
&= 0 && n \text{ even}
\end{aligned}
\right\} \quad \text{(half-wave sym.)} \quad (24)
$$

It should be noted that half-wave symmetry may be present in a waveform which also shows odd symmetry or even symmetry. The waveform sketched in Fig. 17-5a, for example, possesses both even symmetry and half-wave symmetry. When a waveform possesses half-wave symmetry and either even or odd symmetry, then it is possible to reconstruct the waveform if the function is known over any quarter-period interval. The value of a_n or b_n may also be found by integrating over any quarter period. Thus,

$$
\left.
\begin{aligned}
a_n &= \frac{8}{T} \int_0^{T/4} f(t) \cos n\omega_0 t \, dt && n \text{ odd} \\
a_n &= 0 && n \text{ even} \\
b_n &= 0 && \text{all } n
\end{aligned}
\right\} \quad \begin{matrix}\text{(half-wave and} \\ \text{even sym.)}\end{matrix} \quad (25)
$$

and

$$
\left.
\begin{aligned}
a_n &= 0 && \text{all } n \\
b_n &= \frac{8}{T} \int_0^{T/4} f(t) \sin n\omega_0 t \, dt && n \text{ odd} \\
b_n &= 0 && n \text{ even}
\end{aligned}
\right\} \quad \begin{matrix}\text{(half-wave and} \\ \text{odd sym.)}\end{matrix} \quad (26)
$$

It is always worthwhile spending a few moments investigating the symmetry of a function for which a Fourier series is to be determined.

Drill Problems

17-4. Sketch each of the functions described below, state whether or not even symmetry, odd symmetry, and half-wave symmetry are present, and give the period: (a) $v = 0$, $-1 < t < 1$ and $3 < t < 5$; $v = 20$, $1 < t < 3$; $v = -20$, $5 < t < 7$; repeats; (b) $v = 15$, $0 < t < 2$; $v = 0$, $2 < t < 6$; $v = -15$, $6 < t < 8$; repeats; (c) $v = 10|t|$, $-1 < t < 1$; $v = 0$, $1 < t < 3$; repeats.

Ans: No, yes, yes, 8; no, yes, no, 8; yes, no, no, 4

17-5. Determine the Fourier series for each waveform described in Drill Prob. 17-4.

$$Ans: \frac{80}{\pi} \sum_{n=1,odd}^{\infty} \frac{1}{n} \cos \frac{n\pi}{4} \sin \frac{n\pi t}{4}; \quad \frac{30}{\pi} \sum_{n=1}^{\infty} \frac{1}{n} \left(1 - \cos \frac{n\pi}{2}\right) \sin \frac{n\pi t}{4};$$

$$2.5 + \frac{40}{\pi^2} \sum_{n=1}^{\infty} \frac{1}{n^2} \left(\frac{n\pi}{2} \sin \frac{n\pi}{2} + \cos \frac{n\pi}{2} - 1\right) \cos \frac{n\pi t}{2}$$

17-4
Complete response to periodic forcing functions

Through the use of the Fourier series, we may now express an arbitrary periodic forcing function as the sum of an infinite number of sinusoidal forcing functions. The forced response to each of these functions may be determined by conventional steady-state analysis, and the form of the natural response may be determined from the poles of the appropriate network transfer function. The initial conditions existing throughout the network, including the initial value of the forced response, enable the amplitude of the natural response to be selected; and then the complete response is obtained as the sum of the forced and natural responses. Let us illustrate this general procedure by a specific example.

We shall apply the square wave of Fig. 17-6a, which includes a dc component, to the series RL circuit, as shown in Fig. 17-6b. The forcing function is applied at $t = 0$, and the current is the desired response. Its initial value is zero. The forcing function has a fundamental frequency $\omega_0 = 2$ rad/s, and its Fourier series may be written down by comparison with the solution of Drill Prob. 17-3a:

$$v_s(t) = 5 + \frac{20}{\pi} \sum_{n=1,odd}^{\infty} \frac{\sin 2nt}{n}$$

We shall find the forced response of the nth harmonic by working in that frequency domain. Thus,

$$v_{sn}(t) = \frac{20}{\pi n} \sin 2nt$$

and

$$\mathbf{V}_{sn} = \frac{20}{\pi n} (-j1)$$

The impedance offered by the RL circuit at this frequency is

$$\mathbf{Z}_n = 4 + j(2n)2 = 4 + j4n$$

Fig. 17-6: (a) A square-wave voltage forcing function. (b) The forcing function of (a) is applied to this series RL circuit at $t = 0$; the complete response $i(t)$ is desired.

(a)

(b)

and thus the component of the forced response at this frequency is

$$\mathbf{I}_{fn} = \frac{\mathbf{V}_{sn}}{\mathbf{Z}_n} = \frac{-j5}{\pi n(1 + jn)}$$

Transforming to the time domain, we have

$$i_{fn}(t) = \frac{5}{\pi n} \frac{1}{\sqrt{1 + n^2}} \cos(2nt - 90° - \tan^{-1} n)$$

$$= \frac{5}{\pi(1 + n^2)} \left(\frac{\sin 2nt}{n} - \cos 2nt \right)$$

Since the response to the dc component is obviously 1.25 A, the forced response may be expressed as the summation

$$i_f(t) = 1.25 + \frac{5}{\pi} \sum_{n=1,\text{odd}}^{\infty} \left[\frac{\sin 2nt}{n(1 + n^2)} - \frac{\cos 2nt}{1 + n^2} \right]$$

The familiar natural response of this simple circuit is the single exponential term [characterizing the single pole of the transfer function, $\mathbf{I}_f/\mathbf{V}_s = 1/(4 + 2s)$]

$$i_n(t) = Ae^{-2t}$$

The complete response is therefore the sum

$$i(t) = i_f(t) + i_n(t)$$

and, since $i(0) = 0$, it is necessary to select A so that

$$A = -i_f(0)$$

Letting $t = 0$, we find that $i_f(0)$ is given by

$$i_f(0) = 1.25 - \frac{5}{\pi} \sum_{n=1,\text{odd}}^{\infty} \frac{1}{1 + n^2}$$

Although we could express A in terms of this summation, it is more convenient to use the numerical value of the summation. The sum of the first five terms of $\sum 1/(1 + n^2)$ is 0.671, the sum of the first ten terms is 0.695, the sum of the first twenty terms is 0.708, and the exact sum[3] is 0.720 to three significant figures. Thus

$$A = -1.25 + \frac{5}{\pi} 0.720 = -0.104$$

and

$$i(t) = -0.104e^{-2t} + 1.25 + \frac{5}{\pi} \sum_{n=1,\text{odd}}^{\infty} \left[\frac{\sin 2nt}{n(1 + n^2)} - \frac{\cos 2nt}{1 + n^2} \right]$$

[3] The sum of this series is known in closed form: $\displaystyle\sum_{n=1,\text{odd}}^{\infty} \frac{1}{1 + n^2} = \frac{\pi}{4} \tanh \frac{\pi}{2}.$

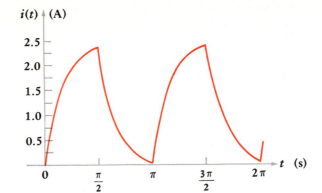

Fig. 17-7: The initial portion of the complete response of the circuit of Fig. 17-6b to the forcing function of Fig. 17-6a.

In obtaining this solution, we have had to use many of the most general concepts introduced in this and the preceding sixteen chapters. Some we did not have to use because of the simple nature of this particular circuit, but their places in the general analysis were indicated above. In this sense, we may look upon the solution of this problem as a significant achievement in our introductory study of circuit analysis. In spite of this glorious feeling of accomplishment, however, it must be pointed out that the complete response, as obtained above in analytical form, is not of much value as it stands; it furnishes no clear picture of the nature of the response. What we really need is a sketch of $i(t)$ as a function of time. This may be obtained by a laborious calculation at a sufficient number of instants of time; an available digital computer or a programmable hand calculator can be of great assistance here. The sketch may be approximated by the graphical addition of the natural response, the dc term, and the first few harmonics; this is an unrewarding task. When all is said and done, the most informative solution of this problem is probably obtained by making a repeated transient analysis. That is, the form of the response can certainly be calculated in the interval from $t = 0$ to $t = \pi/2$ s; it is an exponential rising toward 2.5 A. After determining the value at the end of this first interval, we have an initial condition for the next $\pi/2$-s interval. The process is repeated until the response assumes a generally periodic nature. The method is eminently suitable to this example, for there is negligible change in the current waveform in the successive periods $\pi/2 < t < 3\pi/2$ and $3\pi/2 < t < 5\pi/2$. The complete current response is sketched in Fig. 17-7.

Drill Problem

17-6. Use the methods of Chap. 6 to determine the value of the current sketched in Fig. 17-7 at $t = $: (a) $\pi/2$; (b) π; (c) $3\pi/2$.

Ans: 2.392; 0.1034; 2.396 A

17-5
**Complex form of
the Fourier series**

In obtaining a frequency spectrum, we have seen that the amplitude of each frequency component depends on both a_n and b_n; that is, the sine term and the cosine term both contribute to the amplitude. The exact expression for this amplitude is $\sqrt{a_n^2 + b_n^2}$. It is also possible to obtain the amplitude directly by using a form of Fourier series in which each term is a cosine function with a phase angle; the amplitude and phase angle are functions of $f(t)$ and n. An even more convenient and concise form of the Fourier series is obtained if the sines and cosines are expressed as exponential functions with complex multiplying constants.

Let us first take the trigonometric form of the Fourier series:

$$f(t) = a_0 + \sum_{n=1}^{\infty} (a_n \cos n\omega_0 t + b_n \sin n\omega_0 t)$$

and then substitute the exponential forms for the sine and cosine. After rearranging,

$$f(t) = a_0 + \sum_{n=1}^{\infty} \left(e^{jn\omega_0 t} \frac{a_n - jb_n}{2} + e^{-jn\omega_0 t} \frac{a_n + jb_n}{2} \right)$$

We now define a complex constant $\mathbf{c}_n$:

$$\mathbf{c}_n = \tfrac{1}{2}(a_n - jb_n) \qquad (n = 1,2,3, \ldots) \tag{27}$$

The values of a_n, b_n, and $\mathbf{c}_n$ all depend on n and $f(t)$. Suppose we now replace n by $(-n)$; how do the values of the constants change? The coefficients a_n and b_n are defined by (10) and (11), and it is evident that

$$a_{-n} = a_n$$

but

$$b_{-n} = -b_n$$

From (27), then,

$$\mathbf{c}_{-n} = \tfrac{1}{2}(a_n + jb_n) \qquad (n = 1,2,3, \ldots) \tag{28}$$

Thus,

$$\mathbf{c}_n = \mathbf{c}_{-n}^*$$

We also let

$$\mathbf{c}_0 = a_0$$

We may therefore express $f(t)$ as

$$f(t) = \mathbf{c}_0 + \sum_{n=1}^{\infty} \mathbf{c}_n e^{jn\omega_0 t} + \sum_{n=1}^{\infty} \mathbf{c}_{-n} e^{-jn\omega_0 t}$$

or

$$f(t) = \sum_{n=0}^{\infty} \mathbf{c}_n e^{jn\omega_0 t} + \sum_{n=1}^{\infty} \mathbf{c}_{-n} e^{-jn\omega_0 t}$$

Finally, instead of summing the second series over the positive integers from 1 to ∞, let us sum over the negative integers, from -1 to $-\infty$:

$$f(t) = \sum_{n=0}^{\infty} \mathbf{c}_n e^{jn\omega_0 t} + \sum_{n=-1}^{-\infty} \mathbf{c}_n e^{jn\omega_0 t}$$

or
$$f(t) = \sum_{n=-\infty}^{n=\infty} \mathbf{c}_n e^{jn\omega_0 t} \tag{29}$$

By agreement, a summation from $-\infty$ to ∞ is understood to include a term for $n = 0$.

Equation (29) is the complex form of the Fourier series for $f(t)$; its conciseness is one of the most important reasons for its use. In order to obtain the expression by which a particular complex coefficient $\mathbf{c}_n$ may be evaluated, we substitute (10) and (11) into (27):

$$\mathbf{c}_n = \frac{1}{T} \int_{-T/2}^{T/2} f(t) \cos n\omega_0 t \, dt - j\frac{1}{T} \int_{-T/2}^{T/2} f(t) \sin n\omega_0 t \, dt$$

use the exponential equivalents of the sine and cosine, and simplify:

$$\mathbf{c}_n = \frac{1}{T} \int_{-T/2}^{T/2} f(t) e^{-jn\omega_0 t} \, dt \tag{30}$$

Thus, a single concise equation serves to replace the two equations required for the trigonometric form of the Fourier series. Instead of evaluating two integrals to find the Fourier coefficients, only one integration is required; moreover, it is almost always a simpler integration. It should be noted that the integral (30) contains the multiplying factor $1/T$, whereas the integrals for a_n and b_n both contain the factor $2/T$.

Collecting the two basic relationships for the exponential form of the Fourier series, we have

$$f(t) = \sum_{n=-\infty}^{\infty} \mathbf{c}_n e^{jn\omega_0 t} \tag{29}$$

$$\mathbf{c}_n = \frac{1}{T} \int_{-T/2}^{T/2} f(t) e^{-jn\omega_0 t} \, dt \tag{30}$$

where $\omega_0 = 2\pi/T$ as usual.

The amplitude of the component of the exponential Fourier series at $\omega = n\omega_0$, where $n = 0, \pm 1, \pm 2, \ldots$, is $|\mathbf{c}_n|$. We may plot a discrete frequency spectrum giving $|\mathbf{c}_n|$ versus $n\omega_0$ or nf_0 with an abscissa that shows both positive and negative values, and when we do the graph is symmetrical about the origin since (27) and (28) show that $|\mathbf{c}_n| = |\mathbf{c}_{-n}|$.

We note also from (29) and (30) that the amplitude of the sinusoidal component at $\omega = n\omega_0$, where $n = 1, 2, 3, \ldots$, is $\sqrt{a_n^2 + b_n^2} = 2|\mathbf{c}_n| = 2|\mathbf{c}_{-n}| = |\mathbf{c}_n| + |\mathbf{c}_{-n}|$.

For the dc component, $a_0 = \mathbf{c}_0$.

Let us consider a numerical example to illustrate the use of the complex form of the Fourier series. A train of rectangular pulses of amplitude V_0 and duration τ, recurring periodically every T seconds, is shown in Fig. 17-8a.

(a)

(b)

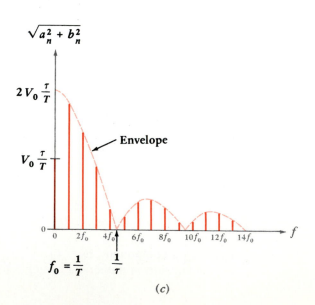

Fig. 17-8: (a) A periodic sequence of rectangular pulses. (b) The corresponding discrete line spectrum for $|\mathbf{c}_n|$, $f = nf_0$, $n = 0, \pm1, \pm2, \ldots$ (c) $\sqrt{a_n^2 + b_n^2}$ versus $f = nf_0$, $n = 0, 1, 2, \ldots$

(c)

The fundamental frequency is, therefore,

$$f_0 = \frac{1}{T}$$

The value of a general complex coefficient is first found from (30):

$$\mathbf{c}_n = \frac{1}{T} \int_{-T/2}^{T/2} f(t) e^{-jn\omega_0 t} \, dt$$

$$= \frac{V_0}{T} \int_{t_0}^{t_0+\tau} e^{-jn\omega_0 t} \, dt$$

$$= \frac{V_0}{-jn\omega_0 T} (e^{-jn\omega_0(t_0+\tau)} - e^{-jn\omega_0 t_0})$$

$$= \frac{2V_0}{n\omega_0 T} e^{-jn\omega_0(t_0+\tau/2)} \sin\left(\tfrac{1}{2}n\omega_0\tau\right)$$

$$= \frac{V_0\tau}{T} \frac{\sin\left(\tfrac{1}{2}n\omega_0\tau\right)}{\tfrac{1}{2}n\omega_0\tau} e^{-jn\omega_0(t_0+\tau/2)}$$

The magnitude of $\mathbf{c}_n$ is therefore

$$|\mathbf{c}_n| = \frac{V_0\tau}{T} \left| \frac{\sin\left(\tfrac{1}{2}n\omega_0\tau\right)}{\tfrac{1}{2}n\omega_0\tau} \right| \tag{31}$$

and the angle of $\mathbf{c}_n$ is

$$\text{ang } \mathbf{c}_n = -n\omega_0\left(t_0 + \frac{\tau}{2}\right) \qquad \text{(possibly plus 180°)} \tag{32}$$

The trigonometric factor in (31) occurs frequently in modern communication theory, and it is called the *sampling function*. The "sampling" refers to the time function of Fig. 17-8a from which the sampling function is derived. The product of this sequence of pulses and any other function $f(t)$ represents *samples* of $f(t)$ every T seconds if τ is small and $V_0 = 1$. We define

$$\text{Sa}(x) = \frac{\sin x}{x}$$

Because of the way in which it helps to determine the amplitude of the various frequency components in $v(t)$, it is worthwhile discovering the important characteristics of this function. First, we note the $\text{Sa}(x)$ is zero whenever x is an integral multiple of π; that is,

$$\text{Sa}(n\pi) = 0 \qquad n = 1, 2, 3, \ldots$$

When x is zero, the function is indeterminate, but it is easy to show that its value is unity:

$$\text{Sa}(0) = 1$$

The magnitude of $\text{Sa}(x)$ therefore decreases from unity at $x = 0$ to zero at $x = \pi$. As x increases from π to 2π, $|\text{Sa}(x)|$ increases from zero to a maximum less than unity, and then decreases to zero once again. As x continues to increase, the successive maxima continually become smaller because the numerator of Sa(x) cannot exceed unity and the denominator is continually increasing. Also, Sa(x) shows even symmetry.

Now let us construct the line spectrum. We first consider $|\mathbf{c}_n|$, writing (31) in terms of the fundamental cyclic frequency f_0:

$$|\mathbf{c}_n| = \frac{V_0\tau}{T} \left| \frac{\sin(n\pi f_0\tau)}{n\pi f_0\tau} \right| \tag{33}$$

The amplitude of any $\mathbf{c}_n$ is obtained from (33) by using the known values τ and $T = 1/f_0$ and selecting the desired value of n, $n = 0, \pm 1, \pm 2, \ldots$. Instead of evaluating (33) at these discrete frequencies, let us sketch the *envelope* of $|\mathbf{c}_n|$ by considering the frequency nf_0 to be a continuous variable. That is, f, which is nf_0, can actually take on only the discrete values of the harmonic frequencies $0, \pm f_0, \pm 2f_0, \pm 3f_0$, and so forth, but we may think of n for the moment as a continuous variable. When f is zero, $|\mathbf{c}_n|$ is evidently $V_0\tau/T$, and when f has increased to $1/\tau$, $|\mathbf{c}_n|$ is zero. The resultant envelope is sketched as the broken line of Fig. 17-8b. The line spectrum is then obtained by simply erecting vertical lines at each harmonic frequency, as shown in the sketch. The amplitudes shown are those of the $\mathbf{c}_n$. The particular case sketched applies to the case where $\tau/T = 1/(1.5\pi) = 0.212$. In this example, it happens that there is no harmonic exactly at that frequency at which the envelope amplitude is zero; another choice of τ or T could produce such an occurrence, however.

In Fig. 17-8c, the amplitude of the sinusoidal component is plotted as a function of frequency. Note again that $a_0 = \mathbf{c}_0$ and $\sqrt{a_n^2 + b_n^2} = |\mathbf{c}_n| + |\mathbf{c}_{-n}|$.

There are several observations and conclusions which we may make about the line spectrum of a periodic sequence of rectangular pulses, as given in Fig. 17-8c. With respect to the envelope of the discrete spectrum, it is evident that the "width" of the envelope depends upon τ, and not upon T. As a matter of fact, the shape of the envelope is not a function of T. It follows that the bandwidth of a filter which is designed to pass the periodic pulses is a function of the pulse width τ, but not of the pulse period T; an inspection of Fig. 17-8c indicates that the required bandwidth is about $1/\tau$ Hz. If the pulse period T is increased (or the pulse repetition frequency f_0 is decreased), the bandwidth $1/\tau$ does not change, but the number of spectral lines between zero frequency and $1/\tau$ Hz increases, albeit discontinuously; the amplitude of each line is inversely proportional to T. Finally, a shift in the time origin does not change the line spectrum; that is, $|\mathbf{c}_n|$ is not a function of t_0. The relative phases of the frequency components do change with the choice of t_0.

Drill Problem

17-7. Determine the general coefficient $\mathbf{c}_n$ in the complex Fourier series for each of the three waveforms shown in Fig. 17-3.

$$Ans: -j\frac{1}{n\pi}(1 - \cos n\pi); \quad \frac{2}{n\pi}\sin\frac{n\pi}{2}; \quad -j\frac{4}{n^2\pi^2}\sin\frac{n\pi}{2}$$

Problems

1 If $f(t) = 2 + 3 \cos (10\pi t + 30°) + 4 \cos (20\pi t + 60°) + \cos (30\pi t + 90°)$, find: (a) the average value of $f(t)$; (b) the effective value of $f(t)$; (c) the period of $f(t)$; (d) the value of $f(t)$ at $t = 0.05$ s.

2 (a) Sketch $f(t) = 2 \cos 2\pi t - \sin 4\pi t + 2 \sin 6\pi t$ over the interval $0 < t < T$. (b) What is the maximum (positive) value of $f(t)$ in this interval?

3 The function shown in Fig. 17-9 is periodic with $T = 10$ s. Find: (a) the average value; (b) the effective value; (c) the amplitude of the third harmonic.

Fig. 17-9: See Prob. 3.

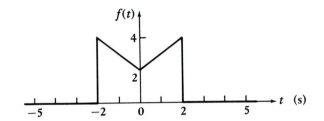

4 For the periodic voltage shown in Fig. 17-10, find: (a) T; (b) f_0; (c) ω_0; (d) a_0; (e) b_1.

Fig. 17-10: See Prob. 4.

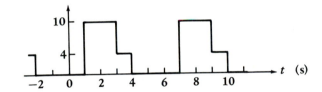

5 Find a_3, b_3, and $\sqrt{a_3^2 + b_3^2}$ for the periodic waveform sketched in Fig. 17-11.

Fig. 17-11: See Prob. 5.

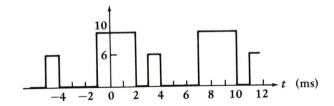

6 Obtain the trigonometric form of the Fourier series, give the value of T, and determine the average value for each of these periodic functions of time: (a) $2.7 \sin^2 90\pi t$; (b) $2.7 \sin^3 90\pi t$; (c) $2.7 \cos 89\pi t - 1.2 \sin 90\pi t$.

7 Given the periodic function, $v(t) = -50t$, $-0.02 < t < 0$ s, and $v(t) = 100t$, $0 < t < 0.01$ s, with $T = 0.03$ s, find: (a) a_0; (b) a_n.

8 Let $f(t) = 20 + 8 \cos 5t - 6 \cos 10t + 3 \cos 15t - 5 \sin 5t + 6 \sin 10t - 4 \sin 15t + 2 \sin 20t$. Find: (a) the period of $f(t)$; (b) the bandwidth in hertz required for the signal; (c) the average value of $f(t)$; (d) the effective value of $f(t)$; (e) the discrete amplitude and phase spectra of the signal.

9 A periodic function is defined as $g(t) = 10 \sin 20t$, $0 < t < \pi/60$ s, with $T = \pi/60$ s. Determine values for: (a) a_0; (b) a_3; (c) b_3.

Fig. 17-12: See Prob. 10.

10 (a) Identify the symmetry of the waveform shown in Fig. 17-12. (b) Which of the a_n, b_n, or a_0 are zero? (c) Find a_0, a_1, b_1, a_2, and b_2.

11 A certain periodic function $f(t)$ is known to have even-function symmetry and the amplitude spectrum shown in Fig. 17-13. (a) Determine the Fourier series for $f(t)$ if a_n and $b_n \geq 0$ for all n. (b) Find the effective value of $f(t)$. (c) Specify the instantaneous value of $f(t)$ at $t = 0.04$ s.

Fig. 17-13: See Prob. 11.

12 A function $f(t)$ is shown in Fig. 17-14 over the interval $0 < t < 4$. Sketch a function $g(t)$ over $0 < t < 20$ that equals $f(t)$ for $0 < t < 4$ and also has: (a) $T = 8$ and even symmetry; (b) $T = 8$ and odd symmetry; (c) $T = 16$, even, and half-wave symmetry; (d) $T = 16$, odd, and half-wave symmetry.

Fig. 17-14: See Prob. 12.

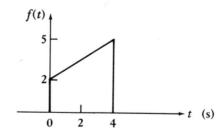

Fig. 17-15: See Prob. 13.

13 Assume that Fig. 17-15 shows one pulse of a series of pulses that repeats every 4 s. (a) Determine the average value of $f(t)$. (b) Calculate the rms value of $f(t)$. (c) Define a continuation of this function in the interval $4 < t < 8$ s so that the new function, $f_{new}(t)$, is odd and has $T = 8$ s. (d) Express $f_{new}(t)$ as a Fourier series. (e) Determine the rms value of the third harmonic of $f_{new}(t)$.

14 Make use of symmetry as much as possible to obtain numerical values of a_0, a_n, and b_n, $1 \leq n \leq 10$, for the waveform shown in Fig. 17-16.

15 In the interval, $2 < t < 5$ ms, a function is given as $10 \sin 1000\pi t$ for $2 < t < 3$ ms and zero for $0 < t < 2$ ms. The function has a period of 12 ms and possesses both even and half-wave symmetry. Find numerical values for a_0, a_1, a_2, a_3, a_4, and a_5.

Fig. 17-16: See Prob. 14.

16 After investigating the symmetry shown by the waveform in Fig. 17-17, write its trigonometric Fourier series, giving complete numerical values for all harmonics up to the eighth.

Fig. 17-17: See Prob. 16.

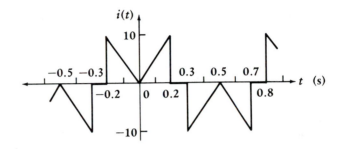

17 For each of the waveforms sketched in Fig. 17-18, state whether or not it has even symmetry, odd symmetry, or half-wave symmetry, or some combination of these symmetries.

Fig. 17-18: See Prob. 17.

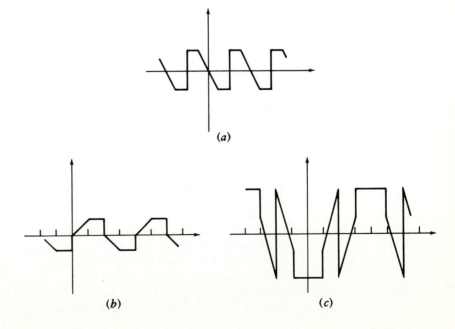

(a)

(b)   (c)

Fig. 17-19: See Prob. 18.

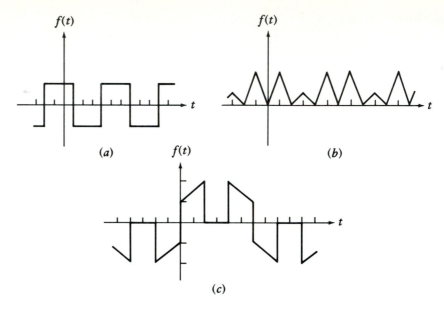

(a)

(b)

(c)

18 For each of the waveforms shown in Fig. 17-19, answer all of these questions: (1) Is even symmetry present? (2) Is odd symmetry present? (3) Is half-wave symmetry present? (4) Is $a_0 = 0$? (5) Are all $a_{\text{even}} = 0$? (6) Are all $a_{\text{odd}} = 0$? (7) Are all $b_{\text{even}} = 0$? (8) Are all $b_{\text{odd}} = 0$?

19 Replace the square wave of Fig. 17-6a with that shown in Fig. 17-20 and repeat the analysis of Sec. 17-4 to obtain a new expression for: (a) $i_f(t)$; (b) $i(t)$.

Fig. 17-20: See Probs. 19 and 20.

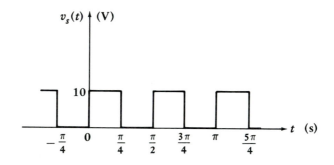

20 The waveform for $v_s(t)$ shown in Fig. 17-20 is applied to the circuit of Fig. 17-6b. Use the standard methods of transient analysis to calculate $i(t)$ at $t = $: (a) $\pi/4$ s; (b) $\pi/2$ s; (c) $3\pi/4$ s.

21 An ideal voltage source, an open switch, a 2-Ω resistor, and a 2.5-F capacitor are in series. The voltage source provides the voltage given in Fig. 17-6a. The switch is closed at $t = 0$ and the capacitor voltage is the desired response. (a) Work in the frequency domain of the nth harmonic to find the forced response as a trigonometric Fourier series. (b) Specify the functional form of the natural response. (c) Determine the complete response.

22 Determine c_3, c_{-3}, $|c_3|$, a_3, b_3, and $\sqrt{a_3^2 + b_3^2}$ for the waveform of Fig. 17-21 knowing that $T = 6$ ms.

Fig. 17-21: See Prob. 22.

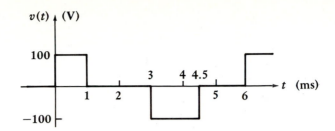

23 Find the complex Fourier series for the periodic waveform shown in Fig. 17-22. Give numerical values for c_n, $n = 0, \pm 1, \pm 2, \pm 3$.

Fig. 17-22: See Prob. 23.

24 A periodic sequence of positive pulses is composed of pulses having an amplitude of 20 V, a duration of 0.4 μs, and a repetition rate of 3000 pulses per second. (a) At what frequency is the envelope of the frequency spectrum equal to zero? (b) What is the frequency separation of the spectral lines? (c) What is the value of $|c_n|$ for the frequency component nearest to 10 kHz? (d) . . . nearest to 1 MHz? (e) What nominal bandwidth is required in an amplifier to transmit this pulse train with reasonable fidelity? (f) How many spectral components lie between 1 and 1.1 Mrad/s? (g) Find $|c_{227}|$ and the frequency to which it corresponds.

25 The complex Fourier coefficients for a periodic voltage are: $c_0 = 10$, $c_1 = 6 - j6$, $c_2 = 4 + j2$, $c_3 = -2 - j3$, and $c_n = 0$ for $|n| \geq 4$. Let $T = 10$ ms. (a) Evaluate $v(0.002\ s)$. (b) What is the effective value of this voltage?

26 A periodic wave with $T = 20$ ms has complex Fourier coefficients given by $c_0 = 1$ A and $c_n = (1/n^2) - j(2/n^3)$ A, for $|n| \geq 1$. (a) Sketch a line spectrum for $|c_n|$. (b) What average power does this current deliver to a 5-Ω resistor?

27 Let $f(t) = 1$ for $-3 < t < -2\ \mu$s and $2 < t < 3\ \mu$s, and $f(t) = 0$ elsewhere in the period interval, $-12.5 < t < 12.5\ \mu$s; $T = 25\ \mu$s. This series of pulses might represent the decimal number 3 being transmitted in binary form by a digital computer. After determining the complex Fourier series for $f(t)$: (a) give the value of c_4; (b) find c_0; (c) find $|c_n|_{max}$; (d) find N so that $|c_n| \leq 0.01|c_n|_{max}$ for all $n > N$. What bandwidth is required to transmit this portion of the spectrum?

28 The signal shown in Fig. 17-23 is periodic with $T = 1$ ms. (a) Find c_n. (b) Evaluate c_5.

Fig. 17-23: See Prob. 28.

29 A certain periodic signal, having a period of 9 s, is described as follows: $f(t) = 2t + 4, -2 < t < -1$ s; $f(t) = 2, -1 < t < 1$ s; $f(t) = 4 - 2t, 1 < t < 2$ s; $f(t) = 0$ elsewhere. (*a*) Find the general coefficient $\mathbf{c}_n$ of the complex Fourier series. (*b*) Evaluate $\mathbf{c}_0$ and $\mathbf{c}_1$.

30 A voltage waveform $v_s(t)$ has a period of $\frac{1}{12}$ s and is defined by: $v_s = 60$ V, $0 < t < \frac{1}{96}$ s; $v_s = 0, \frac{1}{96} < t < \frac{1}{12}$ s. (*a*) Find $\mathbf{c}_4$. (*b*) The voltage v_s is applied as the source in the circuit shown in Fig. 17-24. What average power is delivered to the load?

Fig. 17-24: See Prob. 30.

Chapter 18
Fourier Transforms

<table>
<tr><td>

**18-1
Introduction**

</td><td>

In this chapter we begin to consider the use of transform methods in studying the behavior of linear circuits through a discussion of the Fourier transform. The Laplace transform is the subject of the next chapter.

The Fourier transform and the Laplace transform are operations that convert a function of time into a function of $j\omega$ (Fourier transform) or $\mathbf{s}$ (Laplace transform). Given a suitable function of time, there is one and only one Fourier transform that corresponds to it, and a similar unique correspondence exists for the Laplace transform. Moreover, a one-to-one relationship also exists for the inverse transform; that is, given a suitable function of $j\omega$, or $\mathbf{s}$ for the Laplace transform, there is one and only one function of time that corresponds to it.

The Fourier and Laplace transforms are *integral transforms* which are extremely important in the study of many types of engineering systems, including linear electric circuits. Up to now we have not really needed them, because most of our circuits, along with their forcing and response functions, have been rather simple. The use of relatively powerful integral-transform techniques on them could be likened to using an electric locomotive to crack walnuts; we might lose sight of the meat. But now we have progressed to a point where our forcing functions, and perhaps our circuits, are getting too complicated for the tools we have developed so far. We want to look at forcing functions that are *not* periodic, and in later studies at (random) forcing functions or signals that are not expressible at all as time functions. This desire is often gratified through use of the integral-transform methods which now burst upon us.

</td></tr>
<tr><td>

**18-2
Definition of the
Fourier transform**

</td><td>

Let us proceed to define the Fourier transform by first recalling the spectrum of the periodic train of rectangular pulses we obtained at the end of Chap. 17. That was a *discrete* line spectrum, which is the type that we must always obtain for periodic functions of time. The spectrum was discrete in the sense that it was not a smooth or continuous function of frequency; instead, it had nonzero values only at specific frequencies.

</td></tr>
</table>

There are many important forcing functions, however, that are not periodic functions of time, such as a single rectangular pulse, a step function, a ramp function, or a rather strange type of function called the *impulse function* to be defined later in this chapter. Frequency spectra may be obtained for such nonperiodic functions, but they will be *continuous* spectra in which some energy, in general, may be found in any nonzero frequency interval, no matter how small.

We shall develop this concept by beginning with a periodic function and then letting the period become infinite. Our experience with the periodic rectangular pulses at the end of Chap. 17 should indicate that the envelope will decrease in amplitude without otherwise changing shape, and that more and more frequency components will be found in any given frequency interval. In the limit we should expect an envelope of vanishingly small amplitude, filled with an infinite number of frequency components, separated by vanishingly small frequency intervals. The number of frequency components between, say, 0 and 100 Hz becomes infinite, but the amplitude of each one approaches zero. At first thought, a spectrum of zero amplitude is a puzzling concept. We know that the line spectrum of a periodic forcing function shows the amplitude of each frequency component. But what does the zero-amplitude continuous spectrum of a nonperiodic forcing function signify? That question will be answered in the following section; now we proceed to carry out the limiting procedure suggested above.

We begin with the exponential form of the Fourier series:

$$f(t) = \sum_{n=-\infty}^{n=\infty} \mathbf{c}_n e^{jn\omega_0 t} \tag{1}$$

where

$$\mathbf{c}_n = \frac{1}{T} \int_{-T/2}^{T/2} f(t) e^{-jn\omega_0 t} \, dt \tag{2}$$

and

$$\omega_0 = \frac{2\pi}{T} \tag{3}$$

We now let

$$T \to \infty$$

and thus, from (3), ω_0 must become vanishingly small. We represent this limit by a differential

$$\omega_0 \to d\omega$$

Thus

$$\frac{1}{T} = \frac{\omega_0}{2\pi} \to \frac{d\omega}{2\pi} \tag{4}$$

Finally, the frequency of any "harmonic" $n\omega_0$ must now correspond to the general frequency variable which describes the continuous spectrum. In other words, n must tend to infinity as ω_0 approaches zero, such that the product is finite:

$$n\omega_0 \rightarrow \omega \tag{5}$$

When these four limiting operations are applied to (2), we find that $\mathbf{c}_n$ must approach zero, as we had previously presumed. If we multiply each side of (2) by the period T and then undertake the limiting process, a nontrivial result is obtained:

$$\mathbf{c}_n T \rightarrow \int_{-\infty}^{\infty} f(t)e^{-j\omega t}\, dt$$

The right side of this expression is a function of ω (and *not* of t), and we represent it by $\mathbf{F}(j\omega)$:

$$\mathbf{F}(j\omega) = \int_{-\infty}^{\infty} f(t)e^{-j\omega t}\, dt \tag{6}$$

Now let us apply the limiting process to (1). We begin by multiplying and dividing the summation by T,

$$f(t) = \sum_{n=-\infty}^{n=\infty} \mathbf{c}_n T e^{jn\omega_0 t}\frac{1}{T}$$

next replacing $\mathbf{c}_n T$ by the new quantity $\mathbf{F}(j\omega)$, and then making use of (4) and (5). In the limit, the summation becomes an integral, and

$$f(t) = \frac{1}{2\pi}\int_{-\infty}^{\infty} \mathbf{F}(j\omega)e^{j\omega t}\, d\omega \tag{7}$$

Equations (6) and (7) are collectively called the *Fourier transform pair*. The function $\mathbf{F}(j\omega)$ is the *Fourier transform* of $f(t)$, and $f(t)$ is the *inverse* Fourier transform of $\mathbf{F}(j\omega)$.

This transform-pair relationship is most important! We should memorize it, draw arrows pointing to it, and mentally keep it on the conscious level henceforth and forevermore.[1] We emphasize the importance of these relations by repeating and blocking them off below:

$$\mathbf{F}(j\omega) = \int_{-\infty}^{\infty} e^{-j\omega t}f(t)\, dt \tag{8a}$$

$$f(t) = \frac{1}{2\pi}\int_{-\infty}^{\infty} e^{j\omega t}\mathbf{F}(j\omega)\, d\omega \tag{8b}$$

The exponential terms in these two equations carry opposite signs for the exponents. To keep them straight, it may help to note that the positive sign is associated with the expression for $f(t)$, just as it is with the complex Fourier series (1).

---

[1] Future used-car dealers and politicians may forget it.

One question is appropriate to raise at this time. For the Fourier transform relationships given above, can we obtain the Fourier transform of *any* arbitrarily chosen $f(t)$? It turns out that the answer is affirmative for essentially any voltage or current that we can actually produce. A sufficient condition for the existence of $\mathbf{F}(j\omega)$ is that

$$\int_{-\infty}^{\infty} |f(t)| \, dt < \infty$$

This condition is not necessary, however, because some functions that do not meet it still have a Fourier transform; the step function is one such example. Furthermore, we shall see later that $f(t)$ does not even need to be nonperiodic in order to have a Fourier transform; the Fourier series representation for a periodic time function is just a special case of the more general Fourier transform representation.

As we indicated earlier, the Fourier transform-pair relationship is unique. For a given $f(t)$ there is one specific $\mathbf{F}(j\omega)$; and for a given $\mathbf{F}(j\omega)$ there is one specific $f(t)$.

Let us use the Fourier transform to obtain the continuous spectrum of a single rectangular pulse. We shall select that pulse in Fig. 17-8a (repeated as Fig. 18-1a) which occurs in the interval $t_0 < t < t_0 + \tau$. Thus

$$f(t) = \begin{cases} V_0 & t_0 < t < t_0 + \tau \\ 0 & t < t_0 \text{ and } t > t_0 + \tau \end{cases}$$

The Fourier transform of $f(t)$ is found from (8a):

$$\mathbf{F}(j\omega) = \int_{t_0}^{t_0+\tau} V_0 e^{-j\omega t} \, dt$$

Fig. 18-1: (a) A periodic sequence of rectangular pulses. (b) The corresponding discrete line spectrum for $|\mathbf{c}_n|$.

(a)

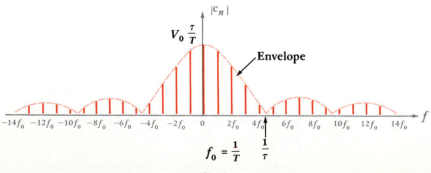

(b)

and this may be easily integrated and simplified:

$$\mathbf{F}(j\omega) = V_0\tau \frac{\sin \frac{1}{2}\omega\tau}{\frac{1}{2}\omega\tau} e^{-j\omega(t_0+\tau/2)}$$

The magnitude of $\mathbf{F}(j\omega)$ yields the continuous frequency spectrum, and it is obviously of the form of the sampling function. The value of $\mathbf{F}(0)$ is $V_0\tau$. The shape of the spectrum is identical with the broken-line curve of Fig. 18-1b. A plot of $|\mathbf{F}(j\omega)|$ as a function of ω does *not* indicate the magnitude of the voltage present at any given frequency. What is it, then? Examination of (7) shows that, if $f(t)$ is a voltage waveform, then $\mathbf{F}(j\omega)$ is dimensionally "volts per unit frequency," a concept that may be strange to many of us. In order to understand this a little better, we next look into some of the properties of $\mathbf{F}(j\omega)$.

Drill Problems

18-1. Let $f(t) = 100$ V for $-0.1 < t < -0.05$ and $0.05 < t < 0.1$, and $f(t) = 0$ for all other t. Evaluate $\mathbf{F}(j\omega)$ for $\omega = :$ (a) 0; (b) 5π rad/s; (c) -5π rad/s; (d) 10π rad/s; (e) 20π rad/s.　　　　*Ans:* 10.00; 3.73; 3.73; -6.37; 0 V/(rad/s)

18-2. Let $\mathbf{F}(j\omega) = 100$ V/(rad/s) for $-2 < \omega < -1$ rad/s and also for $1 < \omega < 2$ rad/s, and $\mathbf{F}(j\omega) = 0$ for all other ω. Evaluate $f(t)$ at $t = :$ (a) 0; (b) $\pi/4$ s; (c) $-\pi/4$ s; (d) $\pi/2$ s; (e) π s.　　　　*Ans:* 31.8; 11.87; 11.87; -20.3; 0 V

18-3
Some properties of the Fourier transform

Our object in this section is to establish several of the mathematical properties of the Fourier transform and, even more important, to understand its physical significance. We begin by using Euler's identity to replace $e^{-j\omega t}$ in (8a),

$$\mathbf{F}(j\omega) = \int_{-\infty}^{\infty} f(t) \cos \omega t \, dt - j \int_{-\infty}^{\infty} f(t) \sin \omega t \, dt \tag{9}$$

Since $f(t)$, $\cos \omega t$, and $\sin \omega t$ are all real functions of time, both the integrals in (9) are real functions of ω. Thus, by letting

$$\mathbf{F}(j\omega) = A(\omega) + jB(\omega) = |\mathbf{F}(j\omega)| e^{j\phi(\omega)} \tag{10}$$

we have

$$A(\omega) = \int_{-\infty}^{\infty} f(t) \cos \omega t \, dt \tag{11}$$

$$B(\omega) = -\int_{-\infty}^{\infty} f(t) \sin \omega t \, dt \tag{12}$$

$$|\mathbf{F}(j\omega)| = \sqrt{A^2(\omega) + B^2(\omega)} \tag{13}$$

and　　　　$$\phi(\omega) = \tan^{-1} \frac{B(\omega)}{A(\omega)} \tag{14}$$

Replacing ω by $-\omega$ shows that $A(\omega)$ and $|\mathbf{F}(j\omega)|$ are both even functions of ω, while $B(\omega)$ and $\phi(\omega)$ are both odd functions of ω.

Now, if $f(t)$ is an even function of t, then the integrand of (12) is an odd function of t, and the symmetrical limits force $B(\omega)$ to be zero; thus, if $f(t)$ is even, its Fourier transform $\mathbf{F}(j\omega)$ is a real, even function of ω, and the phase function $\phi(\omega)$ is zero or π for all ω. However, if $f(t)$ is an odd function of t, then $A(\omega) = 0$, $\mathbf{F}(j\omega)$ is both odd and a pure imaginary function of ω; $\phi(\omega)$ is $\pm\pi/2$. In general, however, $\mathbf{F}(j\omega)$ is a complex function of ω.

Finally, we note that the replacement of ω by $-\omega$ in (9) forms the conjugate of $\mathbf{F}(j\omega)$. Thus,

$$\mathbf{F}(-j\omega) = A(\omega) - jB(\omega) = \mathbf{F}^*(j\omega)$$

and we have

$$\mathbf{F}(j\omega)\mathbf{F}(-j\omega) = \mathbf{F}(j\omega)\mathbf{F}^*(j\omega) = A^2(\omega) + B^2(\omega) = |\mathbf{F}(j\omega)|^2$$

With these basic mathematical properties of the Fourier transform in mind, we are now ready to consider its physical significance. Let us suppose that $f(t)$ is either the voltage across or the current through a 1-Ω resistor, so that $f^2(t)$ is the instantaneous power delivered to the 1-Ω resistor by $f(t)$. Integrating this power over all time, we obtain the total energy delivered by $f(t)$ to the 1-Ω resistor,

$$W_{1\Omega} = \int_{-\infty}^{\infty} f^2(t)\ dt \tag{15}$$

Now let us resort to a little trickery. Thinking of the integrand above as $f(t)$ times itself, we replace one of those functions by (8b),

$$W_{1\Omega} = \int_{-\infty}^{\infty} f(t) \left[\frac{1}{2\pi} \int_{-\infty}^{\infty} e^{j\omega t}\ \mathbf{F}(j\omega)\ d\omega \right] dt$$

Since $f(t)$ is not a function of the variable of integration ω, we may move it inside the bracketed integral, and then interchange the order of integration,

$$W_{1\Omega} = \frac{1}{2\pi} \int_{-\infty}^{\infty} \left[\int_{-\infty}^{\infty} \mathbf{F}(j\omega)e^{j\omega t}f(t)\ dt \right] d\omega$$

Next we shift $\mathbf{F}(j\omega)$ outside the inner integral, causing that integral to become $\mathbf{F}(-j\omega)$,

$$W_{1\Omega} = \frac{1}{2\pi} \int_{-\infty}^{\infty} \mathbf{F}(j\omega)\mathbf{F}(-j\omega)\ d\omega = \frac{1}{2\pi} \int_{-\infty}^{\infty} |\mathbf{F}(j\omega)|^2\ d\omega$$

Collecting these results,

$$\int_{-\infty}^{\infty} f^2(t)\ dt = \frac{1}{2\pi} \int_{-\infty}^{\infty} |\mathbf{F}(j\omega)|^2\ d\omega \tag{16}$$

Equation (16) is a very useful expression known as Parseval's theorem.[2] This theorem, along with (15), tells us that the energy associated with $f(t)$ can be obtained either from an integration over all time in the time domain or by $1/(2\pi)$ times an integration over all frequency in the frequency domain.

---

[2] Marc-Antoine Parseval-Deschenes was a rather obscure French mathematician, geographer, and occasional poet who published these results in 1805, seventeen years before Fourier published his theorem.

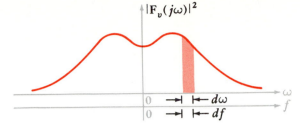

Parseval's theorem also leads us to a greater understanding and interpretation of the meaning of the Fourier transform. Consider a voltage $v(t)$ with Fourier transform $\mathbf{F}_v(j\omega)$ and 1-Ω energy W_{1v},

$$W_{1v} = \frac{1}{2\pi} \int_{-\infty}^{\infty} |\mathbf{F}_v(j\omega)|^2 \, d\omega = \frac{1}{\pi} \int_{0}^{\infty} |\mathbf{F}_v(j\omega)|^2 \, d\omega$$

where the right-most equality follows from the fact that $|\mathbf{F}_v(j\omega)|^2$ is an even function of ω. Then, since $\omega = 2\pi f$, we can write

$$W_{1v} = \int_{-\infty}^{\infty} |\mathbf{F}_v(j\omega)|^2 \, df = 2 \int_{0}^{\infty} |\mathbf{F}_v(j\omega)|^2 \, df \tag{17}$$

Figure 18-2 illustrates a typical plot of $|\mathbf{F}_v(j\omega)|^2$ as a function of both ω and f. If we divide the frequency scale up into vanishingly small increments df, (17) shows us that the area of a differential slice under the $|\mathbf{F}_v(j\omega)|^2$ curve, having a width df, is $|\mathbf{F}_v(j\omega)|^2 \, df$. This area is shown shaded. The sum of all such areas, as f ranges from minus to plus infinity, is the total 1-Ω energy contained in $v(t)$. Thus, $|\mathbf{F}_v(j\omega)|^2$ is the (1-Ω) *energy density* or energy per unit bandwidth (J/Hz) of $v(t)$, and this energy density is always a real, even, nonnegative function of ω. By integrating $|\mathbf{F}_v(j\omega)|^2$ over an appropriate frequency interval, we are able to calculate that portion of the total energy lying within the chosen interval. Note that the energy density is not a function of the phase of $\mathbf{F}_v(j\omega)$, and thus there are an infinite number of time functions and Fourier transforms that possess identical energy-density functions.

As an example, let us assume that $v(t)$ is the voltage forcing function applied to the input of an ideal bandpass filter,[3] and that $v_o(t)$ is the output voltage. For $v(t)$ we select the one-sided [i.e., $v(t) = 0$ for $t < 0$] exponential pulse,

$$v(t) = 4e^{-3t}u(t)$$

and let the filter passband be defined by $1 < |f| < 2$. The energy in $v_o(t)$ will therefore be equal to the energy of that part of $v(t)$ having frequency components in the intervals, $1 < f < 2$ and $-2 < f < -1$. We determine the Fourier transform of $v(t)$,

---

[3] An ideal bandpass filter is a two-port network which allows all those frequency components of the input signal for which $\omega_1 < |\omega| < \omega_2$ to pass unattenuated from the input terminals to the output terminals; all components having frequencies outside this so-called *passband* are completely attenuated.

$$\mathbf{F}_v(j\omega) = 4 \int_{-\infty}^{\infty} e^{-j\omega t} e^{-3t} u(t)\, dt = 4 \int_0^{\infty} e^{-(3+j\omega)t}\, dt$$

$$= \frac{4}{3 + j\omega}$$

and then we may calculate the total 1-Ω energy in the input signal by either

$$W_{1v} = \frac{1}{2\pi} \int_{-\infty}^{\infty} |\mathbf{F}_v(j\omega)|^2\, d\omega = \frac{8}{\pi} \int_{-\infty}^{\infty} \frac{d\omega}{9 + \omega^2}$$

$$= \frac{16}{\pi} \int_0^{\infty} \frac{d\omega}{9 + \omega^2} = \frac{8}{3} \quad \text{J}$$

or $$W_{1v} = \int_{-\infty}^{\infty} v^2(t)\, dt = 16 \int_0^{\infty} e^{-6t}\, dt = \frac{8}{3} \quad \text{J}$$

The total energy in $v_o(t)$, however, is smaller:

$$W_{1o} = \frac{1}{2\pi} \int_{-4\pi}^{-2\pi} \frac{16 d\omega}{9 + \omega^2} + \frac{1}{2\pi} \int_{2\pi}^{4\pi} \frac{16 d\omega}{9 + \omega^2} = \frac{16}{\pi} \int_{2\pi}^{4\pi} \frac{d\omega}{9 + \omega^2}$$

$$= \frac{16}{3\pi} \left(\tan^{-1} \frac{4\pi}{3} - \tan^{-1} \frac{2\pi}{3} \right) = 0.358 \quad \text{J}$$

Thus we see that an ideal bandpass filter enables us to remove energy from prescribed frequency ranges while still retaining the energy contained in other frequency ranges. The Fourier transform helps us to describe the filtering action quantitatively without actually evaluating $v_o(t)$, although we shall see later that the Fourier transform can also be used to obtain the expression for $v_o(t)$ if we wish to do so.

Drill Problems

18-3. Let $i(t) = 10e^{20t}u(-t)$ A, and find: (a) $\mathbf{F}_i(j0)$; (b) A_i (10); (c) B_i (10); (d) $|\mathbf{F}_i(j10)|$; (e) ϕ_i (10).
Ans: 0.500 A/(rad/s); 0.400 A/(rad/s); 0.200 A/(rad/s); 0.447 A/(rad/s); 26.6°

18-4. Find the 1-Ω energy associated with the current $i(t) = 10e^{20t}u(-t)$ A in the interval: (a) $-0.1 < t < 0.1$ s; (b) $-10 < \omega < 10$ rad/s; (c) $10 < \omega < \infty$ rad/s.   *Ans:* 2.45; 0.738; 0.881 J

**18-4
The unit-impulse
function**

Before continuing our discussion of the Fourier transform, we need to pause briefly to define a new singularity function called the *unit-impulse* or *delta function.* We shall see that the unit-impulse function enables us to confirm our previous statement that periodic, as well as nonperiodic, time functions possess Fourier transforms. In addition, our analysis of general *RLC* circuits can be enhanced by utilization of the unit impulse. We have been avoiding the possibility that the voltage across a capacitor or the current through an inductor might change by a finite amount in zero time, for in these cases the capacitor current or the inductor voltage would have had to assume infinite values. Although this is not physically possible, it is mathematically possible. Thus, if a step voltage $V_0u(t)$ is applied directly across an uncharged capacitor C, the time constant is zero, since R_{eq} is zero and C is finite. This means that a charge CV_0 must be established on the capacitor in a vanishingly small time. We might say, with more insight than rigor, that "infinite current flowing for zero time produces a finite charge" on the capacitor. This type of phenomenon can be described through use of the unit-impulse function.

We shall define the unit impulse as a function of time which is zero when its argument, generally $(t - t_0)$, is less than zero; which is also zero when its argument is greater than zero; which is infinite when its argument is zero; and which has unit area. Mathematically, the defining statements are

$$\delta(t - t_0) = 0 \qquad t \neq t_0 \tag{18}$$

and

$$\int_{-\infty}^{\infty} \delta(t - t_0)\, dt = 1 \tag{19}$$

where the symbol δ ("curly" delta) is used to represent the unit impulse. In view of the functional values expressed by (18), it is apparent that the limits on the integral in (19) may be any values which are less than t_0 and greater than t_0. In particular, we may let t_0^- and t_0^+ represent values of time which are arbitrarily close to t_0 and then express (19) as

$$\int_{t_0^-}^{t_0^+} \delta(t - t_0)\, dt = 1 \tag{20}$$

For the most part, we shall concern ourselves with signals having only a single discontinuity, and we shall select our time scale so that the switching operation occurs at $t = 0$. For this special case, the defining equations are

$$\delta(t) = 0 \qquad t \neq 0 \tag{21}$$

and

$$\int_{-\infty}^{\infty} \delta(t)\, dt = 1 \tag{22}$$

or

$$\int_{0^-}^{0^+} \delta(t)\, dt = 1 \tag{23}$$

The unit impulse may be multiplied by a constant also; this cannot, of course, affect (18) or (21), because the value must still be zero when the argument is not zero. However, multiplication of any of the integral expressions by a constant shows that the area under the impulse is now equal to the constant multiplying factor; this area is called the *strength* of the impulse. Thus, the impulse $5\delta(t)$ has a strength of 5, and the impulse $-10\delta(t - 2)$ has a strength of -10. If the unit impulse is multiplied by a function of time, then the strength of the impulse must be the value of that function at the time for which the impulse *argument* is zero. In other words, the strength of the impulse $e^{-t/2}\delta(t - 2)$ is 0.368, and the strength of the impulse $\sin(5\pi t + \pi/4)\delta(t)$ is 0.707. It is therefore possible to write the following integrals[4] which make the same statement in mathematical form:

$$\int_{-\infty}^{\infty} f(t)\, \delta(t)\, dt = f(0) \tag{24}$$

or

$$\int_{-\infty}^{\infty} f(t)\, \delta(t - t_0)\, dt = f(t_0) \tag{25}$$

[4] They are called *sifting integrals* because the integral sifts out a particular value of $f(t)$.

Fig. 18-4: A rectangular pulse of unit area which approaches a unit impulse as $\Delta \to 0$.

Fig. 18-3: A positive and negative impulse are plotted graphically as functions of time. The strengths of the impulses are 4 and -3, respectively, and thus $f(t) = 4\delta(t + 1) - 3\delta(t)$.

The graphical symbol for an impulse is shown in Fig. 18-3, where $f(t) = 4\delta(t + 1) - 3\delta(t)$ is plotted as a function of time. It is customary to indicate the strength of the impulse in parentheses adjacent to the impulse. Note that no attempt should be made to indicate the strength of an impulse by adjusting its amplitude; each spike has infinite amplitude, and all impulses should be drawn as arrows having the same convenient amplitude. Positive and negative impulses should be drawn above and below the time axis, respectively. In order to avoid confusion with the ordinate, the lines and arrowheads used to form the impulses are drawn thicker than are the axes.

We shall find it convenient to be familiar with several other interpretations of the unit impulse. We seek graphical forms which do not have infinite amplitudes, but which will approximate an impulse as the amplitude increases. Let us first consider a rectangular pulse, such as that shown in Fig. 18-4. The pulse width is selected as Δ and its amplitude as $1/\Delta$, thus forcing the area of the pulse to be unity, regardless of the magnitude of Δ. As Δ decreases, the amplitude $1/\Delta$ increases, and the rectangular pulse becomes a better approximation to a unit impulse. The response of a circuit element to a unit impulse may be determined by finding its response to this rectangular pulse and then letting Δ approach zero. However, since the impulse response is easily found, we should also realize that this response may in itself be an acceptable approximation to the response produced by a narrow rectangular pulse.

A triangular pulse, sketched in Fig. 18-5, may also be used to approximate the unit impulse. Since we again prefer unit area, a pulse of amplitude $1/\Delta$ must possess an overall width of 2Δ. As Δ approaches zero, the triangular pulse approaches the unit impulse. There are many other pulse shapes which, in the limit, approach the unit impulse, but we shall take the decaying

Fig. 18-5: A triangular pulse of unit area which approaches a unit impulse as $\Delta \to 0$.

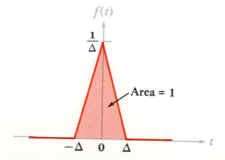

exponential as our last limiting form. We first construct such a waveform with unit area by finding the area under the general exponential,

$$f(t) = \begin{cases} 0 & t < 0 \\ Ae^{-t/\tau} & t > 0 \end{cases}$$

or

$$f(t) = Ae^{-t/\tau}u(t)$$

Thus

$$\text{Area} = \int_0^\infty Ae^{-t/\tau} \, dt = -\tau Ae^{-t/\tau}\Big|_0^\infty = \tau A$$

and we thus must set $A = 1/\tau$. The time constant will be very short, and we shall again use Δ to represent this short time. Thus, the exponential function

$$f(t) = \frac{1}{\Delta} e^{-t/\Delta}u(t)$$

approaches the unit impulse as $\Delta \rightarrow 0$. This representation of the unit impulse indicates that the exponential decay of a current or voltage in a circuit approaches an impulse (but not necessarily a unit impulse) as the time constant is reduced.

For our final interpretation of the unit impulse, let us try to establish a relationship with the unit-step function. The function shown in Fig. 18-6a is almost a unit step; however, Δ seconds are required for it to complete the linear change from zero to unit amplitude. Beneath this modified unit-step function, Fig. 18-6b shows its derivative; since the linear portion of the modified step rises at the rate of 1 unit every Δ seconds, the derivative must be a rectangular pulse of amplitude $1/\Delta$ and width Δ. This, however, was the first function we considered as an approximation to the unit impulse, and we

(a)

Fig. 18-6: (a) A modified unit-step function; the transition from zero to unity is linear over a Δ-s time interval. (b) The derivative of the modified unit step. As $\Delta \rightarrow 0$, (a) becomes the unit step and (b) approaches the unit impulse.

(b)

know that it approaches the unit impulse as Δ approaches zero. But the modified unit step approaches the unit step itself as Δ approaches zero, and we conclude that the unit impulse may be regarded as the time derivative of the unit-step function.[5] Mathematically,

$$\delta(t) = \frac{du(t)}{dt} \tag{26}$$

and conversely,

$$u(t) = \int_{0^-}^{t} \delta(t)\, dt \qquad t > 0 \tag{27}$$

where the lower limit may in general be any value of t less than zero. Either (26) or (27) could be used as the definition of the unit impulse if we wished.

We thus see another method suggesting itself for the determination of the response to a unit impulse. If we can find the unit-step response, then the linear nature of our circuits requires the response to a unit impulse to be the derivative of the response to the unit step. From the opposite point of view, if the response to a unit impulse is known, then the integral of this response from $-\infty$ to t must be the unit-step response.

Let us apply (27) to show how a current impulse can place an immediate charge on a capacitor. Since the strength of an impulse may be interpreted as the area under the impulse, a current impulse must have a strength that has the units of current times time, or charge. Thus, we apply the impulse

$$i(t) = Q_0\, \delta(t - t_0)$$

to a capacitor C. Assuming the passive sign convention, the capacitor voltage is:

$$v(t) = \frac{1}{C} \int_{-\infty}^{t} i(t)\, dt$$

$$= \frac{Q_0}{C} \int_{-\infty}^{t} \delta(t - t_0)\, dt$$

$$= \frac{Q_0}{C}\, u(t - t_0)$$

We see that the capacitor voltage jumps discontinuously from zero to Q_0/C at $t = t_0$.

Drill Problems

18-5. Find the strength of an impulse defined by: (*a*) the limit as $a \to 0$ of $(1/a) \sin(\pi x/a)[u(x) - u(x - a)]$; (*b*) $8e^{-2t}[\cos(4t - 0.5)]\, \delta(t - 0.5)$; (*c*) one part of df/dt where $f = 8e^{-2t} \cos(4t - 0.5)\, u(t)$.   *Ans:* 0.637; 0.208; 7.02

18-6. Evaluate: (*a*) $\int_{-\infty}^{\infty} [(\sin x)/x]\, \delta(x - 1)\, dx$; (*b*) $\int_{-1}^{1} e^{-5t}[\delta(t - 0.1) + \delta(t + 0.1)]\, dt$; (*c*) $\int_{0.1}^{1.3} \frac{1}{7} [\delta(\tau - \frac{1}{4}) + \delta(\tau - \frac{3}{4}) + \delta(\tau - \frac{5}{4})]\, d\tau.$.   *Ans:* 0.841; 2.26; 6.13

---

[5] The fact that the derivative of the unit step does not exist at the point of the discontinuity creates quite a skeptical attitude in many mathematicians; nevertheless, the impulse is a useful analytical function.

18-5
Fourier transform pairs for some simple time functions

We now seek the Fourier transform of the unit impulse $\delta(t - t_0)$. That is, we are interested in the spectral properties or frequency-domain description of this singularity function. If we use the notation $\mathcal{F}\{\ \}$ to symbolize "Fourier transform of $\{\ \}$," then

$$\mathcal{F}\{\delta(t - t_0)\} = \int_{-\infty}^{\infty} e^{-j\omega t}\,\delta(t - t_0)\,dt$$

From our earlier discussion of this type of integral, and by (25) in particular, we have

$$\mathcal{F}\{\delta(t - t_0)\} = e^{-j\omega t_0} = \cos \omega t_0 - j \sin \omega t_0 \tag{28}$$

This complex function of ω leads to the 1-Ω energy density function,

$$|\mathcal{F}\{\delta(t - t_0)\}|^2 = \cos^2 \omega t_0 + \sin^2 \omega t_0 = 1$$

This remarkable result says that the (1-Ω) energy per unit bandwidth is unity *at all frequencies,* and that the total energy in the unit impulse is infinitely large.[6] No wonder, then, that we must conclude that the unit impulse is "impractical" in the sense that it cannot be generated in the laboratory. Moreover, even if one were available to us, it must appear distorted after being subjected to the finite bandwidth of any practical laboratory instrument.

Since there is a unique one-to-one correspondence between a time function and its Fourier transform, we can say that the inverse Fourier transform of $e^{-j\omega t_0}$ is $\delta(t - t_0)$. Utilizing the symbol $\mathcal{F}^{-1}\{\ \}$ for the inverse transform, we have

$$\mathcal{F}^{-1}\{e^{-j\omega t_0}\} = \delta(t - t_0)$$

Thus, we now know that

$$\frac{1}{2\pi} \int_{-\infty}^{\infty} e^{j\omega t} e^{-j\omega t_0}\,d\omega = \delta(t - t_0)$$

even though we would fail in an attempt at the direct evaluation of this improper integral. Symbolically, we may write

$$\delta(t - t_0) \Leftrightarrow e^{-j\omega t_0} \tag{29}$$

where $\Leftrightarrow$ indicates that the two functions constitute a Fourier transform pair.

Continuing with our consideration of the unit-impulse function, let us consider a Fourier transform of that form,

$$\mathbf{F}(j\omega) = \delta(\omega - \omega_0)$$

---

[6] Note, for example, from Fig. 18-4 that the total energy in the unit impulse is

$$\lim_{\Delta \to 0} \int_{-\Delta/2}^{\Delta/2} \left(\frac{1}{\Delta}\right)^2 dt = \lim_{\Delta \to 0} \left(\frac{1}{\Delta}\right) = \infty$$

which is a unit impulse *in the frequency domain* located at $\omega = \omega_0$. Then $f(t)$ must be

$$f(t) = \mathcal{F}^{-1}\{\mathbf{F}(j\omega)\} = \frac{1}{2\pi} \int_{-\infty}^{\infty} e^{j\omega t} \, \delta(\omega - \omega_0) \, d\omega = \frac{1}{2\pi} e^{j\omega_0 t}$$

where we have used the sifting property of the unit impulse. Thus we may now write

$$\frac{1}{2\pi} e^{j\omega_0 t} \Leftrightarrow \delta(\omega - \omega_0)$$

or

$$e^{j\omega_0 t} \Leftrightarrow 2\pi\delta(\omega - \omega_0) \tag{30}$$

Also, by a simple sign change we obtain

$$e^{-j\omega_0 t} \Leftrightarrow 2\pi\delta(\omega + \omega_0) \tag{31}$$

Clearly, the time function is complex in both (30) and (31), and does not exist in the real world of the laboratory. Time functions such as $\cos \omega_0 t$, for example, can be produced with laboratory equipment, but a function like $e^{j\omega_0 t}$ cannot.

However, we know that

$$\cos \omega_0 t = \tfrac{1}{2} e^{j\omega_0 t} + \tfrac{1}{2} e^{-j\omega_0 t}$$

and it is easily seen from the definition of the Fourier transform that

$$\mathcal{F}\{f_1(t)\} + \mathcal{F}\{f_2(t)\} = \mathcal{F}\{f_1(t) + f_2(t)\} \tag{32}$$

Therefore,

$$\begin{aligned}
\mathcal{F}\{\cos \omega_0 t\} &= \mathcal{F}\{\tfrac{1}{2} e^{j\omega_0 t}\} + \mathcal{F}\{\tfrac{1}{2} e^{-j\omega_0 t}\} \\
&= \pi\delta(\omega - \omega_0) + \pi\delta(\omega + \omega_0)
\end{aligned}$$

which indicates that the frequency-domain description of $\cos \omega_0 t$ shows a *pair* of impulses, located at $\omega = \pm\omega_0$. This should not be a great surprise, for in our first discussion of complex frequency in Sec. 13-2, we noted that a sinusoidal function of time was always represented by a pair of imaginary frequencies located at $\mathbf{s} = \pm j\omega_0$. We have, therefore,

$$\cos \omega_0 t \Leftrightarrow \pi[\delta(\omega + \omega_0) + \delta(\omega - \omega_0)] \tag{33}$$

Before establishing the Fourier transforms of any more time functions, we should understand just where we are heading, and why. Thus far we have determined several Fourier transform pairs. As we build up our knowledge of such pairs, they, in turn, can be used to obtain more pairs and eventually we will have a catalog of most of the familiar time functions encountered in circuit analysis, along with the corresponding Fourier transforms. Thus we will have not only the time-domain descriptions of such functions, but their frequency-domain descriptions as well. Then, just as the use of phasor transforms simplified the determination of the steady-state sinusoidal response, we shall find that the use of the Fourier transforms of various forcing functions can simplify the determination of the complete response, both the natural and the forced components. When we extend our thinking to the use of the Laplace transform in the following chapter, we shall even be able to account for the troublesome initial conditions that have plagued us in the past. With these thoughts in mind, let us look at just a few more transform pairs with the goal of listing our findings in a form that will be useful for quick reference later.

The first forcing function that we considered many chapters ago was a dc voltage or current. To find the Fourier transform of a constant function of time, $f(t) = K$, our first inclination might be to substitute this constant into the defining equation for the Fourier transform and evaluate the resulting integral. If we did, we would find ourselves with an indeterminate expression on our hands. Fortunately, however, we have already solved this problem, for from (31),

$$e^{-j\omega_0 t} \Leftrightarrow 2\pi\delta(\omega + \omega_0)$$

We see that, if we simply let $\omega_0 = 0$, then the resulting transform pair is

$$1 \Leftrightarrow 2\pi\delta(\omega) \tag{34}$$

from which it follows that

$$K \Leftrightarrow 2\pi K\delta(\omega) \tag{35}$$

and our problem is solved. The frequency spectrum of a constant function of time consists only of a component at $\omega = 0$, which we knew all along.

As another example, let us obtain the Fourier transform of a singularity function known as the *signum function*, sgn (t), defined by

$$\text{sgn}(t) = \begin{cases} -1 & t < 0 \\ 1 & t > 0 \end{cases} \tag{36}$$

or
$$\text{sgn}(t) = u(t) - u(-t)$$

Again, if we should try to substitute this time function into the defining equation for the Fourier transform, we would face an indeterminate expression upon substitution of the limits of integration. This same problem will arise every time we attempt to obtain the Fourier transform of a time function that does not approach zero as $|t|$ approaches infinity. Eventually, we will avoid this situation by using the *Laplace transform*, because it contains a built-in convergence factor that will cure many of the inconvenient ills associated with the evaluation of certain Fourier transforms.

The signum function under consideration can be written as

$$\text{sgn}(t) = \lim_{a \to 0} [e^{-at}u(t) - e^{at}u(-t)]$$

Notice that the expression within the brackets *does* approach zero as $|t|$ gets very large. Using the definition of the Fourier transform, we obtain

$$\mathscr{F}\{\text{sgn}(t)\} = \lim_{a \to 0} \left[\int_0^\infty e^{-j\omega t}e^{-at}\, dt - \int_{-\infty}^0 e^{-j\omega t}e^{at}\, dt \right]$$

$$= \lim_{a \to 0} \frac{-j2\omega}{\omega^2 + a^2} = \frac{2}{j\omega}$$

Table 18-1: Some familiar Fourier transform pairs

$f(t)$	$f(t)$	$\mathscr{F}\{f(t)\} = \mathbf{F}(j\omega)$	$\lvert\mathbf{F}(j\omega)\rvert$
	$\delta(t - t_0)$	$e^{-j\omega t_0}$	
	$e^{j\omega_0 t}$	$2\pi\delta(\omega - \omega_0)$	
	$\cos \omega_0 t$	$\pi[\delta(\omega + \omega_0) + \delta(\omega - \omega_0)]$	
	1	$2\pi\delta(\omega)$	
	$\text{sgn}(t)$	$\dfrac{2}{j\omega}$	
	$u(t)$	$\pi\delta(\omega) + \dfrac{1}{j\omega}$	
	$e^{-\alpha t}u(t)$	$\dfrac{1}{\alpha + j\omega}$	
	$e^{-\alpha t}\cos \omega_d t \cdot u(t)$	$\dfrac{\alpha + j\omega}{(\alpha + j\omega)^2 + \omega_d^2}$	
	$u(t + \tfrac{1}{2}T) - u(t - \tfrac{1}{2}T)$	$T\,\dfrac{\sin \dfrac{\omega T}{2}}{\dfrac{\omega T}{2}}$	

The real component is zero since sgn (t) is an odd function of t. Thus,

$$\text{sgn } (t) \Leftrightarrow \frac{2}{j\omega} \qquad (37)$$

As a final example in this section, let us look at the familiar unit-step function, $u(t)$. Making use of our work on the signum function above, we represent the unit step by

$$u(t) = \tfrac{1}{2} + \tfrac{1}{2} \text{ sgn } (t)$$

and obtain the Fourier transform pair

$$u(t) \Leftrightarrow \left[\pi\delta(\omega) + \frac{1}{j\omega} \right] \qquad (38)$$

Table 18-1 presents the conclusions drawn from the examples discussed in this section, along with a few others that have not been detailed here.

Drill Problems

18-7. Find $\mathbf{F}(j2)$ if $f(t) = :$ (a) $2u(t) - \delta(t - 1)$; (b) $5 \text{ sgn } (t) - 3$; (c) $3e^{-4t}u(t)$; (d) lim as $a \rightarrow 0$ of $e^{-at}u(t - 1)$, $a > 0$.

Ans: $0.426\underline{/-12.30°}$; $5\underline{/-90°}$; $0.671\underline{/-26.6°}$; $0.500\underline{/155.4°}$

18-8. Find $f(2)$ if $\mathbf{F}(j\omega) = :$ (a) $3[\delta(\omega + 1) + \delta(\omega - 1)]$; (b) $2/(j\omega - 0.1)$; (c) $(1 + j2\omega)/(0.5 - \omega^2 + j\omega)$.

Ans: -0.397; 2.44; 0.398

18-6
The Fourier transform of a general periodic time function

In Sec. 18-2 we remarked that we would be able to show that periodic time functions, as well as nonperiodic functions, possess Fourier transforms. Since a promise made is a debt unpaid, let us now establish this fact on a rigorous basis. Consider a periodic time function $f(t)$ with period T and Fourier series expansion, as outlined by (1), (2), and (3),

$$f(t) = \sum_{n=-\infty}^{\infty} \mathbf{c}_n e^{jn\omega_0 t} \qquad (1)$$

where

$$\mathbf{c}_n = \frac{1}{T} \int_{-T/2}^{T/2} f(t)e^{-jn\omega_0 t} \, dt \qquad (2)$$

and

$$\omega_0 = \frac{2\pi}{T} \qquad (3)$$

Bearing in mind that the Fourier transform of a sum is just the sum of the transforms of the terms in the sum, and that $\mathbf{c}_n$ is not a function of time, we can write

$$\mathscr{F}\{f(t)\} = \mathscr{F} \left\{ \sum_{n=-\infty}^{\infty} \mathbf{c}_n e^{jn\omega_0 t} \right\} = \sum_{n=-\infty}^{\infty} \mathbf{c}_n \mathscr{F}\{e^{jn\omega_0 t}\}$$

After obtaining the transform of $e^{jn\omega_0 t}$ from (30), we have

$$f(t) \Leftrightarrow 2\pi \sum_{n=-\infty}^{\infty} \mathbf{c}_n \delta(\omega - n\omega_0) \qquad (39)$$

This shows that $f(t)$ has a discrete spectrum consisting of impulses located at points on the ω axis given by $\omega = n\omega_0$, $n = \ldots, -2, -1, 0, 1, \ldots$. The strength of each impulse is 2π times the value of the corresponding Fourier coefficient appearing in the complex form of the Fourier series expansion for $f(t)$.

As a check on our work, let us see whether the inverse Fourier transform of the right side of (39) is once again $f(t)$. This inverse transform can be written as

$$\mathscr{F}^{-1}\{\mathbf{F}(j\omega)\} = \frac{1}{2\pi} \int_{-\infty}^{\infty} e^{j\omega t} \left[2\pi \sum_{n=-\infty}^{\infty} \mathbf{c}_n \delta(\omega - n\omega_0) \right] d\omega \stackrel{?}{=} f(t)$$

Since the exponential term does not contain the index of summation n, we can interchange the order of the integration and summation operations,

$$\mathscr{F}^{-1}\{\mathbf{F}(j\omega)\} = \sum_{n=-\infty}^{\infty} \int_{-\infty}^{\infty} \mathbf{c}_n e^{j\omega t} \delta(\omega - n\omega_0) \, d\omega \stackrel{?}{=} f(t)$$

Because it is not a function of the variable of integration ω, $\mathbf{c}_n$ can be treated as a constant. Then, using the sifting property of the impulse, we obtain

$$\mathscr{F}^{-1}\{\mathbf{F}(j\omega)\} = \sum_{n=-\infty}^{\infty} \mathbf{c}_n e^{jn\omega_0 t} \stackrel{?}{=} f(t)$$

which is exactly the same as (1), the complex Fourier series expansion for $f(t)$. The question marks in the preceding equations can now be removed, and the existence of the Fourier transform for a periodic time function is established. This should come as no great surprise, however. In the last section we evaluated the Fourier transform of a cosine function, which is certainly periodic, although we made no direct reference to its periodicity. However, we did use a backhanded approach in getting the transform. But now we have a mathematical tool by which the transform can be obtained more directly. To demonstrate this procedure, consider $f(t) = \cos \omega_0 t$ once more. First we evaluate the Fourier coefficients $\mathbf{c}_n$,

$$\mathbf{c}_n = \frac{1}{T} \int_{-T/2}^{T/2} \cos \omega_0 t \; e^{-jn\omega_0 t} \, dt = \begin{cases} \frac{1}{2} & n = \pm 1 \\ 0 & \text{otherwise} \end{cases}$$

Then

$$\mathscr{F}\{f(t)\} = 2\pi \sum_{n=-\infty}^{\infty} \mathbf{c}_n \delta(\omega - n\omega_0)$$

This expression only has values that are nonzero when $n = \pm 1$, and it follows, therefore, that the entire summation reduces to

$$\mathscr{F}\{\cos \omega_0 t\} = \pi[\delta(\omega - \omega_0) + \delta(\omega + \omega_0)]$$

which is precisely the expression that we obtained before. What a relief!

Drill Problem

18-9. Find: (*a*) $\mathscr{F}\{3 \cos^2 2t\}$; (*b*) $\mathscr{F}\{\sin \omega_0 t\}$; (*c*) $\mathscr{F}\{4 \cos(2t + 60°)\}$.
Ans: $3\pi\delta(\omega) + 1.5\pi[\delta(\omega + 4) + \delta(\omega - 4)]$; $j\pi[\delta(\omega + \omega_0) - \delta(\omega - \omega_0)]$;
$(4\pi\underline{/-60°})\delta(\omega + 2) + (4\pi\underline{/60°})\delta(\omega - 2)$

18-7
Convolution and circuit response in the time domain

Before continuing our discussion of the Fourier transform, let us note again where all this is leading. Our goal is a technique for simplifying those problems in linear circuit analysis that involve the determination of explicit expressions for response functions caused by the application of one or more forcing functions. We accomplish this by utilizing a transfer function called the *system function* of the circuit. It turns out that this system function is the Fourier transform of the unit-impulse response of the circuit.

The specific analytical technique that will be used requires the evaluation of the Fourier transform of the forcing function, the multiplication of this transform by the system function to obtain the transform of the response function, and then the inverse-transform operation to obtain the response function. By these means some relatively complicated integral expressions will be reduced to simple functions of ω, and the mathematical operations of integration and differentiation will be replaced by the simpler operations of algebraic multiplication and division. With these remarks in mind, let us now proceed to examine the unit-impulse response of a circuit and eventually establish its relation to the system function. Then we can look at some specific analysis problems.

Consider a linear electric network N without initial stored energy to which a forcing function $x(t)$ is applied. At some point in this circuit, a response function $y(t)$ is present. We show this in block diagram form in Fig. 18-7a along with general sketches of typical time functions. The forcing function is arbitrarily shown to exist only over the interval $a < t < b$. Thus, $y(t)$ can exist only for $t > a$.

Fig. 18-7: A conceptual development of the convolution integral,

$$y(t) = \int_{-\infty}^{\infty} x(\lambda)h(t - \lambda)\,d\lambda.$$

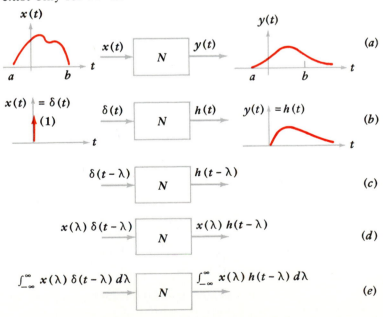

The question that we now wish to answer is this: if we know the form of $x(t)$, then how is $y(t)$ described? To answer this question, it is obvious that we need to know something about N. Suppose, therefore, that our knowledge of N is the way it responds when the forcing function is a unit impulse. That is, we are assuming that we know $h(t)$, the response function resulting from a unit impulse being supplied as the forcing function at $t = 0$, as shown in Fig. 18-7b. The function $h(t)$ is commonly called the unit-impulse response function or the *impulse response*. This is a very important descriptive property for an electric circuit.

Instead of applying the unit impulse at time $t = 0$, let us now suppose that it is applied at time $t = \lambda$ (lambda). We see that the only change in the output is a time delay. Thus, the output becomes $h(t - \lambda)$ when the input is $\delta(t - \lambda)$, as shown in Fig. 18-7c. Next, suppose that the input impulse has some strength other than unity. Specifically, let the strength of the impulse be numerically equal to the value of $x(t)$ when $t = \lambda$. This value $x(\lambda)$ is a constant; we know that the multiplication of a single forcing function in a linear circuit by a constant simply causes the response to change proportionately. Thus, if the input is changed to $x(\lambda)\delta(t - \lambda)$, then the response becomes $x(\lambda)h(t - \lambda)$, as shown in Fig. 18-7d.

Now let us sum this latest input over all possible values of λ and use the result as a forcing function for N. Linearity decrees that the output must be equal to the sum of the responses resulting from the use of all possible values of λ. Loosely speaking, the integral of the input produces the integral of the output, as shown in Fig. 18-7e. But what is the input now? Given the sifting property of the unit impulse, we see that the input is simply $x(t)$, the original input.

Our question is now answered. When $x(t)$, the input to N, is known, and when $h(t)$, the impulse response of N, is known, then $y(t)$, the output or response function, is expressed by

$$y(t) = \int_{-\infty}^{\infty} x(\lambda)h(t - \lambda)\,d\lambda \tag{40}$$

as shown in Fig. 18-7f. This important relationship is known far and wide as the *convolution integral*. In words, this last equation states that "the output is equal to the input convolved with the impulse response." It is often abbreviated by means of

$$y(t) = x(t) * h(t) \tag{41}$$

where the asterisk is read "convolved with."

Equation (40) sometimes appears in a slightly different but equivalent form. If we let $z = t - \lambda$, $d\lambda = -dz$, the expression for $y(t)$ becomes

$$y(t) = \int_{\infty}^{-\infty} -x(t - z)h(z)\,dz = \int_{-\infty}^{\infty} x(t - z)h(z)\,dz$$

and, since the symbol that we use for the variable of integration is unimportant, we can modify (40) to write

$$y(t) = x(t) * h(t) = \int_{-\infty}^{\infty} x(z)h(t - z)\, dz = \int_{-\infty}^{\infty} x(t - z)h(z)\, dz \quad (42)$$

The two forms of the convolution integral given by (42) are worth memorizing.

The result that we have in (42) is very general. It applies to any linear system. However, we are usually interested in *physically realizable* systems, those that *do* exist or *could* exist, and such systems have a property that modifies the convolution integral slightly. That is, the response of the system cannot begin before the forcing function is applied. In particular, $h(t)$ is the response of the system resulting from the application of a unit impulse at $t = 0$. Therefore, $h(t)$ cannot exist for $t < 0$. It follows that, in the second integral of (42), the integrand is zero when $z < 0$; in the first integral, the integrand is zero when $t - z$ is negative or when $z > t$. Therefore, for realizable systems the limits of integration change in the convolution integrals:

$$y(t) = x(t) * h(t) = \int_{-\infty}^{t} x(z)h(t - z)\, dz = \int_{0}^{\infty} x(t - z)h(z)\, dz \quad (43)$$

Equations (42) and (43) are both valid, but the latter is more specific when we are speaking of *realizable* linear systems.

Before discussing the significance of the impulse response of a circuit any further, let us consider a numerical example which will give us some insight into just how the convolution integral can be evaluated. Although the expression itself is simple enough, the evaluation is sometimes troublesome, especially with regard to the values used as the limits of integration.

Suppose that the input is a rectangular voltage pulse that starts at $t = 0$, has a duration of 1 s, and is 1 V in amplitude,

$$x(t) = v_i(t) = u(t) - u(t - 1)$$

Suppose also that the impulse response of this circuit is known to be an exponential function of the form:[7]

$$h(t) = 2e^{-t}u(t)$$

We wish to evaluate the output voltage $v_o(t)$, and we can write the answer immediately in integral form,

$$y(t) = v_o(t) = v_i(t) * h(t) = \int_{0}^{\infty} v_i(t - z)h(z)\, dz$$
$$= \int_{0}^{\infty} [u(t - z) - u(t - z - 1)][2e^{-z}u(z)]\, dz$$

[7] A description of one possible circuit to which this impulse response might apply is developed in Prob. 35.

Obtaining this expression for $v_o(t)$ was simple enough, but the presence of the many unit-step functions tends to make its evaluation confusing. Careful attention must be paid to determine those portions of the range of integration in which the integrand is zero.

Let us first use some graphical assistance to help us understand what the convolution integral says. We begin by drawing several z axes lined up one above the other, as shown in Fig. 18-8. We know what $v_i(t)$ looks like, so we know what $v_i(z)$ looks like also; this is plotted as Fig. 18-8a. The function $v_i(-z)$ is simply $v_i(z)$ run backwards with respect to z, or rotated about the ordinate axis; it is shown in Fig. 18-8b. Next we wish to represent $v_i(t - z)$, which is $v_i(-z)$ after it is shifted to the right by an amount $z = t$ as shown in Fig. 18-8c. On the next z axis in Fig. 18-8d, the hypothetical impulse response is plotted.

Fig. 18-8: Graphical concepts in evaluating a convolution integral.

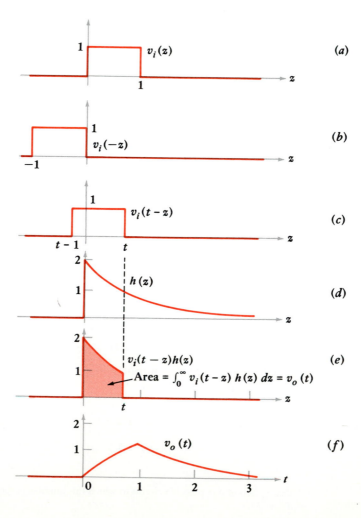

Finally, we multiply the two functions $v_i(t - z)$ and $h(z)$. The result is shown in Fig. 18-8e. Since $h(z)$ does not exist prior to $t = 0$ and $v_i(t - z)$ does not exist for $z > t$, notice that the product of these two functions has nonzero values only in the interval $0 < z < t$ for the case shown where $t < 1$; when $t > 1$, then the nonzero values for the product are obtained in the interval, $(t - 1) < z < t$. The *area* under the product curve (shown shaded in the figure) is numerically equal to the value of v_o corresponding to the specific value of t selected in Fig. 18-8c. As t increases from zero to unity, the area under the product curve continues to rise, and thus $v_o(t)$ continues to rise. But as t increases beyond $t = 1$, the area under the product curve, which is equal to $v_o(t)$, starts decreasing and approaches zero. For $t < 0$, the curves representing $v_i(t - z)$ and $h(z)$ do not overlap at all, so the area under the product curve is obviously zero. Now let us use these graphical concepts to obtain an explicit expression for $v_o(t)$.

For values of t that lie between zero and unity, we must integrate from $z = 0$ to $z = t$; for values of t that exceed unity, the range of integration is $t - 1 < z < t$. Thus, we may write

$$
v_o(t) = \begin{cases}
0 & t < 0 \\[2mm]
\displaystyle\int_0^t 2e^{-z}\, dz = 2(1 - e^{-t}) & 0 < t < 1 \\[2mm]
\displaystyle\int_{t-1}^t 2e^{-z}\, dz = 2(e - 1)e^{-t} & t > 1
\end{cases}
$$

This function is shown plotted versus the time variable t in Fig. 18-8f, and our solution is completed.

As a second convolution example, let us apply a unit-step function, $x(t) = u(t)$, as the input to a system whose impulse response is $h(t) = u(t) - 2u(t - 1) + u(t - 2)$. These two functions of time are shown in Figs. 18-9a and b.

We arbitrarily choose to evaluate the first integral of (43),

$$
y(t) = \int_{-\infty}^t x(z)\, h(t - z)\, dz
$$

and prepare a sequence of sketches to help select the correct limits of inte-

Fig. 18-9: Sketches of (a) the input signal, $x(t) = u(t)$, and (b) the unit-impulse response, $h(t) = u(t) - 2u(t - 1) + u(t - 2)$, for a linear system.

(a)

(b)

Fig. 18-10: (a) The input signal, and (b) the impulse response are plotted as functions of z. (c) h(−z) is obtained by flipping h(z) about the vertical axis, and (d) h(t − z) results when h(−z) is slid t units to the right.

gration. Figure 18-10 shows these functions in order: the input $x(z)$ as a function of z; the impulse response $h(z)$; the curve of $h(-z)$, which is just $h(z)$ rotated about the vertical axis; and $h(t - z)$, obtained by sliding $h(-z)$ to the right t units. For this sketch, we have selected t in the range, $0 < t < 1$.

It is now easy to visualize the product of the first graph, $x(z)$, and the last, $h(t - z)$, for the various ranges of t. When t is less than zero, there is no overlap, and

$$y(t) = 0 \qquad t < 0$$

For the case sketched in Fig. 18-10d, the curves overlap from $z = 0$ to $z = t$, and each is unity in value. Thus,

$$y(t) = \int_0^t 1 \times 1 \; dz = t \qquad 0 < t < 1$$

When t lies between 1 and 2, $h(t - z)$ has slid far enough to the right to bring under the step function that part of the negative square wave extending from $z = 0$ to $z = t - 1$. We have

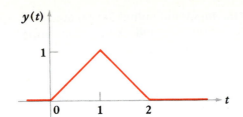

Fig. 18-11: The result of convolving the $x(t)$ and $h(t)$ shown in Fig. 18-9.

$$y(t) = \int_0^{t-1} 1 \times (-1)\, dz + \int_{t-1}^t 1 \times 1\, dz$$

$$= -z\big|_{z=0}^{z=t-1} + z\big|_{z=t-1}^{z=t}$$

Therefore,

$$y(t) = -(t-1) + t - (t-1) = 2 - t \qquad 1 < t < 2$$

Finally, when t is greater than 2, $h(t-z)$ has slid far enough to the right to lie entirely to the right of $z = 0$. The intersection with the unit step is complete, and

$$y(t) = \int_{t-2}^{t-1} 1 \times (-1)\, dz + \int_{t-1}^t 1 \times 1\, dz$$

$$= -z\big|_{z=t-2}^{z=t-1} + z\big|_{z=t-1}^{z=t}$$

or $\qquad y(t) = -(t-1) + (t-2) + t - (t-1) = 0 \qquad t > 2$

These four segments of $y(t)$ are collected as a continuous curve in Fig. 18-11.

There is a great deal of information and technique wrappped up in these two examples. The drill problems below offer an opportunity to make sure that the procedure is understood sufficiently well to make it worthwhile to pass on to new material. In other words, pay attention to the old Chinese proverb: I hear, I forget; I see, I remember; I do, I understand.

Drill Problems

18-10. A linear network is characterized by the impulse response, $h(t) = 0.3u(t-0.2)$. An input signal, $x(t) = 4[u(t) - u(t-1)]$, is applied. Find $y(t)$ at $t = :$ (a) -0.1; (b) 0.1; (c) 0.6; (d) 1.1; (e) 1.6; (f) 2.3.
Ans: 0; 0; 0.480; 1.080; 1.200; 1.200

18-11. If $h(t) = 0.6e^{-2t}u(t)$ and $x(t) = 10e^{-5t}u(t)$, find $y(t)$ at $t = :$ (a) -1; (b) 1; (c) 2.
Ans: 0; 0.257; 0.0365

18-8
The system function and response in the frequency domain

In the previous section the problem of determining the output of a physical system in terms of the input and the impulse response was solved by using the convolution integral and working entirely in the time domain. The input, the output, and the impulse response are all *time* functions. Now let us see whether some analytical simplification can be wrought by working with frequency-domain descriptions of these three functions.

To do this we examine the Fourier transform of the system output. Assuming that the input and output are voltages, we apply the basic definition of the Fourier transform and express the output by the convolution integral (42),

$$\mathcal{F}\{v_o(t)\} = \mathbf{F}_o(j\omega) = \int_{-\infty}^{\infty} e^{-j\omega t} \left[\int_{-\infty}^{\infty} v_i(t - z)h(z)\, dz \right] dt$$

where we again assume no initial energy storage. At first glance this expression may seem rather formidable, but it can be reduced to a result that is surprisingly simple. We may move the exponential term inside the inner integral because it does not contain the variable of integration z. Next we reverse the order of integration, obtaining

$$\mathbf{F}_o(j\omega) = \int_{-\infty}^{\infty} \int_{-\infty}^{\infty} e^{-j\omega t} v_i(t - z)h(z)\, dt\, dz$$

Since it is not a function of t, we can extract $h(z)$ from the inner integral and simplify the integration with respect to t by a change of variable, $t - z = x$,

$$\mathbf{F}_o(j\omega) = \int_{-\infty}^{\infty} h(z) \int_{-\infty}^{\infty} e^{-j\omega(x+z)} v_i(x)\, dx\, dz$$

$$= \int_{-\infty}^{\infty} e^{-j\omega z} h(z) \int_{-\infty}^{\infty} e^{-j\omega x} v_i(x)\, dx\, dz$$

But now the sum is starting to break through, for the inner integral is merely the Fourier transform of $v_i(t)$. Furthermore, it contains no z terms and can be treated as a constant in any integration involving z. Thus, we can move this transform, $\mathbf{F}_i(j\omega)$, completely outside all the integral signs,

$$\mathbf{F}_o(j\omega) = \mathbf{F}_i(j\omega) \int_{-\infty}^{\infty} e^{-j\omega z} h(z)\, dz$$

Finally, the remaining integral exhibits our old friend once more, another Fourier transform! This one is the Fourier transform of the impulse response, which we shall designate by the notation $\mathbf{H}(j\omega)$. Therefore, all our work has boiled down to the simple result:

$$\mathbf{F}_o(j\omega) = \mathbf{F}_i(j\omega)\mathbf{H}(j\omega) = \mathbf{F}_i(j\omega)\mathcal{F}\{h(t)\} \tag{44}$$

This is another important result: it defines the *system function* $\mathbf{H}(j\omega)$ as the ratio of the Fourier transform of the response function to the Fourier transform of the forcing function. Moreover, the system function and the impulse response comprise a Fourier transform pair:

$$h(t) \Leftrightarrow \mathbf{H}(j\omega) \tag{45}$$

The development in the preceding paragraph also serves to prove the general statement that the Fourier transform of the convolution of two time functions is the product of their Fourier transforms,

$$\mathcal{F}\{f(t) * g(t)\} = \mathbf{F}_f(j\omega)\mathbf{F}_g(j\omega) \tag{46}$$

To recapitulate, if we know the Fourier transforms of the forcing function and the impulse response, then the Fourier transform of the response function can be obtained as their product. The result is a description of the response function in the frequency domain; if we wish to do so, we can obtain the time-domain description of the response function by taking the inverse Fourier transform. Thus we see that the process of convolution in the time domain is equivalent to the relatively simple operation of multiplication in the frequency domain. This is one fact that makes the use of integral transforms so attractive.

The foregoing comments might make us wonder once again why we would ever choose to work in the time domain, but we must always remember that we seldom get something for nothing. A poet once said, "Our sincerest laughter, with some pain is fraught."[8] The pain herein is the occasional difficulty in obtaining the inverse Fourier transform of a response function, for reasons of mathematical complexity. On the other hand, a modern digital computer can convolve two time functions with magnificent celerity. For that matter, it can also obtain an FFT (fast Fourier transform)[9] quite rapidly. Consequently there is no clear-cut advantage between working in the time domain or the frequency domain. A decision must be made each time a new problem arises; it should be based on the given information available and on the computational facilities at hand.

Now let us attempt a frequency-domain analysis of the problem that we worked in the preceding section with the convolution integral. We had a forcing function of the form

$$v_i(t) = u(t) - u(t - 1)$$

and a unit-impulse response defined by

$$h(t) = 2e^{-t}u(t)$$

We first obtain the corresponding Fourier transforms. The forcing function is the difference between two unit-step functions. These two functions are identical, except that one is initiated 1 s after the other. We shall evaluate the response due to $u(t)$, and the response due to $u(t - 1)$ is the same, but delayed in time by 1 s. The difference between these two partial responses will be the total response due to $v_i(t)$.

The Fourier transform of $u(t)$ was obtained in Sec. 18-5,

$$\mathcal{F}\{u(t)\} = \pi\delta(\omega) + \frac{1}{j\omega}$$

---

[8] A cultural message from P. B. Shelley, "To a Skylark," 1821.

[9] The fast Fourier transform is a type of *discrete* Fourier transform, which is a numerical approximation to the (continuous) Fourier transform we have been considering. Many references concerning it are given in G. D. Bergland, A Guided Tour of the Fast Fourier Transform, *IEEE Spectrum*, vol. 6, no. 7, pp. 41–52, July 1969.

The system function is obtained by taking the Fourier transform of $h(t)$, listed in Table 18-1,

$$\mathcal{F}\{h(t)\} = \mathbf{H}(j\omega) = \mathcal{F}\{2e^{-t}u(t)\} = \frac{2}{1 + j\omega}$$

The inverse transform of the product of these two functions yields that component of $v_o(t)$ caused by $u(t)$,

$$v_{o1}(t) = \mathcal{F}^{-1}\left\{\frac{2\pi\delta(\omega)}{1 + j\omega} + \frac{2}{j\omega(1 + j\omega)}\right\}$$

Using the sifting property of the unit impulse, the inverse transform of the first term is just a constant equal to unity.

The second term contains a product of terms in the denominator, each of the form $(\alpha + j\omega)$, and its inverse transform is found most easily by making use of the partial-fraction expansion that we all should have seen in our introductory calculus course. Let us select a technique for obtaining a partial-fraction expansion that has one big advantage—it always works, although faster methods are usually available for most situations.[10] We assign an unknown quantity in the numerator of each fraction, here two in number,

$$\frac{2}{j\omega(1 + j\omega)} = \frac{A}{j\omega} + \frac{B}{1 + j\omega}$$

and then substitute a corresponding number of simple values for $j\omega$. Here we let $j\omega = 1$,

$$1 = A + B/2$$

and then $j\omega = -2$,

$$1 = -A/2 - B$$

This leads to $A = 2$ and $B = -2$. Thus,

$$\mathcal{F}^{-1}\left\{\frac{2}{j\omega(1 + j\omega)}\right\} = \mathcal{F}^{-1}\left\{\frac{2}{j\omega} - \frac{2}{1 + j\omega}\right\} = \text{sgn}\,(t) - 2e^{-t}u(t)$$

so that

$$v_{o1}(t) = 1 + \text{sgn}\,(t) - 2e^{-t}u(t)$$

$$= 2u(t) - 2e^{-t}u(t)$$

$$= 2(1 - e^{-t})u(t)$$

It follows that $v_{o2}(t)$, the component of $v_o(t)$ produced by $u(t - 1)$, is

$$v_{o2}(t) = 2(1 - e^{-(t-1)})u(t - 1)$$

Therefore,

$$v_o(t) = v_{o1}(t) + v_{o2}(t)$$

$$= 2(1 - e^{-t})u(t) - 2(1 - e^{-t+1})u(t - 1)$$

[10] See that old calculus text!

The discontinuities at $t = 0$ and $t = 1$ dictate a separation into three time intervals:

$$v_o(t) = \begin{cases} 0 & t < 0 \\ 2(1 - e^{-t}) & 0 < t < 1 \\ 2(e - 1)e^{-t} & t > 1 \end{cases}$$

This is the same result we obtained by working this problem with the convolution integral in the time domain.

It appears that some ease of solution is realized by working this particular problem in the frequency domain rather than in the time domain. We are able to trade a relatively complicated time-domain integration for a simple multiplication of two Fourier transforms in the frequency domain, and (this is a most noteworthy point) we did not have to pay the price of having a difficult Fourier transform to invert in determining $v_o(t)$.

Drill Problems

18-12. Let $v_1(t) = 10e^{-2t}u(t)$ V in a linear network for which $h(t) = 2.5e^{-8t}u(t)$. Find $\mathbf{H}(j\omega)$, $\mathbf{V}_i(j\omega)$, $\mathbf{V}_o(j\omega)$, and then: (a) $v_o(0.5)$; (b) $v_o(1.5)$; (c) $v_{o,\max}$. *Ans:* 1.457; 0.207; 1.969 V

18-13. Use frequency-domain methods to find the time-domain response of a network having a system function $(3 - j\omega)/(2 + j\omega)$ if the input is: (a) a unit impulse; (b) a unit-step function; (c) $\cos t$.
Ans: $-\delta(t) + 5e^{-2t}u(t)$; $(1.5 - 2.5e^{-2t})u(t)$; $\sqrt{2} \cos (t - 45°)$

18-9
The physical significance of the system function

In this section we shall try to connect several aspects of the Fourier transform with work we completed in earlier chapters.

Given a general linear two-port network N without any initial energy storage, we assume sinusoidal forcing and response functions, arbitrarily assumed to be voltages, as shown in Fig. 18-12. We let the input voltage be simply $A \cos (\omega_x t + \theta)$, and the output can be described in general terms as $v_o(t) = B \cos (\omega_x t + \phi)$, where the amplitude B and phase angle ϕ are functions of ω_x. In phasor form, we can write the forcing and response functions as $\mathbf{V}_i = Ae^{j\theta}$ and $\mathbf{V}_o = Be^{j\phi}$. The ratio of the phasor response to the phasor forcing function is a complex number that is a function of ω_x.

$$\frac{\mathbf{V}_o}{\mathbf{V}_i} = \mathbf{G}(\omega_x) = \frac{B}{A} e^{j(\phi - \theta)}$$

where B/A is the amplitude of $\mathbf{G}$ and $\phi - \theta$ is its phase angle. This transfer function $\mathbf{G}(\omega_x)$ could be obtained in the laboratory by varying ω_x over a large range of values, and measuring the amplitude B/A and phase $\phi - \theta$ for each value of ω_x. If we would then plot each of these parameters as a function of frequency, the resultant pair of curves would completely describe the transfer function.

$v_i(t) = A \cos (\omega_x t + \theta)$ | N | $v_o(t) = B \cos (\omega_x t + \phi)$

Fig. 18-12: Sinusoidal analysis can be used to determine the transfer function, $\mathbf{H}(j\omega_x) = (B/A)e^{j(\phi-\theta)}$, where B and ϕ are functions of ω_x.

Now let us hold these comments in the backs of our minds for a moment as we consider a slightly different aspect of the same analysis problem.

For the circuit with sinusoidal input and output shown in Fig. 18-12, what is the system function $\mathbf{H}(j\omega)$? To answer this question, we begin with the definition of $\mathbf{H}(j\omega)$ as the ratio of the Fourier transforms of the output and the input. Both of these time functions involve the functional form, $\cos(\omega_x t + \beta)$, whose Fourier transform we have not evaluated as yet, although we can handle $\cos \omega_x t$. The transform we need is

$$\mathcal{F}\{\cos(\omega_x t + \beta)\} = \int_{-\infty}^{\infty} e^{-j\omega t} \cos(\omega_x t + \beta)\, dt$$

If we make the substitution, $\omega_x t + \beta = \omega_x \tau$, then

$$\mathcal{F}\{\cos(\omega_x t + \beta)\} = \int_{-\infty}^{\infty} e^{-j\omega\tau + j\omega\beta/\omega_x} \cos \omega_x \tau\, d\tau$$

$$= e^{j\omega\beta/\omega_x}\mathcal{F}\{\cos \omega_x t\}$$

$$= \pi e^{j\omega\beta/\omega_x}[\delta(\omega - \omega_x) + \delta(\omega + \omega_x)]$$

This is a new Fourier transform pair,

$$\cos(\omega_x t + \beta) \Leftrightarrow \pi e^{j\omega\beta/\omega_x}[\delta(\omega - \omega_x) + \delta(\omega + \omega_x)] \tag{47}$$

which we can now use to evaluate the desired system function,

$$\mathbf{H}(j\omega) = \frac{\mathcal{F}\{B \cos(\omega_x t + \phi)\}}{\mathcal{F}\{A \cos(\omega_x t + \theta)\}}$$

$$= \frac{\pi B e^{j\omega\phi/\omega_x}[\delta(\omega - \omega_x) + \delta(\omega + \omega_x)]}{\pi A e^{j\omega\theta/\omega_x}[\delta(\omega - \omega_x) + \delta(\omega + \omega_x)]}$$

$$= \frac{B}{A} e^{j\omega(\phi - \theta)/\omega_x}$$

Now we recall the expression for $\mathbf{G}(\omega_x)$,

$$\mathbf{G}(\omega_x) = \frac{B}{A} e^{j(\phi - \theta)}$$

where B and ϕ were evaluated at $\omega = \omega_x$, and we see that evaluating $\mathbf{H}(j\omega)$ at $\omega = \omega_x$ gives

$$\mathbf{H}(j\omega_x) = \mathbf{G}(\omega_x) = \frac{B}{A} e^{j(\phi - \theta)}$$

Since there is nothing special about the x subscript, we conclude that the system function and the transfer function are identical:

$$\mathbf{H}(j\omega) = \mathbf{G}(\omega) \tag{48}$$

The fact that one argument is ω while the other is indicated by $j\omega$ is immaterial and arbitrary; the j merely makes possible a more direct comparison between the Fourier and Laplace transforms.

Equation (48) represents a direct connection between Fourier transform techniques and sinusoidal steady-state analysis. Our previous work on steady-state sinusoidal analysis using phasors was but a special case of the more general techniques of Fourier transform analysis. It was "special" in the sense that the inputs and outputs were sinusoids, whereas the use of Fourier transforms and system functions enables us to handle nonsinusoidal forcing functions and responses.

Thus, to find the system function $\mathbf{H}(j\omega)$ for a network, all we need to do is to determine the corresponding sinusoidal transfer function as a function of ω (or $j\omega$).

As an example of the application of these profound generalities, let us look at the simple RL series circuit shown in Fig. 18-13a. We seek the voltage across the inductor when the input voltage is a simple exponentially decaying pulse. We need the system function; but it is not necessary to apply an impulse, find the impulse response, and then determine its inverse transform. Instead we use (48) to obtain the system function $\mathbf{H}(j\omega)$ by assuming that the input and output voltages are both sinusoids described by their corresponding phasors, as shown in Fig. 18-13b. Using voltage division, we have

$$\mathbf{H}(j\omega) = \frac{\mathbf{V}_o}{\mathbf{V}_i} = \frac{j2\omega}{4 + j2\omega}$$

The transform of the forcing function is

$$\mathcal{F}\{v_i(t)\} = \frac{5}{3 + j\omega}$$

and thus the transform of $v_o(t)$ is given as

$$\mathcal{F}\{v_o(t)\} = \mathbf{H}(j\omega)\mathcal{F}\{v_i(t)\}$$

$$= \frac{j2\omega}{4 + j2\omega} \frac{5}{3 + j\omega}$$

$$= \frac{15}{3 + j\omega} - \frac{10}{2 + j\omega}$$

where the partial fractions appearing in the last step help to determine the inverse Fourier transform

$$v_o(t) = \mathcal{F}^{-1} \left\{ \frac{15}{3 + j\omega} - \frac{10}{2 + j\omega} \right\}$$

$$= 15e^{-3t}u(t) - 10e^{-2t}u(t)$$

$$= 5(3e^{-3t} - 2e^{-2t})u(t)$$

Fig. 18-13: (a) The response $v_o(t)$ caused by $v_i(t)$ is desired. (b) The system function $\mathbf{H}(j\omega)$ may be determined by sinusoidal steady-state analysis: $\mathbf{H}(j\omega) = \mathbf{V}_o/\mathbf{V}_i$.

Our problem is completed without fuss, convolution, or differential equations.

Returning again to (48), the identity between the system function $\mathbf{H}(j\omega)$ and the sinusoidal steady-state transfer function $\mathbf{G}(\omega)$, we may now consider the system function as the ratio of the output phasor to the input phasor. Suppose that we hold the input-phasor amplitude at unity and the phase angle at zero. Then the output phasor is $\mathbf{H}(j\omega)$. Under these conditions, if we record the output amplitude and phase as functions of ω, for all ω, we have recorded the system function $\mathbf{H}(j\omega)$ as a function of ω, for all ω. We thus have examined the system response under the condition that an infinite number of sinusoids, all with unity amplitude and zero phase, were successively applied at the input. Now suppose that our input is a single unit impulse and look at the impulse response $h(t)$. Is the information we examine really any different from that we just obtained? The Fourier transform of the unit impulse is a constant equal to unity, indicating that all frequency components are present, all with the same magnitude, and all with zero phase. Our system response is the sum of the responses to all these components. The result might be viewed at the output on a cathode-ray oscilloscope. It is evident that the system function and the impulse-response function contain equivalent information regarding the response of the system.

We therefore have two different methods of describing the response of a system to a general forcing function; one is a time-domain description, and the other a frequency-domain description. Working in the time domain, we convolve the forcing function with the impulse response of the system to obtain the response function. As we saw when we first considered convolution, this procedure may be interpreted by thinking of the input as a continuum of impulses of different strengths and times of application; the output which results is a continuum of impulse responses.

In the frequency domain, however, we determine the response by multiplying the Fourier transform of the forcing function by the system function. In this case we interpret the transform of the forcing function as a frequency spectrum, or a continuum of sinusoids. Multiplying this by the system function, we obtain the response function, also as a continuum of sinusoids.

Whether we choose to think of the output as a continuum of impulse responses or as a continuum of sinusoidal responses, the linearity of the network and the superposition principle enable us to determine the total output as a time function by summing over all frequencies (the inverse Fourier transform), or as a frequency function by summing over all time (the Fourier transform).

Unfortunately, both of these techniques have some difficulties or limitations associated with their use. In using convolution, the integral itself can often be rather difficult to evaluate when complicated forcing functions or impulse response functions are present. Furthermore, from the experimental point of view, we cannot really measure the impulse response of a system because we cannot actually generate an impulse. Even if we approximate the impulse by a narrow high-amplitude pulse, we would probably drive our system into saturation and out of its linear operating range.

With regard to the frequency domain, we encounter one absolute limitation in that we may easily hypothesize forcing functions that we would like to apply theoretically that do not possess Fourier transforms. Moreover, if we wish to find the time-domain description of the response function, we must evaluate an inverse Fourier transform, and some of these inversions can be extremely difficult.

Finally, neither of these techniques offers a very convenient method of handling initial conditions.

The greatest benefits derived from the use of the Fourier transform arise through the abundance of useful information it provides about the spectral properties of a signal, particularly the energy or power per unit bandwidth.

Most of the difficulties and limitations associated with the Fourier transform are overcome with the use of the Laplace transform. We shall see that it is defined in such a way that there is a built-in convergence factor that enables transforms to be determined for a much wider range of input time functions than the Fourier transform can accommodate. Moreover, we shall find that initial conditions may be handled in a manner that may make us wonder why this marvelous technique was not introduced before the last chapter. Finally, the spectral information that is so readily available in the Fourier transform can also be obtained from the Laplace transform, at least for most of the time functions that arise in real engineering problems.

Well, why *has* this all been withheld until now? The best answer is probably because these powerful techniques can overcomplicate the solution of simple problems and tend to obscure the physical interpretation of the performance of the simpler networks. For example, if we are interested only in the forced response, then there is little point in using the Laplace transform and obtaining both the forced and natural response after laboring through a difficult inverse transform operation.

So much for generalities; on to Laplace.

Drill Problem

18-14. Use Fourier-transform techniques on the circuit of Fig. 18-14 to find the value of $i_o(t)$ at $t = 1$ s if $v_s = :$ (*a*) $\delta(t)$; (*b*) $u(t)$; (*c*) $\cos t$.

Ans: -0.541; 0.271; -0.457 mA

Fig. 18-14: See Drill Prob. 18-14.

Problems

1 Use Eq. (8*a*) to determine the Fourier transform of: (*a*) $f(t) = e^{-at}u(t - t_0)$, $a > 0$; (*b*) $g(t) = te^{-at}u(t)$, $a > 0$; (*c*) $v(t) = (t - t_0)e^{-a(t - t_0)}u(t - t_0)$, $a > 0$.

2 Let $f(t) = 100u(t + 2) + 100u(t + 1) - 100u(t - 1) - 100u(t - 2)$. (*a*) Sketch $f(t)$. (*b*) Find $\mathbf{F}(j\omega)$ by using the defining equations for the Fourier transform.

Fig. 18-15: See Prob. 3.

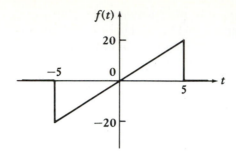

3 Find the Fourier transform of the single sawtooth pulse sketched in Fig. 18-15.
4 Find the Fourier transforms for the waveforms illustrated in Fig. 18-16.

(a)

(b)

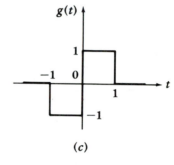

(c)

Fig. 18-16: See Prob. 4.

5 Determine the time function that corresponds to each of these continuous frequency spectra and then evaluate it at $t = 0.5$: (a) $\mathbf{F}(j\omega) = 1$, $1 < |\omega| < 2$; $\mathbf{F}(j\omega) = 0$, $|\omega| < 1$ and $|\omega| > 2$; (b) $\mathbf{F}(j\omega) = e^{-|\omega|}$; (c) $\mathbf{F}(j\omega) = j \sin \omega$, $|\omega| < \pi$; $\mathbf{F}(j\omega) = 0$, $|\omega| > \pi$.

6 A single pulse is defined by $f(t) = 12 \cos 200\pi t$ V for $-2.5 \leq t \leq 2.5$ ms, and $f(t) = 0$ for $t < -2.5$ ms and $t > 2.5$ ms. Evaluate $\mathbf{F}(j\omega)$ for $\omega = $: (a) 0; (b) 100π rad/s; (c) -100π rad/s; (d) 200π rad/s.

7 The voltage $v_s(t)$ is present across a 5-Ω resistor and has a Fourier transform whose magnitude is $|\mathbf{F}_s(j\omega)| = 9 - \omega^2$ V/(rad/s) for $|\omega| < 3$ and 0 for $|\omega| > 3$ rad/s. (a) What is the total energy present in the signal? (b) In what range, $|\omega| < \omega_1$, is 50 percent of this energy located?

8 Let $\mathbf{F}(j\omega) = \pi/\sqrt{|\omega|}$ be the Fourier transform of $f(t)$. (a) Given the definite integral, $\int_0^\infty [(\cos mx)/\sqrt{x}] dx = \sqrt{\pi/2m}$, find $f(t)$. (b) Calculate the 1-Ω energy in $f(t)$ that is found in the frequency band, $1 < |\omega| < 3$ rad/s. (c) Calculate the 1-Ω energy in $f(t)$ that is found in the time interval, $1 < t < 3$ s.

9 Let $i(t) = 10te^{-5t}u(t)$ A be the current in a 1-Ω resistor. Calculate: (a) the total energy absorbed by the resistor over all time, and (b) the percentage of the energy that lies in the frequency range, $|\omega| < 5$ rad/s.

10 If $f(t) = 2e^{-8t}u(t)$, find: (a) the 1-Ω energy associated with this signal; (b) the amplitude spectrum of $\mathbf{F}(j\omega)$; and (c) the frequency range $|\omega| < \omega_1$ in which 90 percent of the 1-Ω energy lies.

11 Find the strength of the impulse that is defined by the limit as $\varepsilon \to 0$ of: (a) $(1/\varepsilon)[\cos (\pi t/2\varepsilon)^2 [u(t + \varepsilon) - u(t - \varepsilon)]$; (b) $(1/\varepsilon)[1 - (t/\varepsilon)^2][u(t + \varepsilon) - u(t - \varepsilon)]$.

12 (a) Find $i_R(t)$, $i_C(t)$, and $i_s(t)$ for the circuit of Fig. 18-17. Express all currents in terms of singularity functions. (b) Specify values for the differences, $i_R(0^+) - i_R(0^-)$, $i_C(0^+) - i_C(0^-)$, and $q_C(0^+) - q_C(0^-)$.

Fig. 18-17: See Prob. 12.

13 (a) Find $v_L(t)$, $i_x(t)$, $v_R(t)$, and $v_s(t)$ for the circuit of Fig. 18-18. Express each quantity in terms of singularity functions. (b) Show that $v_R(t) = 0.8 di_x/dt$. Remember that $(d/dt)(uv) = u\, dv/dt + v\, du/dt$.

Fig. 18-18: See Prob. 13.

14 (a) If $v_s(t) = \delta(t)$ V in the circuit shown in Fig. 18-19, find $v_{20,a}(t)$. (b) Find $v_{20,b}(t)$ if $v_s(t) = u(t)$ V. (c) Show that $v_{20,a}(t) = (d/dt)[v_{20,b}(t)]$.

Fig. 18-19: See Prob. 14.

15 The voltage $v(t) = (20/a)[u(t) - u(t - a)]$ V is applied to the series combination of a 10-kΩ resistor and a 5-μF capacitor. Find the magnitude of the capacitor voltage at $t = 30$ ms if: (a) $a = 40$ ms; (b) $a = 20$ ms; (c) $a = 2$ ms; (d) in the limit as $a \to 0$.

16 Use the definition of the Fourier transform to prove the following results, where $\mathcal{F}\{f(t)\} = \mathbf{F}(j\omega)$: (a) $\mathcal{F}\{f(t - t_0)\} = e^{-j\omega t_0}\mathcal{F}\{f(t)\}$; (b) $\mathcal{F}\{df(t)/dt\} = j\omega\mathcal{F}\{f(t)\}$; (c) $\mathcal{F}\{f(kt)\} = (1/|k|)\mathbf{F}(j\omega/k)$; (d) $\mathcal{F}\{f(-t)\} = \mathbf{F}(-j\omega)$; (e) $\mathcal{F}\{tf(t)\} = jd[\mathbf{F}(j\omega)]/d\omega$.

17 Find the Fourier transform of $f(t) = :$ (a) $5[\text{sgn}\ (t - 1)]\ \delta(t)$; (b) $5[\text{sgn}\ (t - 1)]$ $\delta(t - 2)$; (c) $5 \sin 2t\ u(t)$; (d) $5 \sin 2t[u(t + \pi/2) - u(t - \pi/2)]$.

18 Find $\mathbf{F}(j\omega)$ if $f(t) = :$ (a) $A \sin\ (\omega_0 t + \theta)$; (b) $2\delta(t) - 4\ \text{sgn}\ (t) + 3u(t)$; (c) $\cosh 5t\ u(t)$.

19 Find $f(t)$ at $t = 2$ if $\mathbf{F}(j\omega) = :$ (a) $2u(\omega + 2) - 2u(\omega - 2)$; (b) $2u(-2 - \omega) + 2u(-2 + \omega)$; (c) $\delta(\omega) + 2u(-2 - \omega) + 2u(-2 + \omega)$.

20 Find the function of time that corresponds to each of the following Fourier transforms: (a) $2 - 2\delta(\omega - 2) + j2/\omega + 2/(2 + j\omega)$; (b) $(2 + j\omega)/[(2 + j\omega)^2 + 2^2]$; (c) $(\sin 2\omega)/(2\omega)$.

21 Find the Fourier transform of the periodic waveform shown in Fig. 18-20.

Fig. 18-20: See Prob. 21.

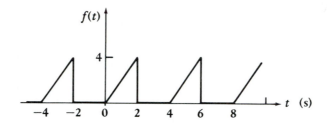

22 A periodic waveform $f(t)$ is described over the period $0 < t < 5$ by $f_1(t) = 2u(t) + 3u(t - 2) - 5u(t - 4)$. Find $\mathbf{F}(j\omega)$.

23 Let $\mathbf{F}(j\omega) = 100 \sum_{n=-\infty}^{\infty} \dfrac{1}{|n|!} \delta(\omega - 0.1n)$. Find the value of the corresponding $f(t)$ at $t = 2$.

24 Given $x(t) = h(t) = u(t) - u(t - 1)$: (a) use convolution in the time domain to find $y(t) = x(t) * h(t)$ and (b) sketch the result.

25 Let $x(t) = 2[u(t) - u(t - 2)]$ and $h(t) = u(t)$. Find $y(t) = x(t) * h(t)$ by using: (a) the first integral in (43); (b) the second integral in (43).

26 Use convolution in the time domain to find $f(t) = f_1(t) * f_2(t)$, if $f_1(t) = (1 + 2e^{-t})u(t)$ and $f_2(t) = 4[u(t + 1) - u(t - 2)]$.

27 The unit-impulse response and the input to a certain linear system are shown in Fig. 18-21. (a) Obtain an integral expression for the output that is valid in the range, $4 < t < 6$, and does not contain any singularity functions. (b) Evaluate the output at $t = 5$.

Fig. 18-21: See Prob. 27.

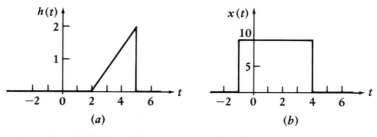

28 The input signal for a certain linear system is $x(t) = 4e^{-(t-1)}u(t - 1)$, while the impulse response is $h(t) = (16 - 2t)[u(t - 5) - u(t - 8)]$. Find the output signal value at: (a) $t = 5$; (b) $t = 8$.

29 If the input to a linear system is $x_1(t) = \delta(t)$, then the output is $y_1(t) = 8 - 2t$ for $0 < t < 4$, and zero elsewhen. A different input signal, $x_2(t) = 10\delta(t - 2) + 6u(t)$, is now applied. Calculate values for the output $y_2(t)$ at $t = $: (a) 1; (b) 2^-; (c) 2^+; (d) 3; (e) 5; (f) 7.

30 If $y(t) = x(t) * h(t)$, where $x(t) = 5\delta(t - 4) + 6[u(t) - u(t - 4)]$ and $h(t) = 9e^{-0.2t}u(t)$: (a) sketch $y(t)$ for $0 < t < 8$, and (b) find values for $y(3)$ and $y(5)$.

31 If $h(t) = u(t) - u(t - 1)$ and $x(t) = 6[u(t - 2) - u(t - 5)]$, use convolution in the time domain to evaluate $y(t) = x(t) * h(t)$ at $t =$: (a) 1; (b) 2.5; (c) 4; (d) 5.6; (e) 7.

32 The impulse response of a linear system is found to be $h(t) = 9e^{-0.2t}u(t)$. When a unit impulse is applied, what percentage of the 1-Ω energy output lies within the band, $-0.02 < f < 0.02$ Hz?

33 If $\mathbf{F}(j\omega) = (2 - j\omega)/[(1 + j\omega)(2 + j\omega)]$, find: (a) the total 1-$\Omega$ energy present in the signal, and (b) the maximum value of $f(t)$.

34 Find the inverse Fourier transform of $\mathbf{F}(j\omega) =$: (a) $1/[(1 + j\omega)(2 + j\omega)(3 + j\omega)]$; (b) $j\omega/[(1 + j\omega)(2 + j\omega)(3 + j\omega)]$; (c) $(j\omega)^2/[(1 + j\omega)(2 + j\omega)(3 + j\omega)]$; (d) $(j\omega)^3/[(1 + j\omega)(2 + j\omega)(3 + j\omega)]$.

35 In Sec. 18-7 the impulse response, $h(t) = 2e^{-t}u(t)$, was hypothesized. In order to develop one network which has such a response: (a) determine $\mathbf{H}(j\omega) = \mathbf{V}_o(j\omega)/\mathbf{V}_i(j\omega)$. (b) Either by inspecting $h(t)$ or $\mathbf{H}(j\omega)$, note that the network has a single energy storage element. Arbitrarily selecting an RC circuit with $R = 1\ \Omega$, $C = 1$ F, to provide the necessary time constant, determine the form of the circuit to give $\frac{1}{2}h(t)$ or $\frac{1}{2}\mathbf{H}(j\omega)$. (c) Place an ideal-voltage amplifier in cascade with the network to provide the proper multiplicative constant. What is the gain of the amplifier?

36 A linear system is characterized by $h(t) = 3e^{-4(t-2)}u(t - 2)$. If a unit impulse, $x(t) = \delta(t)$, is applied to the system input, use the statements of Prob. 16 (if necessary) to help find: (a) the 1-Ω energy density of the system output, and (b) the total energy in the output.

37 Consider a linear system for which $\mathbf{H}(j\omega) = 2/(3 + j\omega)$. (a) Find the unit-impulse response of the system. (b) If the input, $x(t) = 5 \cos 4t$, is applied to the system, find the output $y(t)$.

38 Find $i_x(t)$ for the circuit of Fig. 18-22 if $i_s =$: (a) $\delta(t)$; (b) $\delta(t - 1)$; (c) $u(t)$.

Fig. 18-22: See Prob. 38.

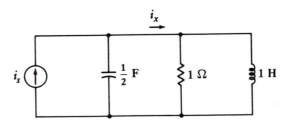

39 (a) If $f(t) = 3e^{-t}u(t)$ and $g(t) = 4e^{-2t}u(t)$, find $f(t) * g(t)$ by working in the frequency domain. (b) Check your results by convolving in the time domain.

40 Find $v_o(t)$ for the circuit of Fig. 18-23 if $v_s(t) = \delta(t)$.

Fig. 18-23: See Prob. 40.

Chapter 19
Laplace Transform Techniques

19-1
Introduction

As we begin this last chapter on the Laplace transform, it may be beneficial to pause long enough to review our progress to date, first because we have covered a considerable amount of ground since page 1 and some tying together is advisable, and second because we need to see how the Laplace transform fits into our hierarchy of analytical methods.

Our goal has constantly been one of analysis: given some forcing function at one point in a linear circuit, determine the response at some other point. For the first several chapters, we played only with dc forcing functions and responses of the form $V_0 e^0$. However, after the introduction of inductance and capacitance, the sudden dc excitation of simple RL and RC circuits produced responses varying exponentially with time $V_0 e^{\sigma t}$. When we considered the RLC circuit, the responses took on the form of the exponentially varying sinusoid $V_0 e^{\sigma t} \cos (\omega t + \theta)$. All of this work was accomplished in the time domain, and the dc forcing function was the only one we considered.

As we advanced to the use of the sinusoidal forcing function, the tedium and complexity of solving the integrodifferential equations caused us to begin casting about for an easier way to work problems. The phasor transform was the result, and we might remember that we were led to it through consideration of a complex forcing function of the form $V_0 e^{j\theta} e^{j\omega t}$. As soon as we concluded that we did not need the factor containing t, we were left with the phasor $V_0 e^{j\theta}$; we had arrived at the frequency domain.

After gaining some facility in sinusoidal steady-state analysis, a modicum of pleasant cerebration led us to apply a forcing function of the form $V_0 e^{j\theta} e^{(\sigma + j\omega)t}$, and we thereby invented the complex frequency $\mathbf{s}$, relegating all our previous functional forms to special cases: dc ($\mathbf{s} = 0$), exponential ($\mathbf{s} = \sigma$), sinusoidal ($\mathbf{s} = j\omega$), and exponential sinusoid ($\mathbf{s} = \sigma + j\omega$). Such was our status at the end of Chap. 16.

Continuing to strive for mastery over an ever greater number of different types of forcing functions, we turned next to nonsinusoidal periodic functions. Here the infinite series developed by Fourier, $a_0 + \Sigma(a_n \cos n\omega_0 t + b_n \sin n\omega_0 t)$, was found to be capable of representing almost any periodic function in which we might be interested. Thanks to linearity and superposition, the response could then be found as the sum of the responses to the individual sinusoidal terms.

Remembering our success with the complex exponential function, we next developed the complex form of the Fourier series, $\Sigma \mathbf{c}_n e^{jn\omega_0 t}$, suitable again only for periodic functions. However, by letting the period of a periodic sequence of pulses increase without limit, we arrived at a form applicable to a single pulse, the inverse Fourier transform:

$$v(t) = \frac{1}{2\pi} \int_{-\infty}^{\infty} e^{j\omega t} \mathbf{V}(j\omega) \, d\omega$$

This then is our current status. What more could we possibly need? It turns out that there are several things. First, there are a few time functions for which a Fourier transform does not exist, such as the increasing exponential, many random signals, and other time functions that are not absolutely integrable. Second, we have not yet permitted initial-energy storage in the networks whose transient or complete response we desired to compute by Fourier-transform methods. Both of these objections are overcome with the Laplace transform, and in addition we shall note a simpler nomenclature and ease of manipulation.

19-2
Definition of the
Laplace transform

We shall present the Laplace transform as a development or evolution of the Fourier transform, but, instead, we might define it directly. For those who prefer to look ahead, (5) and (6) are such definitions of the two-sided Laplace transform, while (7) and (8) give us the one-sided Laplace transform that is used in the remainder of this chapter.

We begin our development of the Laplace transform by using the inverse Fourier transform to interpret $v(t)$ as the sum (integral) of an infinite number of terms, each having the form

$$\left[\frac{\mathbf{V}(j\omega) \, d\omega}{2\pi} \right] e^{j\omega t}$$

Upon comparing this with that of the complex forcing function,

$$[V_0 e^{j\theta}] e^{j\omega t}$$

which led us to the phasor $V_0 e^{j\theta}$, we should note that both bracketed terms are complex quantities in general; therefore, $(1/2\pi)\mathbf{V}(j\omega) \, d\omega$ can also be interpreted as some kind of a phasor. Of course, the frequency differential leads to a vanishingly small amplitude, but we add an infinite number of such terms together when we perform the integration.

Our final step, therefore, parallels that which we took when we introduced phasors in the complex-frequency domain; we now let the time variation have the form, $e^{(\sigma + j\omega)t}$.

To do this, consider the Fourier transform of $e^{-\sigma t}v(t)$, rather than $v(t)$ itself. For $v(t)$ alone we have

$$\mathbf{V}(j\omega) = \int_{-\infty}^{\infty} e^{-j\omega t} v(t) \, dt \tag{1}$$

and

$$v(t) = \frac{1}{2\pi} \int_{-\infty}^{\infty} e^{j\omega t} \mathbf{V}(j\omega) \, d\omega \tag{2}$$

Letting

$$g(t) = e^{-\sigma t} v(t) \tag{3}$$

we see that

$$\mathbf{G}(j\omega) = \int_{-\infty}^{\infty} e^{-j\omega t} e^{-\sigma t} v(t) \, dt$$

$$= \int_{-\infty}^{\infty} e^{-(\sigma + j\omega)t} v(t) \, dt$$

or

$$\mathbf{G}(j\omega) = \mathbf{V}(\sigma + j\omega) = \int_{-\infty}^{\infty} e^{-(\sigma + j\omega)t} v(t) \, dt \tag{4}$$

through comparison with (1). Taking the inverse Fourier transform, we obtain

$$g(t) = \frac{1}{2\pi} \int_{-\infty}^{\infty} e^{j\omega t} \mathbf{G}(j\omega) \, d\omega$$

$$= \frac{1}{2\pi} \int_{-\infty}^{\infty} e^{j\omega t} \mathbf{V}(\sigma + j\omega) \, d\omega$$

With the help of (3) we have

$$e^{-\sigma t} v(t) = \frac{1}{2\pi} \int_{-\infty}^{\infty} e^{j\omega t} \mathbf{V}(\sigma + j\omega) \, d\omega$$

or, upon transferring $e^{-\sigma t}$ inside the integral,

$$v(t) = \frac{1}{2\pi} \int_{-\infty}^{\infty} e^{(\sigma + j\omega)t} \mathbf{V}(\sigma + j\omega) \, d\omega$$

We now replace $\sigma + j\omega$ by the single complex variable $\mathbf{s}$, and, since σ is a constant, $d\mathbf{s} = j \, d\omega$, we have

$$v(t) = \frac{1}{2\pi j} \int_{\sigma_0 - j\infty}^{\sigma_0 + j\infty} e^{\mathbf{s}t} \mathbf{V}(\mathbf{s}) \, d\mathbf{s} \tag{5}$$

where the real constant σ_0 is included in the limits to assure convergence of this improper integral. In terms of $\mathbf{s}$, (4) may be written

$$\mathbf{V(s)} = \int_{-\infty}^{\infty} e^{-\mathbf{s}t} v(t) \, dt \tag{6}$$

Equation (6) defines the *two-sided* or *bilateral Laplace transform* of $v(t)$. The terms "two-sided" or "bilateral" are used to emphasize the fact that both positive and negative times are included in the range of integration. Equation (5) is the inverse Laplace transform, and the two equations constitute the two-sided Laplace transform pair.

  If we continue the process we began earlier in this section, we may note that $v(t)$ is now represented as the sum (integral) of terms of the form

$$\left[\frac{\mathbf{V(s)} \, d\mathbf{s}}{2\pi j}\right] e^{\mathbf{s}t} = \left[\frac{\mathbf{V(s)} \, d\omega}{2\pi}\right] e^{\mathbf{s}t}$$

Such terms have the same form as those we encountered when we used phasors to represent exponentially varying sinusoids,

$$[\mathbf{V}_0 e^{j\theta}]e^{\mathbf{s}t}$$

where the bracketed term was a function of $\mathbf{s}$. Thus, the two-sided Laplace transform may be interpreted as expressing $v(t)$ as the sum (integral) of an infinite number of vanishingly small terms with complex frequency, $\mathbf{s} = \sigma + j\omega$. The variable is $\mathbf{s}$ or ω, and σ should be thought of as governing the convergence factor $e^{-\sigma t}$. That is, by including this exponential term in (6), more positive values of σ ensure that the function $e^{-\sigma t} v(t) \, u(t)$ is absolutely integrable for almost any $v(t)$ that we might meet. The more negative values of σ are needed when $t < 0$. Thus, the two-sided Laplace transform exists for a wider class of functions $v(t)$ than does the Fourier transform. The exact conditions required for the existence of the (one-sided) Laplace transform are given in the following section.

  In many of our circuit analysis problems, the forcing and response functions do not exist forever in time, but rather they are initiated at some specific instant that we usually select as $t = 0$. Thus, for time functions that do not exist for $t < 0$, or for those time functions whose behavior for $t < 0$ is of no interest, the time-domain description can be thought of as $v(t)u(t)$. The defining integral for the Laplace transform is taken with the lower limit at $t = 0^-$ in order to include the effect of any discontinuity at $t = 0$, such as an impulse or higher-order singularity. The corresponding Laplace transform is then

$$\mathbf{V(s)} = \int_{-\infty}^{\infty} e^{-\mathbf{s}t} v(t) u(t) \, dt = \int_{0^-}^{\infty} e^{-\mathbf{s}t} v(t) \, dt$$

This defines the *one-sided Laplace transform* of $v(t)$, or simply the *Laplace transform* of $v(t)$, one-sided being understood. The inverse transform expression remains unchanged, but when evaluated, it is understood to be valid only for $t > 0$. Here then is the definition of the Laplace transform pair that will be used from now on:

$$\mathbf{V(s)} = \int_{0^-}^{\infty} e^{-st} v(t) \, dt \tag{7}$$

$$v(t) = \frac{1}{2\pi j} \int_{\sigma_0 - j\infty}^{\sigma_0 + j\infty} e^{st} \mathbf{V(s)} \, ds \tag{8}$$

$$v(t) \Leftrightarrow \mathbf{V(s)}$$

These are memorable expressions.

The script $\mathcal{L}$ may also be used to indicate the direct or inverse Laplace transform operation:

$$\mathbf{V(s)} = \mathcal{L}\{v(t)\}$$

and

$$v(t) = \mathcal{L}^{-1}\{\mathbf{V(s)}\}$$

Drill Problems

19-1. If $f(t) = 4e^{-3t}[u(t + 2) - u(t - 2)]$, find its: (a) Fourier transform; (b) two-sided Laplace transform; (c) Laplace transform.

$$Ans: \frac{4}{3 + j\omega} [e^{6 + j2\omega} - e^{-6 - j2\omega}]; \frac{4}{3 + \mathbf{s}} (e^{6 + 2\mathbf{s}} - e^{-6 - 2\mathbf{s}}); \frac{4}{3 + \mathbf{s}} (1 - e^{-6 - 2\mathbf{s}})$$

19-2. Assuming that a is real and positive, find the Laplace transform of each of these functions and state the minimum value of σ for which the transform exists: (a) e^{-at}; (b) $e^{-at}u(t + 2)$; (c) $e^{-at}u(t - 2)$; (d) e^{at}; (e) $e^{at}u(t - 2)$; (f) $\delta(t)$; (g) $u(t)$.

$$Ans: \frac{1}{\mathbf{s} + a}, > -a; \frac{1}{\mathbf{s} + a}, > -a; \frac{1}{\mathbf{s} + a} e^{-2\mathbf{s} - 2a}, > -a;$$

$$\frac{1}{\mathbf{s} - a}, > a; \frac{1}{\mathbf{s} - a} e^{2a - 2\mathbf{s}}, > a; 1, \text{any finite value of } \sigma;$$

$$\frac{1}{\mathbf{s}}, > 0$$

19-3
Laplace transforms of some simple time functions

In this section we shall begin to build up a catalog of Laplace transforms for those time functions most frequently encountered in circuit analysis, just as we did for Fourier transforms. This will be done, at least initially, by utilizing the definition.

$$\mathbf{V(s)} = \int_{0^-}^{\infty} e^{-st} v(t) \, dt = \mathcal{L}\{v(t)\}$$

which, along with the expression for the inverse transform,

$$v(t) = \frac{1}{2\pi j} \int_{\sigma_0 - j\infty}^{\sigma_0 + j\infty} e^{st} \mathbf{V(s)} \, ds = \mathcal{L}^{-1}\{\mathbf{V(s)}\}$$

establishes a one-to-one correspondence between $v(t)$ and $\mathbf{V}(\mathbf{s})$. That is, for every $v(t)$ for which $\mathbf{V}(\mathbf{s})$ exists, there is a unique $\mathbf{V}(\mathbf{s})$. At this point, we may be looking with some trepidation at the rather ominous form given for the inverse transform. Fear not! As we shall see shortly, an introductory study of Laplace transform theory does not require actual evaluation of this integral. By going from the time domain to the frequency domain and taking advantage of the uniqueness mentioned above, we shall be able to generate a catalog of transform pairs such that nearly every transform that we wish to invert will already have the corresponding time function listed in our catalog.

Before we start evaluating some transforms, however, we must pause to consider whether there is any chance that the transform may not even exist for some $v(t)$ that concerns us. In our study of Fourier transforms, we were forced to take a kind of back-door approach in finding several transforms. This was true with time functions that were not absolutely integrable, such as the unit step function. In the case of the Laplace transform, the use of realizable linear circuits and real-world forcing functions almost never leads to troublesome time functions. Technically, a set of conditions sufficient to ensure the absolute convergence of the Laplace integral for $\mathrm{Re}(\mathbf{s}) > \sigma_0$ is:

1 The function $v(t)$ is integrable in every finite interval $t_1 < t < t_2$, where $0 \leq t_1 < t_2 < \infty$.
2 The limit, $\lim\limits_{t \to \infty} e^{-\sigma_0 t}|v(t)|$, exists for some value of σ_0.

Time functions that do not satisfy these conditions are seldom encountered by the circuit analyst.[1]

Now let us look at some specific transforms. In a somewhat vengeful mood, we first examine the Laplace transform of the unit-step function $u(t)$ which caused us some earlier trouble. From the defining equation, we may write

$$\mathcal{L}\{u(t)\} = \int_{0^-}^{\infty} e^{-\mathbf{s}t}u(t)\,dt = \int_{0}^{\infty} e^{-\mathbf{s}t}\,dt$$

$$= \left(-\frac{1}{\mathbf{s}}e^{-\mathbf{s}t}\right)_{0}^{\infty} = \frac{1}{\mathbf{s}}$$

since $\mathrm{Re}(\mathbf{s}) > 0$ to satisfy condition 2 above. Thus,

$$u(t) \Leftrightarrow \frac{1}{\mathbf{s}} \tag{9}$$

and our first Laplace transform pair has been established with great ease.

---

[1] Examples of such functions are e^{t^2} and e^{e^t}, but not t^n or n^t. For a somewhat more detailed discussion of the Laplace transform and its applications, refer to Clare D. McGillem and George R. Cooper, "Continuous and Discrete Signal and System Analysis," 2d ed., chap. 5, Holt, Rinehart, and Winston, New York, 1984.

Another singularity function whose transform is of considerable interest is the unit impulse function $\delta(t - t_0)$, where $t_0 > 0^-$:

$$\mathcal{L}\{\delta(t - t_0)\} = \int_{0^-}^{\infty} e^{-st}\delta(t - t_0)\, dt = e^{-st_0}$$

$$\delta(t - t_0) \Leftrightarrow e^{-st_0} \tag{10}$$

In particular, note that we obtain

$$\delta(t) \Leftrightarrow 1 \tag{11}$$

for $t_0 = 0$.

Recalling our past interest in the exponential function, we examine its transform,

$$\mathcal{L}\{e^{-\alpha t}u(t)\} = \int_{0^-}^{\infty} e^{-\alpha t}e^{-st}\, dt$$

$$= \left[-\frac{1}{s + \alpha}\, e^{-(s + \alpha)t} \right]_0^{\infty}$$

$$= \frac{1}{s + \alpha}$$

and

$$e^{-\alpha t}u(t) \Leftrightarrow \frac{1}{s + \alpha} \tag{12}$$

It is understood that $\text{Re}\,(\mathbf{s}) > -\alpha$.

As a final example, for the moment, let us consider the ramp function $tu(t)$. We obtain

$$\mathcal{L}\{tu(t)\} = \int_{0^-}^{\infty} te^{-st}\, dt = \frac{1}{s^2}$$

$$tu(t) \Leftrightarrow \frac{1}{s^2} \tag{13}$$

either by a straightforward integration by parts or by use of a table of definite intergrals.

In order to accelerate the process of deriving more Laplace transform pairs, we shall pause to develop several useful theorems in the following section.

Drill Problems

19-3. Find $\mathcal{L}\{f(t)\}$ if $f(t) = $: (a) $3\delta(t - 5)$; (b) $4[\delta(t - 1) + \delta(t - 2)]$; (c) $2u(t - 5)$; (d) $2(t - 3)u(t - 5)$.   *Ans:* $3e^{-5s}$; $4(e^{-s} + e^{-2s})$; $2e^{-5s}/s$; $e^{-5s}(4s + 2)/s^2$

19-4. Find $\mathcal{L}^{-1}\{F(s)\}$ if $F(s) = $: (a) -5; (b) $-5e^{-5s}$; (c) $-5/(s + 5)$; (d) $(-5s + 5)/(s + 5)$; (e) $(-5s + 5)e^{-5s}/(s + 5)$.   *Ans:* $-5\delta(t)$; $-5\delta(t - 5)$; $-5e^{-5t}u(t)$; $-5\delta(t) + 30e^{-5t}u(t)$; $-5\delta(t - 5) + 30e^{-5(t - 5)}u(t - 5)$

19-4
Several basic
theorems for the
Laplace transform

The further evaluation of Laplace transforms is facilitated by applying several basic theorems. One of the simplest and most obvious is the linearity theorem: the Laplace transform of the sum of two or more time functions is equal to the sum of the transforms of the individual time functions. For two time functions we have

$$\mathcal{L}\{f_1(t) + f_2(t)\} = \int_{0^-}^{\infty} e^{-st}[f_1(t) + f_2(t)] \, dt$$

$$= \int_{0^-}^{\infty} e^{-st}f_1(t) \, dt + \int_{0^-}^{\infty} e^{-st}f_2(t) \, dt$$

$$= \mathbf{F}_1(\mathbf{s}) + \mathbf{F}_2(\mathbf{s})$$

As an example of the use of this theorem, suppose that we have a Laplace transform $\mathbf{V(s)}$ and want to know the corresponding time function $v(t)$. It will often be possible to decompose $\mathbf{V(s)}$ into the sum of two or more functions, say $\mathbf{V}_1(\mathbf{s})$ and $\mathbf{V}_2(\mathbf{s})$, whose inverse transforms, $v_1(t)$ and $v_2(t)$, are already tabulated. It then becomes a simple matter to apply the linearity theorem and write

$$v(t) = \mathcal{L}^{-1}\{\mathbf{V(s)}\} = \mathcal{L}^{-1}\{\mathbf{V}_1(\mathbf{s}) + \mathbf{V}_2(\mathbf{s})\}$$
$$= \mathcal{L}^{-1}\{\mathbf{V}_1(\mathbf{s})\} + \mathcal{L}^{-1}\{\mathbf{V}_2(\mathbf{s})\} = v_1(t) + v_2(t)$$

As a specific example, let us determine the inverse Laplace transform of

$$\mathbf{V(s)} = \frac{1}{(\mathbf{s} + \alpha)(\mathbf{s} + \beta)}$$

Although it is possible to substitute this expression into the defining equation for the inverse transform, it is much easier to utilize the linearity theorem. Using the partial-fraction expansion, we can split the given transform into the sum of two simpler transforms,

$$\mathbf{V(s)} = \frac{A}{\mathbf{s} + \alpha} + \frac{B}{\mathbf{s} + \beta}$$

where A and B may be found by any of several methods. Perhaps the quickest solution is obtained by recognizing that

$$A = \lim_{s \to -\alpha} \left[(\mathbf{s} + \alpha)\mathbf{V(s)} - \frac{\mathbf{s} + \alpha}{\mathbf{s} + \beta} B \right]$$

$$= \lim_{s \to -\alpha} \left[\frac{1}{\mathbf{s} + \beta} - 0 \right] = \frac{1}{\beta - \alpha}$$

Similarly,

$$B = \frac{1}{\alpha - \beta}$$

and therefore,

$$\mathbf{V(s)} = \frac{1/(\beta - \alpha)}{\mathbf{s} + \alpha} + \frac{1/(\alpha - \beta)}{\mathbf{s} + \beta}$$

We have already evaluated inverse transforms of the form shown on the right, and thus

$$v(t) = \frac{1}{\beta - \alpha} e^{-\alpha t} u(t) + \frac{1}{\alpha - \beta} e^{-\beta t} u(t)$$

$$= \frac{1}{\beta - \alpha} (e^{-\alpha t} - e^{-\beta t}) u(t)$$

If we wished we could now include this as a new entry in our catalog of Laplace pairs,

$$\frac{1}{\beta - \alpha} (e^{-\alpha t} - e^{-\beta t}) u(t) \Leftrightarrow \frac{1}{(\mathbf{s} + \alpha)(\mathbf{s} + \beta)} \tag{14}$$

It is noteworthy that a transcendental function of t in the time domain transforms into a simpler rational function of $\mathbf{s}$ in the frequency domain. Simplifications of this kind are of paramount importance in transform theory. In this example also note that we made use of the fact that

$$kv(t) \Leftrightarrow k\mathbf{V(s)} \tag{15}$$

where k is a constant of proportionality. This result is obviously a direct consequence of the definition of the Laplace transform.

We are now able to consider two theorems that might be considered collectively as the raison d'être for Laplace transforms in circuit analysis—the time differentiation and integration theorems. These will help us transform the derivatives and integrals appearing in the time-domain circuit equations.

Let us look at time differentiation first by considering a time function $v(t)$ whose Laplace transform $\mathbf{V(s)}$ is known to exist. We want the transform of the first derivative of $v(t)$,

$$\mathscr{L}\left\{\frac{dv}{dt}\right\} = \int_{0^-}^{\infty} e^{-\mathbf{s}t} \frac{dv}{dt} dt$$

This can be integrated by parts,

$$U = e^{-\mathbf{s}t} \qquad dV = \frac{dv}{dt} dt$$

with the result,

$$\mathscr{L}\left\{\frac{dv}{dt}\right\} = \left[v(t)e^{-\mathbf{s}t}\right]_{0^-}^{\infty} + \mathbf{s}\int_{0^-}^{\infty} e^{-\mathbf{s}t} v(t) dt$$

The first term on the right must approach zero as t increases without limit; otherwise $\mathbf{V(s)}$ would not exist. Hence,

$$\mathcal{L}\left\{\frac{dv}{dt}\right\} = 0 - v(0^-) + \mathbf{s}\mathbf{V}(\mathbf{s})$$

and
$$\frac{dv}{dt} \Leftrightarrow \mathbf{s}\mathbf{V}(\mathbf{s}) - v(0^-) \tag{16}$$

Similar relationships may be developed for higher-order derivatives,

$$\frac{d^2v}{dt^2} \Leftrightarrow \mathbf{s}^2\mathbf{V}(\mathbf{s}) - \mathbf{s}v(0^-) - v'(0^-) \tag{17}$$

$$\frac{d^3v}{dt^3} \Leftrightarrow \mathbf{s}^3\mathbf{V}(\mathbf{s}) - \mathbf{s}^2v(0^-) - \mathbf{s}v'(0^-) - v''(0^-) \tag{18}$$

where $v'(0^-)$ is the value of the first derivative of $v(t)$ evaluated at $t = 0^-$, $v''(0^-)$ is the initial value of the second derivative of $v(t)$, and so forth. When all initial conditions are zero, we see that differentiating once with respect to t in the time domain corresponds to one multiplication by $\mathbf{s}$ in the frequency domain; differentiating twice in the time domain corresponds to multiplication by $\mathbf{s}^2$ in the frequency domain, and so on. Thus, differentiation in the time domain is equivalent to multiplication in the frequency domain. This is a substantial simplification! We should also begin to see that, when the initial conditions are not zero, their presence is still accounted for. A simple example will serve to demonstrate this.

Suppose that we have the series RL circuit shown in Fig. 19-1. The network is driven by a unit-step voltage, and we assume an initial value of the current (at $t = 0^-$) of 5 A.[2] Using KVL to write the single loop equation in the time domain, we have

$$2\frac{di}{dt} + 4i = 3u(t)$$

Instead of solving this differential equation as we have done previously, we first transform to the frequency domain by taking the Laplace transform of each term,

$$2[\mathbf{s}\mathbf{I}(\mathbf{s}) - i(0^-)] + 4\mathbf{I}(\mathbf{s}) = \frac{3}{\mathbf{s}}$$

[2] We could have established this current by letting the source be $20u(-t) + 3u(t)$ V, or $20 - 17u(t)$ V, or other expressions of this nature.

Fig. 19-1: A circuit which is analyzed by transforming the differential equation $2di/dt + 4i = 3u(t)$ into $2[\mathbf{s}\mathbf{I}(\mathbf{s}) - i(0^-)] + 4\mathbf{I}(\mathbf{s}) = 3/\mathbf{s}$.

2 H

$i(t)$

$3u(t)$ V

4 Ω

$i(0^-) = 5$ A

We next solve for $\mathbf{I(s)}$, substituting $i(0^-) = 5$,

$$(2s + 4)\mathbf{I(s)} = \frac{3}{s} + 10$$

and

$$\mathbf{I(s)} = \frac{1.5}{s(s + 2)} + \frac{5}{s + 2}$$

Now,

$$\lim_{s \to 0} [s\mathbf{I(s)}] = 0.75$$

and

$$\lim_{s \to -2} [(s + 2)\mathbf{I(s)}] = -0.75 + 5$$

and thus

$$\mathbf{I(s)} = \frac{0.75}{s} + \frac{4.25}{s + 2}$$

We then use our known transform pairs to invert:

$$i(t) = 0.75u(t) + 4.25e^{-2t}u(t)$$
$$= (0.75 + 4.25e^{-2t})u(t)$$

Our solution for $i(t)$ is complete. Both the forced response $0.75u(t)$ and the natural response $4.25e^{-2t}u(t)$ are present, and the initial condition was automatically incorporated into the solution. The method illustrates a very painless way of obtaining the complete solution of many differential equations.

The same kind of simplification can be accomplished when we meet the operation of integration with respect to time in our circuit equations. Let us determine the Laplace transform of the time function described by $\int_{0^-}^{t} v(x)\, dx$,

$$\mathcal{L}\left\{\int_{0^-}^{t} v(x)\, dx\right\} = \int_{0^-}^{\infty} e^{-st}\left[\int_{0^-}^{t} v(x)\, dx\right] dt$$

Integrating by parts, we let

$$u = \int_{0^-}^{t} v(x)\, dx \qquad dw = e^{-st}\, dt$$

$$du = v(t)\, dt \qquad w = -\frac{1}{s} e^{-st}$$

Then

$$\mathcal{L}\left\{\int_{0^-}^{t} v(x)\, dx\right\} = \left\{\left[\int_{0^-}^{t} v(x)\, dx\right]\left[-\frac{1}{s} e^{-st}\right]\right\}_{t=0^-}^{t=\infty} - \int_{0^-}^{\infty} -\frac{1}{s} e^{-st}v(t)\, dt$$

$$= \left[-\frac{1}{s} e^{-st}\int_{0^-}^{t} v(x)\, dx\right]_{0^-}^{\infty} + \frac{1}{s} \mathbf{V(s)}$$

But, since $e^{-st} \to 0$ as $t \to \infty$, the first term on the right vanishes at the upper limit, and when $t \to 0^-$, the integral in this term likewise vanishes. This leaves only the $\mathbf{V(s)/s}$ term, so

$$\int_{0^-}^{t} v(x)\, dx \Leftrightarrow \frac{\mathbf{V(s)}}{\mathbf{s}} \tag{19}$$

and thus integration in the time domain corresponds to division by $\mathbf{s}$ in the frequency domain. Once more, a relatively complicated calculus operation in the time domain simplifies to an algebraic operation in the frequency domain.

As an example of how this helps us in circuit analysis, we shall determine $i(t)$ and $v(t)$ for $t > 0^-$ in the series RC circuit shown in Fig. 19-2. It is assumed that there was some initial energy stored in the capacitor prior to $t = 0^-$ so that $v(0^-) = 9$ V. We first write the single loop equation,

$$u(t) = 4i(t) + 16 \int_{-\infty}^{t} i(t)\, dt$$

In order to apply the time-integration theorem, we must arrange for the lower limit of integration to be 0^-. Thus, we set

$$16 \int_{-\infty}^{t} i(t)\, dt = 16 \int_{-\infty}^{0^-} i(t)\, dt + 16 \int_{0^-}^{t} i(t)\, dt = v(0^-) + 16 \int_{0^-}^{t} i(t)\, dt$$

Therefore,

$$u(t) = 4i(t) + v(0^-) + 16 \int_{0^-}^{t} i(t)\, dt$$

We next take the Laplace transform of both sides of this equation. Since we are utilizing the one-sided transform, $\mathscr{L}\{v(0^-)\}$ is simply $\mathscr{L}\{v(0^-)u(t)\}$, and thus

$$\frac{1}{\mathbf{s}} = 4\mathbf{I(s)} + \frac{9}{\mathbf{s}} + \frac{16}{\mathbf{s}}\,\mathbf{I(s)}$$

and solving for $\mathbf{I(s)}$,

$$\mathbf{I(s)} = \frac{-2}{\mathbf{s} + 4}$$

the desired result is immediately obtained,

$$i(t) = -2e^{-4t}u(t)$$

Fig. 19-2: A circuit illustrating the use of the Laplace transform pair $\int_{0^-}^{t} i(t)dt \Leftrightarrow (1/\mathbf{s})\mathbf{I(s)}$.

If $v(t)$ had been our desired response, we would simply have written a single nodal equation,

$$\frac{v(t) - u(t)}{4} + \frac{1}{16}\frac{dv}{dt} = 0$$

Taking the Laplace transform, we obtain

$$\frac{\mathbf{V}(s)}{4} - \frac{1}{4s} + \frac{1}{16}s\mathbf{V}(s) - \frac{v(0^-)}{16} = 0$$

or

$$\mathbf{V}(s)\left(1 + \frac{s}{4}\right) = \frac{1}{s} + \frac{9}{4}$$

Thus,

$$\mathbf{V}(s) = \frac{4}{s(s+4)} + \frac{9}{s+4} = \frac{1}{s} - \frac{1}{s+4} + \frac{9}{s+4}$$

$$= \frac{1}{s} + \frac{8}{s+4}$$

and taking the inverse transform,

$$v(t) = (1 + 8e^{-4t})u(t)$$

we quickly obtain the desired capacitor voltage without recourse to the usual differential equation solution. To check this result, we note that $(\frac{1}{16})dv/dt$ should yield the previous expression for $i(t)$. For $t > 0$,

$$\frac{1}{16}\frac{dv}{dt} = \frac{1}{16}(-32)e^{-4t} = -2e^{-4t}$$

which is correct.

To illustrate the use of both the linearity theorem and the time-differentiation theorem, not to mention the addition of a most important pair to our forthcoming Laplace transform table, let us establish the Laplace transform of $\sin \omega t\, u(t)$. We could use the defining integral expression with integration by parts, but this is needlessly difficult. Instead, we use the relationship,

$$\sin \omega t = \frac{1}{2j}(e^{j\omega t} - e^{-j\omega t})$$

The transform of the sum of these two terms is just the sum of the transforms, and each term is an exponential function for which we already have the transform. We may immediately write

$$\mathcal{L}\{\sin\omega t\, u(t)\} = \frac{1}{2j}\left(\frac{1}{s - j\omega} - \frac{1}{s + j\omega}\right) = \frac{\omega}{s^2 + \omega^2}$$

$$\sin \omega t\, u(t) \Leftrightarrow \frac{\omega}{s^2 + \omega^2} \tag{20}$$

We next use the time-differentiation theorem to determine the transform of $\cos \omega t\, u(t)$, which is proportional to the derivative of $\sin \omega t$. That is,

$$\mathcal{L}\{\cos \omega t\, u(t)\} = \mathcal{L}\left\{\frac{1}{\omega}\frac{d}{dt}\left[\sin \omega t\, u(t)\right]\right\}$$

$$= \frac{1}{\omega}\,\mathbf{s}\,\frac{\omega}{\mathbf{s}^2 + \omega^2}$$

and
$$\cos \omega t\, u(t) \Leftrightarrow \frac{\mathbf{s}}{\mathbf{s}^2 + \omega^2} \qquad (21)$$

Drill Problems

19-5. Use the method of partial-fraction expansion to find the inverse transform of: (a) $(s + 1)/[(s + 2)(s + 3)]$; (b) $(s + 1)^2/[(s + 2)(s + 3)]$; (c) $1/[(s + 1)s^2]$.

Ans: $(-e^{-2t} + 2e^{-3t})u(t)$; $\delta(t) + (e^{-2t} - 4e^{-3t})u(t)$; $(e^{-t} - 1 + t)u(t)$

19-6. Use Laplace-transform methods to find $v(t)$ if: (a) $2\,dv/dt + 8v = 6u(t)$, $v(0^-) = 1$ V; (b) $d^2v/dt^2 + 3\,dv/dt + 2v = 4\delta(t)$, $v'(0^-) = 5$ V/s, $v(0^-) = 0$; (c) $v(t)$ is the voltage indicated in Fig. 19-3a.

Ans: $(0.75 + 0.25e^{-4t})u(t)$; $9(e^{-t} - e^{-2t})u(t)$; $(20 + 80e^{-5t})u(t)$ V

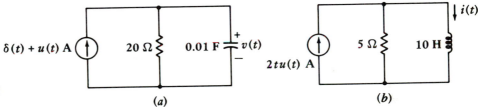

Fig. 19-3: (a) See Drill Prob. 19-6. (b) See Drill Prob. 19-7.

19-7. Find $i(t)$ at $t = 0.8$ s if: (a) $1.5\int_{0^-}^{t} i(t)\,dt + 6\,di/dt = \delta(t)$, $i(0^-) = 0.5$ A; (b) $10\int_{-\infty}^{t} i(t)\,dt + 4i(t) = 8 + 2u(t)$, $\int_{-\infty}^{0^-} i(t)\,dt = 0.8$ C; (c) $i(t)$ is the current defined in Fig. 19-3b. *Ans:* 0.614; 0.0677; 0.281 A

19-5
Convolution Again

In our study of the Fourier transform, we discovered that the Fourier transform of $f_1(t) * f_2(t)$, the convolution of two time functions, was simply the product of the transforms of the individual functions. This then led to the very useful concept of the system function, defined as the ratio of the transform of the system output to the transform of the system input. Exactly the same fortunate circumstances exist when we work with the Laplace transform, as we shall now see.

Convolution was first defined by Eq. (40) of Chap. 18, and we therefore may write

$$f_1(t) * f_2(t) = \int_{-\infty}^{\infty} f_1(\lambda)f_2(t - \lambda)\,d\lambda$$

Now we let $\mathbf{F}_1(\mathbf{s})$ and $\mathbf{F}_2(\mathbf{s})$ be the Laplace transforms of $f_1(t)$ and $f_2(t)$, respectively, and consider the Laplace transform of $f_1(t) * f_2(t)$,

$$\mathcal{L}\{f_1(t) * f_2(t)\} = \mathcal{L}\left\{\int_{-\infty}^{\infty} f_1(\lambda)f_2(t - \lambda)\,d\lambda\right\}$$

As we discovered in our earlier look at convolution, one of these time functions will often be the forcing function that is applied at the input terminals of a linear circuit, and the other will be the unit-impulse response of the circuit. That is, the response of a linear circuit is just the convolution of the input and the impulse response.

Since we are now dealing with time functions that do not exist prior to $t = 0^-$ (the definition of the Laplace transform forces us to assume this), the lower limit of integration can be changed to 0^-. Then, using the definition of the Laplace transform, we get

$$\mathcal{L}\{f_1(t) * f_2(t)\} = \int_{0^-}^{\infty} e^{-st}\left[\int_{0^-}^{\infty} f_1(\lambda)f_2(t - \lambda)\, d\lambda\right] dt$$

Since e^{-st} does not depend upon λ, we can move this factor inside the inner integral. If we do this and also reverse the order of integration, the result is

$$\mathcal{L}\{f_1(t) * f_2(t)\} = \int_{0^-}^{\infty}\left[\int_{0^-}^{\infty} e^{-st} f_1(\lambda)f_2(t - \lambda)\, dt\right] d\lambda$$

Continuing with the same type of trickery, we note that $f_1(\lambda)$ does not depend upon t, so it can be moved outside the inner integral,

$$\mathcal{L}\{f_1(t) * f_2(t)\} = \int_{0^-}^{\infty} f_1(\lambda)\left[\int_{0^-}^{\infty} e^{-st}f_2(t - \lambda)\, dt\right] d\lambda$$

We then make the substitution, $x = t - \lambda$, in the bracketed integral (where we may treat λ as a constant) and, while we are at it, remove the factor $e^{-s\lambda}$,

$$\mathcal{L}\{f_1(t) * f_2(t)\} = \int_{0^-}^{\infty} e^{-s\lambda}f_1(\lambda)\left[\int_{0^-}^{\infty} e^{-sx}f_2(x)\, dx\right] d\lambda$$

The bracketed term is $\mathbf{F}_2(\mathbf{s})$, which is not a function of t and can therefore be moved outside both integral signs. What remains is the Laplace transform $\mathbf{F}_1(\mathbf{s})$, and we have

$$f_1(t) * f_2(t) \Leftrightarrow \mathbf{F}_1(\mathbf{s})\mathbf{F}_2(\mathbf{s}) \tag{22}$$

which is the desired result. Stated slightly differently, we may conclude that the inverse transform of the product of two transforms is the convolution of the individual inverse transforms, a result that is sometimes useful in obtaining inverse transforms.

As an example of the use of the convolution theorem, let us reconsider the first example of Sec. 19-4, in which we were given the transform,

$$\mathbf{V}(\mathbf{s}) = \frac{1}{(\mathbf{s} + \alpha)(\mathbf{s} + \beta)}$$

and obtained the inverse transform by a partial-fraction expansion. We now identify $\mathbf{V}(\mathbf{s})$ as the product of two transforms,

$$\mathbf{V}_1(\mathbf{s}) = \frac{1}{\mathbf{s} + \alpha}$$

$$\mathbf{V}_2(\mathbf{s}) = \frac{1}{\mathbf{s} + \beta}$$

where

$$v_1(t) = e^{-\alpha t}u(t)$$

and

$$v_2(t) = e^{-\beta t}u(t)$$

The desired $v(t)$ can be immediately expressed as

$$v(t) = \mathcal{L}^{-1}\{\mathbf{V}_1(\mathbf{s})\mathbf{V}_2(\mathbf{s})\} = v_1(t) * v_2(t)$$

$$= \int_{0^-}^{\infty} v_1(\lambda)v_2(t - \lambda)\, d\lambda$$

$$= \int_{0^-}^{\infty} e^{-\alpha\lambda}u(\lambda)e^{-\beta(t-\lambda)}u(t - \lambda)\, d\lambda$$

$$= \int_{0^-}^{t} e^{-\alpha\lambda}e^{-\beta t}e^{\beta\lambda}\, d\lambda$$

$$= e^{-\beta t}\int_{0^-}^{t} e^{(\beta-\alpha)\lambda}\, d\lambda$$

$$= e^{-\beta t}\frac{e^{(\beta-\alpha)t} - 1}{\beta - \alpha}u(t)$$

and finally,

$$v(t) = \frac{1}{\beta - \alpha}(e^{-\alpha t} - e^{-\beta t})u(t)$$

which is the same result that we obtained before using the partial-fraction expansion. Note that it is necessary to insert the unit step $u(t)$ in the result because all (one-sided) Laplace transforms are valid only for nonnegative time.

Was the result easier to obtain by this method? Not unless one is in love with convolution integrals. The partial-fraction-expansion method is usually simpler, assuming that the expansion itself is not too cumbersome.

As we have noted several times before, the output $v_o(t)$ at some point in a linear circuit can be obtained by convolving the input $v_i(t)$ with the unit-impulse response $h(t)$. However, we must remember that the impulse response results from the application of a unit impulse at $t = 0$ *with all initial conditions zero.* Under these conditions, the Laplace transform of the output is

$$\mathcal{L}\{v_o(t)\} = \mathbf{V}_o(\mathbf{s}) = \mathcal{L}\{v_i(t) * h(t)\} = \mathbf{V}_i(\mathbf{s})[\mathcal{L}\{h(t)\}]$$

Thus, the ratio $\mathbf{V}_o(\mathbf{s})/\mathbf{V}_i(\mathbf{s})$ is equal to the transform of the impulse response, which we shall denote by $\mathbf{H}(\mathbf{s})$,

$$\mathcal{L}\{h(t)\} = \mathbf{H}(\mathbf{s}) = \frac{\mathbf{V}_o(\mathbf{s})}{\mathbf{V}_i(\mathbf{s})} \tag{23}$$

The expression $\mathbf{H(s)}$ was first introduced in Sec. 13-7 as the ratio of the forced response to the forcing function that caused it, and we termed it a *transfer function*. We now use the same symbol $\mathbf{H(s)}$ and the same name to represent the ratio of the Laplace transform of the output (or response) to the Laplace transform of the input (or forcing function) when all initial conditions are zero. The equivalence of these two descriptions of the transfer function $\mathbf{H(s)}$ is demonstrated in the next-to-last section of this chapter.

From (23) we see that the impulse response and the transfer function make up a Laplace transform pair,

$$h(t) \Leftrightarrow \mathbf{H(s)} \tag{24}$$

This is an important fact that we shall utilize later to analyze the behavior of some circuits that would previously have baffled us.

It is evident that there is considerable similarity between the transfer function of Laplace-transform theory,

$$\mathbf{H(s)} = \frac{\mathbf{V}_o(\mathbf{s})}{\mathbf{V}_i(\mathbf{s})} = \frac{\mathscr{L}\{v_o(t)\}}{\mathscr{L}\{v_i(t)\}}$$

and the system function defined earlier for Fourier transforms,

$$\mathbf{H}(j\omega) = \frac{\mathbf{V}_o(j\omega)}{\mathbf{V}_i(j\omega)} = \frac{\mathscr{F}\{v_o(t)\}}{\mathscr{F}\{v_i(t)\}}$$

More advanced treatments of this subject[3] show that $\mathbf{H}(j\omega) = \mathbf{H(s)}|_{s=j\omega}$ if all the poles of $\mathbf{H(s)}$ lie in the left half of the $\mathbf{s}$ plane. If any poles of $\mathbf{H(s)}$ are located on the $j\omega$ axis, then $\mathbf{H}(j\omega)$ also contains two delta functions for each of these poles. And, if $\mathbf{H(s)}$ has any poles in the RHP, then $\mathbf{H}(j\omega)$ does not exist.

Drill Problems

19-8. Use convolution in the time domain to find the inverse Laplace transform of $\mathbf{F}_1(\mathbf{s})\mathbf{F}_2(\mathbf{s}) = $: (*a*) $[1/\mathbf{s}][1/(\mathbf{s} + 2)]$; (*b*) $[1/\mathbf{s}^2][1/(\mathbf{s} + 2)]$. Find $f_1(t) * f_2(t)$ by any method of your choice if $f_1(t) = \sin 2t \, u(t)$ and $f_2(t) = $: (*c*) $u(t)$; (*d*) $tu(t)$.
Ans: $0.5(1 - e^{-2t})u(t)$; $(0.5t + 0.25e^{-2t} - 0.25)u(t)$; $0.5(1 - \cos 2t)u(t)$;
$(0.5t - 0.25 \sin 2t)u(t)$

19-9. Find $\mathbf{H(s)}$ and $h(t)$ for the circuit of Fig. 19-4 if the output is taken as: (*a*) $v_C(t)$; (*b*) $i_C(t)$; (*c*) $i_s(t)$.
Ans: $1/(\mathbf{s} + 1.25)$, $e^{-1.25t}u(t)$; $0.5\mathbf{s}/(\mathbf{s} + 1.25)$ A/V, $0.5\delta(t) -$
$0.625e^{-1.25t}u(t)$ A/V;
$0.5(\mathbf{s} + 0.25)/(\mathbf{s} + 1.25)$ A/V, $0.5\delta(t) - 0.5e^{-1.25t}u(t)$ A/V

Fig. 19-4: See Drill Prob. 19-9.

[3] McGillem and Cooper, op. cit., pp. 262–265.

**19-6
Time-shift and
periodic functions**

By now we have obtained a number of entries for the catalog of Laplace transform pairs that we agreed to construct earlier. Included are the transforms of the impulse function, the step function, the exponential function, the ramp function, the sine and cosine functions, and the sum of two exponentials. In addition, we have noted the consequences in the s-domain of the time-domain operations of differentiation, integration, and convolution; and we have used the Laplace transform to define what we mean by a transfer function. These results, plus several others, are collected together for quick reference in Tables 19-1 and 19-2. Some of the tabulated relationships are unfamiliar, however, and we shall now develop a few additional useful theorems to help us obtain them.

As we have seen in some of our earlier transient problems, not all forcing functions begin at $t = 0$. What happens to the transform of a time function if that function is simply shifted in time by some known amount? In particular,

Table 19-1: Laplace transform pairs

$f(t) = \mathcal{L}^{-1}\{\mathbf{F(s)}\}$	$\mathbf{F(s)} = \mathcal{L}\{f(t)\}$
$\delta(t)$	1
$u(t)$	$\dfrac{1}{s}$
$tu(t)$	$\dfrac{1}{s^2}$
$\dfrac{t^{n-1}}{(n-1)!}u(t), n = 1, 2, \ldots$	$\dfrac{1}{s^n}$
$e^{-\alpha t}u(t)$	$\dfrac{1}{s + \alpha}$
$te^{-\alpha t}u(t)$	$\dfrac{1}{(s + \alpha)^2}$
$\dfrac{t^{n-1}}{(n-1)!}e^{-\alpha t}u(t), n = 1, 2, \ldots$	$\dfrac{1}{(s + \alpha)^n}$
$\dfrac{1}{\beta - \alpha}(e^{-\alpha t} - e^{-\beta t})u(t)$	$\dfrac{1}{(s + \alpha)(s + \beta)}$
$\sin \omega t\, u(t)$	$\dfrac{\omega}{s^2 + \omega^2}$
$\cos \omega t\, u(t)$	$\dfrac{s}{s^2 + \omega^2}$
$\sin (\omega t + \theta)u(t)$	$\dfrac{s \sin \theta + \omega \cos \theta}{s^2 + \omega^2}$
$\cos (\omega t + \theta)u(t)$	$\dfrac{s \cos \theta - \omega \sin \theta}{s^2 + \omega^2}$
$e^{-\alpha t} \sin \omega t\, u(t)$	$\dfrac{\omega}{(s + \alpha)^2 + \omega^2}$
$e^{-\alpha t} \cos \omega t\, u(t)$	$\dfrac{s + \alpha}{(s + \alpha)^2 + \omega^2}$

Table 19-2: Laplace transform operations

Operation	$f(t)$	$F(s)$
Addition	$f_1(t) \pm f_2(t)$	$F_1(s) \pm F_2(s)$
Scalar multiplication	$kf(t)$	$kF(s)$
Time differentiation	$\dfrac{df}{dt}$	$sF(s) - f(0^-)$
	$\dfrac{d^2f}{dt^2}$	$s^2F(s) - sf(0^-) - f'(0^-)$
	$\dfrac{d^3f}{dt^3}$	$s^3F(s) - s^2f(0^-) - sf'(0^-) - f''(0^-)$
Time integration	$\displaystyle\int_{0^-}^{t} f(t)\,dt$	$\dfrac{1}{s}F(s)$
	$\displaystyle\int_{-\infty}^{t} f(t)\,dt$	$\dfrac{1}{s}F(s) + \dfrac{1}{s}\displaystyle\int_{-\infty}^{0^-} f(t)\,dt$
Convolution	$f_1(t) * f_2(t)$	$F_1(s)F_2(s)$
Time shift	$f(t-a)u(t-a),$ $a \geq 0$	$e^{-as}F(s)$
Frequency shift	$f(t)e^{-at}$	$F(s+a)$
Frequency differentiation	$-tf(t)$	$\dfrac{dF(s)}{ds}$
Frequency integration	$\dfrac{f(t)}{t}$	$\displaystyle\int_{s}^{\infty} F(s)\,ds$
Scaling	$f(at), a \geq 0$	$\dfrac{1}{a}F\left(\dfrac{s}{a}\right)$
Initial value	$f(0^+)$	$\displaystyle\lim_{s\to\infty} sF(s)$
Final value	$f(\infty)$	$\displaystyle\lim_{s\to 0} sF(s)$, all poles of $sF(s)$ in LHP
Time periodicity	$f(t) = f(t + nT),$ $n = 1, 2, \ldots$	$\dfrac{1}{1 - e^{-Ts}} F_1(s),$ where $F_1(s) = \displaystyle\int_{0^-}^{T} f(t)e^{-st}\,dt$

if the transform of $f(t)u(t)$ is the known function $\mathbf{F}(s)$, then what is the transform of $f(t - a)u(t - a)$, the original time function delayed by a seconds (and not existing for $t < a$)? Working directly from the definition of the Laplace transform, we get

$$\mathcal{L}\{f(t - a)u(t - a)\} = \int_{0^-}^{\infty} e^{-st}f(t - a)u(t - a)\,dt = \int_{a^-}^{\infty} e^{-st}f(t - a)\,dt$$

for $t \geq a^-$. Choosing a new variable of integration, $\tau = t - a$, we obtain

$$\mathcal{L}\{f(t - a)u(t - a)\} = \int_{0^-}^{\infty} e^{-s(\tau+a)}f(\tau)\,d\tau = e^{-as}\mathbf{F}(s)$$

Therefore,

$$f(t - a)u(t - a) \Leftrightarrow e^{-as}\mathbf{F(s)} \qquad (a \geq 0) \qquad (25)$$

This result is known as the *time-shift theorem,* and it simply states that, if a time function is delayed by a time a in the time domain, the result in the frequency domain is a multiplication by e^{-as}.

As an example of the application of this theorem, consider the rectangular pulse described by

$$v(t) = u(t - 2) - u(t - 5)$$

which has unit value for the time interval, $2 < t < 5$, and has zero value elsewhere. We know that the transform of $u(t)$ is just $1/\mathbf{s}$, and since $u(t - 2)$ is simply $u(t)$ delayed by 2 s, the transform of this delayed function is $e^{-2s}/\mathbf{s}$. Similarly, the transform of $u(t - 5)$ is $e^{-5s}/\mathbf{s}$. It follows, then, that the desired transform is

$$\mathbf{V(s)} = \frac{e^{-2s} - e^{-5s}}{\mathbf{s}}$$

It was not necessary to revert to the definition of the Laplace transform in order to determine $\mathbf{V(s)}$.

The time-shift theorem is also useful in evaluating the transform of periodic time functions. Suppose that $f(t)$ is periodic with a period T for positive values of t. The behavior of $f(t)$ for $t < 0$ has no effect on the (one-sided) Laplace transform, as we know. Thus, $f(t)$ can be written as

$$f(t) = f(t - nT) \qquad n = 0, 1, 2, \ldots$$

If we now define a new time function which is nonzero only in the first period of $f(t)$,

$$f_1(t) = [u(t) - u(t - T)]f(t)$$

then the original $f(t)$ can be represented as the sum of an infinite number of such functions, delayed by integral multiples of T. That is,

$$f(t) = [u(t) - u(t - T)]f(t) + [u(t - T) - u(t - 2T)]f(t)$$
$$+ [u(t - 2T) - u(t - 3T)]f(t) + \cdots$$
$$= f_1(t) + f_1(t - T) + f_1(t - 2T) + \cdots$$

or

$$f(t) = \sum_{n=0}^{\infty} f_1(t - nT)$$

The Laplace transform of this sum is just the sum of the transforms,

$$\mathbf{F(s)} = \sum_{n=0}^{\infty} \mathscr{L}\{f_1(t - nT)\}$$

so that the time-shift theorem leads to

$$\mathbf{F(s)} = \sum_{n=0}^{\infty} e^{-nTs}\mathbf{F}_1(\mathbf{s})$$

where

$$\mathbf{F}_1(\mathbf{s}) = \mathscr{L}\{f_1(t)\} = \int_{0^-}^{T} e^{-st}f(t)\ dt$$

Since $\mathbf{F}_1(\mathbf{s})$ is not a function of n, it can be removed from the summation, and $\mathbf{F}(\mathbf{s})$ becomes

$$\mathbf{F}(\mathbf{s}) = \mathbf{F}_1(\mathbf{s})[1 + e^{-Ts} + e^{-2Ts} + \cdots]$$

When we apply the binomial theorem to the bracketed expression, it simplifies to $1/(1 - e^{-Ts})$. Thus, we conclude that the periodic function $f(t)$, with period T, has a Laplace transform expressed by

$$\mathbf{F}(\mathbf{s}) = \frac{\mathbf{F}_1(\mathbf{s})}{1 - e^{-Ts}} \qquad (26)$$

where

$$\mathbf{F}_1(\mathbf{s}) = \mathscr{L}\{[u(t) - u(t - T)]f(t)\} \qquad (27)$$

is the transform of the first period of the time function.

To illustrate the use of this transform theorem for periodic functions, let us apply it to the familiar rectangular pulse train, Fig. 19-5. We may describe this periodic function analytically:

$$v(t) = \sum_{n=0}^{\infty} V_0[u(t - nT) - u(t - nT - \tau)] \qquad t > 0$$

The function $\mathbf{V}_1(\mathbf{s})$ is simple to calculate,

$$\mathbf{V}_1(\mathbf{s}) = V_0 \int_{0^-}^{\tau} e^{-st}\ dt = \frac{V_0}{\mathbf{s}}(1 - e^{-s\tau})$$

Now, to obtain the desired transform, we just divide by $(1 - e^{-sT})$,

$$\mathbf{V}(\mathbf{s}) = \frac{V_0}{\mathbf{s}}\frac{1 - e^{-s\tau}}{1 - e^{-sT}} \qquad (28)$$

We should note how the two theorems described in this section show up in the transform (28). The $(1 - e^{-sT})$ factor in the denominator accounts for the periodicity of the function, the $e^{-s\tau}$ term in the numerator arises from the time delay of the negative square wave that turns off the pulse, and the $V_0/\mathbf{s}$ factor is, of course, the transform of the step functions involved in $v(t)$.

Fig. 19-5: A periodic train of rectangular pulses for which $\mathbf{F}(\mathbf{s}) = (V_0/\mathbf{s})(1 - e^{-s\tau})/(1 - e^{-sT})$.

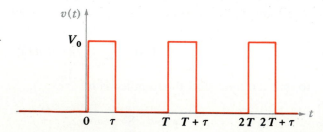

Drill Problems

19-10. Find the Laplace transform of the time function shown in: (*a*) Fig. 19-6*a*; (*b*) Fig. 19-6*b*; (*c*) Fig. 19-6*c*.

Ans: $[0.2(1 - e^{-10s})/s^2] - [2e^{-10s}/s] + [5e^{-10s}]$; $(4e^{-s} + 6e^{-2s} - 10e^{-6s})/s$;
$$8(se^{-s} - e^{-2s} + e^{-3s})/s^2$$

Fig. 19-6: See Drill Prob. 19-10.

(*a*)   (*b*)   (*c*)

19-11. Determine $f_1(t)$ and the Laplace transform of the periodic function shown in: (*a*) Fig. 19-7*a*; (*b*) Fig. 19-7*b*; (*c*) Fig. 19-7*c*.

Ans: $2\delta(t - 2) + 2\delta(t - 6)$, $2(e^{-2s} + e^{-6s})/(1 - e^{-7s})$; $4u(t) - 8u(t - 1) + 8u(t - 3) - 4u(t - 4)$, $(4 - 8e^{-s} + 8e^{-3s} - 4e^{-4s})/[s(1 - e^{-4s})]$; $4\sin 2t[u(t) - u(t - \pi/2)]$, $[8/(s^2 + 4)][(1 + e^{-\pi s/2})/(1 - e^{-\pi s/2})]$

Fig. 19-7: See Drill Prob. 19-11.

(*a*)   (*b*)   (*c*)

19-7 Shifting, differentiation, integration, and scaling in the frequency domain

Other theorems appearing in Table 19-2 specify the results in the time domain of simple operations on **F(s)** in the frequency domain. We shall obtain several easily in this section and then see how they may be applied to derive additional transform pairs.

The first new theorem establishes a relationship between $\mathbf{F(s)} = \mathcal{L}\{f(t)\}$ and $\mathbf{F(s} + a)$. We consider the Laplace transform of $e^{-at}f(t)$,

$$\mathcal{L}\{e^{-at}f(t)\} = \int_{0^-}^{\infty} e^{-st}e^{-at}f(t)\,dt = \int_{0^-}^{\infty} e^{-(s+a)t}f(t)\,dt$$

Looking carefully at this result, we note that the integral on the right is identical to that defining **F(s)** with one exception: $(\mathbf{s} + a)$ appears in place of **s**. Thus,

$$e^{-at}f(t) \Leftrightarrow \mathbf{F(s} + a) \qquad (29)$$

We conclude that replacing **s** by $(\mathbf{s} + a)$ in the frequency domain corresponds to multiplication by e^{-at} in the time domain. This is known as the *frequency-shift* theorem. It can be put to immediate use in evaluating the transform of the exponentially damped cosine function that we used extensively in previous work. Beginning with the known transform of the cosine function,

$$\mathcal{L}\{\cos \omega_0 t\} = \mathbf{F(s)} = \frac{\mathbf{s}}{\mathbf{s}^2 + \omega_0^2}$$

then the transform of $e^{-at} \cos \omega_0 t$ must be $\mathbf{F(s + a)}$,

$$\mathcal{L}\{e^{-at} \cos \omega_0 t\} = \mathbf{F(s + a)} = \frac{(\mathbf{s + a})}{(\mathbf{s + a})^2 + \omega_0^2}$$

and

$$e^{-at} \cos \omega_0 t \Leftrightarrow \frac{\mathbf{s + a}}{(\mathbf{s + a})^2 + \omega_0^2} \qquad (30)$$

Next let us examine the consequences of differentiating $\mathbf{F(s)}$ with respect to $\mathbf{s}$. The result is

$$\frac{d}{d\mathbf{s}} \mathbf{F(s)} = \frac{d}{d\mathbf{s}} \int_{0^-}^{\infty} e^{-\mathbf{s}t} f(t) \, dt$$

$$= \int_{0^-}^{\infty} -te^{-\mathbf{s}t} f(t) \, dt$$

$$= \int_{0^-}^{\infty} e^{-\mathbf{s}t} [-tf(t)] \, dt$$

which is clearly the Laplace transform of $[-tf(t)]$. We therefore conclude that differentiation with respect to $\mathbf{s}$ in the frequency domain results in multiplication by $-t$ in the time domain, or

$$-tf(t) \Leftrightarrow \frac{d}{d\mathbf{s}} \mathbf{F(s)} \qquad (31)$$

Suppose now that $f(t)$ is the unit-ramp function $tu(t)$ whose transform we know is $1/\mathbf{s}^2$. We can use our newly acquired frequency-differentiation theorem to determine the inverse transform of $1/\mathbf{s}^3$ as follows:

$$\frac{d}{d\mathbf{s}} \left(\frac{1}{\mathbf{s}^2}\right) = -\frac{2}{\mathbf{s}^3} \Leftrightarrow -t\mathcal{L}^{-1} \left\{\frac{1}{\mathbf{s}^2}\right\} = -t^2 \, u(t)$$

and

$$\frac{t^2 u(t)}{2} \Leftrightarrow \frac{1}{\mathbf{s}^3} \qquad (32)$$

Continuing with the same procedure, we find

$$\frac{t^3}{3!} u(t) \Leftrightarrow \frac{1}{\mathbf{s}^4} \qquad (33)$$

and in general

$$\frac{t^{n-1}}{(n-1)!} u(t) \Leftrightarrow \frac{1}{\mathbf{s}^n} \qquad (34)$$

The effect on $f(t)$ of integrating $\mathbf{F(s)}$ with respect to $\mathbf{s}$ may be shown by beginning with the definition once more,

$$\mathbf{F(s)} = \int_{0^-}^{\infty} e^{-\mathbf{s}t} f(t) \, dt$$

performing the frequency integration from **s** to ∞,

$$\int_s^\infty \mathbf{F(s)} \, d\mathbf{s} = \int_s^\infty \left[\int_{0^-}^\infty e^{-st}f(t) \, dt \right] \, d\mathbf{s}$$

interchanging the order of integration,

$$\int_s^\infty \mathbf{F(s)} \, d\mathbf{s} = \int_{0^-}^\infty \left[\int_s^\infty e^{-st} \, d\mathbf{s} \right] f(t) \, dt$$

and performing the inner integration,

$$\int_s^\infty \mathbf{F(s)} \, d\mathbf{s} = \int_{0^-}^\infty \left[-\frac{1}{t} e^{-st} \right]_s^\infty f(t) \, dt = \int_{0^-}^\infty \frac{f(t)}{t} e^{-st} \, dt$$

Thus,

$$\frac{f(t)}{t} \Leftrightarrow \int_s^\infty \mathbf{F(s)} \, d\mathbf{s} \tag{35}$$

For example, we have already established the transform pair,

$$\sin \omega_0 t \, u(t) \Leftrightarrow \frac{\omega_0}{\mathbf{s}^2 + \omega_0^2}$$

Therefore,

$$\mathscr{L}\left\{ \frac{\sin \omega_0 t \, u(t)}{t} \right\} = \int_s^\infty \frac{\omega_0 \, d\mathbf{s}}{\mathbf{s}^2 + \omega_0^2} = \tan^{-1} \frac{\mathbf{s}}{\omega_0} \Big|_s^\infty$$

and we have

$$\frac{\sin \omega_0 t \, u(t)}{t} \Leftrightarrow \frac{\pi}{2} - \tan^{-1} \frac{\mathbf{s}}{\omega_0} \tag{36}$$

We next develop the time-scaling theorem of Laplace transform theory by evaluating the transform of $f(at)$, assuming that $\mathscr{L}\{f(t)\}$ is known. The procedure is very simple:

$$\mathscr{L}\{f(at)\} = \int_{0^-}^\infty e^{-st}f(at) \, dt = \frac{1}{a} \int_{0^-}^\infty e^{-(s/a)\lambda}f(\lambda) \, d\lambda$$

where the change of variable, $at = \lambda$, has been employed. The last integral is recognizable as $1/a$ times the Laplace transform of $f(t)$, except that **s** is replaced by **s**/a in the transform. It follows that

$$f(at) \Leftrightarrow \frac{1}{a} \mathbf{F}\left(\frac{\mathbf{s}}{a}\right) \tag{37}$$

As an elementary example of the use of this time-scaling theorem, consider the determination of the transform of a 1-kHz cosine wave. Assuming we know the transform of a 1-rad/s cosine wave,

$$\cos t \, u(t) \Leftrightarrow \frac{\mathbf{s}}{\mathbf{s}^2 + 1}$$

the result is

$$\mathscr{L}\{\cos 2000\pi t \; u(t)\} = \frac{1}{2000\pi} \frac{s/2000\pi}{(s/2000\pi)^2 + 1} = \frac{s}{s^2 + (2000\pi)^2}$$

which is correct.

The time-scaling theorem offers us some computational advantages, for it enables us to work initially in a slowed-down world, where time functions may extend for several seconds and periodic functions possess periods of the order of magnitude of a second. Practical engineering work, however, usually involves time functions that vary much more rapidly with time than our examples might have indicated. The time slowdown is used only to simplify the computational aspects of our problems. These results may then be translated easily to the real world through the use of the time-scaling theorem.

Time scaling is also useful, even essential, in dealing with the analog computer. Here we must work with slowed-down time functions because of the frequency-response limitations of the electronic circuitry in the analog computer.

Drill Problems

19-12. Find: (a) $\mathscr{L}\{e^{-at} \cos(\omega t + \theta)u(t)\}$; (b) $\mathscr{L}\{t \cos(\omega t + \theta)u(t)\}$; (c) $\mathscr{L}\{\sin^2 2t \; u(t)/t\}$.

$$Ans: \; \frac{(s + a)\cos\theta - \omega\sin\theta}{(s + a)^2 + \omega^2}; \; \frac{-2\omega s \sin\theta + s^2 \cos\theta - \omega^2 \cos\theta}{(s^2 + \omega^2)^2};$$

$$0.25 \ln\left(\frac{s^2 + 16}{s^2}\right)$$

19-13. Find the inverse Laplace transform of: (a) $s/(s^2 + 6s + 10)$; (b) $\ln (s + 1) - \ln s$; (c) $s/(s^2 + 1)^2$.

$$Ans: \; e^{-3t}(\cos t - 3 \sin t)u(t); \; (1 - e^{-t})u(t)/t; \; 0.5t \sin t \; u(t)$$

19-14. (a) Find $\mathscr{L}\{f(5t)\}$ if $\mathscr{L}\{f(t)\} = 2/(s^4 + 16)$. Now let $F(s) = 8/(s + 2)^3$ and find: (b) $f(t)$; (c) $f(2t)$.

$$Ans: \; 250/(s^4 + 10^4); \; 4t^2 e^{-2t}u(t); \; 16t^2 e^{-4t}u(t)$$

19-8
The initial-value and final-value theorems

The last two fundamental theorems that we shall discuss are known as the initial-value and final-value theorems. They will enable us to evaluate $f(0^+)$ and $f(\infty)$ by examining the limiting values of $F(s)$.

To derive the initial-value theorem, we consider the Laplace transform of the derivative once again,

$$\mathscr{L}\left\{\frac{df}{dt}\right\} = sF(s) - f(0^-) = \int_{0^-}^{\infty} e^{-st} \frac{df}{dt} \, dt$$

We now let s approach infinity. By breaking the integral into two parts,

$$\lim_{s\to\infty} [sF(s) - f(0^-)] = \lim_{s\to\infty} \left(\int_{0^-}^{0^+} e^0 \frac{df}{dt} \, dt + \int_{0^+}^{\infty} e^{-st} \frac{df}{dt} \, dt \right)$$

we see that the second integral must approach zero in the limit since the integrand itself approaches zero. Also, $f(0^-)$ is not a function of s, and it may be removed from the left limit,

$$-f(0^-) + \lim_{s \to \infty} [sF(s)] = \lim_{s \to \infty} \int_{0^-}^{0^+} df = \lim_{s \to \infty} [f(0^+) - f(0^-)]$$
$$= f(0^+) - f(0^-)$$

and finally

$$f(0^+) = \lim_{s \to \infty} [sF(s)]$$

or

$$\lim_{t \to 0^+} f(t) = \lim_{s \to \infty} [sF(s)] \tag{38}$$

This is the mathematical statement of the *initial-value theorem*. It states that the initial value of the time function $f(t)$ can be obtained from its Laplace transform $F(s)$ by first multiplying the transform by s and then letting s approach infinity. Note that the initial value of $f(t)$ that is obtained is the limit from the right.

The initial-value theorem, along with the final-value theorem that we shall consider in a moment, is useful in checking the results of a transformation or an inverse transformation. For example, when we first calculated the transform of $\cos \omega_0 t\, u(t)$, we obtained $s/(s^2 + \omega_0^2)$. After noting that $f(0^+) = 1$, a partial check on the validity of this result is achieved by applying the initial-value theorem,

$$\lim_{s \to \infty} \left(s\, \frac{s}{s^2 + \omega_0^2} \right) = 1$$

and the check is accomplished.

The final-value theorem is not quite as useful as the initial-value theorem, for it can be used only with a certain class of transforms, those whose poles lie entirely within the left half of the s plane, except for a simple pole at $s = 0$. We again consider the Laplace transform of df/dt,

$$\int_{0^-}^{\infty} e^{-st} \frac{df}{dt}\, dt = sF(s) - f(0^-)$$

this time in the limit as s approaches zero,

$$\lim_{s \to 0} \int_{0^-}^{\infty} e^{-st} \frac{df}{dt}\, dt = \lim_{s \to 0} [sF(s) - f(0^-)] = \int_{0^-}^{\infty} \frac{df}{dt}\, dt$$

We assume that both $f(t)$ and its first derivative are transformable. Now, the last term of this equation is readily expressed as a limit,

$$\int_{0^-}^{\infty} \frac{df}{dt}\, dt = \lim_{t \to \infty} \int_{0^-}^{t} \frac{df}{dt}\, dt = \lim_{t \to \infty} [f(t) - f(0^-)]$$

By recognizing that $f(0^-)$ is a constant, a comparison of the last two equations shows us that

$$\lim_{t \to \infty} f(t) = \lim_{s \to 0} [sF(s)] \tag{39}$$

which is the *final-value theorem*. In applying this theorem, it is necessary to know that $f(\infty)$, the limit of $f(t)$ as t becomes infinite, exists, or, what amounts to the same thing, that the poles of $\mathbf{F(s)}$ all lie *within* the left half of the s plane except for a simple pole at the origin. The product $\mathbf{sF(s)}$ thus has *all* of its poles lying *within* the left-half plane.

As a straightforward example of the application of this theorem, let us consider the function $f(t) = (1 - e^{-at})u(t)$, where $a > 0$. We see immediately that $f(\infty) = 1$. The transform of $f(t)$ is

$$\mathbf{F(s)} = \frac{1}{\mathbf{s}} - \frac{1}{\mathbf{s} + a} = \frac{a}{\mathbf{s(s} + a)}$$

Multiplying by $\mathbf{s}$ and letting $\mathbf{s}$ approach zero, we obtain

$$\lim_{s \to 0} [\mathbf{sF(s)}] = \lim_{s \to 0} \frac{a}{\mathbf{s} + a} = 1$$

which agrees with $f(\infty)$.

If $f(t)$ is a sinusoid, however, so that $\mathbf{F(s)}$ has poles on the $j\omega$ axis, then a blind use of the final-value theorem might lead us to conclude that the final value is zero. We know, however, that the final value of either $\sin \omega_0 t$ or $\cos \omega_0 t$ is indeterminate. So, beware of $j\omega$-axis poles.

We now have all the tools to apply the Laplace transform to the solution of problems that we either could not solve previously, or that we solved with considerable stress and strain.

Drill Problem

19-15. Without finding $f(t)$ first, determine $f(0^+)$ and $f(\infty)$ for each of the following transforms: (a) $[(2s + 1)^2 + 4]/\{[(s + 3)^2 + 16](s + 5)\}$; (b) $(s^3 + 7s^2 + 5)/[s(s^3 + 3s^2 + 4s + 2)]$; (c) $e^{-2s}(s + 2)/(s^3 + 5s)$.

Ans: 4, 0; 1, 2.5; 0, unknown

19-9
The transfer
function H(s)

Earlier in this chapter we defined the transfer function $\mathbf{H(s)}$ as the Laplace transform of the impulse response $h(t)$, initial energy being zero throughout the circuit. Before we can make the most effective use of the transfer function, we need to show that it may be obtained very simply for any linear circuit by frequency-domain analysis. We do this by an argument similar to that presented for the Fourier transform and $\mathbf{H}(j\omega)$ in Sec. 18-9. But we shall be brief.

We work in the complex-frequency domain and apply an input, $v_i(t) = Ae^{\sigma_x t} \cos(\omega_x t + \theta)$, which is $\mathbf{V}_i(\mathbf{s}_x) = Ae^{j\theta}$ in phasor form. Both A and θ are functions of $\mathbf{s}_x = \sigma_x + j\omega_x$, the complex frequency of the excitation. The response is $v_o(t) = Be^{\sigma_x t} \cos(\omega_x t + \phi)$, $\mathbf{V}_o(\mathbf{s}_x) = Be^{j\phi}$. Thus,

$$\frac{\mathbf{V}_o(\mathbf{s}_x)}{\mathbf{V}_i(\mathbf{s}_x)} = \mathbf{G}(\mathbf{s}_x) = \frac{B}{A} e^{j(\phi - \theta)} \tag{40}$$

To determine $\mathbf{H(s)}$, we need to find the ratio of the transform of the output to the transform of the input,

$$\mathbf{H(s)} = \frac{\mathcal{L}\{Be^{\sigma_x t} \cos(\omega_x t + \phi)\}}{\mathcal{L}\{Ae^{\sigma_x t} \cos(\omega_x t + \theta)\}}$$

The required transform is obtained from Table 19-1 by replacing s by $s - \sigma_x$ in the transform for $\cos(\omega t + \theta)$. We have

$$\mathbf{H(s)} = \frac{B \left\{ \dfrac{[(s - \sigma_x) \cos \phi - \omega_x \sin \phi]}{[(s - \sigma_x)^2 + \omega_x^2]} \right\}}{A \left\{ \dfrac{[(s - \sigma_x) \cos \theta - \omega_x \sin \theta]}{[(s - \sigma_x)^2 + \omega_x^2]} \right\}}$$

At $s = s_x$, this simplifies to

$$\mathbf{H(s_x)} = \frac{B}{A} \frac{j\omega_x \cos \phi - \omega_x \sin \phi}{j\omega_x \cos \theta - \omega_x \sin \theta}$$

$$= \frac{B}{A} \frac{\cos \phi + j \sin \phi}{\cos \theta + j \sin \theta} = \frac{B}{A} e^{j(\phi - \theta)}$$

which is identical to (40). Since there is no special significance to the x subscript, it follows that

$$\mathbf{H(s)} = \mathbf{G(s)} \tag{41}$$

Thus, we may find $\mathbf{H(s)}$ by using normal frequency-domain methods with all elements expressed in terms of their impedances at a complex frequency s.

Let us illustrate how this technique may be used to find both $h(t)$ and an output voltage for the circuit shown in Fig. 19-8a. At this time we assume that there is no initial energy storage in the network. We first construct the frequency-domain circuit, shown in Fig. 19-8b. The ratio of $\mathbf{V}_o(s)/\mathbf{V}_i(s)$ may be found by determining $\mathbf{Z}_i(s)$, the impedance of the three parallel branches at the right,

$$\mathbf{Z}_i(s) = \frac{1}{s/24 + \frac{1}{30} + 1/(24 + 48/s)} = \frac{120(s + 2)}{5s^2 + 19s + 8}$$

and then using voltage division,

$$\frac{\mathbf{V}_o(s)}{\mathbf{V}_i(s)} = \frac{\mathbf{Z}_i(s)}{20 + \mathbf{Z}_i(s)} = \frac{6(s + 2)}{5s^2 + 25s + 20}$$

Fig. 19-8: (a) An example in which the transfer function $\mathbf{H(s)} = \mathbf{V}_o(s)/\mathbf{V}_i(s)$ is to be obtained by frequency-domain analysis. Initial conditions are all zero. (b) The frequency-domain circuit.

(a)

(b)

Thus,
$$H(s) = \frac{1.2(s + 2)}{(s + 1)(s + 4)}$$

To find $h(t)$, we need $\mathcal{L}^{-1}\{H(s)\}$:

$$\mathcal{L}^{-1}\{H(s)\} = \mathcal{L}^{-1}\left\{\frac{0.4}{s + 1} + \frac{0.8}{s + 4}\right\}$$

and
$$h(t) = (0.4e^{-t} + 0.8e^{-4t})u(t)$$

Thus, if $v_i(t) = \delta(t)$, $v_o(t) = h(t) = (0.4e^{-t} + 0.8e^{-4t})u(t)$.

For a different input, say $v_i(t) = 50 \cos 2t\, u(t)$ V, we may make use of the transfer function concept,

$$V_o(s) = H(s)V_i(s)$$

where

$$V_i(s) = \mathcal{L}\{50 \cos 2t\, u(t)\} = \frac{50s}{s^2 + 4}$$

and
$$V_o(s) = \frac{1.2(s + 2)}{(s + 1)(s + 4)} \frac{50s}{s^2 + 4}$$

Expanding in partial fractions,

$$V_o(s) = \frac{-4}{s + 1} + \frac{-8}{s + 4} + \frac{6 + j6}{s + j2} + \frac{6 - j6}{s - j2}$$

$$= \frac{-4}{s + 1} + \frac{-8}{s + 4} + \frac{12s + 24}{s^2 + 4}$$

and
$$v_o(t) = (-4e^{-t} - 8e^{-4t} + 12 \cos 2t + 12 \sin 2t)u(t)$$

The solution is straightforward, and we should feel confident of being able to find the response for any input that is Laplace-transformable. We consider the presence of initial-energy storage in the next (and final) section.

Drill Problem

19-16. Determine $h(t)$ for the circuit of Fig. 19-9 if the output is: (a) $i_s(t)$; (b) $v_L(t)$.

Ans: $\frac{1}{16}\delta(t) + 3e^{-12t}u(t) - 9te^{-12t}u(t)$ A; $0.75\delta(t) - 12e^{-12t}u(t) + 36te^{-12t}u(t)$ V

Fig. 19-9: See Drill Prob. 19-16.

**19-10
The complete
response**

When initial energy is present in a circuit, the Laplace transform method may be used to obtain the complete response by any of several different methods. We shall consider two of them.

The first is the more fundamental, for it involves writing the differential equations for the network and then taking the Laplace transform of those equations. The initial conditions appear when a derivative or integral is transformed. The second technique requires each initial capacitor voltage or inductor current to be replaced by an equivalent dc source, often called an *initial-condition generator*. The elements themselves then carry no initial energy, and the transfer function procedure of the preceding section can be followed.

Let us illustrate the differential equation approach by considering the same circuit we just analyzed, but with nonzero initial conditions this time, as shown in Fig. 19-10. We let $v_1(0^-) = 10$ V and $v_2(0^-) = 25$ V.

The differential equations for this circuit may be obtained by writing nodal equations in terms of v_1 and v_2. At the v_1 node,

$$\frac{v_1 - v_2}{24} + \frac{1}{48} v_1' = 0$$

or
$$2v_2 = 2v_1 + v_1' \tag{42}$$

while at the v_2 node,

$$\frac{v_2 - 50 \cos 2t \, u(t)}{20} + \frac{v_2 - v_1}{24} + \frac{v_2}{30} + \frac{1}{24} v_2' = 0$$

or
$$v_1 = v_2' + 3v_2 - 60 \cos 2t \, u(t) \tag{43}$$

Identifying v_2 as the desired response, we eliminate v_1 and v_1' by taking the derivative of (43), remembering that $du(t)/dt = \delta(t)$,

$$v_1' = v_2'' + 3v_2' + 120 \sin 2t \, u(t) - 60\delta(t) \tag{44}$$

and substituting (43) and (44) into (42),

$$2v_2 = 2[v_2' + 3v_2 - 60 \cos 2t \, u(t)] + [v_2'' + 3v_2' + 120 \sin 2t \, u(t) - 60\delta(t)]$$

or
$$v_2'' + 5v_2' + 4v_2 = (120 \cos 2t - 120 \sin 2t)u(t) + 60\delta(t)$$

We now take the Laplace transform,

$$s^2 V_2(s) - sv_2(0^-) - v_2'(0^-) + 5sV_2(s) - 5v_2(0^-) + 4V_2(s) = \frac{120s - 240}{s^2 + 4} + 60$$

Fig. 19-10: The response $v_2(t)$ is obtained for this network with the initial conditions, $v_1(0^-) = 10$ V, $v_2(0^-) = 25$ V.

collect terms,

$$(s^2 + 5s + 4)\mathbf{V}_2(s) = sv_2(0^-) + v_2'(0^-) + 5v_2(0^-) + \frac{120s - 240}{s^2 + 4} + 60$$

let $v_2(0^-) = 25$,

$$(s^2 + 5s + 4)\mathbf{V}_2(s) = 25s + 125 + v_2'(0^-) + \frac{120s - 240}{s^2 + 4} + 60$$

and need a value for $v_2'(0^-)$. This we may obtain from the two circuit equations (42) and (43) by evaluating each term at $t = 0^-$. Actually, we need use only (43) in this problem:

$$v_1(0^-) = v_2'(0^-) + 3v_2(0^-) - 0$$

and

$$v_2'(0^-) = -65$$

Thus,

$$\begin{aligned}
\mathbf{V}_2(s) &= \frac{25s + 120 + 120[(s - 2)/(s^2 + 4)]}{(s + 1)(s + 4)} \\
&= \frac{25s^3 + 120s^2 + 220s + 240}{(s + 1)(s + 4)(s^2 + 4)} \\
&= \frac{\frac{23}{3}}{s + 1} + \frac{\frac{16}{3}}{s + 4} + \frac{12s + 24}{s^2 + 4}
\end{aligned} \tag{45}$$

from which the time-domain response is obtained,

$$v_2(t) = (\tfrac{23}{3}e^{-t} + \tfrac{16}{3}e^{-4t} + 12 \cos 2t + 12 \sin 2t)u(t)$$

Now let us determine the frequency-domain equivalent of an inductor L with initial current $i(0^-)$. The time-domain network of Fig. 19-11a is described by

$$v(t) = Li'$$

and, therefore,

$$\mathbf{V}(s) = sL\mathbf{I}(s) - Li(0^-) \tag{46}$$

or

$$\mathbf{I}(s) = \frac{\mathbf{V}(s)}{sL} + \frac{i(0^-)}{s} \tag{47}$$

The frequency-domain equivalents can be read directly from (46) and (47),

Fig. 19-11: (a) An inductor L with initial current $i(0^-)$ is shown in the time domain. (b) and (c) Frequency-domain networks that are equivalent to (a) for Laplace-transform analysis.

Fig. 19-12: (*a*) A capacitor *C* with initial voltage $v(0^-)$ is shown in the time domain. (*b*) and (*c*) Frequency-domain networks that are equivalent to (*a*) for Laplace-transform analysis.

(*a*)   (*b*)   (*c*)

and they are shown in Figs. 19-11*b* and *c*, respectively. It may be helpful to note that the voltage source in Fig. 19-11*b* is the transform of an impulse, while the current source in *c* is the transform of a step.

Equivalent networks for an initially charged capacitor are obtained by a similar procedure; the results are shown in Fig. 19-12.

We may now use these results to construct a frequency-domain equivalent of the circuit shown in Fig. 19-10, including the effect of the initial conditions. The result is shown in Fig. 19-13. Current sources are used for the initial conditions to expedite the writing of nodal equations.

We must now use both the superposition principle and the concept of the transfer function to see that $V_2(s)$ is composed of the sum of three terms, one due to each source acting alone. Moreover, each of these sources has a transfer function to $V_2(s)$. Each transfer function could be obtained by applying the standard frequency-domain analysis methods. We need not solve three little problems, however, for we can accomplish the frequency-domain analysis with all three sources operating. We do this by writing two nodal equations:

$$\frac{V_1(s) - V_2(s)}{24} + \frac{sV_1(s)}{48} - \frac{10}{48} = 0$$

or

$$(s + 2)V_1 - 2V_2 = 10 \tag{48}$$

and

$$\frac{V_2(s) - 50s/(s^2 + 4)}{20} + \frac{V_2(s) - V_1(s)}{24} + \frac{V_2(s)}{30} - \frac{25}{24} + \frac{sV_2(s)}{24} = 0$$

or

$$(s + 3)V_2 - V_1 = \frac{60s}{s^2 + 4} + 25 \tag{49}$$

Using (48) to eliminate V_1 in the above equation, we have

$$V_1 = \frac{10 + 2V_2}{s + 2}$$

Fig. 19-13: The frequency-domain equivalent of the circuit of Fig. 19-10. The current sources 10/48 and 25/24 provide initial time-domain voltages of 10 and 25 V across the 1/48-F and 1/24-F capacitors, respectively.

and then

$$\mathbf{V}_2 = \frac{25s^3 + 120s^2 + 220s + 240}{(s + 1)(s + 4)(s^2 + 4)}$$

This agrees with Eq. (45) obtained two pages back, and we need not repeat the inverse transform operation. The two methods check.

In comparing their use, it is probably true that the transfer function method with initial-condition generators is a little faster. This becomes a safer statement to make as the complexity of the network increases. However, we must also bear in mind the fact that the closer we stay to fundamentals, the less apt we are to become confused with special techniques and procedures. Certainly the differential equation approach is the more basic. The use of the Laplace transform may then be thought of simply as a handy method of solving linear differential equations.

In finishing our discussion of the Laplace transform, it is helpful to compare its uses with those of the Fourier transform. The latter, of course, does not exist for as wide a variety of time functions; there are also more j's scattered about in the transform expressions, and the Fourier transform is more difficult to apply to circuits containing initial-energy storage. Thus, transient problems are more easily handled with the Laplace transform. However, when spectral information about a signal is desired, such as the distribution of energy across the frequency band, the Fourier transform is the more convenient.

And now we approach the last few paragraphs of the last section of the last chapter. Looking back at six hundred pages of linear circuit analysis, we should ask ourselves what we have accomplished. Are we prepared to tackle a really practical problem, or have we just been tilting at quixotic windmills? Can we analyze a multistage active filter, a complicated telemetry receiver, or a large interconnected power grid? Before we confess that we cannot, or at least admit that there are other things we would rather do, let us consider how far we could go in such a complex problem.

We certainly have developed substantial skills in writing accurate sets of equations to describe the behavior of increasingly complicated linear circuits. This in itself is no mean accomplishment, for we have done so in both the frequency and time domains. Our only real limitation lies in the difficulty of obtaining numerical results as the complexity of the circuit increases. For that reason we concentrated on those examples which suited computational abilities that were only human. However, we did develop an increased understanding of the most important analysis techniques in the process.

In the last several chapters we have been developing those techniques which facilitate the process of obtaining specific numerical answers. We could now write a set of time-domain or frequency-domain equations for any large-scale circuit we choose, but the numerical solution would strain our accuracy and stretch our patience. If our discussion were to proceed any further, we would have to resort to inhuman means to solve these equations. This is the point at which we turn to the computer. Its patience is virtually unlimited, its accuracy is astonishing, and its speed is almost beyond comprehension. The detailed procedure by which the computer arrives at its numerical results, however, is generated by us, and if we make an error in the program, the computer will proceed to generate erroneous results, also with speed, accuracy, and patience.

In summary, then, we can now produce sets of accurate descriptive equations that characterize almost any given linear circuit configuration, and we have some modest abilities for solving the simpler sets of these equations. Our next steps, if we were to continue, would involve learning how to describe circuits by state-variable equations and studying the procedures by which large-scale computer programs can turn the state-variable equations or a set of transform equations into useful data. But we must stop sometime, so let it be here.

Drill Problems

19-17. For the circuit shown in Fig. 19-14, write a suitable differential equation, take the Laplace transform of both sides, and find $v_o(t)$ if $i_s(t) = $: (a) $10u(t)$ A; (b) $10 \cos t\, u(t)$ A; (c) $10e^{-t}u(t)$ A.
Ans: $(12 - 8e^{-0.5t})u(t)$; $(1.6e^{-0.5t} + 2.4 \cos t + 4.8 \sin t)u(t)$; $(16e^{-0.5t} - 12e^{-t})u(t)$ V

Fig. 19-14: See Drill Probs. 19-17 and 19-18.

$i_L(0^-) = 2$ A

19-18. Rework Drill Prob. 19-17 by first installing a suitable initial-condition generator in the circuit and then working in the frequency domain.
Ans: Same

Problems

1 Give the range of σ over which the Laplace transform exists if $f(t) = $: (a) $tu(t)$; (b) t; (c) $e^{-2t}u(t)$; (d) $e^{2t}u(t)$; (e) $e^{-2t}u(t - 3)$; (f) $e^{-2t}u(t + 3)$.

2 Determine the Fourier transform, the two-sided Laplace transform, and the Laplace transform of $f(t) = $: (a) $4e^{-2t}[u(t + 3) - u(t - 2)]$; (b) $4e^{2t}[u(t + 3) - u(t - 2)]$; (c) $4e^{-2|t|}[u(t + 3) - u(t - 2)]$.

3 Use the definition of the Laplace transform to find $\mathbf{F(s)}$ if $f(t) = $: (a) $[u(t - 1)][u(4 - t)][u(t)]$; (b) $2\delta(t - 4)$; (c) $2(\sin \pi t)\delta(t - 0.2)$; (d) $2u(t - 4)$; (e) $2e^{-4t}u(t - 1)$.

4 Determine the Laplace transform of each of the following functions, giving the range of σ over which the transform exists: (a) $[u(t + 3)][u(t)][u(5 - t)]$; (b) $5\delta(t - 5)$; (c) $5\delta(t - 0.1) \cos 2\pi t$; (d) $5u(t - 5)$; (e) $5(t - 2)u(t - 1)$.

5 Find $f(t)$ if $\mathbf{F(s)} = $: (a) $1 + (1/s) + [1/(s + 1)]$; (b) $(s + 1)e^{-s}/s$; (c) $(1 + e^{-s})^2$; (d) $2e^{-3s} \sinh 2\mathbf{s}$.

6 Apply the defining expression for the Laplace transform to calculate $\mathbf{F}(0.1 + j1)$ if

$f(t) = : (a) \ u(t - 1); \ (b) \ tu(t - 1); \ (c) \ (t - 1)u(t - 1); \ (d) \ e^{-2t}u(t - 1); \ (e) \ \delta(t - 1)u(t).$

7 Find $f(t)$ if $\mathbf{F(s)} = : (a) \ 5 - (2/s); \ (b) \ (5/s) + [3/(s + 1)]; \ (c) \ (s + 2)/(s + 1); \ (d) \ (s^2 + 5)/s^2; \ (e) \ 6/[(s + 1)(s + 2)].$

8 Find the inverse Laplace transform of: $(a) \ (s + 6)/(s + 4); \ (b) \ (2 + 3s)/(s^2 + 4); \ (c) \ (4s + 100)/(10s^2 + 20s); \ (d) \ 60/[(s + 2)(s + 3)(s + 5)].$

9 (a) Write a single, nodal, time-domain equation in terms of $v_C(t)$ for the circuit of Fig. 19-15. (b) Take the Laplace transform of the equation and use it to find $v_C(t)$.

Fig. 19-15: See Prob. 9.

10 Let $i_s(t) = 5 + 5u(t)$ A in the circuit shown in Fig. 19-16. Write a single differential equation in terms of $i_L(t)$, take the Laplace transform of the equation, solve for $\mathbf{I}_L(\mathbf{s})$, and then find $i_L(t)$.

Fig. 19-16: See Prob. 10.

11 If $d^3y/dt^3 + 2d^2y/dt^2 = 0$ and $y(0^-) = 0$, $y'(0^-) = 1$, and $y''(0^-) = -1$, take the Laplace transform of the equation, solve for $\mathbf{Y(s)}$, and then obtain $y(t)$.

12 Given the equation, $\int_{-\infty}^{t} 25f(x) \ dx = 30u(t) - df/dt$, where $f(0^-) = 2$ and $\int_{-\infty}^{0^-} f(t) \ dt = 0.2$, find $f(t)$.

13 An ideal voltage source, $v_s(t) = 100e^{-5t}u(t)$ V, a resistor R, an inductor L, and a capacitor C are in series. Let $i(t)$ leave the positive terminal of v_s. Then $v_s = i' + 4i + 3\int_0^t i \ dt$. (a) Find R, L, and C. (b) Use Laplace-transform methods to find $i(t)$.

14 Two signals, $x(t)$ and $y(t)$, are related by the equations, $y'(t) + 2x(t) = t$ and $y + \int_0^t x(z) \ dz = 2t$, where $x(0^-) = 0$ and $\int_{-\infty}^{0} x(t) \ dt = 0$. Find $x(t)$ and $y(t)$ for $t > 0$.

15 Write a single integrodifferential equation in terms of $v_L(t)$ for the circuit of Fig. 19-17, take its Laplace transform, solve for $\mathbf{V}_L(\mathbf{s})$, and obtain $v_L(t)$ by the inverse transform.

Fig. 19-17: See Prob. 15.

16 Let $x(t) = 4tu(t)$ and $h(t) = 3u(t)$. (a) Find $\mathbf{X(s)}$, $\mathbf{H(s)}$, and $\mathscr{L}^{-1}\{\mathbf{X(s)H(s)}\}$. (b) Find $x(t) * h(t)$ by working entirely in the time domain.

17 The input $x(t)$ and the impulse response of a certain linear system are shown in Fig. 19-18. Find the output $y(t)$ as: (a) $x(t) * h(t)$ in the time domain; (b) $\mathscr{L}^{-1}\{\mathbf{X(s)H(s)}\}$ in the frequency domain.

Fig. 19-18: See Prob. 17.

18 Let $f(t) = 2e^{-10t}u(t)$ and $g(t) = 5e^{-3t}u(t)$. Find $f(t) * g(t)$ by two different methods.

19 Find $h(t)$ and $\mathbf{H(s)}$ for the circuit shown in Fig. 19-19 if the output is: (a) $v_R(t)$; (b) $v_L(t)$. (c) Find $v_L(t)$ if $v_s(t) = 0.1\delta(t) - 3u(t)$ V.

Fig. 19-19: See Prob. 19.

20 If $x(t) = 2[u(t) - u(t-1)]$ and $h(t) = u(t) - u(t-2)$: (a) find $\mathscr{L}^{-1}\{\mathscr{L}[x(t)]\mathscr{L}[h(t)]\}$, and (b) sketch the result.

21 Given the waveforms for $x(t)$ and $h(t)$ shown in Fig. 19-20: (a) find $\mathbf{X(s)}$, $\mathbf{H(s)}$, and $y(t) = \mathscr{L}^{-1}\{\mathbf{X(s)H(s)}\}$; (b) find $y(t) = x(t) * h(t)$ by working entirely in the time domain; (c) evaluate $y(2.5)$.

Fig. 19-20: See Prob. 21.

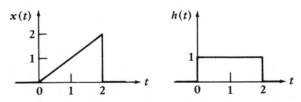

22 Use the concept of time shifting to help in finding: (a) $\mathbf{F(s)}$, if $f(t)$ is zero for $t < 4$, 20 sin $(\pi t/2)$, for $4 < t < 5$, and 20 for $t > 5$; (b) $\mathscr{L}^{-1}\{4e^{-s}/[(s+1)(s+3)]\}$.

23 Use the time-shift theorem to help determine: (a) $\mathbf{F(s)}$ if $f(t) = \cos \pi t[u(t) - u(t-1)]$; (b) $\mathscr{L}^{-1}\{e^{-3s}/[(s+2)^2(s+1)]\}$.

24 For the periodic waveform shown in Fig. 19-21: (a) find $\mathbf{F(s)}$; (b) evaluate $\mathbf{F}(0.5)$.

Fig. 19-21: See Prob. 24.

25 Determine the Laplace transform of the rectified sine wave shown in Fig. 19-22.

Fig. 19-22: See Prob. 25.

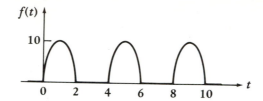

26 Sketch $f(t)$ vs. t if $\mathbf{F(s)} =$: (a) $5s/(s^2 + 4)$; (b) $5se^{-\pi s/2}/(s^2 + 4)$; (c) $5s(1 + e^{-\pi s/2})/(s^2 + 4)$; (d) $5s(1 + e^{-\pi s/2})/[(s^2 + 4)(1 - e^{-1.25\pi s})]$.

27 Find $\mathbf{F(s)}$ if $f(t) =$: (a) $50 \cos (10t - 0.8)u(t)$; (b) $(t - 1)u(t - 2)$; (c) $\sum_{n=0}^{\infty} 13(t - 4n)[u(t - 4n) - u(t - 2 - 4n)]$.

28 Find the Laplace transform of: (a) $5t^3 e^{-2t}u(t)$; (b) $(1/t)(e^{-5t} - e^{-4t})u(t)$; (c) $f(2t)$ if $\mathcal{L}\{f(t)\} = s/[5 + \ln(s + 1)]$.

29 Find the inverse Laplace transform of: (a) $(s^3 + 6s^2 + 13s + 11)/[(s + 2)^3(s + 3)]$; (b) $(s^2 - 4)/(s^2 + 4)^2$; (c) $\mathbf{F}(2s)$ if $\mathcal{L}\{f(t)\} = e^{-4s}/(s^2 + 100)$.

30 Find $f(t)$ if $\mathbf{F(s)} =$: (a) $3/(s + 10)^4$; (b) $e^{-2(s+1)}/(s + 1)^2$; (c) $2/[s^2(s + 2)]$.

31 Let $\mathbf{F_0(s)} = 10/(s^2 + 25)$. (a) Find $f_0(t)$. (b) Find $\mathbf{F_{-1}(s)} = \int_s^\infty \mathbf{F_0(s)}\, ds$ and $f_{-1}(t)$. (c) Find $\mathbf{F_1(s)} = d\mathbf{F_0(s)}/ds$ and $f_1(t)$.

32 If $di/dt + 4i + 3\int_0^t i(z)dz = 12(t - 1)u(t - 1)$ and $i(0^-) = 0$, find $i(t)$.

33 Find $f(0^+)$ and $f(\infty)$ for each of the following transforms, without finding $f(t)$ first: (a) $(3 - e^{-2s})/[s(s^2 + 2s + 5)]$; (b) $3s/(s^2 + 2s + 5)$; (c) $3/(1 - e^{-2s})$.

34 (a) Synthesize an $\mathbf{F(s)}$ as the quotient of two polynomials so that $\lim_{s\to 0} [s\mathbf{F(s)}] = 10$ and $\lim_{s\to\infty} [s\mathbf{F(s)}] = 25$. (b) Find $f(t)$. (c) Find $f(0^+)$ and $f(\infty)$ by limiting operations, and also by applying the initial- and final-value theorems.

35 For each of the following transforms, find $f(0^+)$ and $f(\infty)$; then find the corresponding time functions and show that your results are correct: (a) $50(s + 10)/[s(s + 20)]$; (b) $25(1 + e^{-10s})/s$; (c) $(2s + 6)/(s^2 + 8s + 20)$.

36 The Laplace transform $\mathbf{F(s)}$ has zeros at $\mathbf{s} = 2$ and $\mathbf{s} = -2 \pm j4$; poles at $\mathbf{s} = 0, -3, -5$, and $\mathbf{s} = -1 \pm j1$; and $\mathbf{F}(1) = -5$. (a) Find $f(0^+)$. (b) Find $f(\infty)$. (c) Find $\mathbf{F}(1 + j5)$.

37 Find $h(t)$ for the circuit of Fig. 19-23 by drawing the frequency-domain circuit, finding $\mathbf{V_0(s)}/\mathbf{V_s(s)}$, and then evaluating the inverse transform.

Fig. 19-23: See Prob. 37.

38 The transfer function for a certain linear circuit is $\mathbf{H(s)} = \mathbf{I}_{out}/\mathbf{V}_{in} = 2(s + 10)/[(s + 4)(s^2 + 4)]$. Find $i_{out}(t)$ if $v_{in}(t) = 8u(t)$ V and all initial conditions are zero.

39 (a) Find the impulse response of the circuit shown in Fig. 19-24 by working first in the frequency domain and then taking $\mathcal{L}^{-1}\{\mathbf{V}_{out}(s)\}$. (b) Find $v_{out}(t)$ if $v_{in}(t) = 10u(t)$.

Fig. 19-24: See Prob. 39.

40 Find $i_s(t)$ for the circuit shown in Fig. 19-25.

Fig. 19-25: See Prob. 40.

41 Two identical linear systems are connected in cascade. The input signal to the first stage is $x(t)$, its output is $y(t)$, and the output of the second stage is $z(t)$. If an input $tu(t)$ is applied to a single stage, the output of that stage is $(e^{-2t} - 1)u(t)$. What is $z(t)$ if $x(t) = u(t)$?

42 Determine $i_L(t)$ for the circuit of Fig. 19-26 if $i_s(t) = 16te^{-t}u(t)$ A.

Fig. 19-26: See Prob. 42.

43 (a) Find $\mathbf{V}_C(\mathbf{s})$ for the circuit of Fig. 19-27 by writing one or more differential equations, reducing them to obtain one second-order differential equation, and taking the Laplace transform. (b) Obtain $v_C(t)$.

Fig. 19-27: See Probs. 43 and 44.

44 (a) Draw the transform circuit for Prob. 43 and Fig. 19-27, including two initial-condition generators. (b) Write one or more frequency-domain equations and solve for $\mathbf{V}_C(\mathbf{s})$. (c) Find $v_C(t)$.

45 The sawtooth voltage pulse, $v_s(t) = 40t[u(t) - u(t - 0.1)]$ V, is applied to the series combination of 2 Ω and 0.05 F. Find the two instants of time at which the capacitor voltage is 1 V.

46 Sources that have been switched out of the circuit of Fig. 19-28 prior to $t = 0^-$ cause $i_L(0^-)$ to be 5 A and $v_C(0^-)$ to be 10 V. Supply two initial-condition generators for the transform circuit and find $v_C(t)$. What are $i_L(0^+)$ and $v_C(0^+)$?

Fig. 19-28: See Prob. 46.

47 Find $v_C(t)$ for all t in the circuit of Fig. 19-29.

Fig. 19-29: See Prob. 47.

PART SEVEN
APPENDICES

Appendix 1
Determinants

In Sec. 3-2 of Chap. 3 we obtained a system of three equations for the four-node circuit of Fig. 3-2:

$$7v_1 - 3v_2 - 4v_3 = -11 \qquad (3\text{-}3)$$

$$-3v_1 + 6v_2 - 2v_3 = 3 \qquad (3\text{-}4)$$

$$-4v_1 - 2v_2 + 11v_3 = 25 \qquad (3\text{-}5)$$

This set of equations could have been solved by a systematic elimination of the variables. This procedure is lengthy, however, and may never yield answers if done unsystematically for a greater number of simultaneous equations. A much more orderly method involves using determinants and Cramer's rule, as discussed in most courses in college algebra. The use of determinants has the additional advantages that it leads naturally into the expression of the circuit elements in terms of matrices, and it establishes a method of analyzing a general circuit which will be helpful in proving general theorems. It should be pointed out that the number of arithmetic steps required to solve a large set of simultaneous equations by determinants is excessive; a digital computer would be programmed to use another method. This appendix consists of a brief review of the determinant method and nomenclature.

Consider (3-3), (3-4), and (3-5). The array of the constant coefficients of the equations is called a *matrix*,

$$[\mathbf{G}] = \begin{bmatrix} 7 & -3 & -4 \\ -3 & 6 & -2 \\ -4 & -2 & 11 \end{bmatrix}$$

where the symbol **G** has been selected since each element of the matrix is a conductance value. A matrix has no "value"; it is merely an ordered array of elements. We use boldface type to represent a matrix, and we enclose the array itself by square brackets. Basic algebraic operations using matrices are discussed in Appendix 2.

The *determinant* of a square matrix has a value, however. To be precise, we should say that the determinant of a matrix *is* a value, but common usage enables us to speak of both the array itself and its value as the determinant. We shall symbolize a determinant by Δ, and employ a suitable subscript to denote the matrix to which the determinant refers. Thus,

$$\Delta_G = \begin{vmatrix} 7 & -3 & -4 \\ -3 & 6 & -2 \\ -4 & -2 & 11 \end{vmatrix}$$

Note that simple vertical lines are used to enclose the determinant.

The value of any determinant is obtained by expanding it in terms of its minors. To do this, we select any row j or any column k, multiply each element in that row or column by its minor and by $(-1)^{j+k}$, and then add these products. The *minor* of the element appearing in both row j and column k is the determinant which is obtained when row j and column k are removed; it is indicated by Δ_{jk}.

As an example, let us expand the determinant Δ_G along column 3. We first multiply the (-4) at the top of this column by $(-1)^{3+1} = 1$ and then by its minor,

$$(-4)(-1)^{3+1}\begin{vmatrix} -3 & 6 \\ -4 & -2 \end{vmatrix}$$

and then repeat for the other two elements in column 3, adding the results,

$$\Delta_G = (-4)\begin{vmatrix} -3 & 6 \\ -4 & -2 \end{vmatrix} - (-2)\begin{vmatrix} 7 & -3 \\ -4 & -2 \end{vmatrix} + 11\begin{vmatrix} 7 & -3 \\ -3 & 6 \end{vmatrix}$$

The minors now contain only two rows and columns. They are of *order* two, and their values are easily determined by expanding in terms of minors again, here a trivial operation. Thus, for the first determinant, we expand along the first column by multiplying (-3) by $(-1)^{1+1}$ and its minor, which is merely the element (-2), and then multiplying (-4) by $(-1)^{2+1}$ and by 6. Thus,

$$\begin{vmatrix} -3 & 6 \\ -4 & -2 \end{vmatrix} = (-3)(-2) - 6(-4) = 30$$

It is usually easier to remember the result for a second-order determinant as "upper left times lower right minus upper right times lower left." Finally,

$$\begin{aligned} \Delta_G &= -4[(-3)(-2) - 6(-4)] + 2[7(-2) - (-3)(-4)] + 11[7(6) - (-3)(-3)] \\ &= -4(30) + 2(-26) + 11(33) \\ &= 191 \end{aligned}$$

For practice, let us expand this same determinant along the first row,

$$\begin{aligned} \Delta_G &= 7\begin{vmatrix} 6 & -2 \\ -2 & 11 \end{vmatrix} - (-3)\begin{vmatrix} -3 & -2 \\ -4 & 11 \end{vmatrix} + (-4)\begin{vmatrix} -3 & 6 \\ -4 & -2 \end{vmatrix} \\ &= 7(62) + 3(-41) - 4(30) \\ &= 191 \end{aligned}$$

The expansion by minors is valid for a determinant of any order.

Repeating these rules for evaluating a determinant in more general terms, we would say, given a matrix [**a**],

$$[\mathbf{a}] = \begin{bmatrix} a_{11} & a_{12} & \cdots & a_{1N} \\ a_{21} & a_{22} & \cdots & a_{2N} \\ \cdots\cdots\cdots\cdots\cdots\cdots\cdots \\ a_{N1} & a_{N2} & \cdots & a_{NN} \end{bmatrix}$$

then $\Delta_{\mathbf{a}}$ may be obtained by expansion in terms of minors along any row j:

$$\Delta_{\mathbf{a}} = a_{j1}(-1)^{j+1}\,\Delta_{j1} + a_{j2}(-1)^{j+2}\,\Delta_{j2} + \cdots + a_{jN}(-1)^{j+N}\,\Delta_{jN}$$

$$= \sum_{n=1}^{N} a_{jn}(-1)^{j+n}\,\Delta_{jn}$$

or along any column k:

$$\Delta_{\mathbf{a}} = a_{1k}(-1)^{1+k}\,\Delta_{1k} + a_{2k}(-1)^{2+k}\,\Delta_{2k} + \cdots + a_{Nk}(-1)^{N+k}\,\Delta_{Nk}$$

$$= \sum_{n=1}^{N} a_{nk}(-1)^{n+k}\,\Delta_{nk}$$

The *cofactor* C_{jk} of the element appearing in both row j and column k is simply $(-1)^{j+k}$ times the minor Δ_{jk}. Thus, $C_{11} = \Delta_{11}$, but $C_{12} = -\Delta_{12}$. We may now write

$$\Delta_{\mathbf{a}} = \sum_{n=1}^{N} a_{jn} C_{jn} = \sum_{n=1}^{N} a_{nk} C_{nk}$$

As an example, let us consider this fourth-order determinant:

$$\Delta = \begin{vmatrix} 2 & -1 & -2 & 0 \\ -1 & 4 & 2 & -3 \\ -2 & -1 & 5 & -1 \\ 0 & -3 & 3 & 2 \end{vmatrix}$$

We find

$$\Delta_{11} = \begin{vmatrix} 4 & 2 & -3 \\ -1 & 5 & -1 \\ -3 & 3 & 2 \end{vmatrix} = 4(10 + 3) + 1(4 + 9) - 3(-2 + 15) = 26$$

$$\Delta_{12} = \begin{vmatrix} -1 & 2 & -3 \\ -2 & 5 & -1 \\ 0 & 3 & 2 \end{vmatrix} = -1(10 + 3) + 2(4 + 9) + 0 = 13$$

and $C_{11} = 26$, whereas $C_{12} = -13$. Finding the value of Δ for practice, we have

$$\Delta = 2C_{11} + (-1)C_{12} + (-2)C_{13} + 0$$

$$= 2(26) + (-1)(-13) + (-2)(3) + 0 = 59$$

We next consider Cramer's rule, which enables us to find the values of the unknown variables. Let us again consider (3-3), (3-4), and (3-5); we define the determinant Δ_1 as that determinant which is obtained when the first column of Δ_G is replaced by the three constants on the right sides of the three equations. Thus,

$$\Delta_1 = \begin{vmatrix} -11 & -3 & -4 \\ 3 & 6 & -2 \\ 25 & -2 & 11 \end{vmatrix}$$

We expand along the first column:

$$\Delta_1 = -11 \begin{vmatrix} 6 & -2 \\ -2 & 11 \end{vmatrix} - 3 \begin{vmatrix} -3 & -4 \\ -2 & 11 \end{vmatrix} + 25 \begin{vmatrix} -3 & -4 \\ 6 & -2 \end{vmatrix}$$

$$= -682 + 123 + 750 = 191$$

Cramer's rule then states that

$$v_1 = \frac{\Delta_1}{\Delta_G} = \frac{191}{191} = 1 \quad \text{V}$$

and

$$v_2 = \frac{\Delta_2}{\Delta_G} = \frac{\begin{vmatrix} 7 & -11 & -4 \\ -3 & 3 & -2 \\ -4 & 25 & 11 \end{vmatrix}}{191} = \frac{581 - 63 - 136}{191} = 2 \quad \text{V}$$

and finally,

$$v_3 = \frac{\Delta_3}{\Delta_G} = \frac{\begin{vmatrix} 7 & -3 & -11 \\ -3 & 6 & 3 \\ -4 & -2 & 25 \end{vmatrix}}{191} = \frac{1092 - 291 - 228}{191} = 3 \quad \text{V}$$

Cramer's rule is applicable to a system of N simultaneous linear equations in N unknowns; for the ith variable v_i:

$$v_i = \frac{\Delta_i}{\Delta_G}$$

Drill Problem

A1-1. Evaluate: (a) $\begin{vmatrix} 2 & -1 \\ -3 & 7 \end{vmatrix}$; (b) $\begin{vmatrix} 1 & 2 & 0 \\ 3 & 2 & -1 \\ 4 & -3 & 7 \end{vmatrix}$; (c) $\begin{vmatrix} -2 & 0 & 3 & 1 \\ 1 & 1 & 2 & 0 \\ 2 & 3 & 2 & -1 \\ -1 & 4 & -3 & 7 \end{vmatrix}$

(d) Find i_2 if $4i_1 - 2i_2 - 3i_3 = 100$, $-2i_1 + 5i_2 - i_3 - 2i_4 = 0$, $-i_2 + 4i_3 + i_4 = 50$, and $-5i_1 - 2i_2 - 3i_3 + 2i_4 = 0$.   *Ans:* 11; -39; 105; 149.2

Appendix 2
Matrices

We defined a matrix in Appendix 1 as an ordered array of elements, such as the conductance matrix $[\mathbf{G}]$,

$$[\mathbf{G}] = \begin{bmatrix} 7 & -3 & -4 \\ -3 & 6 & -2 \\ -4 & -2 & 11 \end{bmatrix}$$

We use boldface type to represent a matrix and enclose the symbol within square brackets, as we do the array itself.

A matrix having m rows and n columns is called an (m by n) matrix. Thus,

$$[\mathbf{A}] = \begin{bmatrix} 2 & 0 & 5 \\ -1 & 6 & 3 \end{bmatrix}$$

is a (2 by 3) matrix, and the $[\mathbf{G}]$ matrix above is a (3 by 3) matrix. An (n by n) matrix is a *square matrix* of order n.

An (m by 1) matrix is called a *column matrix* or a *vector*. Thus,

$$[\mathbf{V}] = \begin{bmatrix} \mathbf{V}_1 \\ \mathbf{V}_2 \end{bmatrix}$$

is a (2 by 1) column matrix of phasor voltages, and

$$[\mathbf{I}] = \begin{bmatrix} \mathbf{I}_1 \\ \mathbf{I}_2 \end{bmatrix}$$

is a (2 by 1) phasor-current vector.

A (1 by n) matrix is known as a *row vector*.

Two (m by n) matrices are equal if their corresponding elements are equal. Thus, if a_{jk} is that element of $[\mathbf{A}]$ located in row j and column k, and b_{jk} is the element at row j and column k in matrix $[\mathbf{B}]$, then $[\mathbf{A}] = [\mathbf{B}]$ if and only if $a_{jk} = b_{jk}$ for all $1 \leq j \leq m$ and $1 \leq k \leq n$. Thus, if

$$\begin{bmatrix} \mathbf{V}_1 \\ \mathbf{V}_2 \end{bmatrix} = \begin{bmatrix} \mathbf{z}_{11}\mathbf{I}_1 + \mathbf{z}_{12}\mathbf{I}_2 \\ \mathbf{z}_{21}\mathbf{I}_1 + \mathbf{z}_{22}\mathbf{I}_2 \end{bmatrix}$$

then $\mathbf{V}_1 = \mathbf{z}_{11}\mathbf{I}_1 + \mathbf{z}_{12}\mathbf{I}_2$ *and* $\mathbf{V}_2 = \mathbf{z}_{21}\mathbf{I}_1 + \mathbf{z}_{22}\mathbf{I}_2$.

Two (m by n) matrices may be added by adding corresponding elements. Thus,

$$\begin{bmatrix} 2 & 0 & 5 \\ -1 & 6 & 3 \end{bmatrix} + \begin{bmatrix} 1 & 2 & 3 \\ -3 & -2 & -1 \end{bmatrix} = \begin{bmatrix} 3 & 2 & 8 \\ -4 & 4 & 2 \end{bmatrix}$$

An (n by n) square matrix has a determinant whose value is determined by the rules given for expanding determinants in Appendix 1. The determinant of $[\mathbf{G}]$ is symbolized by $\Delta_\mathbf{G}$ or $\det[\mathbf{G}]$. Therefore, if

$$[\mathbf{G}] = \begin{bmatrix} 7 & -3 & -4 \\ -3 & 6 & -2 \\ -4 & -2 & 11 \end{bmatrix}$$

then

$$\Delta_\mathbf{G} = \det[\mathbf{G}] = \begin{vmatrix} 7 & -3 & -4 \\ -3 & 6 & -2 \\ -4 & -2 & 11 \end{vmatrix}$$

Expanding on the first column, we find

$$\det[\mathbf{G}] = 7(66 - 4) + 3(-33 - 8) - 4(6 + 24) = 191$$

There is only one other matrix operation that is needed in our work, matrix multiplication. Let us consider the matrix product $[\mathbf{A}][\mathbf{B}]$, where $[\mathbf{A}]$ is an (m by n) matrix and $[\mathbf{B}]$ is a (p by q) matrix. If $n = p$, the matrices are said to be *conformal,* and their product exists. That is, matrix multiplication is only defined for the case where the number of columns of the first matrix in the product is equal to the number of rows in the second matrix.

The formal definition of matrix multiplication states that the product of the (m by n) matrix $[\mathbf{A}]$ times the (n by q) matrix $[\mathbf{B}]$ is an (m by q) matrix having elements c_{jk}, $1 \leq j \leq m$ and $1 \leq k \leq q$, where

$$c_{jk} = a_{j1}b_{1k} + a_{j2}b_{2k} + \cdots + a_{jn}b_{nk}$$

That is, to find the element in the second row and third column of the product, we multiply each of the elements in the second row of $[\mathbf{A}]$ by the corresponding element in the third column of $[\mathbf{B}]$ and then add the n results. For example, given the (2 by 3) matrix $[\mathbf{A}]$ and the (3 by 2) matrix $[\mathbf{B}]$,

$$\begin{bmatrix} a_{11} & a_{12} & a_{13} \\ a_{21} & a_{22} & a_{23} \end{bmatrix} \begin{bmatrix} b_{11} & b_{12} \\ b_{21} & b_{22} \\ b_{31} & b_{32} \end{bmatrix}$$

$$= \begin{bmatrix} (a_{11}b_{11} + a_{12}b_{21} + a_{13}b_{31}) & (a_{11}b_{12} + a_{12}b_{22} + a_{13}b_{32}) \\ (a_{21}b_{11} + a_{22}b_{21} + a_{23}b_{31}) & (a_{21}b_{12} + a_{22}b_{22} + a_{23}b_{32}) \end{bmatrix}$$

The result is a (2 by 2) matrix.

As a numerical example of matrix multiplication, we take

$$\begin{bmatrix} 3 & 2 & 1 \\ -2 & -2 & 4 \end{bmatrix} \begin{bmatrix} 2 & 3 \\ -2 & -1 \\ 4 & -3 \end{bmatrix} = \begin{bmatrix} 6 & 4 \\ 16 & -16 \end{bmatrix}$$

where $6 = (3)(2) + (2)(-2) + (1)(4)$, $4 = (3)(3) + (2)(-1) + (1)(-3)$, and so forth.

Matrix multiplication is not commutative. For example, given the (3 by 2) matrix $[\mathbf{C}]$ and the (2 by 1) matrix $[\mathbf{D}]$, it is evident that the product $[\mathbf{C}][\mathbf{D}]$ may be calculated, but the product $[\mathbf{D}][\mathbf{C}]$ is not even defined.

As a final example, let

$$[\mathbf{t}_A] = \begin{bmatrix} 2 & 3 \\ -1 & 4 \end{bmatrix}$$

and

$$[\mathbf{t}_B] = \begin{bmatrix} 3 & 1 \\ 5 & 0 \end{bmatrix}$$

so that both $[\mathbf{t}_A][\mathbf{t}_B]$ and $[\mathbf{t}_B][\mathbf{t}_A]$ are defined. However,

$$[\mathbf{t}_A][\mathbf{t}_B] = \begin{bmatrix} 21 & 2 \\ 17 & -1 \end{bmatrix}$$

while

$$[\mathbf{t}_B][\mathbf{t}_A] = \begin{bmatrix} 5 & 13 \\ 10 & 15 \end{bmatrix}$$

Drill Problem

A2-1. Given $[\mathbf{A}] = \begin{bmatrix} 4 & -1 \\ -2 & 3 \end{bmatrix}$, $[\mathbf{B}] = \begin{bmatrix} 1 & -3 \\ 3 & 5 \end{bmatrix}$, $[\mathbf{C}] = \begin{bmatrix} 30 \\ 50 \end{bmatrix}$, and $[\mathbf{V}] = \begin{bmatrix} \mathbf{V}_1 \\ \mathbf{V}_2 \end{bmatrix}$,

find: (*a*) $[\mathbf{A}] + [\mathbf{B}]$; (*b*) $[\mathbf{A}][\mathbf{B}]$; (*c*) $[\mathbf{B}][\mathbf{A}]$; (*d*) $[\mathbf{A}][\mathbf{V}]$; (*e*) $[\mathbf{A}][\mathbf{V}] + [\mathbf{B}][\mathbf{C}]$; (*f*) $[\mathbf{A}]^2 = [\mathbf{A}][\mathbf{A}]$; (*g*) det $[\mathbf{A}]$.

Ans: $\begin{bmatrix} 5 & -4 \\ 1 & 8 \end{bmatrix}$; $\begin{bmatrix} 1 & -17 \\ 7 & 21 \end{bmatrix}$; $\begin{bmatrix} 10 & -10 \\ 2 & 12 \end{bmatrix}$; $\begin{bmatrix} 4\mathbf{V}_1 - \mathbf{V}_2 \\ -2\mathbf{V}_1 + 3\mathbf{V}_2 \end{bmatrix}$; $\begin{bmatrix} 4\mathbf{V}_1 - \mathbf{V}_2 - 120 \\ -2\mathbf{V}_1 + 3\mathbf{V}_2 + 340 \end{bmatrix}$;

$\begin{bmatrix} 18 & -7 \\ -14 & 11 \end{bmatrix}$; 10

Appendix 3
A Proof of Thévenin's Theorem

We shall prove Thévenin's theorem in the same form in which it is stated in Sec. 3-6 of Chap. 3, repeated here for reference:

> Given any linear circuit, rearrange it in the form of two networks A and B that are connected together by two resistanceless conductors. If either network contains a dependent source, its control variable must be in that same network. Define a voltage v_{oc} as the open-circuit voltage which would appear across the terminals of A if B were disconnected so that no current is drawn from A. Then all the currents and voltages in B will remain unchanged if A is killed (all independent voltage sources and independent current sources in A replaced by short circuits and open circuits, respectively) and an independent voltage source v_{oc} is connected, with proper polarity, in series with the dead (inactive) A network.

We shall effect our proof by showing that the original A network and the Thévenin equivalent of the A network both cause the same current to flow into the terminals of the B network. If the currents are the same, then the voltages must be the same; in other words, if we apply a certain current, which we might think of as a current source, to the B network, then the current source and the B network constitute a circuit which has a specific input voltage as a response. Thus, the current determines the voltage. Alternatively we could, if we wished, show that the terminal voltage at B is unchanged, because the voltage also determines the current uniquely. If the input voltage and current to the B network are unchanged, then it follows that the currents and voltages *throughout* the B network are also unchanged.

Let us first prove the theorem for the case where the B network is inactive (no independent sources). After this step has been accomplished, we may

Fig. A3-1: (a) A general linear network A and a network B that contains no independent sources. Controls for dependent sources must appear in the same part of the network. (b) The Thévenin source is inserted in the circuit and adjusted until $i = 0$. No voltage appears across network B and thus $v_x = v_{oc}$. The Thévenin source thus produces a current $-i$ while network A provides i. (c) The Thévenin source is reversed and network A is killed. The current is therefore i.

then use the superposition principle to extend the theorem to include B networks which contain independent sources. Each network may contain dependent sources, providing their control variables are in the same network.

The current i, flowing in the upper conductor from the A network to the B network in Fig. A3-1a, is therefore caused entirely by the independent sources present in the A network. Suppose now that we add an additional voltage source v_x, which we shall call the Thévenin source, in the conductor in which i is measured, as shown in Fig. A3-1b, and then adjust the magnitude and time variation of v_x until the current is reduced to zero. By our definition of v_{oc}, then, the voltage across the terminals of A must be v_{oc} since $i = 0$. B contains no independent sources, and no current is entering its terminals; therefore, there is no voltage across the terminals of the B network, and by Kirchhoff's voltage law the voltage of the Thévenin source is v_{oc} volts, $v_x = v_{oc}$. Moreover, since the Thévenin source and the A network jointly deliver no current to B, and since the A network by itself delivers a current i, then superposition requires that the Thévenin source acting by itself deliver a current of $-i$ to B. The source acting alone in a reversed direction, as shown in Fig. A3-1c, therefore produces a current i in the upper lead. This situation, however, is the same as the conclusion reached by Thévenin's theorem: the Thévenin source v_{oc} acting in series with the inactive A network is equivalent to the given network.

Now let us consider the case where the B network may be an active network. We now think of the current i, flowing from the A network to the B network in the upper conductor, as being composed of two parts i_A and i_B, where i_A is the current produced by A acting alone and the current i_B is due to B acting alone. Our ability to divide the current into these two components is a direct consequence of the applicability of the superposition principle to

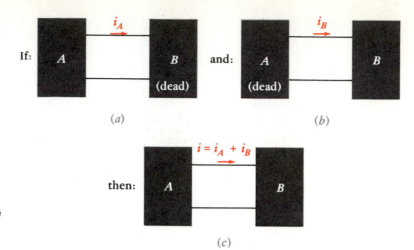

Fig. A3-2: Superposition enables the current i to be considered as the sum of two partial responses.

these two linear networks; the complete response and the two partial responses are indicated by the diagrams of Fig. A3-2.

The partial response i_A has already been considered; if network B is inactive, we know that network A may be replaced by the Thévenin source and the inactive A network. In other words, of the three sources which we must keep in mind, those in A, those in B, and the Thévenin source, the partial response i_A occurs when A and B are dead and the Thévenin source is active. Preparing for the use of superposition, we now let A remain inactive, but turn on B and turn off the Thévenin source; by definition, the partial response i_B is obtained. Superimposing the results, the response when A is dead and both the Thévenin source and B are active is $i_A + i_B$. This sum is the original current i, and the situation wherein the Thévenin source and B are active, but A is dead, is the desired Thévenin equivalent circuit. Thus the active network A may be replaced by its Thévenin source, the open-circuit voltage, in series with the inactive A network, regardless of the status of the B network; it may be either active or inactive.

Appendix 4
Complex Numbers

This appendix includes sections covering the definition of a complex number, the basic arithmetic operations for complex numbers, Euler's identity, and the exponential and polar forms of the complex number. We first introduce the concept of a complex number.

A4-1
The complex number

Our early training in mathematics dealt exclusively with real numbers, such as 4, $-\frac{2}{7}$, and π. Soon, however, we began to encounter algebraic equations, such as $x^2 = -3$, which could not be satisfied by any real number. Such an equation can be solved only through the introduction of the *imaginary unit* or the *imaginary operator,* which we shall designate[1] by the symbol j. By definition, $j^2 = -1$, and thus $j = \sqrt{-1}$, $j^3 = -j$, $j^4 = 1$, and so forth. The product of a real number times the imaginary operator is called an *imaginary number,* and the sum of a real number and an imaginary number is called a *complex number*. Thus, a number having the form $a + jb$, where a and b are real numbers, is a complex number.[2]

We shall designate a complex number by means of a special single symbol; thus, $\mathbf{A} = a + jb$. The complex nature of the number is indicated by the use of boldface type; in handwritten material, a bar over the letter is customary. The complex number $\mathbf{A}$ above is described as having a *real component* or *real part a* and an *imaginary component* or *imaginary part b*. This is also expressed as

$$\text{Re}[\mathbf{A}] = a \qquad \text{Im}[\mathbf{A}] = b$$

The imaginary component of $\mathbf{A}$ is *not jb*. By definition, the imaginary component is a real number.

---

[1] The mathematicians designate the imaginary operator by the symbol i, but it is customary to use j in electrical engineering in order to avoid confusion with the symbol for current.

[2] The choice of the words imaginary and complex is unfortunate. They are used here and in the mathematical literature as technical terms to designate a class of numbers. To interpret imaginary as "not pertaining to the physical world" or complex as "complicated" is neither justified nor intended.

It should be noted that all real numbers may be regarded as complex numbers having imaginary parts equal to zero. The real numbers are therefore included in the system of complex numbers, and we may now consider them as a special case. When we define the fundamental arithmetic operations for complex numbers, we should therefore expect them to reduce to the corresponding definitions for real numbers if the imaginary part of every complex number is set equal to zero.

Since any complex number is completely characterized by a pair of real numbers, such as a and b in the example above, we can obtain some visual assistance by representing a complex number graphically on a rectangular, or cartesian, coordinate system. By providing ourselves with a real axis and an imaginary axis, as shown in Fig. A4-1, we form a *complex plane* or *Argand diagram* on which any complex number can be represented as a single point. The complex numbers $\mathbf{M} = 3 + j1$ and $\mathbf{N} = 2 - j2$ are indicated. It is important to understand that this complex plane is only a visual aid; it is not at all essential to the mathematical statements which follow.

We shall define two complex numbers as being equal if, and only if, their real parts are equal and their imaginary parts are equal. Graphically, then, to each point in the complex plane there corresponds only one complex number, and conversely, to each complex number there corresponds only one point in the complex plane. Thus, suppose we are given the two complex numbers:

$$\mathbf{A} = a + jb \qquad \text{and} \qquad \mathbf{B} = c + jd$$

Then, if

$$\mathbf{A} = \mathbf{B}$$

it is necessary that

$$a = c \qquad \text{and} \qquad b = d$$

A complex number expressed as the sum of a real number and an imagi-

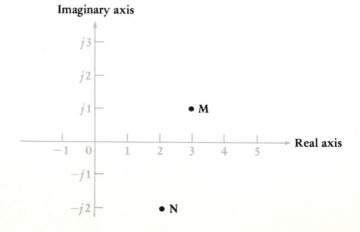

Fig. A4-1: The complex numbers $\mathbf{M} = 3 + j1$ and $\mathbf{N} = 2 - j2$ are shown on the complex plane.

nary number, such as $\mathbf{A} = a + jb$, is said to be in *rectangular* or *cartesian* form. Other forms for a complex number will appear shortly.

Let us now define the fundamental operations of addition, subtraction, multiplication, and division for complex numbers. The sum of two complex numbers is defined as the complex number whose real part is the sum of the real parts of the two complex numbers and whose imaginary part is the sum of the imaginary parts of the two complex numbers. Thus,

$$(a + jb) + (c + jd) = (a + c) + j(b + d)$$

For example,

$$(3 + j4) + (4 - j2) = 7 + j2$$

The difference of two complex numbers is taken in a similar manner; for example,

$$(3 + j4) - (4 - j2) = -1 + j6$$

Addition and subtraction of complex numbers may also be accomplished graphically on the complex plane. Each complex number is represented as a vector, or directed line segment, and the sum is obtained by completing the parallelogram, illustrated by Fig. A4-2a, or by connecting the vectors in a head-to-tail manner, as shown in Fig. A4-2b. A graphical sketch is often useful as a check for a more exact numerical solution.

The product of two complex numbers is defined by

$$(a + jb)(c + jd) = (ac - bd) + j(bc + ad)$$

This result may be easily obtained by a direct multiplication of the two binomial terms, using the rules of the algebra of real numbers, and then simplifying the result by letting $j^2 = -1$. For example,

$$(3 + j4)(4 - j2) = 12 - j6 + j16 - 8j^2$$
$$= 12 + j10 + 8$$
$$= 20 + j10$$

Fig. A4-2: (a) The sum of the complex numbers $\mathbf{M} = 3 + j1$ and $\mathbf{N} = 2 - j2$ is obtained by constructing a parallelogram. (b) The sum of the same two complex numbers is found by a head-to-tail combination.

(a)

(b)

It is easier to multiply the complex numbers by this method, particularly if we immediately replace j^2 by -1, than it is to substitute into the general formula which defines the multiplication.

Before defining the operation of division for complex numbers, we should define the conjugate of a complex number. The *conjugate* of the complex number $\mathbf{A} = a + jb$ is $a - jb$ and is represented as $\mathbf{A}^*$. The conjugate of any complex number is therefore easily obtained by merely changing the sign of the imaginary part of the complex number. Thus, if

$$\mathbf{A} = 5 + j3$$

then

$$\mathbf{A}^* = 5 - j3$$

It is evident that the conjugate of any complicated complex expression may be found by replacing every complex term in the expression by its conjugate, which may be obtained by replacing every j in the expression by $-j$.

The definitions of addition, subtraction, and multiplication show that the following statements are true: the sum of a complex number and its conjugate is a real number; the difference of a complex number and its conjugate is an imaginary number; and the product of a complex number and its conjugate is a real number. It is also evident that if $\mathbf{A}^*$ is the conjugate of $\mathbf{A}$, then $\mathbf{A}$ is the conjugate of $\mathbf{A}^*$; in other words, $\mathbf{A} = (\mathbf{A}^*)^*$. A complex number and its conjugate are said to form a *conjugate complex pair* of numbers.

We now define the quotient of two complex numbers:

$$\frac{\mathbf{A}}{\mathbf{B}} = \frac{(\mathbf{A})(\mathbf{B}^*)}{(\mathbf{B})(\mathbf{B}^*)}$$

and thus

$$\frac{a + jb}{c + jd} = \frac{(ac + bd) + j(bc - ad)}{c^2 + d^2}$$

We multiply numerator and denominator by the conjugate of the denominator in order to obtain a denominator which is real; this process is called *rationalizing the denominator*. As a numerical example,

$$\frac{3 + j4}{4 - j2} = \frac{(3 + j4)(4 + j2)}{(4 - j2)(4 + j2)}$$

$$= \frac{4 + j22}{16 + 4}$$

$$= 0.2 + j1.1$$

The addition or subtraction of two complex numbers which are each expressed in rectangular form is a relatively simple operation; multiplication or division of two complex numbers in rectangular form, however, is a rather unwieldy process. These latter two operations will be found to be much simpler when the complex numbers are given in either exponential or polar form. These forms will be introduced in Secs. A4-3 and A4-4.

Drill Problems

A4-1. Given $\mathbf{A} = -5 + j2$, $\mathbf{B} = 4 - j3$, and $\mathbf{C} = -2 - j4$, find: (a) $\mathbf{C} - \mathbf{B}$; (b) $2\mathbf{A} - 3\mathbf{B} + 4\mathbf{C}$; (c) $j^5\mathbf{C}^2(\mathbf{A} + \mathbf{B})$; (d) $\mathbf{B}$ Re[$\mathbf{A}$] + $\mathbf{A}$ Im[$\mathbf{B}$].

Ans: $-6 - j1$; $-30 - j3$; $4 + j28$; $-5 + j9$

A4-2. For the $\mathbf{A}$, $\mathbf{B}$, $\mathbf{C}$ given above, find: (a) $[(\mathbf{A} - \mathbf{A}^*)(\mathbf{B} + \mathbf{B}^*)^*]^*$; (b) $(1/\mathbf{C}) - (1/\mathbf{B})^*$; (c) $(\mathbf{B} + \mathbf{C})/(2\mathbf{BC})$. *Ans: $-j32$; $-0.260 + j0.320$; $0.0300 + j0.1600$*

A4-2
Euler's identity

In Chap. 9 we begin to encounter functions of time which contain complex numbers, and we are concerned with the differentiation and integration of these functions with respect to the real variable t. We differentiate and integrate such functions with respect to t by exactly the same procedures we use for real functions of time. That is, the complex constants are treated just as though they were real constants when performing the operations of differentiation or integration. If $\mathbf{f}(t)$ is a complex function of time, such as

$$\mathbf{f}(t) = a \cos ct + jb \sin ct$$

then

$$\frac{d\mathbf{f}(t)}{dt} = -ac \sin ct + jbc \cos ct$$

and

$$\int \mathbf{f}(t)\,dt = \frac{a}{c} \sin ct - j\frac{b}{c} \cos ct + \mathbf{C}$$

where the constant of integration $\mathbf{C}$ is a complex number in general.

In Chap. 19 it is necessary to differentiate or integrate a function of a complex variable with respect to that complex variable. In general, the successful accomplishment of either of these operations requires that the function to be differentiated or integrated satisfies certain conditions. All our functions do meet these conditions, and integration or differentiation with respect to a complex variable is achieved by using methods identical to those used for real variables.

At this time we must make use of a very important fundamental relationship known as Euler's identity (pronounced "oilers"). We shall prove this identity, for it is extremely useful in representing a complex number in a form other than rectangular form.

The proof is based on the power-series expansions of $\cos\theta$, $\sin\theta$, and e^z, given towards the back of your favorite college calculus text:

$$\cos\theta = 1 - \frac{\theta^2}{2!} + \frac{\theta^4}{4!} - \frac{\theta^6}{6!} + \cdots$$

$$\sin\theta = \theta - \frac{\theta^3}{3!} + \frac{\theta^5}{5!} - \frac{\theta^7}{7!} + \cdots$$

or
$$\cos \theta + j\sin \theta = 1 + j\theta - \frac{\theta^2}{2!} - j\frac{\theta^3}{3!} + \frac{\theta^4}{4!} + j\frac{\theta^5}{5!} - \cdots$$

and.
$$e^z = 1 + z + \frac{z^2}{2!} + \frac{z^3}{3!} + \frac{z^4}{4!} + \cdots$$

so that
$$e^{j\theta} = 1 + j\theta - \frac{\theta^2}{2!} - j\frac{\theta^3}{3!} + \frac{\theta^4}{4!} + \cdots$$

We conclude that

$$e^{j\theta} = \cos \theta + j \sin \theta \qquad (1)$$

or, if we let $z = -j\theta$, we find that

$$e^{-j\theta} = \cos \theta - j \sin \theta \qquad (2)$$

By adding and subtracting (1) and (2), we obtain the two expressions which we used without proof in our study of the underdamped natural response of the parallel and series RLC circuits,

$$\cos \theta = \tfrac{1}{2}(e^{j\theta} + e^{-j\theta}) \qquad (3)$$

$$\sin \theta = -j\tfrac{1}{2}(e^{j\theta} - e^{-j\theta}) \qquad (4)$$

Drill Problems

A4-3. Use (1) through (4) to evaluate: (a) e^{j2}; (b) e^{1+j2}; (c) $\cos (j2)$; (d) $\sin (j2)$.   *Ans:* $-0.416 + j0.909; -1.131 + j2.47; 3.76 + j0; 0 + j3.63$

A4-4. Evaluate at $t = 0.5$: (a) $(d/dt)(2 \cos 3t - j3 \sin 3t)$; (b) $\int_0^t(2 \cos 3t - j3 \sin 3t)dt$. Evaluate at $s = 1 + j2$: (c) $(d/ds)[2/(s + 3)]$; (d) $\int_s^\infty s^{-2}ds$.   *Ans:* $-5.98 - j0.637; 0.665 - j0.929; -0.0600 + j0.0800; 0.200 - j0.400$

**A4-3
The exponential
form**

Let us now take Euler's identity

$$e^{j\theta} = \cos \theta + j \sin \theta$$

and multiply each side by the real positive number C,

$$Ce^{j\theta} = C \cos \theta + jC \sin \theta \qquad (5)$$

The right side of (5) consists of the sum of a real number and an imaginary number and thus represents a complex number in rectangular form; let us call this complex number $\mathbf{A}$, where $\mathbf{A} = a + jb$. By equating the real parts

$$a = C \cos \theta \qquad (6)$$

and the imaginary parts

$$b = C \sin \theta \qquad (7)$$

squaring and adding (6) and (7),

$$a^2 + b^2 = C^2$$

or
$$C = \pm \sqrt{a^2 + b^2} \qquad (8)$$

and dividing (7) by (6),

$$\frac{b}{a} = \tan \theta$$

or

$$\theta = \tan^{-1} \frac{b}{a} \tag{9}$$

we obtain the relationships (8) and (9), which enable us to determine C and θ from a knowledge of a and b. For example, if $\mathbf{A} = 4 + j2$, then we identify a as 4 and b as 2 and find C and θ:

$$C = \sqrt{4^2 + 2^2} = 4.47$$

$$\theta = \tan^{-1} \tfrac{2}{4} = 26.6°$$

We could use this new information to write $\mathbf{A}$ in the form

$$\mathbf{A} = 4.47 \cos 26.6° + j4.47 \sin 26.6°$$

but it is the form of the left side of (5) which will prove to be the more useful:

$$\mathbf{A} = Ce^{j\theta} = 4.47e^{j26.6°}$$

A complex number expressed in this manner is said to be in *exponential form*. The real positive multiplying factor C is known as the *amplitude* or *magnitude* and the real quantity θ appearing in the exponent is called the *argument* or *angle*. A mathematician would always express θ in radians and would write

$$\mathbf{A} = 4.47e^{j0.464}$$

but electrical engineers customarily work in terms of degrees. The use of the degree symbol (°) in the exponent should make confusion impossible.

To recapitulate, if we have a complex number which is given in rectangular form,

$$\mathbf{A} = a + jb$$

and wish to express it in exponential form,

$$\mathbf{A} = Ce^{j\theta}$$

we may find C and θ by (8) and (9). If we are given the complex number in exponential form, then we may find a and b by (6) and (7).

When $\mathbf{A}$ is expressed in terms of numerical values, the transformation between exponential (or polar) and rectangular forms is available as a built-in operation on most handheld scientific calculators.

One question will be found to arise in the determination of the angle θ by using the arc-tangent relationship (9). This function is multivalued, and an appropriate angle must be selected from various possibilities. One method by which the choice may be made is to select an angle for which the sine and cosine have the proper signs to produce the required values of a and b from (6) and (7). For example, let us convert

$$\mathbf{V} = 4 - j3$$

Fig. A4-3: A complex number may be represented by a point in the complex plane through choosing the correct real and imaginary parts from the rectangular form, or by selecting the magnitude and angle from the exponential form.

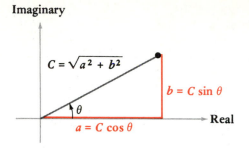

to exponential form. The amplitude is

$$C = \sqrt{4^2 + (-3)^2} = 5$$

and the angle is

$$\theta = \tan^{-1} \frac{-3}{4} \tag{10}$$

A value of θ has to be selected which leads to a positive value for $\cos \theta$, since $4 = 5 \cos \theta$, and a negative value for $\sin \theta$, since $-3 = 5 \sin \theta$. We therefore obtain $\theta = -36.9°$, $323.1°$, $-396.9°$, and so forth. Any of these angles is correct, and we usually select that one which is the simplest, here $-36.9°$. We should note that the solution of (10), $\theta = 143.1°$, is not correct because $\cos \theta$ is negative and $\sin \theta$ is positive.

A simpler method of selecting the correct angle is available if we represent the complex number graphically in the complex plane. Let us first select a complex number, given in rectangular form, $\mathbf{A} = a + jb$, which lies in the first quadrant of the complex plane, as illustrated in Fig. A4-3. If we draw a line from the origin to the point which represents the complex number, we shall have constructed a right triangle whose hypotenuse is evidently the amplitude of the exponential representation of the complex number. In other words, $C = \sqrt{a^2 + b^2}$. Moreover, the counterclockwise angle which the line makes with the positive real axis is seen to be the angle θ of the exponential representation, because $a = C \cos \theta$ and $b = C \sin \theta$. Now if we are given the rectangular form of a complex number which lies in another quadrant, such as $\mathbf{V} = 4 - j3$, which is depicted in Fig. A4-4, the correct angle is graphically

Fig. A4-4: The complex number $\mathbf{V} = 4 - j3 = 5e^{-j36.9°}$ is represented in the complex plane.

evident, either $-36.9°$ or $323.1°$ for this example. The sketch may often be visualized and need not be drawn.

If the rectangular form of the complex number has a negative real part, it is often easier to work with the negative of the complex number, thus avoiding angles greater than 90° in magnitude. For example, given

$$\mathbf{I} = -5 + j2$$

we write

$$\mathbf{I} = -(5 - j2)$$

and then transform $(5 - j2)$ to exponential form,

$$\mathbf{I} = -Ce^{j\theta}$$

where

$$C = \sqrt{29} = 5.39 \quad \text{and} \quad \theta = \tan^{-1}\frac{-2}{5} = -21.8°$$

We therefore have

$$\mathbf{I} = -5.39e^{-j21.8°}$$

The negative sign may be removed from the complex number by increasing or decreasing the angle by 180°, as shown by reference to a sketch in the complex plane. Thus, the result may be expressed in exponential form as

$$\mathbf{I} = 5.39e^{j158.2°} \quad \text{or} \quad \mathbf{I} = 5.39e^{-j201.8°}$$

Note that use of an electronic calculator in the inverse tangent mode always yields angles having magnitudes less than 90°. Thus, $\tan^{-1}[(-3)/4]$ and $\tan^{-1}[3/(-4)]$ both come out as $-36.9°$. Calculators that provide rectangular-to-polar conversion, however, give the correct angle in all cases.

One last remark about the exponential representation of a complex number should be made. Two complex numbers, both written in exponential form, are equal if, and only if, their amplitudes are equal and their angles are equivalent. Equivalent angles are those which differ by multiples of 360°. For example, if $\mathbf{A} = Ce^{j\theta}$ and $\mathbf{B} = De^{j\phi}$, then if $\mathbf{A} = \mathbf{B}$, it is necessary that $C = D$ and $\theta = \phi \pm 360°n$, where $n = 0, 1, 2, 3, \ldots$.

Drill Problems

A4-5. Express each of the following complex numbers in exponential form, using an angle lying in the range, $-180° < \theta \le 180°$: (a) $-47.2 + j28.9$; (b) $-27.6 - j43.2$; (c) $15.62 - j38.7$.   *Ans:* $55.3e^{j148.5°}$; $51.3e^{-j122.6°}$; $41.7e^{-j68.0°}$

A4-6. Express in rectangular form: (a) $20.6e^{-j3}$; (b) $26.1e^{-j102.3°}$; (c) $-30.3e^{j121°}$.   *Ans:* $-20.4 - j2.91$; $-5.56 - j25.5$; $15.61 - j26.0$

A4-4
The polar form

The third (and last) form in which we may represent a complex number is essentially the same as the exponential form, except for a slight difference in symbolism. We use an angle sign ($\underline{/}$) to replace the combination (e^j). Thus, the exponential representation of a complex number $\mathbf{A}$,

$$\mathbf{A} = Ce^{j\theta}$$

may be written somewhat more concisely as

$$\mathbf{A} = C\underline{/\theta}$$

The complex number is now said to be expressed in *polar* form, a name which suggests the representation of a point in a (complex) plane through the use of polar coordinates.

It is apparent that transformation from rectangular to polar form or from polar form to rectangular form is basically the same as transformation between rectangular and exponential form. The same relationships exist between C, θ, a, and b.

The complex number

$$\mathbf{A} = -2 + j5$$

is thus written in exponential form as

$$\mathbf{A} = 5.39e^{j111.8°}$$

and in polar form as

$$\mathbf{A} = 5.39\underline{/111.8°}$$

In order to appreciate the utility of the exponential and polar forms, let us consider the multiplication and division of two complex numbers represented in exponential or polar form. If we are given

$$\mathbf{A} = 5\underline{/53.1°} \quad \text{and} \quad \mathbf{B} = 15\underline{/-36.9°}$$

then the expression of these two complex numbers in exponential form

$$\mathbf{A} = 5e^{j53.1°} \quad \mathbf{B} = 15e^{-j36.9°}$$

enables us to write the product as a complex number in exponential form whose amplitude is the product of the amplitudes and whose angle is the algebraic sum of the angles, in accordance with the normal rules for multiplying two exponential quantities,

$$(\mathbf{A})(\mathbf{B}) = (5)(15)e^{j(53.1°-36.9°)}$$

or

$$\mathbf{AB} = 75e^{j16.2°} = 75\underline{/16.2°}$$

From the definition of the polar form, it is evident that

$$\frac{\mathbf{A}}{\mathbf{B}} = 0.333\underline{/90°}$$

Addition and subtraction of complex numbers are accomplished most easily by operating on complex numbers in rectangular form, and the addition or subtraction of two complex numbers given in exponential or polar form should begin with the conversion of the two complex numbers to rectangular form. The reverse situation applies to multiplication and division; two numbers given in rectangular form should be transformed to polar form, unless the numbers happen to be small integers. For example, if we wish to multiply $(1 - j3)$ by $(2 + j1)$, it is easier to multiply them directly as they stand and obtain $(5 - j5)$. If the numbers can be multiplied mentally, then time is wasted in transforming them to polar form.

We should now endeavor to become familiar with the three different forms in which complex numbers may be expressed and with the rapid conversion from one form to another. The relationships between the three forms seem almost endless, and the following lengthy equation summarizes the various interrelationships:

$$\mathbf{A} = a + jb = \text{Re } [\mathbf{A}] + j \text{ Im } [\mathbf{A}] = Ce^{j\theta} = \sqrt{a^2 + b^2}\, e^{j\tan^{-1}(b/a)}$$

$$= \sqrt{a^2 + b^2}\; \underline{/\tan^{-1} (b/a)} = \frac{b}{\sin [\tan^{-1} (b/a)]}\; \underline{/\tan^{-1} (b/a)}$$

$$= C \cos \theta + jC \sin \theta = C\underline{/\theta}$$

Most of the conversions from one form to another can be done quickly with the help of a calculator.

We shall find that complex numbers are a convenient mathematical artifice which facilitates the analysis of real physical situations. Inevitably in a physical problem a complex number is somehow accompanied by its conjugate.

Drill Problems

A4-7. Express the result of each of these complex number manipulations in polar form, using six significant figures just for the pure joy of calculating: (a) $(25\underline{/112°})/[1 + (3\underline{/-112°})]$; (b) $50/(5.16\underline{/63.2°} - 2.87\underline{/-83.6°})$; (c) $5\underline{/20°} + 4\underline{/30°} + 3\underline{/-115°}$.
Ans: $8.978\ 90\underline{/-155.451°}$; $6.474\ 09\underline{/-74.940\ 6°}$; $6.965\ 59\underline{/8.180\ 74°}$

A4-8. Find $\mathbf{V}$ in rectangular form if: (a) $\mathbf{V} - j1 = 1/\mathbf{V}$; (b) $\mathbf{V} = \ln(4 - j2)$; (c) $\cos \mathbf{V} = 2$. *Ans:* $\pm 0.866 + j0.500$; $1.498 - j(0.464 \pm 2\pi n)$; $\pm j1.317$

Appendix 5
Answers to Odd-Numbered Problems

Chapter 1

1 (a) 1342 mi/h; (b) 196.1 kJ; (c) 22.7 days; (d) Lois Lane
3 (a) 6.97 horses; (b) 0.543 liter
5 15 C
7 (a) -35.6 mC; (b) -34.0 mA; (c) 13.41 μC
9 (a) 4.38 W; (b) 0.400 J
11 (a) 0; (b) 0
13 -32 W, -16 W, 24 W, 56 W, -32 W

Chapter 2

1 (a) 1.667 Ω; (b) 0.144 Ω; (c) 640 W
3 (a) 156.7 ft; (b) 1.410 W
5 (a) -26 V; (b) 320 W; (c) 6 V
7 (a) -96 W; (b) 32 W; (c) 21.3 W; (d) 16.67 W; (e) 10 W
9 (a) 51.1 mW; (b) 52.2 mW; (c) -51.5 mW; (d) -51.0 mW
11 (a) 1.110 Ω; (b) 1.230 Ω; (c) 1.241 Ω
13 (a) $p_{30} = -600$ W, $p_{0.2} = 80$ W, $p_x = 280$ W, $p_{0.6} = 240$ W;
   (b) 0.7 Ω, 14 V (+ at top), 20 A (down); (c-a) 600 W, 80 W, -920 W,
   240 W; (c-b) -2.3 Ω, 46 V (+ at top), 20 A (up)
15 $p_{0.2} = -148.8$ W, $p_{20} = -1090.9$ W, $p_4 = 743.8$ W, $p_6 = 495.9$ W
17 $p_8 = 10.45$ mW, $p_5 = -45.7$ mW, $p_2 = 41.8$ mW, $p_{-3} = -27.4$ mW,
   $p_4 = 20.9$ mW
19 (a) $p_{25} = 576$ W, $p_5 = -600$ W, $p_x = -120$ W, $p_{100} = 144$ W;
   (b) -120 Ω, 1 A (up), 120 V (+ at top); (c-a) 576 W, 600 W, -1320 W,
   144 W; (c-b) -10.91 Ω, 11 A (down), 120 V (+ at bottom)
21 (a) 25 Ω; (b) 20.3 Ω; (c) 22.5 Ω
23 (a) 22.3 Ω; (b) 21.8 Ω; (c) 22.7 Ω
25 5 A, 17.5 V
27 (a) 11.78 m$\mho$; (b) 4.32 $\mho$
29 (a) 2.29 mA, (b) 2.67 A
31 (a) 300 V; (b) 120 V; (c) 0.333 A
33 (a) 1.481 A; (b) 5.39 Ω

35 (*a*) 0.999 950 V; (*b*) -50.0 μV; (*c*) 0.999 950 nW

37 (*a*) 2 V; (*b*) 0.5 V

Chapter 3

1 (*a*) -4.83; (*b*) 200

3 (*a*) -60.9 V; (*b*) 195.7 W

5 25.6 W

7 12.13 W

9 (*a*) 2580 W; (*b*) -32 V

11 1 A

13 1.2 A, 4.2 A, 2 A, 3.2 A

15 (*a*) 3.75 A; (*b*) 4.43 A

17 2.625 A, 43.75 V, 45.9 W

19 4 A

21 1.55 V, 1.7 Ω

23 (*a*) 12 Ω, 0.75 W; (*b*) ∞, 6 V; (*c*) 0, 0.5 A

25 (*a*) -22.5 V, -2.4 A, 9.375 Ω; (*b*) -22.5 V, 9.375 Ω; -2.4 A, 9.375 Ω

27 (*a*) 32 V, 4 Ω; 8 A, 4 Ω; (*b*) 48.6 V, 3.24 Ω; 15 A, 3.24 Ω

29 (*a*) 20 V, -8 Ω; (*b*) 4 A; (*c*) -4 A

31 238 Ω

33 4.999 500 V, 0.199 976 Ω

35 19.02 V

37 7 V

39 (*a*) $8i_1 + 6(i_1 - 2) - 20 - 4i_1 + 30 = 0$; (*b*) 0.2 A

41 1.818 A

Chapter 4

1 (*a*) -1.618 V; (*b*) 0.0711 V, 0.549 s

3 (*a*) 18 A; (*b*) 160 J; (*c*) 320 W

5 (*a*) $-1 + 0.5 \cos 500t + 0.4 \sin 500t$ A; (*b*) 0.1633 J

7 (*a*) 90 μJ; (*b*) 94.2 μs

9 (*a*) $10.009e^{-10t} + 0.001$ V; (*b*) $10e^{-10t}$ V

11 (*a*) 0.0677 J; (*b*) 0; (*c*) -8.83 V

13 (*a*) 6.43 H; (*b*) 2.83 H

15 5 in series in parallel with 2 in series

17 (*a*) 9.66 J; (*b*) 7.12 J; (*c*) 11.24 J

19 (*a*) $10^{-3}v_1 + 2 \times 10^{-6}(v_1' - v_2') = i_s$, $2 \times 10^{-6}(v_2' - v_1') +$

$200\int_0^t v_2 dt + 3 = 0$ (+v_1 top left, +v_2 top right);

(*b*) $1000(i_L - i_s) + 5 \times 10^5 \int_0^t i_L dt + 8 + 0.005i_L' = 0$

21 (*a*) See sketch below; $0.2(v_C - 2e^{-10t}) + 0.1v_C' - 0.04(2e^{-10t} - v_C) +$

$0.05v_R = 0$, $-0.05v_R - 0.1e^{-10t} + 100\int_0^t (-v_R + v_C - 2e^{-10t})dt = 0$;

(*b*) see sketch below; $5i_x + 10\int_0^t(i_x + 0.2i_x + i_L +$

$0.1e^{-10t})dt - 2e^{-10t} = 0$, $0.01i_L' + 20(i_L + 0.1e^{-10t}) +$

$10\int_0^t(1.2i_x + i_L + 0.1e^{-10t})dt - 2e^{-10t} = 0$

P4-21

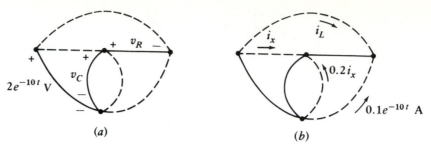

(a) (b)

23 See sketch below; $v_{C1}(0) = 8$ V, $v_{C2}(0) = -4$ V, $i_{L1}(0) = 5$ A,
$i_{L2}(0) = 6$ A, $i_{L3}(0) = 7$ A

P4-23

25 (a) $2 \times 10^{-4}v'_C + 5000\int_0^t v_C dt + 10 = 0$, OK; (b) $v_{C,\text{dual}} =$
10 cos 5000t + 4 sin 5000t V; (c) self-duals

Chapter 5

1 (a) 5 A, 50 J; (b) $5e^{-2.5t}$ A; (c) $-50e^{-2.5t}$ V
3 (a) $5e^{-125t}$ A, $t_1 = 12.88$ ms; (b) $e^{-20.83(t-t_1)}$ A, $t_2 = 344$ ms
5 12.62 nJ
7 (a) 1.134 s; (b) 4.34 s
9 (a) 8 A; (b) 8 A; (c) 20 Ω; (d) 0.5 ms; (e) $8e^{-2000t}$ A
11 8 mA, $t < 0$, and $4.8e^{-480t}$ mA, $t > 0$; 8 mA, 8 mA, 4.8 mA, 1.838 mA,
0.704 mA
13 $5e^{-5.71t}$ A
15 35 A, 35 A; 35 A, 35 A; 35 A, 8.75 A; 21.2 A, 5.31 A; 12.88 A, 3.22 A
17 (a) 4.49 V; (b) 3.59 V; (c) 8.41 V
19 0.248 C
21 (a) 2 mA, 1.540 mA; (b) 0, 56.9 mW
23 $R_1 = 1742$ Ω, $R_2 = 570$ Ω
25 (a) 34.2 V; (b) $34.2e^{-1916t}$ V
27 (a) 100 V, 0; (b) 100 V, 0, 100 V; (c) 0.01 s; (d) $100e^{-100t}$ V;
(e) $40e^{-100t}$ mA; (f) $20 + 80e^{-100t}$ V, $20 - 20e^{-100t}$ V;
(g) 5 + 20 = 25 + 0 mJ
29 $2e^{-400t}$ A
31 500 V, $t < 0$; $250(2e^{-10t} - e^{-2.5t})$ V, $t > 0$

Chapter 6

1 (a) 0; (b) 1; (c) 2; (d) 3; (e) 4; (f) 5; (g) 7 (surprise!)

3 (a) -1; (b) 5; (c) 0; (d) -3.96; (e) $10u(-t) + 2e^{-0.4t}[u(t) - u(t - 5)]$

5 120 V, 100 V, 137.5 V, 300 V, 100 V

7 Let $i = i_1 + i_2$. Then the current due to V is $i_1 = V/R$. The current due to $-Vu(-t)$ is $i_2 = (-V/R)u(-t) - (V/R)e^{-Rt/L}u(t)$. Therefore, $i(t) = (V/R)(1 - e^{-Rt/L})u(t)$

9 $2(1 - e^{-40t})$ A

11 (a) $25(1 - e^{-40t})$ mA; (b) 25 mA, $t < 0$; $50 - 25e^{-40t}$ mA, $t > 0$;
(c) $12.5(\sin 40t - \cos 40t) + 12.5e^{-40t}$ mA, $t > 0$

13 (a) 8.85 A; (b) 120 V

15 $10(1 - e^{-267t})u(t)$ A

17 3 A, -0.304 A, 6 V, 12 V

19 0, $t < 0$; $25(1 - e^{-4000t})$ mA, $0 < t < 0.2$ ms; $55 - 41.2e^{-4000(t-0.0002)}$ mA, $t > 0.2$ ms

21 (a) 1200 V, 1076 V; (b) 999 V

23 $-240u(-t) - (80 + 48e^{-0.2t})u(t)$ V

25 6 V, $t < 0$; $2.4 + 3.6e^{-250t}$ V, $t > 0$

27 (a) $-40 + 100e^{-30\,000t}$ V; (b) 30.5 μs

29 27.7 μs

31 $2.5e^{-5000t}u(t)$ V

33 $5(e^{-t} - 1)u(t)$ V

Chapter 7

1 1 kΩ, 1.429 μF, 7 H

3 $-1.6e^{-400t} - 0.4e^{-1600t}$ A

5 (a) $0.25(e^{-t} + e^{-4t})u(t)$ A; (b) 0.322 s

7 (a) $80e^{-2t} - 20e^{-8t}$ V; (b) $6.4e^{-2t} - 0.4e^{-8t}$ A; (c) 2.45 s; (d) 2.33 s

9 (a) 2.5 kΩ; (b) $24e^{-10^6 t}(1 - 10^6 t)$ mA; (c) 6.27 μs

11 (a) $e^{-4t}(560 \cos 28t - 80 \sin 28t)$ V;
(b) $e^{-4t}(-2.8 \cos 28t - 9.6 \sin 28t)$ A;
(c) $e^{-4t}(2.8 \cos 28t - 0.4 \sin 28t)$ A

13 (a) $e^{-5t}(200 \cos 10t + 100 \sin 10t)$ V;
(b) $20 + e^{-5t}(15 \sin 10t - 20 \cos 10t)$ A

15 (a) $e^{-2t}(-130 \cos 10t - 26 \sin 10t)$ V; (b) 69.4 V, -130 V

17 $e^{-4t}(24 \cos 3t + 32 \sin 3t)$ V

19 (a) $-e^{-6000t} + 19e^{-24\,000t}$ A; (b) 1.091×10^{10} A/s^2

21 (a) $4u(-t) + (10.5e^{-4t} - 6.5e^{-8t})u(t)$ A; (b) $168e^{-4t} - 52e^{-8t}$ V

23 (a) $10 - 10e^{-10t} + 10e^{-40t}$ A; (b) 5.28 A, 10 A

25 (a) $320 + e^{-5t}(-160 \cos 15t + 213 \sin 15t)$ V; (b) 455 V

27 $0.18 + e^{-500t}(-90t - 0.18)$ A

29 (a) $-48 + 64e^{-4t} - 16e^{-8t}$ V; (b) 0 V/s^2, 4096 V/s^3

31 (a) $100[1 - e^{-t}(t + 1)]$ V, $0 < t < 5$ s; see sketch below;
(b) $96.0e^{-2(t-5)}\{\cos[2(t - 5)] - \sin[2(t - 5)]\}$ V, $5 < t < 8$ s; see sketch below

P7-31a,b v_C (V)

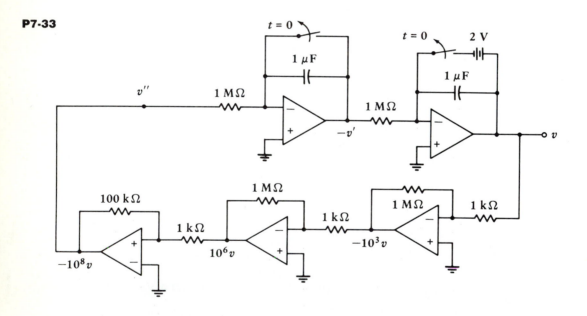

33 See sketch below

P7-33

---

Chapter 8

1 (a) 30.2 Hz; (b) 33.1 ms; (c) 190 rad/s; (d) 31.6; (e) 34.7°; (f) −124.7°;
(g) 25.9, 73.8°

3 (a) −26.8 cos 200πt + 42.2 sin 200πt; (b) 20.4 cos (500πt − 90°)

5 28.0 Ω, 11.47 V, 0.410 cos (500t + 45°) A, 16.38 cos (500t + 135°) V

7 (a) 1.618 W; (b) 1 W; (c) 1 W

9 0.9 cos (400t − 53.1°) A

11 0.4 cos (400t − 53.1°) + 0.693 cos (200t − 33.7°)

13 (a) $-V_m \cos \omega t + Ri + \dfrac{1}{C}\displaystyle\int_{-\infty}^{t} i\,dt = 0,\ \dfrac{di}{dt} + \dfrac{1}{RC} i = -\dfrac{\omega V_m}{R} \sin \omega t$;

(b) $i(t) = \dfrac{V_m}{\sqrt{R^2 + (1/\omega C)^2}} \cos\left(\omega t + \tan^{-1}\dfrac{1}{\omega CR}\right)$

Chapter 9

1 (a) $5.60\underline{/-5.24°}$; (b) $1.585\underline{/78.1°}$; (c) $5.95\underline{/106.9°}$; (d) $1.500 + j8.33$;
   (e) $7.02 + j3.84$

3 $162.8e^{j(20t+47.5°)}$ V

5 (a) $1.25 \cos (1000t - 114°)$ A; (b) $1.25 \cos (1000t - 64°)$ A;
   (c) $1.25e^{j(1000t+3°)}$ A; (d) $0.729e^{j(1000t-55.0°)}$ A

7 (a) 4.81 A; (b) -0.980 A; (c) $10.42\underline{/15.62°}$, $10.42 \cos (100t + 15.62°)$ A

9 (a) -64.0 V; (b) -4.02 A; (c) 257 W

11 $3.54 - j2.83$ A

13 $50 \cos (5000t - 135°)$ V, $69.5 \cos (5000t + 40.7°)$ V,
   $21 \cos (5000t + 38.2°)$ V

15 (a) 1/45 H; (b) 0.896 H

17 (a) $183.5 - j55.0\ \Omega$; (b) 0.717 H; (c) 20.3 and 196.7 rad/s

19 (a) 2.78 mH; (b) 0.750 mH

21 (a) 50 krad/s; (b) 45.9 and 54.5 krad/s

23 (a) 436 rad/s; (b) 816 rad/s; (c) 500 and 2000 rad/s

25 (a) 40 krad/s; (b) 40 and 10 krad/s; (c) 98.0 krad/s; (d) 120 krad/s

27 (a) 90 Ω, 20 mH; (b) 50 Ω, 100 mH

Chapter 10

1 $14.67\underline{/157.6°}$ V

3 $0.285 \cos (10^4 t + 4.09°)$ V

5 $0.808 \cos (10^4 t - 46.0°)$ A

7 (a) Tree: 400 μF, v_s, central 5 Ω; i_L to right in 5 mH; $i_x(t) =$
   $8.98 \cos (1000t + 158.2°)$ A; (b) tree: v_s, 400 μF, right 5 Ω; v_C with
   + at left; $v_x(t) = 44.9 \cos (1000t + 158.2°)$ V

9 $2\underline{/-146.6°}$ and $2 \cos (500t - 146.6°)$ A

11 $0.894\underline{/-63.4°}$ V in series with 0.4 Ω and 1.25 F

13 (a) and (b) $\mathbf{V}_x = -7.5 + j0$ V

15 See sketches below

P10-15

(a)

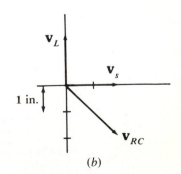

(b)

17 (a) $\mathbf{V}_1 = 80\underline{/30°}(29.69°)$ V, $\mathbf{V}_2 = 50\underline{/-52°}(-52.41°)$ V or $\mathbf{V}_1 = 80\underline{/-30°}$ V,
$\mathbf{V}_2 = 50\underline{/52°}$ V, (b) $\mathbf{V}_1 = 80\underline{/30°}$ V, $\mathbf{V}_2 = 50\underline{/-52°}$ V

19 See sketch below

P10-19

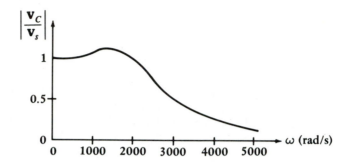

21 See sketch below

P10-21

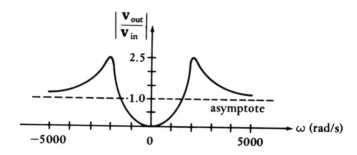

23 See sketches below

P10-23

25 See sketches below

P10-25

Chapter 11

1 (*a*) $-23.1 - 23.1 \cos 40t - 42 \sin 40t$ W; (*b*) -70.6 W

3 -103.9 W, 147.8 W, -219.6 W, 175.7 W

5 (*a*) 0; (*b*) 20; (*c*) 22.5; (*d*) 600

7 -271.2 W, 88.9 W, 0, 182.3 W, 0

9 (*a*) 0.690 W; (*b*) 4.5 W

11 (*a*) 20 W; (*b*) 30 W; (*c*) 40 W; (*d*) 50 W

13 500 Ω, 8 μF, 78.1 W

15 (*a*) $1.965 - j0.262$ Ω; (*b*) 6.25 W

17 (*a*) 0.0474 A; (*b*) 2.14 A

19 (*a*) 21.91; (*b*) 21.21; (*c*) 22.98

21 (*a*) 81.6 V; (*b*) 444 W

23 (*a*) 2.89 kVA, 2.74 kW; (*b*) 0.949 lag; (*c*) 14.98 or 41.6 mH

25 (*a*) 0.8 lag; (*b*) 987 μF

27 (*a*) 0.906 lag; (*b*) 96.3 A rms; (*c*) 322 μF

29 (*a*) $-530.8 + j46.2$ VA, $184.6 + j0$ VA, $0 + j230.8$ VA, $346.2 + j0$ VA, $0 - j276.9$ VA; (*b*) 0.996 lead

31 (*a*) 0.6 lag; (*b*) 0.636 lag; (*c*) 320 μF

Chapter 12

1 5, 6, and -7 A

3 (*a*) $34.0\underline{/21.6°}$ A rms; (*b*) $14.74\underline{/-73.7°}$ A rms

5 (*a*) 23.5 A rms; (*b*) 48.4 A rms; (*c*) 187.7 V rms

7 (*a*) 159.2 μF; (*b*) 3.46 kVA

9 (*a*) $4540\underline{/0.18°}$ V rms; (*b*) $17.93\underline{/-14.04°}$ A rms; (*c*) $246 + j62.2$ Ω; (*d*) 98.7 percent

11 (*a*) $2.84\underline{/-25.1°}$ A rms; (*b*) 1713 W; (*c*) 60.3 W; (*d*) 0.906 lag

13 (*a*) 13.05 μF; (*b*) 1.230 kVAR

15 (*a*) $151.5\underline{/-112.3°}$ V rms; (*b*) $10.18 + j4.03$ kVA

17 $19.83\underline{/117.5°}$, $5.18\underline{/20°}$, and $19.83\underline{/-77.5°}$ A rms; 1496, 423, and 1496 W

19 $16.40\underline{/120.4°}$, $5.90\underline{/20°}$, and $16.40\underline{/-80.4°}$ A rms; 1261, 482, and 1261 W

Chapter 13

1 (a) $-3 \pm j10$, $\pm j2$, and $\pm j18$ s^{-1}; (b) -3 and -11 s^{-1}; (c) 0 and $\pm j7$ s^{-1};
   (d) -1.647 A; (e) 1.569 A; (f) 0.273 A
3 (a) $20s/(s^2 + 250s + 6000)$ A; (b) $0.218\underline{/-73.5°}$ A; (c) 68.4 mA
5 (a) $(4s^2 + 12s + 30)/(s + 3)$ Ω; (b) $8.25\underline{/-14.04°} = 8 - j2$ Ω;
   (c) -0.261 Np/s
7 (a) $(40s^2 + 170s + 40)/(8s^2 + 7s + 20)$ Ω; (b) -1.892 Ω;
   (c) $19.55\underline{/-21.2°}$ Ω; (d) $1/(2 + 8s)$ ℧; (e) $s/(5s + 20)$ ℧; (f) proof
9 (a) $(4s^2 + 65s + 250)/(2s + 25)$ Ω; (b) $(1125\omega + 8\omega^3)/(4\omega^2 + 625)$ Ω;
   (c) $(2s + 25)/(4s^2 + 65s + 250)$ ℧;
   (d) $(6250 + 30\omega^2)/(62\,500 + 2225\omega^2 + 16\omega^4)$ ℧
11 (a) $(40s + 80)/(s^2 + 3s)$ Ω; (b) see sketch below

P13-11b

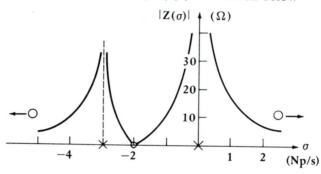

13 (a) $(30s + 50)/(16s + 15)$; (b) -1.667 and -0.938 Np/s;
   (c) see sketch below

P13-13c

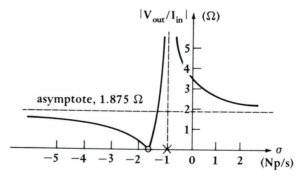

15 (a) $(s^2 + 250s + 10^6)/(20s + 5000)$ Ω;
   (b) pole at $s = -250$ Np/s, zeros at $s = -125 \pm j992$ s^{-1};
   (c) see sketches below

P13-15c

17 (a) See sketch below;
(b) $(8.43s^4 + 50.6s^3 + 194.0s^2 + 422s + 422)/(s^3 + 6s^2 + 10s)$

P13-17a

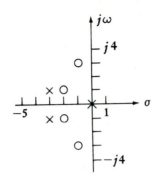

19 (a) See sketch below; (b) 0.360 and 0.400 A/V;
(c) see sketches below; (d) 5.01 in at $\omega = 4.52$ rad/s

P13-19

(a)

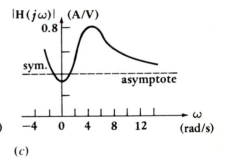

(c)

21 (a) $s(s + 400)/(s^2 + 900s + 230\,000)$; (b) $e^{-450t}(A \cos 165.8t + B \sin 165.8t)$, $t > 0$

23 (a) 20 A; (b) $(20 - 48e^{-10t} + 28e^{-15t})u(t)$ A

25 (a) $250s/(s^2 + 250s + 250\,000)$, $-125 \pm j484$ s^{-1};
(b) 125 Np/s, 500 rad/s, 484 rad/s; (c) 0;
(d) $e^{-125t}(A \cos 484t + B \sin 484t)$; (e) 0, 250 V/s;
(f) $0.516e^{-125t} \sin 484t$ V, $t > 0$

27 (a) 0, $0.05/(s + 0.25)$; (b) $1 + 5(1 - e^{-0.25t})u(t)$ A

29 (a) 10^{-8} F, 10^3 Ω; (b) 10^{-9} F, 10^5 Ω; (c) 0, 10^3 Ω; (d) 0, 10^4 Ω;
(e) 10^{-8} F, 10^3 Ω; 0, 10^4 Ω

31 (a) $-(sC_fR_f + 1)/(sC_fR_1)$; (b) $-(sC_1R_f)/(sC_1R_1 + 1)$;
(c) $-[sR_f + (1/C_f)]/[sR_1 + (1/C_1)]$

Chapter 14

1 (a) 20 krad/s; (b) 3.18 kHz; (c) 30; (d) $-333 \pm j19\,997$ s^{-1}, 0, ∞;
(e) 150 mA; (f) 5 mA

3 (a) 62.5 kΩ, 0.249 mH, 101.9 pF; (b) 8 kΩ, 0.488 H, 513 nF

5 (a) 10 rad/s; (b) 45 Ω

7 $10^{-4} + 10^{-8}s + (2500/s)$ ℧, 500 krad/s, 50

9 (a) 2 H, 545 Ω, 2.73, 36.7 rad/s; (b) 2 H, 444 Ω, 2.22, 45 rad/s

11 (*a*) See sketch below; (*b*) 4.96/−75.6° V

P14-11a

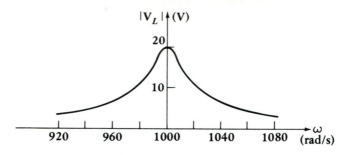

13 (*a*) 89.4 V; (*b*) 483 and 517 Hz; (*c*) 488 Hz
15 (*a*) 998.75 rad/s; (*b*) 250 µ℧; (*c*) 887.4 rad/s; (*d*) 1138.0 rad/s;
    (*e*) 250.6 rad/s; (*f*) 3.99
17 (*a*) 20 pF, 500, 2.5 V; (*b*) 70.7 Mrad/s, 707, 3.54 V; (*c*) 0.249 V
19 $|V_C|_{max}$ = 1000 V, $\mathcal{B}$ = 1 rad/s; ouch!
21 (*a*) 0.990 mH, 927 µF, 0.594 Ω (series); (*b*) 20 mH, 45.9 µF, 12 Ω
    (series); (*c*) 24.8 mH, 37.1 µF, 14.86 Ω (series)
23 (*a*) 8; (*b*) 1500 rad/s;
    (*c*) 5.00 cos (13 000*t* − 53.1°) V (approx.), 5.19 cos (13 000*t* − 53.7°) V
    (exact)
25 (*a*) Series; (*b*) 1000 rad/s, 12.5
27 (*a*) 2.5 Mrad/s, 103.6; (*b*) 25 krad/s, 107.5
29 (*a*) (20s + 100)/(s + 10); (*b*) (40s + 1000)/(s + 50);
    (*c*) see sketch below

P14-29c

31 (*a*) OTSK; (*b*) 60/100° V; (*c*) 600/40° V; (*d*) 12 000/40° V;
    (*e*) 12 000/40° V
33 (*a*) Proof; (*b*) see sketch; (*c*) 30 rad/s; (*d*) 3 rad/s;
    (*e*) 3 Ω → 75 Ω, 600 Ω → 15 kΩ, 1/900 F → 133.3 nF, 1 H → 75 mH

P14-33b

35 (*a*) 0.100 and 100 000 rad/s; (*b*) 0.100 49 and 99 509 rad/s

37 See sketches below

P14-37

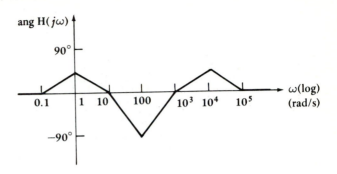

39 See sketches below

P14-39

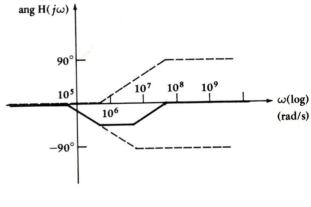

41 $10^6 s^2/(s^3 + 100\,200s^2 + 20\,010\,000s + 10^9)$

Chapter 15

1 (*a*) 1 and 3, or 2 and 4; (*b*) 1 and 4, or 2 and 3;
 (*c*) 1 and 4, or 2 and 3

3 $27.9/\underline{-56.8°}$ A, $9.53/\underline{53.9°}$ A

5 (*a*) $3.58 + j4.16\ \Omega$; (*b*) $2 + j20\ \Omega$

7 (*a*) 0.588 lead; (*b*) 173.0 W

9 50.8 cos ωt V

11 (*a*) $P_{10} = 202$ W, $P_{40} = 1.615$ W; (*b*) $P_{10} = 294$ W, $P_{40} = 235$ W;
 (*c*) see sketch below

P15-11c

13 (*a*) 1.516 cos (40*t* − 145.6°) A, 794 cos (40*t* − 149.7°) V;
    (*b*) to the right
15 (*a*) 576 J; (*b*) 1.248 s
17 (*a*) 3 H; (*b*) 15 H; (*c*) 0.714 H; (*d*) 2.5 H; (*e*) 0.333 H; (*f*) 2 H;
    (*g*) 7 H; (*h*) L_{AB} with *A* to *D* and *B* to *C*, 1.667 H
19 (*a*) $2s/(11s^2 + 145s + 300)$ ℧ ; (*b*) $Ae^{-2.57t} + Be^{-10.61t}$;
    (*c*) $15i_s + 5i_s' - 2i_L' = 100u(t), 20i_L + 3i_L' - 2i_s' = 0$;
    (*d*) 0 A, 18.18 A/s; (*e*) $2.26(e^{-2.57t} - e^{-10.61t})u(t)$ A
21 (*a*) 24.98 Ω and 0.625 mH; (*b*) 24.39 mH; (*c*) 39.00 μF
23 (*a*) 2.499 94 + j0.012 50 Ω; (*b*) 2.5 + j0 Ω
25 P_{10} = 29.4 W, P_5 = 32.7 W, P_3 = 19.59 W
27 (*a*) P_{AB} = 2250 W, P_{CD} = 1000 W; (*b*) P_{AB} = P_{CD} = 1562.5 W
29 (*a*) 4 Ω only; (*b*) 2.61 Ω only
31 1250 W

Chapter 16

1 0.229 ℧
3 $5(3s + 5)/[s(s + 4)]$
5 (*a*) $j\omega CR_1R_3R_4/R_2$; (*b*) proof
7 (*a*) 8.46 Ω; (*b*) 8.46 Ω
9 (*a*) $\begin{bmatrix} 0.712 & -0.394 \\ -0.394 & 0.553 \end{bmatrix}$ (℧); (*b*) same
11 (*a*) 2; (*b*) 180 W

13 (*a*) $\begin{bmatrix} 20 & -5 \\ -20 & 2 \end{bmatrix}$ (m℧); (*b*) $\begin{bmatrix} 10 & -5 \\ -20 & 12 \end{bmatrix}$ (m℧); (*c*) $\begin{bmatrix} 20 & -15 \\ -30 & 12 \end{bmatrix}$ (m℧)

P16-21c

15 (*a*) $\begin{bmatrix} 0.5 & -0.308 \\ 1.75 & 0.821 \end{bmatrix}$ (℧); (*b*) $\begin{bmatrix} 0.1429 & 0.0881 \\ 0.500 & 1.596 \end{bmatrix}$ (℧)

17 (*a*) 32 Ω; (*b*) 40 Ω; (*c*) 28 Ω
19 (*a*) −5; (*b*) 1.667; (*c*) 1.538; (*d*) 7.69; (*e*) 166.7 Ω; (*f*) 65.2 Ω;
    (*g*) 0.667

21 (*a*) $\begin{bmatrix} 1/R_1 & 0 \\ 0 & 0 \end{bmatrix}$; (*b*) $\begin{bmatrix} 1/R_2 & -1/R_2 \\ -1/R_2 & 1/R_2 + 1/R_3 \end{bmatrix}$; (*c*) see sketch; proof

23 (*a*) $\begin{bmatrix} 133.1\underline{/-47.6°} & 94.2\underline{/-2.64°} \\ 9416\underline{/86.8°} & 565\underline{/-3.60°} \end{bmatrix}$ (Ω); (*b*) all four

25 (a) $\begin{bmatrix} 500 & 50 \\ -50\ 000 & 5000 \end{bmatrix}$ (Ω); (b) -28.6, $778\ \Omega$

27 (a) 222 W; (b) 1 Ω; (c) 400 W

29 (a) 5 $\mho$; (b) $-3\ \Omega$; (c) 0.15

31 13.38 W, 476 W, -230 W

33 $\begin{bmatrix} 3.43\ \Omega & 0.1429 \\ -2.86 & 0.214\ \mho \end{bmatrix}$

35 (a) $\begin{bmatrix} 5/3 & 2\ \Omega \\ 3/4\ \mho & 3/2 \end{bmatrix}$; (b) $\begin{bmatrix} 1 & 0 \\ 1/6\ \mho & 1 \end{bmatrix}$; (c) $\begin{bmatrix} 2 & 2\ \Omega \\ 1\ \mho & 1.5 \end{bmatrix}$;

(d) $\begin{bmatrix} 5/3 & 2\ \Omega \\ 37/36\ \mho & 11/6 \end{bmatrix}$

37 $\begin{bmatrix} 0.613 & 3.23\ \Omega \\ 0.0532\ \mho & 0.806 \end{bmatrix}$

39 (a) $\begin{bmatrix} 10/3 & 400/3\ \Omega \\ 1/6\ \mho & 55/6 \end{bmatrix}$; (b) $\begin{bmatrix} 10 & 400/3\ \Omega \\ 5/8\ \mho & 55/6 \end{bmatrix}$

Chapter 17

1 (a) 2; (b) 4.12; (c) 0.2 s; (d) -0.500

3 (a) 1.2; (b) 1.932; (c) 0.703

5 -0.761, 0.337, 0.832

7 (a) 0.500; (b) $3\left(-3 + \cos\dfrac{4\pi n}{3} + 2\cos\dfrac{2\pi n}{3}\right)\Big/(4\pi^2 n^2)$

9 (a) 4.77; (b) -0.0296; (c) -0.922

11 (a) $1 + 3\cos 20\pi t + 2\cos 40\ \pi t$; (b) 2.74; (c) -0.809

13 (a) 3.5; (b) 3.61; (c) $f_{contin.} = -8 + 0.75t$, $4 < t < 8$ s;

(d) $\sum_{1}^{\infty}\left(\dfrac{4}{n\pi} - \dfrac{10}{n\pi}\cos n\pi\right)\sin\dfrac{n\pi t}{4}$; (e) 1.050

15 $a_0 = 0$, $a_1 = 1.091$, $a_2 = 0$, $a_3 = -2.83$, $a_4 = 0$, $a_5 = 3.47$

17 (a) Odd; (b) half-wave; (c) even, half-wave

19 (a) $1.25 + \dfrac{5}{\pi}\sum_{1,odd}^{\infty}\dfrac{1}{1 + 4n^2}\left(\dfrac{1}{n}\sin 4nt - 2\cos 4nt\right)$;

(b) $-0.430\ 26e^{-0.2t} + 1.25 + \dfrac{5}{\pi}\sum_{1,odd}^{\infty}\dfrac{1}{1 + 4n^2}\left(\dfrac{1}{n}\sin 4nt - 2\cos 4nt\right)$

21 (a) $5 + \dfrac{20}{\pi}\sum_{1,odd}^{\infty}\dfrac{1}{1 + 100n^2}\left(\dfrac{1}{n}\sin 2nt - 10\cos 2nt\right)$; (b) $Ae^{-0.2t}$;

(c) $-4.221e^{-0.2t} + 5 + \dfrac{20}{\pi}\sum_{1,odd}^{\infty}\dfrac{1}{1 + 100n^2}\left(\dfrac{1}{n}\sin 2nt - 10\cos 2nt\right)$

23 $10 + \sum_{n=\pm 1}^{\pm\infty}\left[-\dfrac{125}{n^2\pi^2} + \left(\dfrac{125}{n^2\pi^2} + j\dfrac{50}{n\pi}\right)e^{-j0.4\pi n}\right]e^{j400\pi nt}$; $\mathbf{c}_0 = 10$

$\mathbf{c}_{\pm 1} = 6.39 \mp j7.13$, $\mathbf{c}_{\pm 2} = -1.050 \mp j8.30$, $\mathbf{c}_{\pm 3} = -5.66 \mp j3.46$

25 (a) 16.01 V; (b) 17.61 V

27 (a) -0.0620; (b) 0.0800; (c) $|\mathbf{c}_0| = 0.0800$; (d) $N = 740$, 186.0 Mrad/s

29 (a) $\dfrac{9}{n^2\pi^2} \left(\cos \dfrac{2\pi n}{9} - \cos \dfrac{4\pi n}{9} \right)$; (b) 2/3, 0.540

Chapter 18

1 (a) $\dfrac{1}{a + j\omega} e^{-(a+j\omega)t_0}$; (b) $\dfrac{1}{(a + j\omega)^2}$; (c) $\dfrac{1}{(a + j\omega)^2} e^{-j\omega t_0}$

3 $-j\dfrac{8}{\omega^2} \sin 5\omega + j\dfrac{40}{\omega} \cos 5\omega$

5 (a) $\dfrac{1}{\pi t}(\sin 2t - \sin t)$, 0.230; (b) $\dfrac{1}{\pi(1 + t^2)}$, 0.255;

   (c) $\dfrac{-\sin \pi t}{\pi(1 - t^2)}$, -0.424

7 (a) 8.25 J; (b) $\omega_1 = 0.843$ rad/s

9 (a) 0.200 J; (b) 81.8 percent

11 (a) 1.000; (b) 1.333

13 (a) $2\delta(t)$ V, $(10 - 10e^{-5t})u(t)$ A, $40e^{-5t}u(t)$ V, $2\delta(t) + 40e^{-5t}u(t)$ V; (b) proof

15 (a) 225.6 V; (b) 269.9 V; (c) 224.0 V; (d) 219.5 V

17 (a) -5; (b) $5e^{-j2\omega}$; (c) $\dfrac{-10}{\omega^2 - 4}$; (d) $\dfrac{j20}{\omega^2 - 4} \sin \dfrac{\omega\pi}{2}$

19 (a) -0.241; (b) 0.241; (c) 0.400

21 $2\pi \displaystyle\sum_{-\infty}^{\infty} \left[e^{-jn\pi} \left(\dfrac{j2}{n\pi} + \dfrac{2}{n^2\pi^2} \right) - \dfrac{2}{n^2\pi^2} \right] \delta\left(\omega - \dfrac{n\pi}{2} \right)$

23 67.2

25 (a) $t < 0$: 0; $0 < t < 2$: $2t$; $t > 2$: 4; (b) same

27 (a) $\dfrac{20}{3} \displaystyle\int_2^5 (z - 2) \, dz$; (b) 30

29 (a) 42; (b) 72; (c) 152; (d) 150; (e) 116; (f) 96

31 (a) 0; (b) 3; (c) 6; (d) 2.4; (e) 0

33 (a) 0.500 J; (b) 0.5625

35 (a) $\dfrac{2}{1 + j\omega}$; (b) see sketch; (c) 2

37 (a) $2e^{-3t}u(t)$; (b) $2 \cos (4t - 53.1°)$

39 (a) $12(e^{-t} - e^{-2t})u(t)$; (b) same

P18-35b

Chapter 19

1 (a) $\sigma > 0$; (b) $\sigma > 0$; (c) $\sigma > -2$; (d) $\sigma > 2$; (e) $\sigma > -2$; (f) $\sigma > -2$

3 (a) $\dfrac{1}{s}(e^{-s} - e^{-4s})$; (b) $2e^{-4s}$; (c) $1.176e^{-0.2s}$; (d) $\dfrac{2}{s} e^{-4s}$;

   (e) $\dfrac{2}{s + 4} e^{-4-s}$

5 (a) $\delta(t) + (1 + e^{-t})u(t)$; (b) $\delta(t - 1) + u(t - 1)$;
   (c) $\delta(t) + 2\delta(t - 1) + \delta(t - 2)$; (d) $\delta(t - 1) - \delta(t - 5)$

7 (a) $5\delta(t) - 2u(t)$; (b) $(5 + 3e^{-t})u(t)$; (c) $\delta(t) + e^{-t}u(t)$;
   (d) $\delta(t) + 5tu(t)$; (e) $6(e^{-t} - e^{-2t})u(t)$

9 (a) $0.3v_C' + 2.5v_C - 150 = 0$; (b) $20u(-t) + (60 - 40e^{-25t/3})u(t)$ V

11 $(0.25 + 0.5t - 0.25e^{-2t})u(t)$

13 (a) $4\ \Omega$, 1 H, 1/3 F; (b) $(75e^{-3t} - 12.5e^{-t} - 62.5e^{-5t})u(t)$ A

15 $4\delta(t) - 72e^{-20t}u(t)$

17 (a) $t < 1$: $y = 0$; $1 < t < 2$: $y = 5$; $t > 2$: $y = 10$;
   (b) $5u(t - 1) + 5u(t - 2)$

19 (a) $10e^{-10t}u(t)$, $10/(s + 10)$; (b) $\delta(t) - 10e^{-10t}u(t)$, $s/(s + 10)$;
   (c) $0.1\delta(t) - 4e^{-10t}u(t)$ V

21 (a) $\mathbf{X(s)} = (1 - e^{-2s} - 2se^{-2s})/s^2$, $\mathbf{H(s)} = (1 - e^{-2s})/s$,
   $y = \frac{1}{2}t^2u(t) + (-t^2 + 2t)u(t - 2) + (\frac{1}{2}t^2 - 2t)u(t - 4)$;
   (b) same; (c) 1.875

23 (a) $\dfrac{s}{s^2 + \pi^2}(1 + e^{-s})$; (b) $[e^{-(t-3)} - e^{-2(t-3)} - (t - 3)e^{-2(t-3)}]u(t - 3)$

25 $5\pi/[(s^2 + \frac{1}{4}\pi^2)(1 - e^{-2s})]$

27 (a) $(34.8s + 359)/(s^2 + 100)$; (b) $e^{-2s}\left(\dfrac{1}{s^2} + \dfrac{1}{s}\right)$;
   (c) $(13 - 13e^{-2s} - 26se^{-2s})/[s^2(1 - e^{-4s})]$

29 (a) $(\frac{1}{2}t^2e^{-2t} + e^{-3t})u(t)$; (b) $(t \cos 2t)u(t)$; (c) $0.05 \sin [5(t - 8)]\ u(t - 8)$

31 (a) $2 \sin 5t\ u(t)$; (b) $\pi - 2 \tan^{-1}0.2s$, $\dfrac{2}{t} \sin 5t\ u(t)$;
   (c) $-20s/(s^2 + 25)^2$, $-2t \sin 5t\ u(t)$

33 (a) 0, 0.4; (b) 3, 0; (c) ∞, unknown

35 (a) 50, 25, $25(1 + e^{-20t})u(t)$; (b) 25, 50, $25u(t) + 25u(t - 10)$;
   (c) 2, 0, $e^{-4t}(2 \cos 2t - \sin 2t)u(t)$

37 $(e^{-t/4} - \frac{2}{3}e^{-t/6})u(t)$

39 (a) $(1 + t)e^{-2t}u(t)$; (b) $(7.5 - 7.5e^{-2t} - 5te^{-2t})u(t)$

41 $4e^{-2t}(1 - 2t)u(t)$

43 (a) $(7s^3 + 73s^2 + 238s + 292)/[(s + 1)(s + 6)(s^2 + 4)]$;
   (b) $(4.8e^{-t} + 0.1e^{-6t} + 2.1 \cos 2t + 14.7 \sin 2t)u(t)$

45 0.0801 and 0.1386 s

47 $4u(-t) + (2.4e^{-t} + 1.6 \cos 2t + 3.2 \sin 2t)u(t)$

INDEX